Undergraduate Texts in Mathematics

Editors
S. Axler
F.W. Gehring
P.R. Halmos

Springer
New York
Berlin
Heidelberg
Barcelona
Budapest
Hong Kong
London
Milan
Paris
Santa Clara
Singapore
Tokyo

Undergraduate Texts in Mathematics

Anglin: Mathematics: A Concise History and Philosophy.
Readings in Mathematics.

Anglin/Lambek: The Heritage of Thales.
Readings in Mathematics.

Apostol: Introduction to Analytic Number Theory. Second edition.

Armstrong: Basic Topology.

Armstrong: Groups and Symmetry.

Bak/Newman: Complex Analysis.

Banchoff/Wermer: Linear Algebra Through Geometry. Second edition.

Berberian: A First Course in Real Analysis.

Brémaud: An Introduction to Probabilistic Modeling.

Bressoud: Factorization and Primality Testing.

Bressoud: Second Year Calculus.
Readings in Mathematics.

Brickman: Mathematical Introduction to Linear Programming and Game Theory.

Browder: Mathematical Analysis: An Introduction.

Cederberg: A Course in Modern Geometries.

Childs: A Concrete Introduction to Higher Algebra. Second edition.

Chung: Elementary Probability Theory with Stochastic Processes. Third edition.

Cox/Little/O'Shea: Ideals, Varieties, and Algorithms.

Croom: Basic Concepts of Algebraic Topology.

Curtis: Linear Algebra: An Introductory Approach. Fourth edition.

Devlin: The Joy of Sets: Fundamentals of Contemporary Set Theory. Second edition.

Dixmier: General Topology.

Driver: Why Math?

Ebbinghaus/Flum/Thomas: Mathematical Logic. Second edition.

Edgar: Measure, Topology, and Fractal Geometry.

Elaydi: Introduction to Difference Equations.

Fischer: Intermediate Real Analysis.

Flanigan/Kazdan: Calculus Two: Linear and Nonlinear Functions. Second edition.

Fleming: Functions of Several Variables. Second edition.

Foulds: Combinatorial Optimization for Undergraduates.

Foulds: Optimization Techniques: An Introduction.

Franklin: Methods of Mathematical Economics.

Hairer/Wanner: Analysis by Its History.
Readings in Mathematics.

Halmos: Finite-Dimensional Vector Spaces. Second edition.

Halmos: Naive Set Theory.

Hämmerlin/Hoffmann: Numerical Mathematics.
Readings in Mathematics.

Iooss/Joseph: Elementary Stability and Bifurcation Theory. Second edition.

Isaac: The Pleasures of Probability.
Readings in Mathematics.

James: Topological and Uniform Spaces.

Jänich: Linear Algebra.

Jänich: Topology.

Kemeny/Snell: Finite Markov Chains.

Kinsey: Topology of Surfaces.

Klambauer: Aspects of Calculus.

Lang: A First Course in Calculus. Fifth edition.

Lang: Calculus of Several Variables. Third edition.

Lang: Introduction to Linear Algebra. Second edition.

Lang: Linear Algebra. Third edition.

Lang: Undergraduate Algebra. Second edition.

Lang: Undergraduate Analysis.

Lax/Burstein/Lax: Calculus with Applications and Computing. Volume 1.

LeCuyer: College Mathematics with APL.

(continued after index)

John L. Troutman

Variational Calculus and Optimal Control

Optimization with Elementary Convexity

Second Edition

With 87 Illustrations

 Springer

John L. Troutman
Department of Mathematics
Syracuse University
Syracuse, NY 13210
USA

Mathematics Subject Classifications (1991): 49-01

Library of Congress Cataloging-in-Publication Data
Troutman, John L.
    Variational calculus and optimal control: Optimization with elementary convexity
  / John L. Troutman. 2nd edition.
      p.   cm. — (Undergraduate texts in mathematics. Readings in
  mathematics.)
    Includes bibliographical references and index.
    ISBN 0-387-94511-3 (hardcover : alk. paper)
    1. Calculus of variations.   2. Control theory.   3. Mathematical
  optimization.   4. Convex functions.   I. Title.   II. Series.
  QA315.T724   1995
  515'.64—dc20   95-12918

Printed on acid-free paper.

Production coordinated by Brian Howe and managed by Henry Krell; manufacturing super-
vised by Jeffrey Taub.
Typeset by Asco Trade Typesetting Ltd., Hong Kong.
Printed and bound by R.R. Donnelley & Sons, Harrisonburg, VA.
Printed in the United States of America.

9 8 7 6 5 4 3 2 1

ISBN 0-387-94511-3 Springer-Verlag New York Berlin Heidelberg

*This book is dedicated to my parents
and to my teachers*

Preface

Although the calculus of variations has ancient origins in questions of Aristotle and Zenodoros, its mathematical principles first emerged in the postcalculus investigations of Newton, the Bernoullis, Euler, and Lagrange. Its results now supply fundamental tools of exploration to both mathematicians and those in the applied sciences. (Indeed, the macroscopic statements obtained through variational principles may provide the only valid mathematical formulations of many physical laws.) Because of its classical origins, variational calculus retains the spirit of natural philosophy common to most mathematical investigations prior to this century. The original applications, including the Bernoulli problem of finding the brachistochrone, require optimizing (maximizing or minimizing) the mass, force, time, or energy of some physical system under various constraints. The solutions to these problems satisfy related differential equations discovered by Euler and Lagrange, and the variational principles of mechanics (especially that of Hamilton from the last century) show the importance of also considering solutions that just provide stationary behavior for some measure of performance of the system. However, many recent applications do involve optimization, in particular, those concerned with problems in optimal control.

Optimal control is the rapidly expanding field developed during the last half-century to analyze optimal behavior of a constrained process that evolves in time according to prescribed laws. Its applications now embrace a variety of new disciplines, including economics and production planning.[1] In

---

[1] Even the perennial question of how a falling cat rights itself in midair can be cast as a control problem in geometric robotics! See *Dynamics and Control of Mechanical Systems: The Falling Cat and Related Problems*, by Michael Enos, Ed. American Mathematical Society, 1993.

this text we will view optimal control as a special form of variational calculus, although with proper interpretation, these distinctions can be reversed.

In either field, most initial work consisted of finding (necessary) conditions that characterize an optimal solution tacitly assumed to exist. These conditions were not easy to justify mathematically, and the subsequent theories that gave (sufficient) conditions guaranteeing that a candidate solution does optimize were usually substantially harder to implement. (Conditions that ensure existence of an optimizing solution were—and are—far more difficult to investigate, and they cannot be considered at the introductory level of this text. See [Ce].) Now, in any of these directions, the statements of most later theoretical results incorporate some form of convexity in the defining functions (at times in a disguised form). Of course, convexity was to be expected in view of its importance in characterizing extrema of functions in ordinary calculus, and it is natural to employ this central theme as the basis for an introductory treatment.

The present book is both a refinement and an extension of the author's earlier text, *Variational Calculus with Elementary Convexity* (Springer-Verlag, 1983) and its supplement, *Optimal Control with Elementary Convexity* (1986). It is addressed to the same audience of junior to first-year graduate students in the sciences who have some background in multidimensional calculus and differential equations. The goal remains to solve problems completely (and exactly) whenever possible at the mathematical level required to formulate them. To help achieve this, the book incorporates a sliding scale-of-difficulty that allows its user to become gradually more sophisticated, both technically and theoretically. The few starred (*) sections, examples, and problems outside this scheme can usually be overlooked or treated lightly on first reading.

For our purposes, a convex function is a differentiable real-valued function whose graph lies above its tangent planes. In application, it may be enough that a function of several variables have this behavior only in some of the variables, and such "elementary" convexity can often be inferred through pattern recognition. Moreover, with proper formulation, many more problems possess this convexity than is popularly supposed. In fact, using only standard calculus results, we can solve most of the problems that motivated development of the variational calculus, as well as many problems of interest in optimal control.

The paradigm for our treatment is as follows: Elementary convexity suggests simple sufficiency conditions that can often lead to direct solution, and they in turn inform the search for necessary conditions that hold whether or not such convexity is present. For problems that can be formulated on a fixed interval (or set) this statement remains valid even when fixed-endpoint conditions are relaxed, or certain constraints (isoperimetric or Lagrangian) are imposed. Moreover, sufficiency arguments involving elementary convexity are so natural that even multidimensional generalizations readily suggest themselves.

In Part I, we provide the standard results of variational calculus in the context of linear function spaces, together with those in Chapter 3 that use elementary convexity to establish sufficiency. In Part II, we extend this development into more sophisticated areas, including Weierstrass–Hilbert field theory of sufficiency (Chapter 9). We also give an introduction to Hamiltonian mechanics and use it in §8.8 to motivate a different means for recognizing convexity, that leads to new elementary solutions of some classical problems (including that of the brachistochrone). Throughout these parts, we derive and solve many optimization problems of physical significance including some involving optimal controls. But we postpone our discussion of control theory until Part III, where we use elementary convexity to suggest sufficiency of the Pontjragin principle before establishing its necessity in the concluding chapter.

Most of this material has been class-tested, and in particular, that of Part I has been used at Syracuse University over 15 years as the text for one semester of a year-sequence course in applied mathematics. Chapter 8 (on Hamiltonian mechanics) can be examined independently of adjacent chapters, but Chapter 7 is prerequisite to any other subsequent chapters. On the other hand, those wishing primarily an introduction to optimal control could omit both Chapters 8 and 9. The book is essentially self-contained and includes in Chapter 0 a review of optimization in Euclidean space. It does not employ the Lebesque integral, but in the Appendix we develop some necessary results about analysis in Euclidean space and families of solutions to systems of differential equations.

Acknowledgments

I wish once more to express my appreciation to those who first made me aware of the elegance and power of variational methods—Daniel Frederick at V.P.I., M.M. Schiffer at Stanford, and C. Lanczos as author. I must also reacknowledge the contributions that William Hrusa (now at Carnegie-Mellon University) made to the earlier work during his student days. Many of the problems in Part I and II originated with Bill, and his assistance and commentary during that initial production were invaluable. It has been rewarding to hear from those including Terry Rockafellar at the University of Washington, Frank Chorlton at the University of Aston, and Morris Kline at New York University, whose satisfaction with the earlier work helped motivate this extension.

I wish also to express my gratitude to Phil Loewen at the University of British Columbia and Frank Clarke at the Université de Montréal, as well as to Dan Waterman, my friend and colleague at Syracuse, for their suggestions and encouragement. Many thanks are due my other colleagues, especially Gerry Cargo, Phil Church, Phil Griffin, Wolfgang Jurkat, Tadeusz Iwaniec, and Andy Vogel, who taught from this material at Syracuse and made valuable suggestions for its improvement. And of course, I feel deep gratitude to and for my many students over the years, without whose evident enjoyment and expressed appreciation the current work would not have been undertaken. It is a pleasure to recognize those responsible for transforming this work from manuscript to printed page: the principal typists, Louise Capra, Esther Clark, and Steve Everson; the editors and staff at Springer-Verlag, in particular, Jenny Wolkowicki and the late Walter Kauffmann-Bühler. Finally, I wish to thank my wife, Patricia Brookes, for her patience and understanding during the years of revision.

Syracuse, New York   JOHN L. TROUTMAN

Contents

PART THREE
OPTIMAL CONTROL

CHAPTER 10*

Control Problems and Sufficiency Considerations

CHAPTER 0

Review of Optimization in $\mathbb{R}^d$

This chapter presents a brief summary of the standard terminology and basic results related to characterizing the maximal and minimal values of a real valued function f defined on a set D in Euclidean space. With the possible exception of the remarks concerning convexity ((0.8) and (0.9)), this material is covered in texts on multidimensional calculus; the notation is explained at the end of §1.5.

For $d = 1, 2, 3, \ldots$, let $\mathbb{R}^d$ denote d-dimensional real Euclidean space where a typical point or vector $X = (x_1, x_2, \ldots, x_d)$ has the length $|X| = (\sum_{j=1}^{d} |x_j|^2)^{1/2}$ which is positive unless $X = \mathcal{O} = (0, 0, 0, \ldots, 0)$. (We identify $\mathbb{R}^1$ with $\mathbb{R}$.)

On $\mathbb{R}^d$, with $Y = (y_1, y_2, \ldots, y_d)$, we have the vector space operations of componentwise addition

$$X + Y \overset{\text{def}}{=} (x_1 + y_1, x_2 + y_2, \ldots, x_d + y_d),$$

and scalar multiplication:

$$aX \overset{\text{def}}{=} (ax_1, ax_2, \ldots, ax_d), \qquad \forall\, a \in \mathbb{R}.$$

We may also express $|X| = (X \cdot X)^{1/2}$, utilizing the scalar or dot product

$$X \cdot Y \overset{\text{def}}{=} \sum_{j=1}^{d} x_j y_j,$$

which is subject to the *Cauchy* inequality

$$|(X \cdot Y)| \leq |X||Y|. \tag{1}$$

The Cauchy inequality (1) is used to prove the so-called triangle inequality

$$|X + Y| \leq |X| + |Y|, \tag{2a}$$

1

an alternate form of which is

$$||X| - |Y|| \le |X - Y|, \tag{2b}$$

where

$$X - Y \overset{\text{def}}{=} X + (-1)Y; \quad (\text{Problem 0.1}).$$

$|X - Y|$ defines the *Euclidean distance* between X and Y.

When $X_0 \in \mathbb{R}^d$, then for finite $\delta > 0$, the "sphere"

$$S_\delta(X_0) \overset{\text{def}}{=} \{X \in \mathbb{R}^d : |X - X_0| < \delta\}$$

is called an (open) *neighborhood* of X_0, and X_0 is said to be an *interior point* of each set D which contains this neighborhood for some $\delta > 0$. D is *open* when it consists entirely of interior points. An open set D is a *domain* when each pair of its points may be connected by a (polygonal) curve which lies entirely in D. Each open sphere is a domain, as is each open "box"

$$B = \{X \in \mathbb{R}^d : a_j < x_j < b_j, j = 1, 2, \ldots, d\},$$

but the union of disjoint open sets is *not* a domain, although it remains open.

A point *not* in the interior of a set S, and *not* interior to its complement, $\mathbb{R}^d \sim S$, is called a *boundary point* of S. The set of such points, denoted ∂S, is called the *boundary* of S. For example, if $S = \{(X \in \mathbb{R}^d : |X| \le 1\}$, then $\partial S = B = \{X \in \mathbb{R}^d : |X| = 1\}$; also $\partial B = B$. A set $S \subseteq \mathbb{R}^d$ is said to be *bounded* *iff* it is a subset of some sphere.

We suppose that we are given a real valued function f defined on a set $D \subseteq \mathbb{R}^d$ for which we wish to find *extremal* values. That is, we wish to find points in D (called *extremal* points) at which f assumes maximum or minimum values. With such optimization problems we should note the following facts:

(0.0) *f need not have extremal values on D.*

For example, when $D = \mathbb{R}^1$, then the function $f(X) = x_1$ is unbounded in both directions on D. Moreover, on the open interval $D = (-1, 1) \subseteq \mathbb{R}^1$, this same function, although bounded, takes on values as near -1 or 1 as we please but does *not* assume the values ± 1 on D. On the closed interval, $D = [-1, 1]$, this function does assume both maximum and minimum values, but the function

$$f(X) = \frac{1}{x_1}, \quad x_1 \ne 0,$$

$$f(\mathcal{O}) = 0,$$

is again unbounded.

(0.1) *f may assume only one extremal value on D.*

For example, on $D = (-1, 1]$ the function $f(X) = x_1$ assumes a maximum value $(+1)$, but not a minimum value, while on $(-1, 1)$ the function $f(X) = x_1^2$ assumes a minimum value (0) but not a maximum value.

(0.2) *f may assume an extremal value at more than one point.*

On $D = [-1, 1]$, $f(X) = x_1^2$ assumes a maximum value (1) at $x_1 = \pm 1$, while on $D = \mathbb{R}^2$, $f(X) = x_1^2$ assumes its minimum value (0) at every point located on the x_2 axis.

The only reasonable conditions which guarantee the existence of extremal values are contained in the following theorem whose proof is deferred. (See Proposition 5.3.)

(0.3) **Theorem.** *If $D \subseteq \mathbb{R}^d$ is compact and $f: D \to \mathbb{R}$ is continuous, then f assumes both maximum and minimum values on D.*

In $\mathbb{R}^d$, a *compact* set is a bounded set which is *closed* in that it contains each of its boundary points. In particular, each "box" of the form

$$\bar{B} = \{X \in \mathbb{R}^d : a_j \le x_j \le b_j, j = 1, 2, \ldots, d\}$$

for given real numbers $a_j \le b_j$, $j = 1, 2, \ldots, d$ is compact. However, the interval $(-1, +1)$ is *not* compact. (See §A.0.)

$f: D \to \mathbb{R}$ is *continuous* at $X_0 \in D$ iff for each $\varepsilon > 0$, $\exists \delta > 0$, such that when $X \in D$ and $|X - X_0| < \delta$, then $|f(X) - f(X_0)| < \varepsilon$; and f is continuous on D iff it is continuous at each point $X_0 \in D$.

The previous examples show that neither compactness nor continuity can alone assure the existence of extremal values.

(0.4) *The maximum value of f is the minimum value of $-f$ and vice versa.*

Thus it suffices to characterize the *minimum* points, those $X_0 \in D$ for which

$$f(X) \ge f(X_0), \qquad \forall \, X \in D. \tag{3}$$

As we have seen, such points may be present even on a noncompact set.

(0.5) *When D contains a neighborhood of X_0, an extremal point of f, in which f has continuous partial derivatives $f_{x_j} = \partial f / \partial x_j, j = 1, 2, \ldots, d$, then for each vector $U \in \mathbb{R}^d$ of unit length, the (two-sided) directional derivative:*

$$\partial_U f(X_0) \stackrel{\text{def}}{=} \lim_{\varepsilon \to 0} \left[\frac{f(X_0 + \varepsilon U) - f(X_0)}{\varepsilon} \right] = \frac{\partial f}{\partial \varepsilon}(X_0 + \varepsilon U) \Big|_{\varepsilon = 0}$$

$$= 0.$$

[The bracketed quotient reverses sign as the sign of ε is changed. The existence and continuity of the partial derivatives ensures the existence of the limit which must therefore be zero.]

Introducing the *gradient vector* $\nabla f \overset{\text{def}}{=} (f_{x_1}, f_{x_2}, \ldots, f_{x_d})$, we may also express $\partial_U f(X_0) = \nabla f(X_0) \cdot U$, and conclude that at such an interior extremal point X_0,

$$\nabla f(X_0) = \mathcal{O}. \tag{4}$$

(0.6) *The points X_0 at which* (4) *holds, called* stationary points (or critical points) *of f, need not give either a maximum or a minimum value of f.*

For example, on $D = [-1, 1]$, the function $f(X) = x_1^3$ has $x_1 = 0$ as its only stationary point, but its maximum and minimum values occur at the end points 1 and -1, respectively.

On $D = \mathbb{R}^2$, the function $f(X) = x_2^2 - x_1^2$ has $X_0 = (0, 0)$ as its only critical point; at X_0, f has maximal behavior in one direction $(x_2 = 0)$ and minimal behavior in another direction $(x_1 = 0)$.
In such cases, X_0 is said to be a *saddle point* of f.

(0.7) *A stationary point X_0 may be (only) a local extremal point for f; i.e., one for which $f(X) \geq f(X_0)$ (or $f(X) \leq f(X_0)$) for all $X \in D$ which are in some neighborhood of X_0.*

For example, the polynomial $f(X) = x_1^3 - 3x_1$ has on $D = [-3, 3]$, stationary points at $x_1 = -1, 1$; the first is (only) a local maximum point while the second is (only) a local minimum point for f. (See Figure 0.1.)

(0.8) *When f is a convex function on D then it assumes a minimum value at each stationary point in D.*

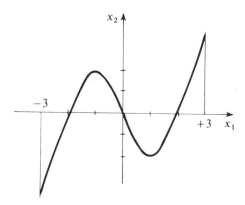

Figure 0.1

For our purposes, f will be said to be *convex* on D when it has continuous partial derivatives in D and satisfies the inequality

$$f(X) \geq f(X_0) + \nabla f(X_0) \cdot (X - X_0), \qquad \forall\, X, X_0 \in D. \tag{5}$$

A convex function need *not* have a stationary point, but obviously when X_0 is a stationary point of the convex function f; then by (4) $\nabla f(X_0) = \mathcal{O}$, so that $f(X) \geq f(X_0)$; i.e., (3) holds.

Observe that we can also express (5) by requiring that at each $X \in D$

$$f(X + V) \geq f(X) + \nabla f(X) \cdot V, \qquad \forall\, V \in \mathbb{R}^d \qquad \text{for which } X + V \in D. \tag{6}$$

For example, in $D = \mathbb{R}^2$, the function $f(x_1, x_2) = x_1^2 + x_2^2$, with gradient $\nabla f(X) = (2x_1, 2x_2)$, satisfies for each $V = (v_1, v_2)$ the inequality

$$\begin{aligned}
f(X + V) = (x_1 + v_1)^2 + (x_2 + v_2)^2 &= x_1^2 + x_2^2 + 2x_1v_1 + 2x_2v_2 + v_1^2 + v_2^2 \\
&= f(X) + \nabla f(X) \cdot V + |V|^2 \\
&\geq f(X) + \nabla f(X) \cdot V
\end{aligned}$$

and so it is convex. At its only stationary point $X_0 = \mathcal{O}$, it assumes its minimum value. [Indeed, trivially, $f(X) = x_1^2 + x_2^2 \geq 0 = f(\mathcal{O})$.]

Moreover, with this example, there is only one minimum point. This property is characteristic of *strictly* convex functions.

(0.9) *When f is strictly convex on D; i.e., when (5) holds at each $X_0 \in D$ with equality iff $X = X_0$, then f can have at most one stationary point, and hence, at most one interior minimum point, in D.*

[When X_0 is a stationary point of a strictly convex function f, then $f(X) > f(X_0)$, $\forall\, X \in D \sim \{X_0\}$. Thus f cannot assume its minimum value $f(X_0)$ at any other point.]

The function of the previous example is strictly convex on each set $D \subseteq \mathbb{R}^2$. For other examples of convex functions and alternate characterizations of convexity, see Problem 0.5, et seq.

(0.10) *f is differentiable at X_0 if for all X in a neighborhood of X_0,*

$$f(X) = f(X_0) + \nabla f(X_0) \cdot (X - X_0) + |X - X_0| \mathfrak{z}(X - X_0), \tag{7}$$

where $\mathfrak{z}(X - X_0)$ is a real valued function (defined for $X \neq X_0$ by (7)) with zero limit as $X \to X_0$. Then the graph of f (in $\mathbb{R}^{d+1}$) has at the point $(X_0, f(X_0))$ a tangent hyperplane, namely, the graph of the affine function

$$T(X) \overset{\text{def}}{=} f(X_0) + \nabla f(X_0) \cdot (X - X_0).$$

[For $d = 1$, this is just the line tangent to the curve representing the graph of f in $\mathbb{R}^2$; for $d = 2$, it is the plane tangent to the surface representing the graph of f in $\mathbb{R}^3$; for the general case, (7) may be used as a definition for tangency.]

We can give the stationarity requirement $\nabla f(X_0) = \mathcal{O}$ the geometric interpretation that the graph of f at X_0 has at the point $(X_0, f(X_0)) \in \mathbb{R}^{d+1}$ a "horizontal" tangent hyperplane; i.e., a d-dimensional subset parallel to $\mathbb{R}^d$. (See § 5.6.) Thus for $d = 2$, a marble "balanced" at $(X_0, f(X_0))$ should not roll but remain "stationary." By (5), we see that a convex differentiable function is one whose graph lies "above" its tangent hyperplanes.

NOTE: The existence of the partial derivatives of f in a neighborhood of X_0 together with their continuity at X_0 (as in (0.5)) are sufficient to guarantee that (7) holds. (See A.7 and [Ed].)

(0.11) *The classification of a stationary point X_0 as a local maximum point, local minimum point, or saddle point may be possible when f has higher-order nonvanishing derivatives at X_0.*

We first observe that *vanishing* higher-order derivatives provide no information. On $D = \mathbb{R}^1$, for each $n = 1, 2, 3, \ldots$ the functions

$$f_n(X) = (x_1)^{2n}, \qquad g_n(X) = (x_1)^{2n+1}$$

each have n derivatives which vanish at their common stationary point $X_0 = \mathcal{O}$, where f_n has a minimum value while g_n has neither a (local) maximum nor a (local) minimum value.

In $\mathbb{R}^1$, it is well known that if at a stationary point, X_0, $f''(X_0) > 0$ ($f''(X_0) < 0$) then f' is strictly increasing (decreasing) at X_0 so that X_0 is a *strict* local minimum (maximum) point for f. The generalization to a higher-dimensional space, where in addition, the possibility of X_0 as saddle point must be permitted, can best be approached through the second directional derivative.

(0.12) **Theorem.** *Let $f: D \to \mathbb{R}$ have continuous second-order partial derivatives*

$$f_{x_i x_j} = (f_{x_i})_{x_j} = \frac{\partial^2 f}{\partial x_j \partial x_i}; \qquad i, j = 1, 2, \ldots, d.$$

If X_0 is a stationary point of f in D, and for each $U = (u_1, u_2, \ldots, u_d) \in \mathbb{R}^d$:

$$\partial_U^2 f(X_0) \overset{\text{def}}{=} \frac{\partial^2}{\partial \varepsilon^2} f(X_0 + \varepsilon U)\Big|_{\varepsilon=0} = q(U) > 0, \qquad \text{when } |U| = 1, \qquad (8)$$

where

$$q(U) \overset{\text{def}}{=} \sum_{i,j=1}^{d} f_{x_i x_j}(X_0) u_i u_j,$$

then X_0 is a strict local minimum point for f.

[Indeed, for each unit vector U, as ε takes on real values in a one-dimensional neighborhood of 0, $\tilde{f}(\varepsilon) \overset{\text{def}}{=} f(X_0 + \varepsilon U)$ takes on all the values of

f in a neighborhood of X_0 in the (two-sided) *direction* of U. Now $\tilde{f}'(0) = (\partial f/\partial \varepsilon)(X_0 + \varepsilon U)|_{\varepsilon=0} = 0$ so that $\tilde{f}$ has a stationary value at 0 and the requirement of the hypothesis is simply that $\tilde{f}''(0) > 0$. Thus f has at X_0 strict local minimal behavior *in the* (two-sided) *direction* U. This extends via a compactness argument to establish that X_0 is a (strict) local minimum point for f [Ed].]

If $\partial_U^2 f(X_0)$ has unlike sign for two directions, U, then f has saddle point behavior at X_0.

(0.13) **Remarks.** If $V = cU \in \mathbb{R}^d$, then $q(V) = c^2 q(U)$. Hence (8) holds *iff* the *quadratic form* $q(V) > 0$, $\forall\, V \in \mathbb{R}^d$, $V \neq \mathcal{O}$.

When (8) holds, the (symmetric) Hessian *matrix* $f_{XX}(X_0)$ whose elements are the second partial derivatives $f_{x_i x_j}(X_0) = f_{x_j x_i}(X_0)$, $i, j = 1, 2, \ldots, d$ (arranged in natural order), is said to be *positive definite*. Conditions which characterize the positive definiteness of such matrices are known. (See Problem 0.10.) For the present, we observe that when (8) holds, the matrix $f_{XX}(X_0)$ is *invertible*. [If $[f_{XX}]V = \mathcal{O}$ for some $V \in \mathbb{R}^d$, then by the laws of matrix multiplication,

$$q(V) = \sum_{i,j=1}^{d} f_{x_i x_j} v_i v_j = \sum_{i=1}^{d} \left(\sum_{j=1}^{d} f_{x_i x_j} v_j \right) v_i = 0,$$

which by (8) is possible at X_0 only if $V = \mathcal{O}$. This condition furnishes the desired invertibility.] When $q(V) \geq 0$, $\forall\, V \in \mathbb{R}^d$, the matrix $f_{XX}(X_0)$ is said to be *positive semidefinite*.

(0.14) *Unless D is open, i.e., has only interior points, then it is also necessary to consider the extremal values of f on ∂D, the boundary of D.*

For example, although on $\mathbb{R}^2$ the function $f(X) = x_2^2 - x_1^2$ has its only stationary point at $X_0 = \mathcal{O}$, X_0 is a saddle point, and so the maximum and minimum values on say $D = \{X \in \mathbb{R}^2 : |X| \leq 2\}$ can be found only along the boundary where $|X| = 2$. In general for this set D, we would have to consider the problem of optimizing f subject to the *constraint* $g(X) \overset{\text{def}}{=} |X|^2 = 4$.

To find the stationary points of f when so constrained to the level sets of one (or more) functions such as g, we can employ the method of Lagrangian multipliers. Since this method will be fully treated in a more abstract setting (§ 5.7) we shall defer its further discussion. (See also [Ed].)

PROBLEMS

0.1. (a) Establish the Cauchy inequality (1). (Hint: If $Y = \mathcal{O}$, the inequality obviously holds; assume $Y \neq \mathcal{O}$, set $\mu = (X \cdot Y)/|Y|^2$, and consider $|X - \mu Y|^2$.)

    (b) What can you deduce about the relationship between X and Y if equality holds in (1)?

(c) Use the Cauchy inequality (1) to prove the triangle inequality (2a).
(d) Conclude that the reverse triangle inequality (2b) holds.

0.2. (a) Derive the inequality

$$|X||Y| - X \cdot Y \le |X - Y|^2 \qquad \text{for } X, Y \in \mathbb{R}^d.$$

(Hint: Show that $|X||Y| \le \frac{1}{2}(|X|^2 + |Y|^2)$ and add $\frac{1}{2}|X - Y|^2$ to the right side of this last inequality.)

(b) Use the result of part (a) to verify that

$$\left| \frac{X}{|X|} - \frac{Y}{|Y|} \right| \le \frac{\sqrt{2}}{\sqrt{|X||Y|}} |X - Y|$$

for $X, Y \in \mathbb{R}^d$, $X \ne 0$, $Y \ne 0$.

0.3. Find the maximum and minimum values (and the points at which they occur) for

$$f(X) = x_1^2 - x_1 x_2 + \tfrac{1}{6}x_2^3$$

on

$$D = \{X \in \mathbb{R}^2 : |x_j| \le 2, j = 1, 2\}.$$

0.4. Find the maximum and minimum values (and the points at which they occur) for

$$f(X) = x_1^2 - 2x_1 x_2 + 1$$

on

$$D = \{X \in \mathbb{R}^2 : |X| \le 1\}.$$

0.5. Which of the following functions are convex on $D = \mathbb{R}^2$? Which are strictly convex?

(a) $f(X) = x_1^2 - x_2^2$.
(b) $f(X) = x_1 - x_2$.
(c) $f(X) = x_1^2 + x_2^2 - 2x_1$.
(d) $f(X) = e^{x_1} + x_2^2$.
(e) $f(X) = x_1 x_2$.

(f) $f(X) = x_1^3 + x_2$.
(g) $f(X) = x_1^4$.
(h) $f(X) = \sin(x_1 + x_2)$.
(i) $f(X) = x_1^2 - 2x_1 x_2 + x_2^2$.
(j) $f(X) = ax_1 + bx_2 + c$, $\qquad a, b, c \in \mathbb{R}$.

0.6. (a) Prove that the sum of two [strictly] convex functions on $D \subseteq \mathbb{R}^d$ is [strictly] convex on D.
(b) Is the sum of a convex function and a strictly convex function strictly convex?
(c) Let f be strictly convex and $c > 0$. Show that cf is strictly convex.
(d) Give an example to show that the product of two convex functions need *not* be convex.

0.7*. Suppose that $f: \mathbb{R}^d \to \mathbb{R}$ is differentiable. Show that f is convex on $\mathbb{R}^d$ iff for each $X, X_0 \in \mathbb{R}^d$,

$$f(tX + (1 - t)X_0) \le tf(X) + (1 - t)f(X_0), \qquad \forall t, \qquad 0 < t < 1.$$

(Hint: To prove the "if" statement, use equation (7) to express

$$f(tX + (1 - t)X_0) = f(X_0) + t\nabla f(X_0) \cdot (X - X_0) + t|X - X_0|_3(tX - tX_0),$$

divide both sides of the resulting inequality by t, and let t approach zero. For the "only if" part, set $Y = tX + (1 - t)X_0$, $0 < t < 1$, and establish the inequalities

$$f(X_0) \geq f(Y) + \nabla f(Y) \cdot (X_0 - Y),$$

$$f(X) \geq f(Y) - \frac{(1-t)}{t} \nabla f(Y) \cdot (X_0 - Y).$$

Then, combine these last two inequalities to get the result.)

0.8*. Assume that $f: \mathbb{R}^d \to \mathbb{R}$ has continuous second-order partial derivatives. Show that f is convex on $\mathbb{R}^d$ iff for each $X_0 \in \mathbb{R}^d$, the matrix of second partial derivatives, $f_{XX}(X_0)$, is positive *semidefinite*, i.e.,

$$\sum_{i,j=1}^d f_{x_i x_j}(X_0) u_i u_j \geq 0, \qquad \forall\, U \in \mathbb{R}^d.$$

(Hint: Use Taylor's theorem for $\tilde{f}(t) \overset{\text{def}}{=} f(X_0 + tU)$ where $U = X - X_0, t \in \mathbb{R}$.)

0.9*. Let $D = \mathbb{R}^2$ and

$$f(X) = \begin{cases} \dfrac{2x_1 x_2}{x_1^2 + x_2^2}, & X \neq \mathcal{O}; \\[2mm] 0, & X = \mathcal{O}. \end{cases}$$

(a) Verify that the partial derivatives $\partial f / \partial x_1$ and $\partial f / \partial x_2$ both exist at $X = \mathcal{O}$.
(b) Show that f is *not* continuous at $X = \mathcal{O}$!

0.10. Let $f: \mathbb{R}^2 \to \mathbb{R}$ have continuous second-order partial derivatives and let $U = (u_1, u_2)$ be a unit vector.

(a) Verify that at X_0:

$$\partial_U^2 f(X_0) = A u_1^2 + 2B u_1 u_2 + C u_2^2,$$

where

$$A = f_{x_1 x_1}(X_0), \qquad B = f_{x_1 x_2}(X_0), \qquad C = f_{x_2 x_2}(X_0).$$

(b) If X_0 is a stationary point of f, with both $AC - B^2 > 0$ and $A > 0$, conclude that X_0 is a strict local minimum point for f. (Hint: let $u_1 = \cos \theta$, $u_2 = \sin \theta$.)

(c) Write the conditions of (b) in terms of subdeterminants of the 2×2 matrix f_{XX}, and conjecture a form for a corresponding set of conditions for the general $d \times d$ matrix.

0.11. Is $f(X) = |X|$ [strictly] convex on $\mathbb{R}^d \sim \{\theta\}$?

PART ONE

BASIC THEORY

Groningen,
January 1, 1697

AN ANNOUNCEMENT

"I, Johann Bernoulli, greet the most clever mathematicians in the world. Nothing is more attractive to intelligent people than an honest, challenging problem whose possible solution will bestow fame and remain as a lasting monument. Following the example set by Pascal, Fermat, etc., I hope to earn the gratitude of the entire scientific community by placing before the finest mathematicians of our time a problem which will test their methods and the strength of their intellect. If someone communicates to me the solution of the proposed problem, I shall then publicly declare him worthy of praise."

CHAPTER 1

Standard Optimization Problems

"Which method is best?" is a question of perennial validity, and through the centuries we have required for it answers of increasing sophistication. When "best" can be assessed numerically, then this assessment may be regarded as a real valued function of the method under consideration which is to be optimized—either maximized or minimized. We are interested not only in the optimum values which can be achieved, but also in the method (or methods) which can produce these values.

When the questions arise from classical science, then it is usually one of the fundamental quantities—length, area, volume, time, work or energy which is to be optimized over an appropriate domain of *functions* describing the particular class of method, process, or configuration so measured. Modern interests have added to this list cost, efficiency, ..., etc.

In this chapter, we shall examine several such problems—chiefly those of classical origin which have been influential in the development of a theory to furnish answers to the above questions. Although we can deal effectively with only two of these in this introductory chapter, the rest will serve in formulating the general class of problems to be considered in this text (§1.5).

§1.1. Geodesic Problems

Whether as a result of inherent laziness, or out of respect for efficiency, we apparently have long wished to know which of the many paths connecting two fixed points A and B is the shortest—i.e., has the least length. Although in $\mathbb{R}^3$ a straight line provides the shortest distance between two points (for reasons substantiated below), in general it may not be reasonable (or possible) to take this route because of natural obstacles, and then it is necessary to consider the more complicated problem of finding the *geodesic* curves (i.e., those of least length) among those constrained to a given "hyper" surface. In particular, we might wish to characterize in $\mathbb{R}^3$, the geodesics on the surface of a sphere, on a cylinder, or on a cone.

(a) Geodesics in $\mathbb{R}^d$

A curve in $\mathbb{R}^d$, joining points A and B may be considered as the range of a vector valued function $Y(t) = (y_1(t), y_2(t), \ldots, y_d(t))$, $t \in [0, 1]$, with components that are *continuous* on $[0, 1]$, such that $Y(0) = A$ and $Y(1) = B$. (When $d = 2$ or 3, we may think of $Y(t)$ as the position at "time" t, and see that the componentwise continuity reflects our desire that the curve not have jumps.) In particular, $Y_0(t) \overset{\text{def}}{=} (1 - t)A + tB$, $t \in [0, 1]$ defines one such curve, namely, the straight line segment determined by A and B (Figure 1.0).

With such generality, the curve may be nonrectifiable; i.e., it need not have finite length. However, if we require more smoothness; if, for example, we require that the component functions have continuous derivatives in $(0, 1)$, then we may think of $Y'(t) = (y_1'(t), y_2'(t), \ldots, y_d'(t))$ as representing the velocity at time t with the associated speed $|Y'(t)|$; hence the length of the curve should be the distance travelled during this motion,

$$L(Y) \overset{\text{def}}{=} \int_0^1 |Y'(t)| \, dt,$$

(considered as a possibly improper Riemann integral). For finiteness of L, we must require that each component of $Y'(t)$ be integrable, and this is most easily obtained by requiring that each component is continuously differentiable on $[0, 1]$. Thus we are led to consider the purely mathematical problem of *minimizing* this length function $L(Y)$ over all vector valued functions Y which meet the above conditions, as in Figure 1.0.

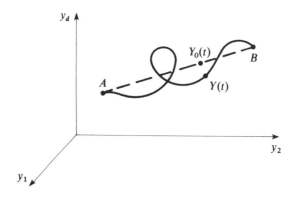

Figure 1.0

Since $Y_0(t)$ is surely admissible, with the derivative $Y_0'(t) = B - A$, we know that if a curve of minimal length $L_{\min}$ exists, then

$$L_{\min} \leq \int_0^1 |Y_0'(t)| \, dt = |B - A|,$$

the Euclidean distance between A and B.

To verify the natural conjecture, that $L_{\min} = |B - A|$, we need only establish the inequality that for any admissible Y,

$$|B - A| \leq L(Y) = \int_0^1 |Y'(t)| \, dt,$$

which is trivial when $A = B$; when $A \neq B$, this is most conveniently done as follows: Observe that from the fundamental theorem of calculus (A.8),

$$B - A = Y(1) - Y(0) = \int_0^1 Y'(t) \, dt;$$

(i.e., $y_j(1) - y_j(0) = \int_0^1 y_j'(t) \, dt, j = 1, 2, \ldots, d$). Thus

$$|B - A|^2 = (B - A) \cdot (B - A) = (B - A) \cdot \int_0^1 Y'(t) \, dt$$

$$= \int_0^1 [(B - A) \cdot Y'(t)] \, dt \leq \int_0^1 |B - A| |Y'(t)| \, dt,$$

or

$$|B - A|^2 \leq |B - A| \int_0^1 |Y'(t)| \, dt,$$

and for $A \neq B$, the desired inequality follows upon division by $|B - A|$. The reader should verify each step in this chain. In obtaining the inequality between the integrals, we utilized at each $t \in (0, 1)$, the Cauchy inequality in the form

$$[(B - A) \cdot Y'(t)] \leq |B - A| |Y'(t)|,$$

and this pointwise inequality in turn implies that of the corresponding integrals. [See A.10.]

Can a nonstraight curve also be of minimal length? (See Problem 6.39.)

(b) Geodesics on a Sphere

For airlines facing fuel shortages, it is essential to know the shortest route linking a pair of cities. Insofar as the earth can be regarded as a sphere, we see that we require knowledge of the geodesics on the surface of a sphere (albeit one which is large enough to be "above" the highest mountain range).

Each point $Y = (y_1, y_2, y_3) \in \mathbb{R}^3$ on the surface of a sphere of radius R centered at $\mathcal{O}$ (except the north and south poles) is specified through its

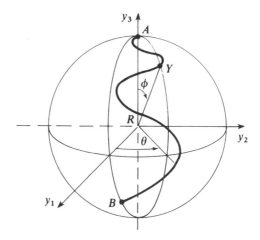

Figure 1.1

spherical coordinates R, φ, and θ as follows:

$$Y = (R \cos \theta \sin \varphi, \, R \sin \theta \sin \varphi, \, R \cos \varphi), \tag{1}$$

for unique $\varphi \in (0, \pi)$ and $\theta \in [0, 2\pi)$. Moreover, given distinct points A and B on this surface, we suppose, as we may, that the axes are chosen so that A is at the north pole ($\varphi = 0$), while $B \neq A$ has the spherical coordinates $(R, 0, \varphi_1)$, for $\varphi_1 > 0$. (See Figure 1.1.)

Then a curve joining A to B on the surface of this sphere is determined through (1) by the pair of continuous functions $(\theta(t), \varphi(t))$, $t \in [0, 1]$, with say $\varphi(0) = 0$; $\theta(1) = 0$, $\varphi(1) = \varphi_1$. (Note: To remain continuous θ and φ may have to exceed their ranges of $[0, 2\pi)$ and $[0, \pi)$, respectively. Also the choice $\theta(1) = 0$ is made for convenience.) By the same considerations as in the previous case we should require the continuous differentiability of θ and φ on $[0, 1]$. Then the resulting curve defined by

$$Y(t) = R(\cos \theta(t) \sin \varphi(t), \sin \theta(t) \sin \varphi(t), \cos \varphi(t)), \qquad t \in [0, 1],$$

[which has at each $t \in (0, 1)$ the derivative

$$Y'(t) = R[-\sin \theta(\sin \varphi)\theta' + \cos \theta(\cos \varphi)\varphi', \, \cos \theta(\sin \varphi)\theta'$$
$$+ \sin \theta(\cos \varphi)\varphi', \, -(\sin \varphi)\varphi'](t)],$$

has the length

$$L(Y) = \int_0^1 |Y'(t)| \, dt = R \int_0^1 \sqrt{\sin^2 \varphi(t)\theta'(t)^2 + \varphi'(t)^2} \, dt$$

$$\geq R \int_0^1 \varphi'(t) \, dt = R\varphi(t) \Big|_0^1 ;$$

thus $L(Y) \geq R\varphi_1$ by our requirements on φ. (See Problem 1.7.)

[Moreover, according to A.10, equality in the above occurs *iff* $(\sin^2 \varphi)\theta'^2 = 0$ and $\varphi' \geq 0$: or for $\varphi \in (0, \pi)$, when $\theta' \equiv 0$ so that $\theta = \text{const.} = \theta(1) = 0$; this corresponds to the smaller great circle arc joining A to B.] Thus we confirm the result known from antiquity that the shortest route joining two points on a spherical surface is precisely along the (shorter) great circle joining these points. Aircraft pilots are well advised to fly such routes, even though they may have to travel over polar regions in doing so.

(c) Other Geodesic Problems

As the last example shows, to characterize the geodesics of a specific hyper-surface in $\mathbb{R}^d$, it may be possible to utilize the properties of special coordinates associated with that surface. For example, in $\mathbb{R}^3$, we may use cylindrical coordinates to search for geodesics on a cylinder, on a cone, and on a general surface of revolution; some of the resulting problems will be examined in the next chapter. Curiously, the original mathematical interest in finding the geodesics on a general surface of revolution (as expressed by Jakob Bernoulli in 1697) arose because of the then recent discovery that our planet is *not* perfectly spherical.

The consideration of geodesics on a general surface S in $\mathbb{R}^3$ was begun by Johann Bernoulli (1698) and his pupil Euler (c. 1728), continued by Lagrange (1760) and treated rather decisively by Gauss (1827). When the surface can be described by $S = \{Y \in \mathbb{R}^3 : g(Y) = 0\}$ for some function g, we are required to minimize a length integral of the form $L(Y) = \int_0^1 |Y'(t)| \, dt$ as above, but now subject to the *Lagrangian* constraint $g(Y(t)) \equiv 0$. This problem forms a branch of differential geometry, and it will be considered again in §6.7.

§1.2. Time-of-Transit Problems

If we travel at constant speed, then the geodesic routes determined in the last section will also provide the least time of transit between given points A, B. However, if we cannot travel with constant speed—and, in particular, if the speed depends upon the path taken, then the problems of least distance and least time in transit must be considered separately. In this section we shall examine several such problems, including that of the brachistochrone (from the Greek βραχιστος ≡ "shortest," χρονος ≡ "time") which has been very significant in the emergence of the calculus of variations.

(a) The Brachistochrone

In falling under the action of gravity an object accelerates quite rapidly. Thus it was natural for Galileo to wonder (c. 1637) whether a wire bent in the shape

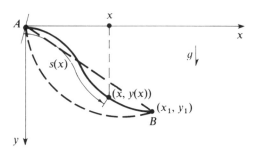

Figure 1.2

of the circular arc shown in Figure 1.2 might not offer a faster time of transit to a bead sliding down it under the action of gravity than would a straight wire joining the same two points.

In 1696, Johann Bernoulli challenged mathematicians to find the brachistochrone, that is, the planar curve which would provide the least time of transit. His own solution was derived from an optical analogy, [see Problem 1.1]; and solutions were provided by his brother Jakob, by Newton, by Euler, and by Leibniz. Although all of these reached the same conclusion—that the brachistochrone is not the circular arc, but a cycloid—none of their solutions is entirely satisfactory; however, that of Jakob Bernoulli admitted refinement and far-reaching generalization: the variational calculus.

If we use the coordinate system shown in Figure 1.2, in which the initial point A is the origin and the positive y axis is taken as vertically *downward*, then a typical curve which might represent the brachistochrone joining A to a lower point $B = (x_1, y_1)$ where x_1 and y_1 are positive, can be represented as the *graph* of a continuous function $y = y(x)$, $x \in [0, x_1]$ with $y(0) = 0$ and $y(x_1) = y_1$. (Here we are abandoning the parametric form of representing curves used in the previous section in favor of one less general but more convenient. Although it is reasonable that the class of curves needed should be so representable (Why?), the reader should consider whether something essential is lost with this restriction.)

Assuming sufficient differentiability, this curve has length l and the time required to travel along it is given through pure kinematic considerations as

$$T = T(y) = \int_0^l \frac{ds}{v},$$

where $v = ds/dt$ is the speed of travel at a distance s along the curve.

Now from calculus, for each $x \in [0, x_1]$, $s = s(x) = \int_0^x \sqrt{1 + y'(\xi)^2} \, d\xi$ is the arc length corresponding to the horizontal position x, and we may regard $v = v(x)$ as the associated speed. Thus with these substitutions,

$$T = T(y) = \int_0^{x_1} \frac{\sqrt{1 + y'(x)^2}}{v(x)} \, dx.$$

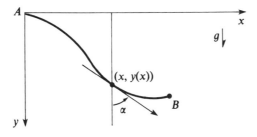

Figure 1.3

To express v in terms of y we resort to Newton's laws of dynamics. Assuming that the gravitational acceleration, g, is constant during the fall, and neglecting the effects of other forces (including that of friction), the acceleration of a bead (of unit mass) along the wire at time t is

$$\dot{v} = g \cos \alpha; \qquad \text{also } \dot{y} = v \cos \alpha,$$

where α is the angle between the tangent line to the curve and the y axis at this point (Figure 1.3). (We use the Newtonian notation of a dot to signify a derivative with respect to time.)

Thus $v\dot{v} = g\dot{y}$, or upon integrating with respect to time,

$$v^2 = 2gy + \text{const.}$$

But at A, $v = y = 0$, so that in general,

$$v = \sqrt{2gy} \qquad \text{or } v(x) = \sqrt{2gy(x)}.$$

Thus finally,

$$T = T(y) = \frac{1}{\sqrt{2g}} \int_0^{x_1} \left(\frac{1 + y'(x)^2}{y(x)} \right)^{1/2} dx, \qquad (2)$$

and we have the problem of minimizing T over all functions $y = y(x)$ which validate the above analysis. However, we may consider also the mathematical problem of minimizing the integral

$$\int_0^{x_1} \left(\frac{1 + y'(x)^2}{y(x)} \right)^{1/2} dx$$

over all functions y continuous on $[0, x_1]$ for which $y(0) = 0$, $y(x_1) = y_1$, and the integral is defined. This last condition requires that y have a derivative integrable on $[0, x_1]$ and that y be ≥ 0 with

$$\int_0^{x_1} (y(x))^{-1/2} dx < +\infty.$$

In any case, there is no obvious answer although we may verify Galileo's conjecture that a circular arc is superior to the straight line. (See Problem 1.2.)

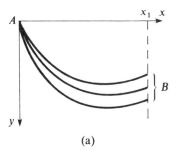 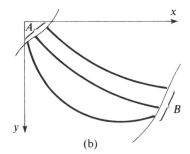

(a)   (b)

Figure 1.4

Alternate forms of this problem soon suggested themselves. For example, in 1697, Jakob Bernoulli challenged mathematicians to find the brachisto-chrone among those curves representing travel over a fixed horizontal distance x_1, but for which the y_1 coordinate may vary, as in Figure 1.4(a). Later, mathematicians investigated the problem of finding the brachistochrone joining fixed *curves* as in Figure 1.4(b).

Newton also considered the problem of finding the brachistochrone associated with tunnels *through* the earth connecting fixed points. Again, the least time is given not by the straight line, but by another type of cycloid [Sm].

It required nearly two more centuries to obtain, for these problems, the mathematical solutions to be presented in this book. See §3.4, §6.2, §8.8, and §9.2. At each stage of development, however, the brachistochrone remained a testing ground for the state of the art.

(b) Steering and Control Problems

Closely related to the problem of the brachistochrone, is that from this century concerning the best course to steer when navigating through a current of varying speed. For example, which path joining fixed points A and B on opposite banks of a river with varying current will provide minimum transit time for a boat which travels with constant speed w *with respect to the water*? As in Figure 1.5, we suppose the river banks parallel and utilize the coordinate system shown, in which the y axis represents one bank and the line $x = x_1$ the other. We also assume that $w = 1$ and that the river current r is directed *downstream* and admits the prescription $r = r(x)$, continuous on $[0, x_1]$. Then the time of transit of a boat travelling between the origin A and the downstream point $B = (x_1, y_1)$ along a smooth path which is the graph of a function $y = y(x)$ on $[0, x_1]$ is given (after some work left to Problem 1.3) by

$$T(y) = \int_0^{x_1} [\alpha(x)\sqrt{1 + (\alpha y')^2(x)} - (\alpha^2 r y')(x)]\, dx, \tag{3}$$

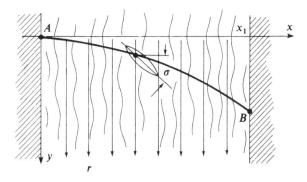

Figure 1.5

where $\alpha(x) = (1 - r^2(x))^{-1/2}$. (In order that the crossing be possible we must have $0 \le r(x) < 1$.) (Why?) We are required to minimize this rather compli- cated integral over all those functions y which are continuously differentiable on $[0, x_1]$, and satisfy $y(0) = 0$, $y(x_1) = y_1$. The methods of Chapter 3 will provide access to a solution. (See Problem 3.20.)

We may also consider the more natural problems in which the river banks are represented by curves, and permit the crossing points to vary. Finally, we can also permit the current to vary with y as well as x. In fact, in 1931, Zermelo investigated the two-dimensional version of this problem which could be equally significant to the piloting of a submarine or a light aircraft [C]. And when we ask how to operate our craft so that it travels along an optimal path, we enter the realm of *optimal control problems*, first consid- ered by Minorsky around 1920.

(Problems 1.1–1.3, 1.8)

§1.3. Isoperimetric Problems

One of the most ancient optimization problems of which we have record is that of finding the maximal area which can be enclosed by a curve of fixed perimeter. According to Virgil, this problem was already of importance to Dido of Carthage (c. 850 B.C.) and she supposedly obtained a solution on heuristic grounds. (See Problem 1.5.) The Greek geometer Zenodoros appar- ently knew that the circle provided a greater area than polygons having this same perimeter, and a few centuries later Pappus (c. A.D. 390) concluded that the circle was maximal among *isoperimetric* curves.

There is a simple physical analogy which supports this conjecture: Sup- pose that a cylinder with thin inextensible impervious, but completely flexi- ble walls is deformed so that its constant cross-sectional shape is that of the area to be maximized. Then a section of it is placed with walls vertical on a

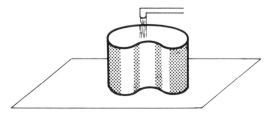

Figure 1.6

smooth horizontal surface and filled with water as in Figure 1.6. Under the
action of gravity, the water will seek its lowest level and if we assume none
lost by leakage at the base, then this will be accomplished by a movement of
the walls of the cylinder until the effects of hydrostatic pressure are equalized.
Since this pressure is exerted uniformly at each depth, the final configuration
must have constant curvature, i.e., it is that of a right circular cylinder.
However, the configuration which provides least depth must clearly be that
having maximal cross-sectional area. Thus the circle encloses a greater area
than any noncircular isoperimetric curve.

One mathematical formulation of this problem is as follows: We suppose
that a smooth simple closed curve of length l is represented parametrically by
$Y(t) = (x(t), y(t))$, $t \in [0, 1]$, with $Y(0) = Y(1)$ for closure (Figure 1.7).

Then according to Green's theorem [Ed], the area of the domain D
bounded by the curve is

$$A(Y) = \iint_D dx\, dy = \int_{\partial D} x\, dy,$$

where ∂D denotes the boundary of D assumed positively oriented by the
parametrization through Y. Utilizing this parametrization we have

$$A(Y) = \int_0^1 x(t) y'(t)\, dt; \tag{4}$$

we must maximize $A(Y)$ over all functions $Y(t)$ having continuously differ-

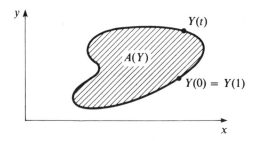

Figure 1.7

entiable components on $[0, 1]$, which meet the closure conditions $Y(0) = Y(1)$, and for which the resulting curve satisfies the *isoperimetric condition*

$$L(Y) \equiv \int_0^1 |Y'(t)| \, dt = l \tag{5}$$

for a *given l*.

We shall return to the solution of this problem in §6.7.

Remark. This problem has received much attention and there have been other less restrictive formulations. In particular, the German geometer Jakob Steiner (c. 1840) devised several techniques to attack it [P]. One well-known analytical solution utilizes properties of Fourier series, (see Problem 1.6.), but in §8.8, we will present a recently discovered solution that seems more natural.

A modern version which combines features of the isoperimetric and steering problems from the previous section is due to Chaplygin (1938). It consists of describing the closed path which an airplane say, on reconnaisance, should fly with constant airspeed in the presence of a wind, in order to enclose the maximum area. When the wind speed is zero, then this problem is equivalent to the classical isoperimetric problem (Problem 1.4).

Zenodoros considered also the higher-dimensional analogue of maximizing the volume of a solid having fixed surface area (Problem 1.9).

A different isoperimetric problem (often attributed to Euler)[1] consists of finding the shape which a thin inextensible long cable or chain of weight/unit length W, and given length L, will assume under its own weight when supported freely from end points separated a fixed horizontal distance H. As we shall show in §3.5, this requires the minimization of the center-of-mass integral $F(y) = W \int_0^L y(s) \, ds$ over all functions $y \geq 0$ continuously differentiable in $[0, L]$, with $y(0) = y(L) = 0$, which satisfy the "isoperimetric" condition

$$G(y) \equiv \int_0^L \sqrt{1 - y'(s)^2} \, ds = H. \tag{6}$$

In general, the term "isoperimetric" is assigned to any optimization problem in which the class of competing functions is subject to integral or global constraints of the form (5) or (6). The resulting isoperimetric problems admit simple abstract formulation (§5.7) in contrast to Lagrangian problems (such as those for geodesics on a general surface) in which intrinsic constraining conditions must be satisfied (§6.7).

(Problems 1.4–1.6)

---

[1] This problem was first proposed by Galileo (who believed that a parabolic shape would be optimal) and it was then attacked mathematically by the Bernoullis in 1701. (See Goldstine.)

§1.4. Surface Area Problems

A higher-dimensional analogue of the geodesic problem discussed in §1.1, might be formulated as follows:

Find the surface of minimum area that spans fixed closed curves in $\mathbb{R}^3$.

(a) Minimal Surface of Revolution

For example, when the curves consist of a pair of "concentric" parallel circles such as those shown in Figure 1.8, then we could ask for the surface of revolution which "joins" them and has minimum area—or, equivalently, we would wish to find the shape of its boundary curve. The problem in this form was first attacked by Euler (c. 1744) who employed his recently developed theory of the calculus of variations in its solution.

If we utilize the coordinate system shown, then we would be led by elementary calculus to minimize the surface area function

$$S(y) = 2\pi \int_a^b y(x)\, ds(x) = 2\pi \int_a^b y(x)\sqrt{1 + y'(x)^2}\, dx \qquad (7)$$

among all functions y which are *nonnegative*, continuously differentiable on $[a, b]$, and, for which

$$y(a) = a_1 \qquad \text{and} \quad y(b) = b_1.$$

Here, a_1 and b_1 denote the given radii of the bounding circles, one of which may degenerate to a point. (However, see Problem 1.7.)

When a_1 and b_1 are comparable to $b - a$ as in Figure 1.8, it is reasonable to expect a minimizing y to be of this form. However, when $b - a$ greatly exceeds a_1 and b_1, as in Figure 1.9, then it is seen that the surface area can be

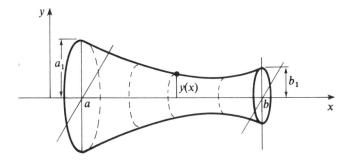

Figure 1.8

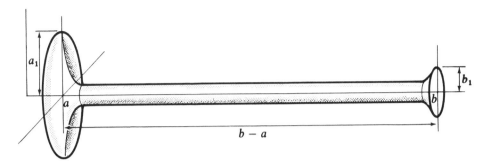

Figure 1.9

made as close as we please to the area of the bounding disks—and that these probably represent the "least" area—but the associated "boundary" curve is not of the form admitted.

We have two alternatives: First, we can simply predict that in this case the problem as posed has no solution and attempt to substantiate this. Or, we can seek a reformulation of the problem in which such cornered curves remain admissible. We shall adopt the second alternative when we return to this problem in §7.5. For the present, we note that a framework large enough to include this alternative must accomodate an accurate description of cornered curves.

(b) Minimal Area Problem

Suppose that the competing surfaces can be represented as the graphs of smooth functions $u = u(x, y)$ defined on a common planar domain D, as in Figure 1.10. Then the associated surface area is given by calculus as

$$S(u) = \int_D \sqrt{1 + u_x^2 + u_y^2} \, dA,$$

where dA denotes the two-dimensional element of integration over D; the boundary of D is denoted by ∂D, and in this text, we suppose it to be so well-behaved that Riemann integration of continuous functions can be defined over D or $\overline{D}$ and over ∂D [Ed].

We would then seek the minimum for $S(u)$ over all functions u which are continuous on $\overline{D} = D \cup \partial D$, continuously differentiable inside D, and have given continuous boundary values $u|_{\partial D} = \gamma$, say. We shall obtain some partial results for this problem, which need *not* have a solution, in §3.4(e), under the assumption that D is a *Green's domain* (one whose boundary smoothness admits use of Green's theorem [F1].)

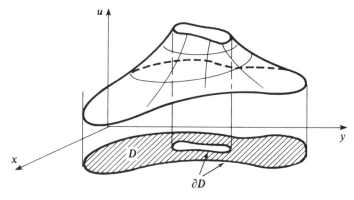

Figure 1.10

(c) Plateau's Problem

A new impetus was given to this class of problems in 1873 when the Belgian mathematical physicist Joseph Plateau noticed that wires bent in the shape of the bounding curves could be made to support a thin film of glycerine which in order to minimize the surface tension should assume the shape associated with the minimal area. These "soap film" experiments have been continued until this day; they show that for some configurations, more than one type of solution is possible, and in some cases the solutions can be observed to change form as the relative geometry is altered. For example, from our discussion of the minimal surface of revolution, we would expect that a soap film joining a pair of circular rings that is initially cylindrical in shape might transform into a pair of disks as the distance between the rings is increased.

The mathematics supporting a general theory for such surfaces is outside the scope of the present text. (See, for example, [Os].)

(Problems 1.7–1.9)

§1.5. Summary: Plan of the Text

These problems exhibit common ingredients. Each requires the optimization of a real valued function F defined on a certain domain $\mathscr{D}$ of functions Y, usually by means of an integral of the form

$$F(Y) \int_a^b f(x, Y(x), Y'(x)) \, dx, \tag{8}$$

for some given real valued function f. Here Y is a real [or possibly a vector] valued function [each component of] which is continuous on $[a, b]$ with a continuous derivative in (a, b); and $\mathscr{D}$, in general, consists of those functions of this class which meet certain specified boundary conditions at a and/or b. However, as in the case of the brachistochrone, it may be necessary to impose further restrictions such as requiring that $Y = y \geq 0$ on $[a, b]$. In addition, there may be constraints of the isoperimetric form in which

$$G(Y) \overset{\text{def}}{=} \int_a^b g(x, Y(x), Y'(x))\, dx = l; \tag{9}$$

or of the Lagrangian form in which

$$g(x, Y(x), Y'(x)) \equiv 0, \qquad x \in (a, b), \tag{10}$$

for given functions g and constants l.

We may also find it necessary to enlarge the domain $\mathscr{D}$ to include functions with corner points; i.e., discontinuous derivatives, and to permit more freedom of the boundary points such as that required to treat the brachistochrone joining fixed curves.

Finally, we should also consider the obvious extensions to integrals involving higher derivatives of the functions and to functions of more than one variable.

In the generality considered it is seen that the theory should be applicable to any problem of finding the "best" process which meets the following criteria:

(1) The processes can be described by a suitable class of functions.
(2) The value of each process can be measured by a weighted averaging usually represented through an integral involving the function and one or more of its derivatives.

Thus, for example, it should be applicable to problems of minimizing the strain energy stored in a deformed elastic body such as a bent beam. And it should also provide some insight as to how an economic process should be operated over a fixed period of time in order to minimize the cost of operation—or maximize the return on investment.

A complete solution to any of these problems requires characterizing those functions in the domain which could produce the extremal values of interest (either $F_{\max}$ or $F_{\min}$) and proving that these are indeed the extremal values sought. In this chapter we have obtained a satisfactory solution to only one of our problems—that of finding the geodesics on a sphere (§1.1(b)). There, a proper (and natural) selection of the domain $\mathscr{D}$ led directly to the inequality

$$L \text{ (curve)} \geq L \text{ (shorter great circle joining same points)},$$

with equality *iff* the curve is the great circle. Moreover, it was not necessary to "know" or guess the answer beforehand as we did when proving that the geodesics in $\mathbb{R}^3$ are the straight lines.

The general theory known as the calculus of variations has been developed over the last three centuries to handle such problems. It arose out of efforts to duplicate, insofar as possible, the analysis of optimization of ordinary functions defined on a set in $\mathbb{R}^d$ (a review of which is given in Chapter 0).

From the theoretical calculus of variations we learn that functions which could provide the extremal values sought must satisfy the differential equations of Euler-Lagrange determined by the integrand f (and g, if constraints of the form (9) or (10) are present). On the other hand, solution to these equations need not be possible within the required class ($\mathscr{D}$), and even when possible, need not furnish the extremal values sought.

In $\mathbb{R}^d$, the minimization of convex functions is simple to analyze, and in Chapter 3 we shall see that suitable convexity of F again provides access to the solution of some of these problems—even when isoperimetric or Lagrangian constraints are present.

However, this convexity is best formulated in terms of the directional derivatives of F (its Gâteaux variations), in the setting of a linear space (Chapter 2).

After the brief technical Chapter 4, we return to this space in Chapter 5, supply it with a norm and obtain conditions (again expressed through Gâteaux variations) which are necessarily satisfied, if a function is minimized or maximized locally on a subset of this normed linear space. In this setting we shall also develop the method of Lagrangian multipliers to treat optimization with constraints of the isoperimetric type (9).

In Chapter 6, we apply the results of Chapter 5 to functions F defined by integrals such as (8) to obtain the classical theory of Euler (1742) and Lagrange (1755). The effects of constraints of various types are also considered. Extensions to integrals involving higher-ordered derivatives or vector valued functions, and to multidimensional integrals will also be given. However, we reserve the generalization to piecewise C^1 (cornered) extremals for presentation in Chapter 7, where it will also be shown that the existence of a *minimizing* function demands some convexity from the integrand f (§7.6).

In Chapter 9, we demonstrate that, conversely, such partial convexity of the integrand ensures that some solutions obtained previously do minimize their integral functions, provided that a suitable family of these solutions is available. These sufficiency arguments, which here incorporate earlier results from Legendre and Jacobi, are now known as field theory and are due largely to Weierstrass (c. 1879) and Hilbert (1900). (The so-called direct methods, also initiated by Hilbert, but developed by Tonelli and his successors from 1915, wherein existence of a minimizing function is established through *a priori* estimates, are considered to lie outside the scope of this text. See [G-F], [Ak].)

However, one of the most satisfying uses of the Euler-Lagrange theory has come from another direction, namely, the recognition that differential equations which describe a physical process might be regarded as the Euler-Lagrange equations for a variational problem. This generated the variational principles of mechanics culminating in the Hamilton (c. 1835)–Jacobi (c. 1840) theory, which has survived the transition from classical physics to quantum mechanics, and which properly interpreted, is still regarded as the correct theoretical foundation for the laws that govern the operation of our universe. We shall examine this aspect of the subject in Chapter 8 (which is essentially independent of Chapters 7 and 9).

In the concluding part (Part III), we turn to problems of optimal control where the vector-state Y of a system at time t is governed by a dynamical system of differential equations of the vector form

$$Y'(t) = G(t, Y(t), U(t))$$

dependent upon a vector "control" function $U(t)$. The task is to determine a "path" Y_0 and a control U_0 that optimizes some performance assessment integral of the form

$$F(Y, U) = \int_a^b F(t, Y(t), U(t)) \, dt$$

perhaps subject to certain endpoint or other constraints, and control restrictions. (For example, an engine throttle can open only so far.)

In Chapter 10 we use partial convexity of some associated integrals to suggest a governing principle (due to Pontjragin, c. 1960) guaranteeing optimality of the choices that it dictates. Then in Chapter 11, we will learn that these Pontjragin conditions are necessary for the optimal choices, whether or not such convexity is present. Finally we use the principle to obtain necessary conditions when other types of Lagrangian or isoperimetric constants are present.

To establish the above assertions, we shall use several results from advanced calculus and the theory of differential equations, proofs for most of which will be found in the Appendix.

Notation: Uses and Abuses

In this book, we use the following abbreviations which are now standard in mathematical literature:

$\in$	for "is an element of" or "is in" or "in";
$\exists$	for "there exists";
$\forall$	for "for each" or "for every";
iff	for "if and only if";
C^1	for "continuously differentiable";
C^2	for "twice continuously differentiable";
C^n	for "n times continuously differentiable";
$\mathbb{R}^d$	for "d-dimensional Euclidean space";
$\square$	for "end of proof";
$x \nearrow a$	for "x approaches a from below";
$x \searrow a$	for "x approaches a from above";
$\overset{\text{def}}{=}$	for "defined as";
$*$	to designate material of more than average difficulty.

In addition, we shall need notation which is not standard in order to handle effectively composite functions such as

$$f(x, y(x), y'(x)) \quad (\text{or } f(x, Y(x), Y'(x)))$$

and their derivatives.

First, we write $f = f(x, y, z)$ for which the partial derivatives are as usual denoted

$$f_x(x, y, z), f_y(x, y, z), f_z(x, y, z), f_{xx}(x, y, z), f_{xy}(x, y, z), f_{xz}(x, y, z), \ldots, \text{ etc.}$$

Then we evaluate at $(x, y(x), y'(x))$, and denote the respective results by

$$f[y(x)], f_x[y(x)], f_y[y(x)], f_z[y(x)], f_{xx}[y(x)], f_{xy}[y(x)], f_{xz}[y(x)], \ldots, \text{ etc.}$$

However, for a *given* function $y = y(x)$, in Chapter 6 we further condense the notation to

$$f(x) = f[y(x)] = f(x, y(x), y'(x)),$$
$$f_x(x) = f_x[y(x)] = f_x(x, y(x), y'(x)),$$
$$f_y(x) = f_y[y(x)] = f_y(x, y(x), y'(x)),$$
$$f_{y'}(x) = f_z[y(x)] = f_z(x, y(x), y'(x)),$$
$$f_{y'y'}(x) = f_{zz}[y(x)]; \ldots, \text{ etc.},$$

where use of the subscript (y') is consistent with regarding $f(x, y, z)$ as $f(x, y, y')$, an approach taken by some authors.

Vectors will always be indicated by capital letters. For vector valued functions $Y = Y(x)$, the corresponding abbreviations when $f = f(x, Y, Z)$, are as follows:

$$f(x) = f[Y(x)] = f(x, Y(x), Y'(x)),$$
$$f_x(x) = f_x[Y(x)] = f_x(x, Y(x), Y'(x)),$$
$$f_Y(x) = f_Y[Y(x)] = f_Y(x, Y(x), Y'(x)),$$
$$f_{Y'}(x) = f_Z[Y(x)] = f_Z(x, Y(x), Y'(x)).$$

where $f_Y(x, Y, Z)$ is the *vector* with jth component $f_{y_j}(x, Y, Z)$; and again use of the subscript (Y') is consistent with regarding $f(x, Y, Z)$ as $f(x, Y, Y')$.

Observe that with this notation, the chain rule takes the following forms

$$\frac{d}{dx} f[Y(x)] = f_x[Y(X)] + f_Y[Y(x)] \cdot Y'(x) + f_Z[Y(x)] \cdot Y''(x)$$

or

$$\frac{d}{dx} f(x) = f_x(x) + f_Y(x) \cdot Y'(x) + f_{Y'}(x) \cdot Y''(x),$$

assuming sufficient differentiability. (In the later chapters, we will utilize a notation for the *matrices* of the *second* partial derivatives (and similar expressions) which is consistent with that above. It will be explained as it is required.)

Finally, because of frequency of occurrence, we simplify $(y'(x))^2$ to $y'(x)^2$, and make corresponding reductions for $(y''(x))^2$, $(y'(x))^3$, ..., etc., while $(y(x))^2$ is given either of the forms $y^2(x)$ or $y(x)^2$, as desired.

PROBLEMS

1.1. **The Brachistochrone.** (The following optical analogy was used by Johann Bernoulli, in 1696, to solve the brachistochrone problem.) In a nonuniform medium, the speed of light is not constant, but varies inversely with the index of refraction. According to Fermat's principle of geometric optics, a light ray travelling between two points in such a medium follows the "quickest" path joining the points. Bernoulli thus concluded that finding the brachistochrone is equivalent to finding the path of a light ray in a planar medium whose index of refraction is inversely proportional to $\sqrt{y}$.

The optics problem can be solved by using Snell's law which states that at each point along the path of a light ray, the sine of the angle which the path makes with the y-axis is inversely proportional to the index of refraction (and hence proportional to the speed). Therefore, the brachistochrone should satisfy

$$c \sin \alpha = \sqrt{y}, \tag{11}$$

where c is a constant and α is as shown in Figure 1.3.

(a) Assuming that the brachistochrone joining $(0, 0)$ to (x_1, y_1) can be represented as the graph of a smooth function $y = y(x)$, use (11) to prove that

$$c^2 = y(x)[1 + y'(x)^2], \qquad 0 < x < x_1.$$

(b) Show that the cycloid given parametrically by

$$x(\theta) = \frac{c^2}{2}(\theta - \sin \theta),$$

$$0 \le \theta \le \theta_1,$$

$$y(\theta) = \frac{c^2}{2}(1 - \cos \theta),$$

satisfies the differential equation found in part (a) and has $(0, 0)$ as one end point. (It will be shown in §3.4 that c and θ_1 can always be chosen to make (x_1, y_1) the other end point, and that the resulting curve is expressible in the form $y = y(x)$.)

(Although this does not constitute a mathematically rigorous solution to the problem, it illustrates an important parallel between geometric optics and particle mechanics which led to the works of Hamilton.)

1.2. **A Brachistochrone.** (See §1.2(a).) Let $x_1 = y_1 = 1$.

(a) Use equation (2) to compute $T(y)$ for the straight line path $y = x$.

(b) Use equation (2) and the trigonometric substitution $1 - x = \cos \theta$ to show that for the circular arc $y = \sqrt{1 - (x - 1)^2}$,

$$T(y) = (2g)^{-1/2} \int_0^{\pi/2} (\sin \theta)^{-1/2} \, d\theta.$$

(c)   Use a table of definite integrals to conclude that the circular arc in part (b) provides a smaller transit time than the line segment in part (a).

(d)   Use the inequality $\sin \theta \le \theta$, $\theta \ge 0$, to obtain a *lower* estimate for the transit time of the circular arc of (b);

(e)* Find similar (but more precise) *upper* estimates which lead to the same conclusion as in (c) *without* obtaining a numerical value for the integral in (b).

1.3. Transit Time of a Boat. (See §1.2(b).) Use the following steps to derive equation (3):

(a)   Show that the x- and y-components of the velocity of the boat are given respectively by $\cos \sigma$ and $r + \sin \sigma$, where σ is the steering angle of the boat relative to the x-axis shown in Figure 1.5.

(b)   Prove that the crossing time is given by

$$T = \int_0^{x_1} \sec \sigma \, dx, \qquad \ge x_1.$$

(c)   Show that $y' = r \sec \sigma + \sqrt{\sec^2 \sigma - 1}$.

(d)   Conclude that

$$T(y) = \int_0^{x_1} [\alpha(x)\sqrt{1 + (\alpha y')^2(x)} - (\alpha^2 r y')(x)] \, dx,$$

where $\alpha(x) = (1 - r^2(x))^{-1/2}$.

1.4. Chaplygin's Problem. A small airplane in level flight with constant unit airspeed flies along a simple smooth closed loop in one hour in the presence of a wind with constant direction and speed $w \le 1$. Suppose that its ground position at time t is given by $Y(t) = (x(t), y(t))$ where the wind is in the direction of the positive x-axis.

(a)   Argue that $(x'(t) - w)^2 + y'(t)^2 \equiv 1$, $t \in [0, 1]$, while $A(Y) = \int_0^1 x(t) y'(t) \, dt$ represents the area enclosed by the flight path.

(b)   Formulate the problem of finding the flight path(s) maximizing the area enclosed. (We return to this formulation in Problem 9.19.)

(c)   When $w = 0$, show why a solution of the problem in (b) would solve the classical isoperimetric problem of §1.3.

(d)   As formulated in (b) Chaplygin's problem is not isoperimetric. (Why not?) Recast it as an isoperimetric problem in terms of $\sigma(t)$, the steering angle at time t between the wind direction and the longitudinal axis of the plane. (Hint: Take $Y(0) = Y(1) = \mathcal{O}$ and conclude that $x(t) = wt + \int_0^t \cos \sigma(\tau) \, d\tau$, while $y(t) = \int_0^t \sin \sigma(\tau) \, d\tau$).

1.5. Queen Dido's Conjecture. According to Virgil, Dido (of Carthage), when told that her province would consist only of as much land as could be enclosed by the hide of a bull, tore the hide into thin strips and used them to form a long rope of length l with ends anchored to the "straight" Mediterranean coast as shown in Figure 1.11. The rope itself was arranged in a semicircle which she believed would result in the maximum "enclosed" province. And thus was Carthage founded—in mythology.

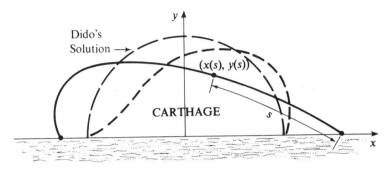

Figure 1.11

For simplicity, suppose that $l = \pi$; then use the arc length s as the parameter.

(a) Show that the conjectured inequality for the vector valued function $Y = (x, y)$ with $y(0) = y(\pi) = 0$, is

$$A(Y) = -\int_0^\pi y(s)x'(s)\,ds \le \frac{\pi}{2} = \left(\frac{1}{2}\right)\int_0^\pi [x'(s)^2 + y'(s)^2]\,ds.$$

Hint: Use Green's theorem.

(b) Prove that the inequality in (a) is satisfied if $\int_0^\pi [y'(s)^2 - y^2(s)]\,ds \ge 0$, when $y(0) = y(\pi) = 0$, and y is continuously differentiable on $[0, \pi]$. (We shall establish this inequality in §9.1.)

(c) Show that equality in (a) and (b) would imply that $x'(s) = -y(s)$.

(d) Verify that equality will hold in (a) and (b) for the trigonometric functions $x(s) = \cos s$, $y(s) = \sin s$, defining the semicircle.

(e) How is Dido's problem related to that considered in §1.3? Could a solution to one of these problems yield a solution to the other?

1.6. The Isoperimetric Inequality. (See §1.3 and §9.5.)

To derive a formulation analogous to that in the preceding problem, use the arc length parametrization of a closed curve of length $l = 2\pi$.

(a) Show that for $Y = (x, y)$ so parametrized, with say $Y(0) = Y(2\pi) = \mathcal{O}$, the isoperimetric inequality $A(Y) \le \pi$, would follow from *Wirtinger's inequality:*

$$\int_0^{2\pi} [y'(s)^2 - y(s)^2]\,ds \ge 0, \quad \text{when } y(0) = y(2\pi) = 0, \quad and \quad \int_0^{2\pi} y(s)\,ds = 0,$$

where y is continuously differentiable on $[0, 2\pi]$. Hint: What is the geometrical significance (for the curve) of the last integral requirement?

(b) Prove that equality throughout in (a) would require that $x'(s) = -y(s)$, with $x(0) = x(2\pi) = 0$.

(c) Verify that the trigonometric functions $x(s) = \cos s - 1$, $y(s) = \sin s$ meet all of the requirements in (a), (b), and give equality throughout. What curve do they parametrize?

(d) Show that the integral inequality in (a) is *violated* for the function $y_1(s) = \sin s/2$. Is $\int_0^{2\pi} y_1(s)\, ds = 0$?

1.7. Degenerate Minimal Surface of Revolution. (See §1.4(a).)

When one of the end circles degenerates to a point, then we can suppose that a typical curve to be rotated is the graph of a continuously differentiable function $y = y(x)$, $0 \le x \le b$, with $0 = y(0) \le y(x)$; $y(b) = b_1 > 0$. Prove that the resulting surface area $S(y) = 2\pi \int_0^b y(x)\sqrt{1 + y'(x)^2}\, dx \ge \pi b_1^2$. Give an interpretation of this inequality. Hint: $\sqrt{1 + c^2} \ge c$.

1.8. A Seismic Wave Problem.

Suppose that a seismic disturbance or wave travels with a speed which is a linear function of its depth η below the earth's surface (assumed flat) along a subterranean path which minimizes its time of transit between fixed points.

Assume that a typical (planar) path joining the points is the graph of a continuously differentiable function $y(x) = \eta(x) + \sigma$, $a \le x \le b$, where σ is a positive absolute *constant*, and η is the positive local depth of the path, as shown in Figure 1.12(a); $y(a) = a_1$ and $y(b) = b_1$.

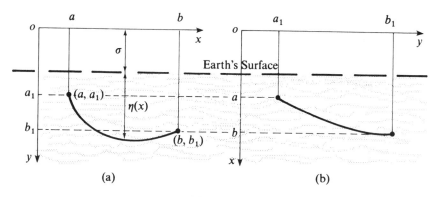

Figure 1.12

(a) Show that the time of transit along this path is for some positive *constant* p, given by $T(y) = p\int_a^b(\sqrt{1 + y'(x)^2}/y(x))\, dx$. For further analysis of this problem, see Problem 8.29 and §9.3.)

(b) When the x- and y-axes are interchanged, if this path can be described as the graph of a function $y = y(x)$ as in Figure 1.12(b), show that the time of transit integral is for $a < h$ given by $T(y) = p\int_a^b(\sqrt{1 + y'(x)^2}/x)\, dx$. This formulation is examined further in Problem 3.23.

1.9. The Zenodoros Problem.

(a) Show that the problem of finding the solid of revolution of maximal volume enclosed by a surface of *fixed* total area $2\pi T$ leads to the consideration of maximizing $V(y) = \pi \int_0^b y^2 \, dx$ where $y \geq 0$, with $y(0) = y(b) = 0$, subject to the isoperimetric constraint

$$2\pi T = S(y) \equiv 2\pi \int_0^b y(x) \, ds(x) = 2\pi \int_0^b y(x)\sqrt{1 + y'(x)^2} \, dx,$$

where $b > 0$ *must also* be varied. (Why?)

(b)* Use Euler's substitution $t = \int y \, ds$, so that $0 \leq t \leq T$, to reduce this to the *non*-isoperimetric problem of maximizing

$$\tilde{V}(y) = \pi \int_0^T y^2(t)x'(t) \, dt = \pi \int_0^T y(t)\sqrt{1 - (yy')^2(t)} \, dt,$$

where $y = y(t)$ is continuously differentiable on $[0, T]$, and $0 \leq y(t)$, with $y(0) = y(T) = 0$.

The analysis is continued in Problems 6.43, 8.27 and 9.12.

CHAPTER 2

Linear Spaces and Gâteaux Variations

Each problem considered previously reduces to that of optimizing (usually minimizing) a real valued function J defined on a subset $\mathscr{D}$ of a linear space $\mathscr{Y}$. In the present chapter we shall view problems in this context and introduce the associated directional derivatives (Gâteaux variations) of the functions which will be required for what follows. We begin with a catalogue of standard linear spaces presupposing some familiarity with vector space operations, with continuity, and with differentiability in $\mathbb{R}^d$.

§2.1. Real Linear Spaces

All functions considered in this text are assumed to be real valued or real vector valued. The principal requirement of a real linear (or vector) space of functions is that it contain the sums and (real) scalar multiples of those functions. We remark without proof that the collection of real valued functions f, g, on a (nonempty) set S forms a real linear space (or vector space) with respect to the operations of pointwise addition:

$$(f + g)(x) = f(x) + g(x), \qquad \forall\, x \in S$$

and scalar multiplication:

$$(cf)(x) = cf(X), \qquad \forall\, x \in S, \quad c \in \mathbb{R}. \quad ([N]).$$

Similarly, for each $d = 1, 2, \ldots$ the collection of all d-dimensional real vector valued functions on this set S forms a linear space with respect to the following operations of componentwise addition and scalar multiplication: if $F = (f_1, f_2, \ldots, f_d)$ and $G = (g_1, \ldots, g_d)$, where f_j and g_j are real valued functions on S for $j = 1, 2, \ldots, d$, so that $F(x) = (f_1(x), f_2(x), \ldots, f_d(x))$, and $G(x) = (g_1(x), g_2(x), \ldots, g_d(x))$, $\forall\, x \in S$, then

$$(F + G)(x) = F(x) + G(x)$$
$$\overset{\text{def}}{=} (f_1(x) + g_1(x), f_2(x) + g_2(x), \ldots, f_d(x) + g_d(x))$$

and

$$(cF)(x) = cF(x) \overset{\text{def}}{=} (cf_1(x), cf_2(x), \ldots, cf_d(x)), \qquad \forall\, x \in S.$$

It follows that each *subspace* of these spaces, i.e., each *subset* which is closed under the defining operations of addition and scalar multiplication, is itself a real linear space.

In particular, if continuity is definable on S, then $C(S)\,(= C^0(S))$, the set of continuous real valued functions on S, will be a real linear space since the sum of continuous functions, or the multiple of a continuous function by a real constant, is again a continuous function. Similarly, for each *open* subset D of Euclidean space and each $m = 1, 2, \ldots$, $C^m(D)$, the set of functions on D having continuous partial derivatives of order $\leq m$, is a real linear space, since the laws of differentiation guarantee that the sum or scalar multiple of such functions will be another. In addition, if D is bounded with boundary ∂D, and $\bar{D} = D \cup \partial D$, then $C^m(\bar{D})$, the subset of $C^m(D) \cap C(\bar{D})$ consisting of those functions whose partial derivatives of order $\leq m$ each admit continuous extension to $\bar{D}$, is a real linear space.

For example, when $a, b \in \mathbb{R}$, then $\overline{(a, b)} = [a, b]$, is a closed and bounded interval. A function y, which is continuous on $[a, b]$, is in $C^1[a, b]$ [1] if it is continuously differentiable in (a, b) and its derivative y' has finite limiting values from the right at a (denoted $y'(a+)$) and from the left at b (denoted $y'(b-)$). When no confusion can arise we shall use the simpler notations $y'(a)$ and $y'(b)$, respectively, for these values, with a similar convention for higher derivatives at a, b, when present. Observe that $y_0(x) = x^{3/2}$ does define a function in $C^1[0, 1]$ while $y_1(x) = x^{1/2}$ does not.

Finally, for $d = 1, 2, \ldots$, $[C(S)]^d$, $[C^m(D)]^d$, and $[C^m(\bar{D})]^d$, the sets of d-dimensional vector valued functions whose components are in $C(S)$, $C^m(D)$, and $C^m(\bar{D})$, respectively, also form real linear spaces.

We know that subsets $\mathscr{D}$ of these spaces provide natural domains for optimization of the real valued functions in Chapter 1. However, these subsets do *not* in general constitute linear spaces themselves. For example,

$$\mathscr{D} = \{y \in C[a, b] : y(a) = 0, y(b) = 1\}$$

is not a linear space since if $y \in \mathscr{D}$ then $2y \notin \mathscr{D}$. ($2y(b) = 2(1) = 2 \neq 1$.) However,

$$\mathscr{D}_0 = \{y \in C[a, b] : y(a) = y(b) = 0\}$$

is a linear space. (Why?)

In the sequel we shall assume the presence of a real linear space $\mathscr{Y}$ consist-

---

[1] We abbreviate $C((a, b))$ by $C(a, b)$, $C^1([a, b])$ by $C^1[a, b]$, etc.

ing of points (or vectors), y, in which are defined the operations of (vector) addition and (real) scalar multiplication obeying the usual commutative, associative, and distributive laws. In particular, there is a unique vector $\mathcal{O}$ such that $c\mathcal{O} = 0y = \mathcal{O}, \forall\, y \in \mathcal{Y}, c \in \mathbb{R}$; we also adopt the standard abbreviations that $1y = y$ and $-1y = -y, \forall\, y \in \mathcal{Y}$.

§2.2. Functions from Linear Spaces

Although we may wish to optimize a real valued function J over a subset $\mathcal{D}$ of a linear space $\mathcal{Y}$, it is frequently the case that the natural domain $\mathcal{D}^*$ of J is larger than $\mathcal{D}$, and may be $\mathcal{Y}$ itself.

Example 1.

$$J(y) = \int_a^b [\sin^3 x + y^2(x)]\, dx$$

is defined on all of $\mathcal{Y} = C[a, b]$, since each continuous function $y \in \mathcal{Y}$ results in a continuous integrand, $\sin^3 x + y^2(x)$, whose integral is finite.

Example 2.

$$J(y) = \int_a^b \rho(x)\sqrt{1 + y'(x)^2}\, dx, \quad \text{with } \rho \in C[a, b],$$

is defined for each $y \in \mathcal{Y} = C^1[a, b]$ since the assumption that y has a derivative on (a, b), which has a continuous extension to $[a, b]$, again results in a continuous integrand.

(Actually, J remains defined on $\mathcal{Y}$, when ρ is (Riemann) integrable over $[a, b]$.)

Example 3. The brachistochrone function of §1.2(a),

$$T(y) = \frac{1}{\sqrt{2g}} \int_0^{x_1} \frac{\sqrt{1 + y'(x)^2}}{\sqrt{y(x)}}\, dx,$$

is not defined on $\mathcal{Y} = C^1[0, x_1]$ because of the presence of the term $\sqrt{y(x)}$ in the denominator of the integrand. It *is* defined on the subset

$$\mathcal{D}^* = \left\{ y \in C^1[0, x_1] : y(x) \geq 0, \forall\, x \in (0, x_1), \text{ and } \int_0^{x_1} (y(x))^{-1/2}\, dx < +\infty \right\},$$

which is not a linear space. (Why not?)

Example 4. When $f \in C([a, b] \times \mathbb{R}^2)$, then

$$F(y) = \int_a^b f(x, y(x), y'(x))\, dx$$

is defined on $\mathcal{Y} \equiv C^1[a, b]$, since for each $y \in \mathcal{Y}$, the composite function

$$f[y(x)] = f(x, y(x), y'(x)) \in C[a, b].$$

However, if $f \in C([a, b] \times D)$ where D is a *domain* in $\mathbb{R}^2$, then F is defined only on a subset of

$$\mathcal{D}^* = \{y \in C^1[a, b] : (y(x), y'(x)) \in D, \forall \, x \in [a, b]\}.$$

Example 5. For each $d = 1, 2, \ldots$, the evaluation function $L(Y) = Y(a)$ is defined on $\mathcal{Y} = (C[a, b])^d$. It is even *linear* in that

$$L(cY + \tilde{c}\tilde{Y}) = cL(Y) + \tilde{c}L(\tilde{Y}), \qquad \forall \, c, \tilde{c} \in \mathbb{R} \text{ and } Y, \tilde{Y} \in \mathcal{Y}.$$

Also linear are

$$L_1(Y) = Y'\left(\frac{a+b}{2}\right) \qquad \text{on } (C^1[a, b])^d;$$

and

$$L(y) = \int_a^b 3xy(x)\, dx \qquad \text{on } C[a, b].$$

However, most functions of interest to us in this text are (highly) nonlinear.

Example 6. If J and $\tilde{J}$ are real valued functions defined on a subset $\mathcal{D}^*$ of any linear space $\mathcal{Y}$, then for $c, \tilde{c} \in \mathbb{R}$

$$cJ, \; cJ + \tilde{c}\tilde{J}, \; J\tilde{J}, \; e^J, \; \sin J,$$

are also defined on $\mathcal{D}^*$; but $1/J$, $\sqrt{J}$, $\tan J$, need *not* be defined. Thus

$$J(y) = \int_a^b \sqrt{1 + y'(x)^2}\, dx + 2y(a)$$

is defined on $C^1[a, b]$, but

$$J_1(y) = \frac{1}{y(a)} \text{ is not.}$$

Example 7. If D is a sufficiently nice bounded domain in $\mathbb{R}^2$, then as in §1.4(b),

$$S(u) = \int_D \sqrt{1 + u_x^2 + u_y^2}\, dA$$

is defined $\forall \, u \in C^1(\bar{D})$.

(Problems 2.1–2.3)

§2.3. Fundamentals of Optimization

When J is a real valued function defined on a subset $\mathcal{D}$ of a linear space $\mathcal{Y}$, as in the previous section, there may be interest in the extremal values of J which would occur at those points $y_0 \in \mathcal{D}$ for which either

$$J(y) \geq J(y_0), \qquad \forall \, y \in \mathcal{D};$$

or

$$J(y) \leq J(y_0), \qquad \forall\, y \in \mathcal{D},$$

and in finding these points if they exist.

Since the latter points are those for which $-J(y) \geq -J(y_0)$, $\forall\, y \in \mathcal{D}$, it will suffice to characterize the former—those *minimum points* y_0 which produce *minimal* values of J on $\mathcal{D}$. We say that a point $y_0 \in \mathcal{D}$ minimizes J on $\mathcal{D}$ *uniquely* when it is the only such point in $\mathcal{D}$, or equivalently, when $J(y) \geq J(y_0)$, $\forall\, y \in \mathcal{D}$ with equality *iff* $y = y_0$.

For example, $J(y) \stackrel{\text{def}}{=} \int_0^1 y^2(x)\, dx \geq 0 = J(y_0)$, if $y_0(x) = 0$, so that y_0 minimizes J on $\mathcal{D} = C[0, 1]$. Moreover, it does so uniquely, because $J(y) = 0$ implies that $p(x) = y^2(x) = 0$ (see Lemma A.9 in the Appendix) so that $y = \mathcal{O} = y_0$.

In this section and in the next chapter, we will be concerned only with such *global* minimum points; (the consideration of *local* minimum points is reserved until Chapter 5, et seq.). We will also consider these minimal points in problems involving isoperimetric or Lagrangian constraints.

First, let's make a restatement utilizing the linearity of $\mathcal{Y}$.

(2.1) **Lemma.** $y_0 \in \mathcal{D}$ *minimizes* J *on* $\mathcal{D}$ [*uniquely*] *iff*

$$J(y_0 + v) - J(y_0) \geq 0, \qquad \forall y_0 + v \in \mathcal{D},$$

[*with equality iff* $v = \mathcal{O}$].[1]

PROOF. For each $y \in \mathcal{D}$, set $v = y - y_0$, so that $y = y_0 + v$, and $y = y_0$ *iff* $v = \mathcal{O}$.   □

Example 1. To minimize $J(y) = \int_a^b y'(x)^2\, dx$ on

$$\mathcal{D} = \{y \in C^1[a, b] : y(a) = 0,\ y(b) = 1\}, \tag{1}$$

we observe that obviously $J(y) \geq 0$, and by inspection, $J(y_1) = 0$ if $y_1' = 0$, but $y_1 = \text{const.}$ is not in $\mathcal{D}$.

However, if we reformulate the problem as suggested by the lemma, then for $y_0 \in \mathcal{D}$, and $y_0 + v \in \mathcal{D}$, we should examine

$$J(y_0 + v) - J(y_0) = \int_a^b [(y_0'(x) + v'(x))^2 - y_0'(x)^2]\, dx$$

$$= \int_a^b v'(x)^2\, dx + 2 \int_a^b y_0'(x)v'(x)\, dx$$

$$\geq 2 \int_a^b y_0'(x)v'(x)\, dx.\ (\text{Why?})$$

Now $y_0(a) = (y_0 + v)(a) = y_0(a) + v(a)$ so that $v(a) = 0$, and similarly, $v(b) =$

---

[1] This is to be considered as two assertions; the first is made by deleting the bracketed expressions throughout, while the second requires their presence.

0. By *inspection*, $y_0' = $ const., makes

$$\int_a^b y_0'(x) v'(x)\, dx = \text{const.} \int_a^b v'(x)\, dx = \text{const. } v(x)\Big|_a^b = 0, \qquad \forall \text{ such } v,$$

so that we have the *inequality* $J(y_0 + v) - J(y_0) \geq 0$. Moreover, $y_0(x) = (x - a)/(b - a)$ is in $\mathscr{D}$ and has $y_0'(x) = $ const. Hence, by the lemma, y_0 minimizes J on $\mathscr{D}$. It also does so uniquely, for *equality* demands that $\int_a^b v'(x)^2\, dx = 0$, which by A.9 requires that $v'(x)^2 = 0$, or that $v(x) = $ const. $= v(a) = 0$: i.e., $v = \mathcal{O}$.

Next, we make a simple observation that permits us to ignore inessential constants.

(2.2) Proposition. y_0 *minimizes* J *on* $\mathscr{D}$ [*uniquely*] *iff for constants* c_0 *and* $c \neq 0$, y_0 *minimizes* $c^2 J + c_0$ *on* $\mathscr{D}$ [*uniquely*].

PROOF. If $y \in \mathscr{D}$, then

$$(c^2 J + c_0)(y) = c^2 J(y) + c_0 \geq c^2 J(y_0) + c_0 = (c^2 J + c_0)(y_0),$$

iff y_0 minimizes J on $\mathscr{D}$ [with equality iff $y = y_0$]. □

Thus from Example 1, we may also say that $y_0(x) = (x - a)/(b - a)$ minimizes

$$J_1(y) \overset{\text{def}}{=} 3 \int_a^b (y'(x)^2 + \sin^3 x)\, dx = 3 \int_a^b y'(x)^2\, dx + 3 \int_a^b \sin^3 x\, dx$$

$$= 3J(y) + c_0, \text{ on } \mathscr{D} \text{ of (1) uniquely.}$$

Constraints

If we seek to minimize J on $\mathscr{D}$, when it is further restricted to a level set of one or more similar functions G, then as was known to Lagrange and Euler, it may suffice to minimize an augmented function *without* constraints.

(2.3) Proposition. *If functions* J *and* $G_1, G_2, \ldots, G_N$ *are defined on* $\mathscr{D}$, *and for some constants* $\lambda_1, \ldots, \lambda_N$, y_0 *minimizes* $\tilde{J} = J + \lambda_1 G_1 + \lambda_2 G_2 + \cdots + \lambda_N G_N$ *on* $\mathscr{D}$ [*uniquely*], *then* y_0 *minimizes* J *on* $\mathscr{D}$ [*uniquely*] *when further restricted to the set* $G_{y_0} \overset{\text{def}}{=} \{ y \in \mathscr{D} : G_j(y) = G_j(y_0), j = 1, 2, \ldots, N \}$.

PROOF. For each $y \in \mathscr{D}$:

$$\tilde{J}(y) = J(y) + \sum_{j=1}^N \lambda_j G_j(y) \geq \tilde{J}(y_0) = J(y_0) + \sum_{j=1}^N \lambda_j G_j(y_0);$$

but when $y \in G_{y_0}$, then $J(y) \geq J(y_0)$, since the terms involving the G_j will have the *same* values on each side of the inequality. Uniqueness is clearly preserved if present. □

Remark. The hope is that the λ_j can be found so that in addition $G_j(y_0) = l_j$ for *prescribed* values l_j, $j = 1, 2, \ldots, N$. Reinforcement for this possibility will be given in the discussion of the method of Lagrangian multipliers (§5.7). Actually, y_0 minimizes J *automatically* on a much larger set.

(2.4) Corollary. y_0 *of Proposition* 2.3 *minimizes* J *on* $\mathscr{D}$ [*uniquely*] *when restricted to the set*

$$G_{y_0}^* = \{y \in \mathscr{D} : \lambda_j G_j(y) \le \lambda_j G_j(y_0), j = 1, 2, \ldots, N\}.$$

PROOF. For $y \in \mathscr{D}$, the previous inequality gives us

$$J(y) - J(y_0) \ge \sum_{j=1}^{N} [\lambda_j G_j(y_0) - \lambda_j G_j(y)] \ge 0, \quad \text{if } y \in G_{y_0}^*.$$

If $J(y) = J(y_0)$ *under these conditions,* then $\lambda_j G_j(y_0) = \lambda_j G_j(y), j = 1, 2, \ldots, N$ (Why?), so that $\tilde{J}(y) = \tilde{J}(y_0)$. [With uniqueness, it follows that $y = y_0$.]   □

These results illustrate an important principle: *The solution to one minimization problem may also provide a solution for other problems.*

The above abstract formulation is suitable for attacking problems involving *isoperimetric* constraints such as the following:

Application: Rotating Fluid Column

A circular column of water of radius l is rotated about its vertical axis at constant angular velocity, ω, inside a smooth-walled cylinder as shown in Figure 2.1. Then the (upper) free surface assumes a shape which preserves the volume of the fluid and minimizes the *potential* energy—given within an

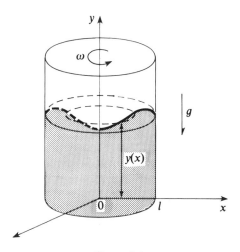

Figure 2.1

inessential additive constant c_0 by

$$J(y) = \rho\pi \int_0^l [gy^2(x) - \omega^2 x^2 y(x)]x \, dx, \tag{2}$$

on

$$\mathscr{D} = \{C[0, l] : y(x) \geq 0\}.$$

Here, ρ is the mass density of the water, and g the standard gravitational constant, while $y(x)$ is the height of the liquid at a radial distance x from the center. To obtain (2) we regard the column as if it were static and maintained in the shape shown in Figure 2.1, by the interaction of the downward gravitational force and the radially directed centrifugal force on each of its particles. (We are invoking Bernoulli's principle for stable equilibrium which is discussed in §8.9.)

The volume of the (static) column is given by

$$G(y) = 2\pi \int_0^l xy(x) \, dx, \tag{3}$$

and hence, according to Proposition 2.3, we should consider minimizing $\tilde{J}(y) = J(y) + \lambda G(y)$, on $\mathscr{D}$. Then by Lemma 2.1, suppressing the subscript 0, we should examine $\tilde{J}(y + v) - \tilde{J}(y)$. Upon replacing λ by $\rho\lambda/2$ and simplifying, we obtain the *inequality*

$$\tilde{J}(y + v) - \tilde{J}(y) = \pi\rho \int_0^l \{g[(y + v)^2(x) - y^2(x)] + (\lambda - \omega^2 x^2)v(x)\}x \, dx$$

$$= \pi\rho \int_0^l \{gv^2(x) + [2gy(x) + (\lambda - \omega^2 x^2)]v(x)\}x \, dx$$

$$\geq \pi\rho \int_0^l [2gy(x) + (\lambda - \omega^2 x^2)]v(x)x \, dx,$$

and this last integral vanishes $\forall \, y + v \in \mathscr{D}$, when the bracketed term $\equiv 0$; i.e., when

$$y(x) = y_0(x) = \frac{\omega^2 x^2 - \lambda}{2g}. \tag{4}$$

Moreover, *equality* is possible only when $\int_0^l v^2(x)x \, dx = 0$, which implies by A.9, that $v = \mathcal{O}$, so that y_0 minimizes $\tilde{J}$ on $\mathscr{D}$ uniquely. It follows from Proposition 2.3, that the free surface will be the *paraboloid* of revolution defined by (4).

Now, by (3),

$$G(y_0) = \frac{\pi}{g} \int_0^l (\omega^2 x^2 - \lambda)x \, dx = \frac{\pi l^2}{g}\left[\frac{\omega^2 l^2}{4} - \frac{\lambda}{2}\right],$$

and λ may be selected to make $G(y_0)$ match the given fluid volume.

Observe that the minimizing shape will depend on ω as we would expect (and is of constant height when $\omega = 0$) but it is *independent* of the density ρ, and thus would be the same for another (perfect) fluid. (The actual minimal value of $J(=J(y_0))$, although calculable, is of little interest.)

The approach taken here extends also to problems involving *Lagrangian* constraints on suitable function spaces. To exemplify this, we recall the notation from §1.5, that when $f = f(x, y, z) \in C([a, b] \times \mathbb{R}^2)$ and $y \in C^1[a, b]$, then $f[y(x)] \stackrel{\text{def}}{=} f(x, y(x), y'(x))$.

(2.5) **Proposition.** *Suppose* $f = f(x, y, z)$ *and* $g = g(x, y, z)$ *are continuous on* $[a, b] \times \mathbb{R}^2$ *and there is a function* $\lambda \in C[a, b]$, *for which* y_0 *minimizes* $\tilde{F}(y) = \int_a^b \tilde{f}[y(x)]\, dx$ *on* $\mathcal{D} \subseteq C^1[a, b]$ *[uniquely] where* $\tilde{f} \stackrel{\text{def}}{=} f + \lambda g$. *Then* y_0 *minimizes* $F(y) = \int_a^b f[y(x)]\, dx$ *on* $\mathcal{D}$ *[uniquely] under the (inequality) constraint*

$$\lambda(x)g[y(x)] \leq \lambda(x)g[y_0(x)], \qquad \forall x \in [a, b]. \tag{5}$$

PROOF. If $y \in \mathcal{D}$, then

$$\tilde{F}(y) = F(y) + \int_a^b \lambda(x)g[y(x)]\, dx \geq \tilde{F}(y_0) = F(y_0) + \int_a^b \lambda(x)g[y_0(x)]\, dx$$

so that

$$F(y) - F(y_0) \geq \int_a^b \lambda(x)(g[y_0(x)] - g[y(x)])\, dx$$

$$\geq 0 \qquad \text{when (5) holds.}$$

Moreover, if $F(y) = F(y_0)$ under conditions (5), then $\int_a^b \lambda(x)(g[y_0(x)] - g[y(x)])\, dx = 0$, and $\tilde{F}(y) = \tilde{F}(y_0)$ [which with uniqueness is possible *iff* $y = y_0$]. $\qquad\square$

Unfortunately, this result, although suggestive, cannot be implemented readily, since it does not provide a method for determining a suitable λ. However, we may be able to use a *known* solution to another problem.

From the analysis used in the previous application, we may claim that $y_0(x) = x^2 - 1$ minimizes $\tilde{F}(y) = \int_0^b [y^2(x) + (2 - 2x^2)y(x)]x\, dx$ uniquely on $\mathcal{D} = C[0, b]$ (Problem 2.12).

Setting $\lambda(x) = x$ in Proposition 2.5, it follows that y_0 also minimizes $F(y) = \int_0^b 2xy(x)\, dx$ on $\mathcal{D}$ uniquely under the Lagrangian constraint:

$$g[y(x)] \stackrel{\text{def}}{=} y^2(x) - 2x^2y(x) = g[y_0(x)] = 1 - x^4,$$

and since $\lambda(x) = x \geq 0$ on $[0, b]$, also under the *inequality* constraint $g[y(x)] \leq 1 - x^4$.

In this section, we have shown that direct inequalities may be of use in finding the minima for some functions. All of the functions analyzed here were actually convex, and the convexity methods developed in the next chapter will provide a more systematic means of obtaining such inequalities for a large and useful class of functions. We shall also return to the basic approach adopted here in the analysis of sufficient conditions for a minimum in Chapter 9.

§2.4. The Gâteaux Variations

A decisive role in the optimization of a real valued function on a subset of $\mathbb{R}^d$ is played by its partial derivatives—or more generally by its directional derivatives—if they exist (§0.5). When J is a real valued function on a subset of a linear space $\mathcal{Y}$, then it is not evident how to define its partial derivatives (unless $\mathcal{Y}$ can be assigned a distinguished coordinate system). However, a definition for its directional derivatives is furnished by a straightforward generalization of that in $\mathbb{R}^d$:

(2.6) **Definition.** For $y, v \in \mathcal{Y}$:

$$\delta J(y; v) \stackrel{\text{def}}{=} \lim_{\varepsilon \to 0} \frac{J(y + \varepsilon v) - J(y)}{\varepsilon},$$

assuming that this limit exists, is called the *Gâteaux variation of J at y in the direction v.*

Observe that both y and the direction v are fixed while the limit as $\varepsilon \to 0$ is taken in 2.6. The existence of the limit presupposes that:

(i) $J(y)$ is defined; (6)

(ii) $J(y + \varepsilon v)$ is defined for all sufficiently small ε; and then

$$\delta J(y; v) = \frac{\partial}{\partial \varepsilon} J(y + \varepsilon v) \bigg|_{\varepsilon = 0}, \tag{7}$$

if this "ordinary" derivative with respect to the real variable ε exists at $\varepsilon = 0$.

The Gâteaux variation of J at y depends only on the local behavior of J near y; however, this variation need not exist in any direction $v \neq \mathcal{O}$, or it may exist in some directions and not in others.

As is to be expected from (7), the Gâteaux variation is a linear operation on the functions J; i.e., if at some $y_0 \in \mathcal{Y}$, functions J and $\tilde{J}$ each have Gâteaux variations in the *same* direction v, then for constants $c, \tilde{c} \in \mathbb{R}$, $\delta(cJ + \tilde{c}\tilde{J})(y_0; v)$ exists and equals $c\delta J(y_0; v) + \tilde{c}\delta\tilde{J}(y_0; v)$. This is a direct consequence of (7) and the linearity of the ordinary derivative. We also see that $\delta J(y; \mathcal{O}) = 0$, $\forall y$ at which $J(y)$ is defined, and that when $c \in \mathbb{R}$, $\delta J(y; cv) = c\delta J(y; v)$ provided that the variation on the right exists (Problem 2.4).

In particular, for $c = -1$:

$$\delta J(y; -v) = -\delta J(y; v) = \delta(-J)(y; v),$$

whenever any of these variations exists.

Example 1. If $J = f \in C^1(\mathbb{R}^d)$ and $Y, V \in \mathcal{Y} = \mathbb{R}^d$, then

$$\delta f(Y; V) = \lim_{\varepsilon \to 0} \frac{f(Y + \varepsilon V) - f(Y)}{\varepsilon}$$

is just the directional derivative of f when V is a *unit* vector (§0.5). Thus we have that

$$\delta f(Y; V) = \nabla f(Y) \cdot V,$$

and this holds for all $V \in \mathscr{Y}$. (Why?)

Example 2. If $\mathscr{Y} = C[a, b]$, then as in §2.2, Example 1,

$$J(y) = \int_a^b [\sin^3 x + y^2(x)] \, dx$$

is defined $\forall \; y \in \mathscr{Y}$. Thus for fixed y, $v \in \mathscr{Y}$ and $\varepsilon \neq 0$; $y + \varepsilon v \in \mathscr{Y}$ (since $\mathscr{Y}$ is a linear space) so that

$$J(y + \varepsilon v) = \int_a^b [\sin^3 x + (y + \varepsilon v)^2(x)] \, dx \quad \text{is defined.}$$

After successive cancellations we obtain the ratio

$$\frac{J(y + \varepsilon v) - J(y)}{\varepsilon} = \frac{1}{\varepsilon} \int_a^b [(y + \varepsilon v)^2(x) - y^2(x)] \, dx$$

$$= \frac{1}{\varepsilon} \int_a^b [y^2(x) + 2\varepsilon y(x)v(x) + \varepsilon^2 v^2(x) - y^2(x)] \, dx$$

$$= 2 \int_a^b y(x)v(x)dx + \varepsilon \int_a^b v^2(x) \, dx.$$

Each of the integrals in this last expression is a constant (Why?), and the limit as $\varepsilon \to 0$ exists. Hence from Definition 2.6, we conclude that

$$\delta J(y; v) = 2 \int_a^b y(x)v(x) \, dx, \qquad \forall \; y, v \in \mathscr{Y}.$$

Alternatively, using (7), we could form:

$$J(y + \varepsilon v) = \int_a^b [\sin^3 x + (y + \varepsilon v)^2(x)] \, dx$$

$$= \int_a^b [\sin^3 x + y^2(x)] \, dt + 2\varepsilon \int_a^b y(x)v(x) \, dx + \varepsilon^2 \int_a^b v^2(x) \, dx,$$

compute for fixed y, v, the derivative

$$\frac{\partial}{\partial \varepsilon} J(y + \varepsilon v) = 2 \int_a^b y(x)v(x) \, dx + 2\varepsilon \int_a^b v^2(x) \, dx,$$

and evaluate at $\varepsilon = 0$ to get

$$\delta J(y; v) = 2 \int_a^b y(x)v(x) \, dx \qquad\qquad (8)$$

as above.

In general, it is easier technically to use the second method for comput-ing δJ in concrete cases simply because we are more familiar with the tech-niques for differentiating standard real valued functions. However, the second method as *carried out here* requires that $(\partial/\partial\varepsilon)J(y + \varepsilon v)$ exist for small $\varepsilon \neq 0$, and that it be continuous at $\varepsilon = 0$. By contrast, the first requires only the existence of this derivative at $\varepsilon = 0$.

Example 3. When $\rho \in C[a, b]$, the function

$$J(y) = \int_a^b \rho(x)\sqrt{1 + y'(x)^2}\, dx$$

is defined $\forall\, y \in \mathcal{Y} = C^1[a, b]$. (Example 2 of §2.2.)

Hence, using the second method, we form for fixed $y, v \in \mathcal{Y}$,

$$J(y + \varepsilon v) = \int_a^b \rho(x)\sqrt{1 + (y + \varepsilon v)'(x)^2}\, dx,$$

compute its derivative by differentiating under the integral sign (as in A.13) to get

$$\frac{\partial}{\partial\varepsilon}J(y + \varepsilon v) = \int_a^b \frac{\partial}{\partial\varepsilon}[\rho(x)\sqrt{1 + (y + \varepsilon v)'(x)^2}]\, dx$$

$$= \int_a^b \frac{\rho(x)(y + \varepsilon v)'(x)v'(x)}{\sqrt{1 + (y + \varepsilon v)^2(x)^2}}\, dx,$$

(which is justified by the continuity of this last integrand on $[a, b] \times \mathbb{R}$), and evaluate at $\varepsilon = 0$ to obtain

$$\delta J(y; v) = \int_a^b \frac{\rho(x)y'(x)v'(x)}{\sqrt{1 + y'(x)^2}}\, dx. \tag{9}$$

Example 4. When $f \in C([a, b] \times \mathbb{R}^2)$, the function

$$F(y) = \int_a^b f(x, y(x), y'(x))\, dx = \int_a^b f[y(x)]\, dx,$$

is defined whenever $y \in \mathcal{Y} = C^1[a, b]$. (Example 4 of §2.2.) However to com-pute $\delta F(y; v)$ by differentiating

$$F(y + \varepsilon v) = \int_a^b f(x, (y + \varepsilon v)(x), (y + \varepsilon v)'(x))\, dx = \int_a^b f[(y + \varepsilon v)(x)]\, dx,$$

with respect to ε under the integral sign, we obviously should require that the function $f(x, y, z)$ have partial derivatives $f_y, f_z \in C([a, b] \times \mathbb{R}^2)$. Then from the chain rule, for fixed x, y, z, v, w:

$$\frac{\partial}{\partial\varepsilon}f(x, y + \varepsilon v, z + \varepsilon w) = f_y(x, y + \varepsilon v, z + \varepsilon w)v + f_z(x, y + \varepsilon v, z + \varepsilon w)w.$$

With this preparation we have by A.13 that

$$\frac{\partial}{\partial\varepsilon}F(y+\varepsilon v) = \int_a^b \frac{\partial}{\partial\varepsilon}f[(y+\varepsilon v)(x)]\,dx$$

$$= \int_a^b \frac{\partial}{\partial\varepsilon}f(x,\,y(x)+\varepsilon v(x),\,y'(x)+\varepsilon v'(x))\,dx$$

$$= \int_a^b (f_y[(y+\varepsilon v)(x)]v(x)+f_z[(y+\varepsilon v)(x)]v'(x))\,dx,$$

(since the last integrand is continuous on $[a,b]\times\mathbb{R}$), so that when $\varepsilon = 0$, we see that $F(y) = \int_a^b f(x,\,y(x),\,y'(x))\,dx$ has the variation

$$\delta F(y;v) = \int_a^b [f_y(x,\,y(x),\,y'(x))v(x)+f_z(x,\,y(x),\,y'(x))v'(x)]\,dx$$

$$= \int_a^b (f_y[y(x)]v(x)+f_z[y(x)]v'(x))\,dx \qquad \text{for } y,\,v \in C^1[a,b]. \quad (10)$$

We shall return to the analysis of this function in Chapters 3 and 6.

Example 5. The function

$$J(y) = \int_a^b \sin y(x)\,dx + y^2(b),$$

for which

$$J(y+\varepsilon v) = \int_a^b \sin(y(x)+\varepsilon v(x))\,dx + (y+\varepsilon v)^2(b),$$

is defined on $\mathscr{Y} = C[a,b]$ and has at each $y \in \mathscr{Y}$ and in each direction $v \in \mathscr{Y}$ the Gâteaux variation

$$\delta J(y;v) = \int_a^b [\cos y(x)]v(x)\,dx + 2y(b)v(b).$$

Example 6*. For $\mathscr{Y} = C[0,\pi]$, the function

$$J(y) = \int_0^\pi \sqrt{1-y(x)^2}\,dx$$

is not defined on $\mathscr{Y}$, but it is defined on

$$\mathscr{D} = \{y \in C[0,\pi]: \|y\|_M \le 1\},$$

where

$$\|y\|_M = \max_{x\in[0,\pi]} |y(x)|,$$

so that, for example, $y_1(x) = \sin x$, $x \in [0,\pi]$ is a function in $\mathscr{D}$, but $y_2(x) = x$ is not.

Moreover, for a given $y \in \mathcal{D}$, only those directions v could be considered for which $y + \varepsilon v \in \mathcal{D}$ for sufficiently small ε: i.e., for which $|y(x) + \varepsilon v(x)| \leq 1$ for such ε. The function $y(x) \equiv 1$ is in $\mathcal{D}$, but its only possible direction is $v = 0$; the function $y_1(x) = \sin x$ has (at most) as its possible directions, those v for which $v(\pi/2) = 0$. On the other hand, for $y(x) = \frac{1}{2}\sin x$ and a given $v \in C[0, \pi]$, then $|v(x)| \leq \|v\|_M = M_1$, say, and so $|\varepsilon v(x)| \leq \frac{1}{2}$ when $\varepsilon \leq 1/2M_1$; thus $y + \varepsilon v \in \mathcal{D}$ when $\varepsilon \leq 1/2M_1$.

To proceed, we assume that for some $y \in \mathcal{D}$, $v \in \mathcal{Y}$, we know that $y + \varepsilon v \in \mathcal{D}$, and consider for small $\varepsilon \neq 0$ the formally obtained derivative

$$\frac{\partial}{\partial \varepsilon} J(y + \varepsilon v) = \frac{\partial}{\partial \varepsilon} \int_0^\pi \sqrt{1 - (y + \varepsilon v)^2}(x)\, dx$$

$$= \int_0^\pi \frac{\partial}{\partial \varepsilon} \sqrt{1 - (y + \varepsilon v)^2}(x)\, dx$$

$$= -\int_0^\pi \frac{(y + \varepsilon v)(x)v(x)}{\sqrt{1 - (y + \varepsilon v)^2(x)}}\, dx,$$

which for $\varepsilon = 0$ should give the value

$$\delta J(y; v) = -\int_0^\pi \frac{y(x)v(x)}{\sqrt{1 - y^2(x)}}\, dx,$$

provided that this process is valid. Now from A.13, it is valid if the integrand

$$\tilde{f}(x, \varepsilon) = \frac{[y(x) + \varepsilon v(x)]v(x)}{\sqrt{1 - (y + \varepsilon v)^2(x)}}$$

is continuous on $[0, \pi] \times [-\varepsilon_0, \varepsilon_0]$ for some $\varepsilon_0 > 0$ and this requires that $\|y\|_M < 1$. However, if $\|y\|_M < 1$, then for each $v \in \mathcal{Y}$, $\|y + \varepsilon v\|_M < 1$ for all ε sufficiently small, by the argument above.

Thus we can at least say that when $\|y\|_M < 1$, $\delta J(y; v)$ exists $\forall\, v \in \mathcal{Y}$ and is given above. At other $y \in \mathcal{D}$ it is more difficult to consider the Gâteaux variation.

Example 7. The area function

$$A(Y) = \int_0^1 x(t)y'(t)\, dt$$

of §1.3 is defined $\forall\, Y = (x, y) \in \mathcal{Y} = (C^1[0, 1])^2$. Since for $V = (u, v) \in \mathcal{Y}$,

$$A(Y + \varepsilon V) = \int_0^1 (x(t) + \varepsilon u(t))(y'(t) + \varepsilon v'(t))\, dt,$$

it follows that

$$\delta A(Y; V) = \int_0^1 [(x(t) + \varepsilon u(t))v'(t) + (y'(t) + \varepsilon v'(t))u(t)]\, dt \Big|_{\varepsilon=0}$$

$$= \int_0^1 [x(t)v'(t) + y'(t)u(t)]\, dt. \tag{11}$$

Example 8. More generally, if $f = f(x, Y, Z) \in C^1([a, b] \times \mathbb{R}^{2d})$, then

$$F(Y) = \int_a^b f(x, Y(x), Y'(x)) \, dx = \int_a^b f[Y(x)] \, dx$$

is defined for all vector valued functions $Y \in \mathcal{Y} = (C^1[a, b])^d$, $(d = 2, 3, \ldots)$. Its Gâteaux variation at $V \in \mathcal{Y}$ is given by

$$\delta F(Y; V) = \int_a^b f_Y[Y(x)] \cdot V(x) + f_Z[Y(x)] \cdot V'(x)) \, dx, \qquad (12)$$

where $f_Y[Y(x)]$ is the *vector valued* function with components

$$f_{y_j}[Y(x)] \stackrel{\text{def}}{=} f_{y_j}(x, Y(x), Y'(x)), \qquad j = 1, 2, \ldots, d;$$

while $f_Z[Y(x)]$ is that with components

$$f_{z_j}[Y(x)] \stackrel{\text{def}}{=} f_{z_j}(x, Y(x), Y'(x)), \qquad j = 1, 2, \ldots, d.$$

(See Problem 2.6.) This result together with the notation used to express it should be compared with the one-dimensional case of Example 4.

Example 9. The surface area function of §2.2, Example 7,

$$S(u) = \int_D \sqrt{1 + u_x^2 + u_y^2} \, dA$$

is defined $\forall u \in \mathcal{Y} = C^1(\overline{D})$, and

$$\delta S(u; v) = \int_D \frac{u_x v_x + u_y v_y}{\sqrt{1 + u_x^2 + u_y^2}} \, dA \qquad (13)$$

exists $\forall u, v \in \mathcal{Y}$, since the denominator can never vanish. (See Problem 2.7.)

(Problems 2.4–2.13)

PROBLEMS

2.1. Give an example of a *nonconstant* function in each of the following sets and determine whether the set is a subspace of $\mathcal{Y} = C^1[a, b]$.
(a) $C[a, b]$.
(b) $\mathcal{D} = \{y \in \mathcal{Y} : y(a) = 0\}$.
(c) $\mathcal{D} = \{y \in \mathcal{Y} : y'(a) = 0, y(b) = 1\}$.
(d) $C^2[a, b]$.
(e) $\mathcal{D} = \{y \in \mathcal{Y} : y(a) = y'(a)\}$.
(f) $(C^1[a, b])^2$.
(g) $\mathcal{D} = \{y \in \mathcal{Y} : \int_a^b y(x) \, dx = 0\}$.
(h) $\mathcal{D} = \{y \in \mathcal{Y} : y'(x) = y(x), x \in (a, b)\}$.

2.2. Which of the following functions are defined on: (a) $C^1[a, b]$? (b) $C[a, b]$?

$$L(y) = \int_a^b (\sin^3 x) y(x)\, dx; \qquad J(y) = \int_a^b \frac{y'(x)}{\sqrt{1 + y^2(x)}}\, dx;$$

$$F(y) = \int_a^b \frac{dx}{\sqrt{1 - y'(x)^2}} + y(a); \qquad G(y) = \int_a^b \log y(x)\, dx.$$

2.3. For each of the following functions give a subset $\mathscr{D}$ of $\mathscr{Y}$ (possibly $\mathscr{Y}$ itself) on which the function is defined and determine whether or not your $\mathscr{D}$ is a subspace of $\mathscr{Y}$:
(a) $F(Y) = \int_a^b |Y'(x)|\, dx$, $\mathscr{Y} = (C^1[a, b])^d$.
(b) $G(y) = \int_a^b \sqrt{1 + xy^2(x)}\, dx$, $\mathscr{Y} = C[a, b]$.
(c) $H(y) = \int_a^b \log y'(x)\, dx$, $\mathscr{Y} = C^1[a, b]$.
(d) $J(u) = \int_D \sqrt{u_x^2 - u_y^2}\, dA$; $\mathscr{Y} = C^1(\bar{D})$, where D is a nice bounded domain of $\mathbb{R}^2$.
(e) $K(y) = \int_a^b (1 + y''(x)^2) y(x)\, dx$, $\mathscr{Y} = C^2[a, b]$.

2.4. Let J and $\tilde{J}$ be real valued functions defined on a subset of a linear space $\mathscr{Y}$, and suppose that for some y, $v \in \mathscr{Y}$, $\delta J(y; v)$ and $\delta \tilde{J}(y; v)$ exist. If $c \in \mathbb{R}$, establish existence and equality as required for the following assertions:
(a) $\delta(cJ)(y; v) = \delta J(y; cv) = c\delta J(y; v)$.
(b) $\delta(J + \tilde{J})(y; v) = \delta J(y; v) + \delta \tilde{J}(y; v)$.
Assuming the existence of the variations involved,
(c) is $\delta J(cy; v) = c\delta J(y; v)$?
(d) is $\delta J(y + \tilde{y}; v) = \delta J(y; v) + \delta J(\tilde{y}; v)$?

2.5. Let $\mathscr{Y} = C^1[a, b]$ and find $\delta J(y; v)$ for y, $v \in \mathscr{Y}$, when
(a) $J(y) = y(a)^3$.
(b) $J(y) = \int_a^b [y(x)^3 + xy'(x)^2]\, dx$.
(c) $J(y) = \int_a^b \sqrt{2 + x^2} - \sin y'(x)\, dx$.
(d) $J(y) = \int_a^b [e^x y(x) - 3y'(x)^4]\, dx + 2y'(a)^2$.
(e) $J(y) = \int_a^b [x^2 y(x)^2 + e^{y'(x)}]\, dx$.
(f) $J(y) = \sin y'(a) + \cos y(b)$.
(g) $J(y) = (\int_a^b [2y'(x) + x^2 y(x)]\, dx)(\int_a^b [1 + y'(x)]^2\, dx)$.
(h) $J(y) = \int_a^b y(x)\, dx / \int_a^b [1 + y'(x)^2]\, dx$.

2.6. In Example 8 of §2.4, verify equation (12) by formal differentiation under the integral sign.

2.7. Let $\mathscr{Y} = C^1(\bar{D})$ where D is a bounded domain in the x–y plane with a nice boundary.
(a) For $J(u) = \frac{1}{2} \int_D (u_x^2 + u_y^2)\, dA$, show that

$$\delta J(u; v) = \int_D (u_x v_x + u_y v_y)\, dA, \quad \forall\, u, v \in \mathscr{Y}.$$

(b) For $S(u) = \int_D \sqrt{1 + u_x^2 + u_y^2}\, dA$, verify that

$$\delta S(u; v) = \int_D \frac{u_x v_x + u_y v_y}{\sqrt{1 + u_x^2 + u_y^2}}\, dA, \quad \forall\, u, v \in \mathscr{Y}.$$

2.8. Let $\mathcal{Y} = (C^1[a, b])^2$, $Y = (y_1, y_2)$, $V = (v_1, v_2)$, and find $\delta J(Y; V)$ for $Y, V \in \mathcal{Y}$ when

(a) $J(Y) = Y(a) \cdot Y(b)$.

(b) $J(y) = \int_a^b [y_1(x)^2 + y_2'(x)^3] \, dx$.

(c) $J(Y) = \int_a^b [e^{y_1(x)} - x^2 y_1(x) y_2'(x)] \, dx$.

(d) $J(y) = \int_a^b [\sin^2 y_1(x) + xy_2(x) + y_1'(x) y_2(x)^2] \, dx$.

2.9. Let $\mathcal{Y} = C^2[a, b]$ and $F(y) = \int_a^b f(x, y(x), y'(x), y''(x)) \, dx = \int_a^b f[y(x)] \, dx$, where $f = f(x, y, z, r)$ in $C^1([a, b] \times \mathbb{R}^3)$ is given. For $v, y \in \mathcal{Y}$, prove that

$$\delta F(y; v) = \int_a^b (f_y[y(x)]v(x) + f_z[y(x)]v'(x) + f_r[y(x)]v''(x)) \, dx.$$

2.10. Assume that $\delta J(y; v)$ and $\delta \tilde{J}(y; v)$ both exist for $y, v \in \mathcal{Y}$.

(a) Verify the product formula

$$\delta(J\tilde{J})(y; v) = \delta J(y; v)\tilde{J}(y) + J(y)\delta \tilde{J}(y; v).$$

(b) Establish the quotient rule

$$\delta\left(\frac{J}{\tilde{J}}\right)(y; v) = \frac{\tilde{J}(y)\delta J(y; v) - J(y)\delta \tilde{J}(y; v)}{\tilde{J}(y)^2},$$

provided that $\tilde{J}(y) \neq 0$.

(c) Supposing that $h \in C^1(\mathbb{R})$, show that

$$\delta(h(J))(y; v) = h'(J(y))\delta J(y; v).$$

2.11. If L is a linear function on $\mathcal{Y}$ (as in Example 5 of §2.2) prove that

$$\delta L(y; v) = L(v), \; \forall \, y, v \in \mathcal{Y}.$$

2.12. (a) For $\rho, \beta \in \mathcal{Y} = C[0, b]$, with $\rho > 0$, find a function y_0 which minimizes

$$F(y) = \int_0^b [\rho(x)y^2(x) + \beta(x)y(x)] \, dx \qquad \text{on } \mathcal{D} = \mathcal{Y}.$$

(b) Show that y_0 is unique.

(c) What can you conclude if $\rho < 0$?

2.13. Show that for the arc length function $L(Y) = \int_0^1 |Y'(t)| \, dt$ of §1.1, formal differentiation produces

$$\partial L(Y; V) = \int_0^1 \frac{Y'(t)}{|Y'(t)|} \cdot V'(t) \, dt, \qquad \forall \, V \in \mathcal{Y} = (C^1[0, 1])^d,$$

if $Y \in \mathcal{D}^* = \{Y \in \mathcal{Y} : |Y'| \neq 0\}$. Is $\mathcal{D}^*$ a subspace of $\mathcal{Y}$?

CHAPTER 3

Minimization of Convex Functions

By utilizing the Gâteaux variations from the preceding chapter, it is straightforward to characterize convexity for a function J on a subset $\mathcal{D}$ of a linear space $\mathcal{Y}$, such that a convex function is automatically minimized by a $y \in \mathcal{D}$ at which its Gâteaux variations vanish.[1] Moreover, in the presence of strict convexity, there can be at most one such y. A large and useful class of functions is shown to be convex. In particular, in §3.2, the role of [strongly] convex integrands f in producing [strictly] convex integral functions F is examined, and a supply of such f is made accessible through the techniques and examples of §3.3. Moreover, the Gâteaux variations of integral functions will, in general, vanish at each solution y of an associated differential equation (of Euler–Lagrange).

The resulting theory extends to problems involving convex constraining functions (§3.5), and it is used in several applications of interest including a version of the brachistochrone (§3.4), and the hanging cable (or catenary) problem of Euler. Additional applications will be found in the problem set together with extensions of this theory to other types of integral functions.

In this chapter only those conditions sufficient for a minimum are considered, and it is shown that in the presence of strict convexity they can supply a complete and satisfactory solution to the problems of interest. In particular, we may be able to ignore the difficult question of *a priori* existence of a minimum by simply exhibiting the (unique) function which minimizes. Actually, the direct approach developed here within the framework of convexity extends in principle to other problems (§3.6).

---

[1] The definitions of functional convexity employed in this book incorporate, for convenience, some presupposed differentiability of the functions. For a *convex* set $\mathcal{D}$, less restrictive formulations are available, but they are more difficult to utilize. (See Problem 0.7, [Fl], and [E–T].)

§3.1. Convex Functions

When $f \in C^1(\mathbb{R}^3)$ then for $Y = (x, y, z)$, $V = (u, v, w) \in \mathbb{R}^3$, we have

$$\delta f(Y; V) = \nabla f(Y) \cdot V; \quad \text{(as in §2.4, Example 1)};$$

moreover, f is defined to be convex (§0.8) provided that $\forall\ Y, V \in \mathbb{R}^3$:

$$f(Y + V) - f(Y) \geq \nabla f(Y) \cdot V = \delta f(Y; V), \tag{1}$$

and strictly convex when equality holds at Y iff $V = \mathcal{O}$ (§0.9). We also observe that minimization of a convex function f may be particularly easy to establish, in that a point Y at which $\nabla f(Y) = \mathcal{O}$ clearly minimizes f. (1) suggests the following:

(3.1) **Definition.** A real valued function J defined on a set $\mathscr{D}$ in a linear space $\mathscr{Y}$ is said to be [strictly] convex on $\mathscr{D}$ provided that when y and $y + v \in \mathscr{D}$ then $\delta J(y; v)$ is defined and $J(y + v) - J(y) \geq \delta J(y; v)$ [with equality iff $v = \mathcal{O}$].[1]

$\mathscr{D}$ itself may be *nonconvex*. (See 3.15)

Although "most" functions are not convex, a surprisingly large number of those of interest to us are convex—even strictly convex—as the applications will show. The following observation will prove valuable:

(3.2) **Proposition.** *If J and $\tilde{J}$ are convex functions on a set $\mathscr{D}$ then for each $c \in \mathbb{R}$, $c^2 J$ and $J + \tilde{J}$ are also convex. Moreover, the latter functions will be strictly convex with J (for $c \neq 0$).*

PROOF.

$$(c^2 J + \tilde{J})(y + v) - (c^2 J + \tilde{J})(y) = c^2(J(y + v) - J(y)) + (\tilde{J}(y + v) - \tilde{J}(y))$$

$$\geq c^2 \delta J(y; v) + \delta \tilde{J}(y; v) = \delta(c^2 J + \tilde{J})(y; v), \quad \text{if } y, y + v \in \mathscr{D} \quad \text{(by 3.1)}.$$

This establishes the convexity of $J + \tilde{J}$ (when $c^2 = 1$) and of $c^2 J$ (when $\tilde{J} = \mathcal{O}$). Moreover, when J is strictly convex and $c \neq 0$, then there must be strict inequality except for the trivial case of $v = \mathcal{O}$. $\qquad\square$

(3.3) **Proposition.** *If J is [strictly] convex on $\mathscr{D}$ then each $y_0 \in \mathscr{D}$ for which $\delta J(y_0; v) = 0$, $\forall\ y_0 + v \in \mathscr{D}$ minimizes J on $\mathscr{D}$ [uniquely].*

PROOF. If $y \in \mathscr{D}$, then with $v = y - y_0$

$$J(y) - J(y_0) = J(y_0 + v) - J(y_0)$$

$$\geq \delta J(y_0; v) = 0, \quad \text{(by 3.1 and hypotheses)}$$

$$\text{[with equality iff } v = \mathcal{O}].$$

[1] Recall the footnote on page 40.

Hence $J(y) \geq J(y_0)$ [with equality *iff* $y = y_0$] and this is the desired conclusion. ∎

Example 1.

$$J(y) = \int_a^b (\sin^3 x + y^2(x))\, dx$$

for which

$$\delta J(y; v) = 2 \int_a^b y(x)v(x)\, dx, \qquad \forall\, y, v \in \mathcal{Y} = C[a, b] \quad (\S 2.4, \text{ Example 2}),$$

is strictly convex on $\mathcal{Y}$ since

$$J(y + v) - J(y) = \int_a^b ((y + v)^2(x) - y^2(x))\, dx$$

$$= 2 \int_a^b y(x)v(x)\, dx + \int_a^b v^2(x)\, dx$$

$$\geq 2 \int_a^b y(x)v(x)\, dx = \delta J(y; v),$$

with equality *iff* $\int_a^b v^2(x)\, dx = 0$ which by A.9 is possible for the continuous function v^2 *iff* $v^2(x) \equiv 0$; i.e., $v = \mathcal{O}$. Thus, by Proposition 3.3, each $y \in \mathcal{Y}$ which makes

$$\delta J(y; v) = 2 \int_a^b y(x)v(x)\, dx = 0,$$

$\forall\, v \in \mathcal{Y}$, minimizes J on $\mathcal{Y}$ uniquely. Clearly, $y = \mathcal{O}$ accomplishes this and hence it is the unique minimizing function.

On the other hand, to minimize J on

$$\mathcal{D} = \{y \in C[a, b]: y(a) = a_1, y(b) = b_1\},$$

we would again try to have $\delta J(y; v) = 0$, but now only for those $y, y + v \in \mathcal{D}$, i.e., only for those $v \in \mathcal{D}_0$ where

$$\mathcal{D}_0 = \{v \in C[a, b]: v(a) = v(b) = 0\}. \quad (\text{Why?})$$

Again $y = \mathcal{O}$ would make $\delta J(y; v) = 0$ but now it is *not* in $\mathcal{D}$ (unless $a_1 = b_1 = 0$).

Example 2.

$$F(y) = \int_a^b y'(x)^2\, dx$$

for which

$$\delta F(y; v) = 2 \int_a^b y'(x)v'(x)\, dx, \qquad \forall\, y, v \in \mathcal{Y} = C^1[a, b],$$

is also convex on $\mathcal{Y}$ (Why?), but now the equality $F(y + v) - F(y) = \delta F(y; v)$

is possible *iff* $\int_a^b v'(x)^2\, dx = 0$ and this occurs whenever $v(x) = \text{const.}$ Thus F is *not* strictly convex on $\mathscr{Y}$. However, F *is* strictly convex on

$$\mathscr{D} = \{y \in C^1[a, b]:\ y(a) = a_1,\ y(b) = b_1\},$$

since now $y,\ y + v \in \mathscr{D} \Rightarrow v \in \mathscr{D}_0$ where

$$\mathscr{D}_0 = \{v \in C^1[a, b]:\ v(a) = v(b) = 0\},$$

and hence equality for v in $\mathscr{D}_0$ is possible *iff* $v(x) = \text{const.} = v(a) = 0$. Thus each $y \in \mathscr{D}$ which makes $\delta F(y; v) = 2\int_a^b y'(x)v'(x)\, dx = 0,\ \forall\, v \in \mathscr{D}_0$ minimizes F on $\mathscr{D}$ uniquely. By inspection, $y' = \text{const.}$ will accomplish this since

$$\int_a^b (\text{const.})\, v'(x)\, dx = (\text{const.}) \int_a^b v'(x)\, dx$$

$$= \text{const.}\ [v(b) - v(a)] = 0, \qquad \forall\, v \in \mathscr{D}_0.$$

The linear function $y_0(x) = [(b_1 - a_1)/(b - a)](x - a) + a_1 \in \mathscr{D}$ and has $y_0' = \text{const.}$ Hence it minimizes F on $\mathscr{D}$ uniquely.

Example 3. A linear function L on $\mathscr{Y}$ is convex, since by Problem 2.11,

$$L(y + v) - L(y) = L(v) = \delta L(y; v), \qquad \forall\, y, v \in \mathscr{Y},$$

but it is not strictly convex. (Why?)

Example 4. The evaluation function from Example 5 of §2.4, $\tilde{J}(y) = y^2(b)$, which for $y,\ v \in \mathscr{Y} = C[a, b]$ has the variation $\delta\tilde{J}(y; v) = 2y(b)v(b)$, is convex since

$$\tilde{J}(y + v) - \tilde{J}(y) = (y + v)^2(b) - y^2(b) = 2y(b)v(b) + v^2(b)$$

$$\geq \delta\tilde{J}(y; v);$$

but not *strictly* convex since equality occurs when $v(b) = 0$. Clearly $\tilde{J}(y) \geq 0$, but $\tilde{J}$ assumes its minimum value of 0 whenever $y(b) = 0$.

§3.2. Convex Integral Functions

If $f = f(x, y, z)$ and its partial derivatives $f_y,\ f_z$ are defined and continuous on $[a, b] \times \mathbb{R}^2$, then as in Example 4 of §2.4, we know that the integral function

$$F(y) = \int_a^b f(x, y(x), y'(x))\, dx = \int_a^b f[y(x)]\, dx,$$

has $\forall\, y,\ v \in C^1[a, b]$, the variation

$$\delta F(y; v) = \int_a^b (f_y[y(x)]v(x) + f_z[y(x)]v'(x))\, dx, \tag{2}$$

where from §1.5 we recall the generic abbreviation

$$f[y(x)] \overset{\text{def}}{=} f(x, y(x), y'(x)), \tag{3}$$

so that

$$f_y[y(x)] = f_y(x, y(x), y'(x)) \quad \text{and} \quad f_z[y(x)] = f_z(x, y(x), y'(x)).$$

Hence, convexity of F requires that $\forall\ y, y + v \in C^1[a, b]$

$$F(y + v) - F(y) \geq \delta F(y; v),$$

or that

$$\int_a^b (f[y(x) + v(x)] - f[y(x)]) \, dx \geq \int_a^b (f_y[y(x)]v(x) + f_z[y(x)]v'(x)) \, dx.$$

Now this would follow from the corresponding pointwise inequality between the integrands in the last expression; i.e., if for each $x \in (a, b)$:

$$f[y(x) + v(x)] - f[y(x)] \geq f_y[y(x)]v(x) + f_z[y(x)]v'(x), \tag{4}$$

or from (3), if

$$f(x, y + v, z + w) - f(x, y, z) \geq f_y(x, y, z)v + f_z(x, y, z)w,$$

$$\forall\ (x, y, z), (x, y + v, z + w) \in (a, b) \times \mathbb{R}^2, \tag{5}$$

where we have incorporated the abbreviations:

$$y = y(x), \qquad v = v(x) \quad \text{and} \quad z = y'(x), \qquad w = v'(x).$$

Inequality (5) simply states that $f = f(x, y, z)$ is convex when x is held fixed. [See (6) of §0.8.] This restricted or *partial* convexity essential to our development is expressed and extended in the following which uses for illustration a function defined on a subset of $\mathbb{R}^3$:

(3.4) **Definition.** $f(\underline{x}, y, z)$ is said to be [strongly] convex on $S \subseteq \mathbb{R}^3$ if $f = f(x, y, z)$ and its partial derivatives f_y and f_z are defined and continuous on this set and there they satisfy the inequality:

$$f(x, y + v, z + w) - f(x, y, z) \geq f_y(x, y, z)v + f_z(x, y, z)w,$$

$$\forall\ (x, y, z) \text{ and } (x, y + v, z + w) \in S, \tag{6}$$

[with equality at (x, y, z) only if $v = 0$ or $w = 0$].

Observe that the underlined variable(s) (if any) are held fixed in the inequality while partial derivatives of f are required only for the remaining variables (y and z). Clearly, if f itself is convex on $\mathbb{R}^3$ as in §0.8 then $f(\underline{x}, y, z)$ will be convex as above. Moreover, if as in §0.9, f is *strictly* convex, then $f(\underline{x}, y, z)$ will be *strongly* convex. However, in general, strong convexity is weaker than strict convexity. (Why?) Also, $f(y, z)$ is [strongly] convex on $D \subseteq \mathbb{R}^2$ precisely when $\tilde{f}(\underline{x}, y, z) = f(y, z)$ is [strongly] convex on $[a, b] \times D$.

For instance, we will see in the next section that $f(y, z) = z^2 + 4y$ is strongly convex on $\mathbb{R}^2$ (even though it is not strictly convex). Therefore $\tilde{f}(\underline{x}, y, z) = z^2 + 4y$ is strongly convex on $[a, b] \times \mathbb{R}^2$ for any interval $[a, b]$.

The significance of strong convexity is seen in the following:

(3.5) Theorem. *Let D be a domain in $\mathbb{R}^2$ and for given a_1, b_1, set*

$$\mathscr{D} = \{y \in C^1[a, b]: y(a) = a_1, y(b) = b_1; (y(x), y'(x)) \in D\}.$$

If $f(\underline{x}, y, z)$ is [strongly] convex on $[a, b] \times D$, then

$$F(y) = \int_a^b f(x, y(x), y'(x)) \, dx$$

is [strictly] convex on $\mathscr{D}$. Hence each $y \in \mathscr{D}$ for which

$$\frac{d}{dx} f_z[y(x)] = f_y[y(x)]$$

on (a, b), minimizes F on $\mathscr{D}$ [uniquely].

PROOF. When $y, y + v \in \mathscr{D}$, then inequality (6) shows that at each $x \in (a, b)$,

$$f[y(x) + v(x)] - f[y(x)] \geq f_y[y(x)]v(x) + f_z[y(x)]v'(x), \qquad (7)$$

[with equality only if $v(x) = 0$ or $v'(x) = 0$ so that $v(x)v'(x) = 0$]. Integrating (7) gives

$$\int_a^b (f[y(x) + v(x)] - f[y(x)]) \, dx \geq \int_a^b (f_y[y(x)]v(x) + f_z[y(x)]v'(x)) \, dx,$$

or with (2) and (3):

$$F(y + v) - F(y) \geq \delta F(y; v),$$

so that F is convex.

Moreover, in the presence of (7), equality between the integrals representing these last functions is possible only when equality holds everywhere in (7) (A.10). [But if $f(\underline{x}, y, z)$ is strongly convex, this in turn is possible only if the product $v(x)v'(x) = \frac{1}{2}(v^2(x))' \equiv 0$; then $v^2(x) = \text{const.} = v^2(a) = 0$ when both y and $y + v \in \mathscr{D}$. Thus $v = \mathcal{O}$ so that F is *strictly* convex on $\mathscr{D}$.]

Finally from (2), each y for which

$$\frac{d}{dx} f_z[y(x)] = f_y[y(x)] \quad \text{on } (a, b) \tag{8}$$

makes

$$\delta F(y; v) = \int_a^b \frac{d}{dx}(f_z[y(x)]v(x)) \, dx = f_z[y(x)]v(x) \Big|_a^b$$

$$= 0 \quad \text{when } y, y + v \in \mathscr{D}. \quad \text{(Why?)} \tag{8'}$$

Thus by Proposition 3.3, y minimizes F on $\mathscr{D}$ [uniquely].   □

Neither of the convexity implications stated in this theorem is reversible. (See Problem 3.16.)

Example 1. To minimize

$$F(y) = \int_0^1 (y'^2(x) + 4y(x)) \, dx$$

on

$$\mathscr{D} = \{y \in C^1(0, 1]: y(0) = 0, y(1) = 1\}$$

we recall that $f(\underline{x}, y, z) = z^2 + 4y$ is strongly convex on $[0, 1] \times \mathbb{R}^2$. Hence according to Theorem 3.5, F is minimized uniquely on $\mathscr{D}$ by a solution y_0 of the equation

$$\frac{d}{dx} f_z[y(x)] = f_y[y(x)] \qquad (0 < x < 1),$$

which for this f is just

$$\frac{d}{dx} [2y'(x)] = 4 \quad \text{or} \quad y'' = 2.$$

Upon integrating twice we obtain the general solution

$$y(x) = x^2 + cx + c_0$$

for constants c, c_0 to be found *if possible* so that $y \in \mathscr{D}$. We require $y(0) = c_0 = 0$ and (then) $y(1) = 1 + c = 1$, or $c = 0$. Consequently $y_0(x) = x^2$ minimizes F on $\mathscr{D}$ and it is the only function which does so!

(3.6) Remarks. The differential equation (8) whose solutions in $\mathscr{D}$ minimize our convex F is known as the Euler–Lagrange equation. It is a fundamental tool of the variational calculus and we will examine it thoroughly in Chapter 6. For the present, note that if $f \in C^2([a, b] \times D)$, then we may use the chain rule (formally) on the left side of (8) and seek a minimizing $y \in \mathscr{D} \cap C^2[a, b]$ which satisfies the second-order differential equation

$$f_{zx}[y(x)] + f_{zy}[y(x)]y'(x) + f_{zz}[y(x)]y''(x) = f_y[y(x)], \tag{9}$$

(with the obvious abbreviations). Although there are standard existence theorems which provide conditions for a solution y to (9) in a neighborhood of $x = a$ which satisfies $y(a) = a_1$, these theorems do *not* gurantee that such solutions can be extended to $[a, b]$, or that when extendable they can meet the second end point condition $y(b) = b_1$. (In Problem 3.20, we have an example for which even the simpler equation $f_z(x, y'(x)) = \text{const.}$ cannot be satisfied in $\mathscr{D}$.) Thus we do *not* have a proof for the existence of a function which minimizes F on $\mathscr{D}$, and indeed as we shall see, such functions need *not* exist. Our condition (8) is at best sufficient, and we must consider each application independently.

There is some simplification when y is not present explicitly, i.e., when $f = f(x, z)$ alone (or $f = f(z)$). Then $f_y \equiv 0$, and for an interval I, the appropriate requirement for the [strong] convexity of $f(\underline{x}, z)$ on $[a, b] \times I$ is that for each $x \in [a, b]$:

$$f(x, z + w) - f(x, z) \geq f_z(x, z)w, \qquad \forall\, z, z + w \in I;$$

$$\text{[with equality at } z \text{ iff } w = 0\text{].} \tag{10}$$

If, in addition, $f_x \equiv 0$, then $f = f(z)$ alone which is [strictly] convex on I precisely when $\tilde{f}(\underline{x}, z) = f(z)$ is [strongly] convex on $[a, b] \times I$. (Why?)

With these reductions, the next results should be apparent.

(3.7) Theorem. *Let I be an interval and set*

$$\mathscr{D} = \{y \in C^1[a, b]: y(a) = a_1, y(b) = b_1; y'(x) \in I\}.$$

Then, if $f(\underline{x}, z)$ is [strongly] convex on $[a, b] \times I$, each $y \in \mathscr{D}$ which makes $f_z(x, y'(x)) = \text{const. on } (a, b)$ minimizes $F(y) = \int_a^b f(x, y'(x))\, dx$ on $\mathscr{D}$ [uniquely].

PROOF. This follows immediately from (10) upon setting $f_y \equiv 0$ in the statement and proof of the previous theorem. $\qquad\square$

(3.8) Corollary. *If $f = f(z)$ is [strictly] convex on I and*

$$m = [(b_1 - a_1)/(b - a)] \in I,$$

then $y_0(x) \equiv m(x - a) + a_1$ minimizes $F(y) = \int_a^b f(y'(x))\}\, dx$ on $\mathscr{D}$ [uniquely].

PROOF. If $y_0'(x) = m \in I$, then $y_0 \in \mathscr{D}$ (Why?) and $f_z(y_0'(x)) = f_z(m)$ is constant on (a, b). Hence Theorem 3.7 is applicable. $\qquad\square$

There are similar simplifications when $f = f(x, y)$, but the associated integrands occur less frequently in application. (See Problem 3.18.)

Free End-Point Problems

When we examine the proof of Theorem 3.5 we see that the end-point specification was used only to conclude that the *constant* $v^2(x) = 0$ and that $f_z[y(x)]v(x)|_a^b = 0$. Hence these end-point conditions on y may be relaxed, if suitable compensation is made in $f_z[y(x)]$.

(3.9) Proposition. *Let D be a domain in $\mathbb{R}^2$ and suppose that $f(\underline{x}, y, z)$ is [strongly] convex on $[a, b] \times D$. Then each solution $y_0 \in \mathscr{D} = \{y \in C^1[a, b]: (y(x), y'(x)) \in D\}$ of the differential equatiion $(d/dx)f_z[y(x)] = f_y[y(x)]$ minimizes*

$$F(y) = \int_a^b f[y(x)]\, dx:$$

(i) *on* $\mathscr{D}^b = \{y \in \mathscr{D}: y(a) = y_0(a)\}$, *if* $f_z[y_0(b)] = 0$ [*uniquely*];
(ii) *on* $\mathscr{D}$, *if* $f_z[y_0(a)] = f_z[y_0(b)] = 0$, [*uniquely within an additive constant*].[1]

(As we shall see in §6.4, these "natural" boundary conditions on $f_z[y_0(x)]$ are also necessary for the minimization.)

PROOF. Only the last assertion in (ii) requires further comment. If $f(\underline{x}, y, z)$ is strongly convex on $[a, b] \times D$, and $y_0 \in \mathscr{D}$ is a solution of the given differential equation, then when $y_0 + v \in \mathscr{D}$ we have from (8') that

$$F(y_0 + v) - F(y_0) \geq \delta F(y_0; v) = f_z[y_0(x)]v(x)|_a^b = 0,$$

with equality only if $v^2(x) = \text{const.}$ ($\Rightarrow 2v^2(x)v'(x) = v(x)(v^2(x))' \equiv 0$) so that $v'(x) \equiv 0$ and $v(x) = \text{const.}$ on $[a, b]$. Thus

$$F(y_0 + v) = F(y_0) \Rightarrow y_0 + v = y_0 + \text{const.} \qquad \square$$

Extension of the results of this section to convex functions defined by improper Riemann integrals is treated in Problem 3.21*.

Example 2. Let's return to the problem in Example 1, where we found that $y_0(x) = x^2$ minimizes. It happens that $f_z[y_0(x)] = 2y_0'(x) = 4x$ vanishes at $x = 0$ and so from Proposition 3.9 we see that y_0 also minimizes F uniquely on the larger set

$$\mathscr{D}_1 = \{y \in C^1[0, 1]: y(1) = 1\}.$$

Similarly, we can show that $y_1(x) = x^2 - 2x$ minimizes F on $\mathscr{D}^1 = \{y \in C^1[0, 1]: y(0) = 0\}$ uniquely. However, our method cannot produce a function of the form $y_2(x) = x^2 + cx + c_0$ that minimizes F on the still larger set $\mathscr{D}_2 = C^1[0, 1]$, since in this case, $f_z[y_2(x)] = 2y_2'(x) = 2(2x + c)$ cannot be zero at *both* end-points. In view of Remarks (3.6) we should not be surprised that we did not get everything we might wish. We were lucky to get what we did so easily.

§3.3. [Strongly] Convex Functions

In order to apply the results of the previous section, we require a supply of functions which are [strongly] convex. In this section techniques for recognizing such convexity will be developed.

We begin with the simpler case $f = f(x, z)$, where as we have seen, the defining inequality for [strong] convexity of $f(\underline{x}, z)$ on $[a, b] \times I$ is (10). Now (10) is in turn guaranteed by a simple condition on f_{zz} which should recall the

---

[1] The choice of constants may be limited. See Problem 3.9.

criterion from elementary calculus for the convexity of a function defined in an interval I.

(3.10) Proposition. *If $f = f(x, z)$ and f_{zz} are continuous on $[a, b] \times I$ and for each $x \in [a, b]$, $f_{zz}(x, z) > 0$ (except possibly at a finite set of z values) then $f(x, z)$ is strongly convex on $[a, b] \times I$.*

PROOF. For fixed $x \in [a, b]$, let $g(z) = f(x, z)$ so that $g''(z) = f_{zz}(x, z) > 0$ on I (with a possible finite set of exceptiional values). Then integrating by parts gives for distinct $z, \zeta \in I$:

$$g(\zeta) - g(z) = \int_z^\zeta g'(t)\, dt = (\zeta - z)g'(z) + \int_z^\zeta (\zeta - t)g''(t)\, dt > (\zeta - z)g'(z),$$

since the last integral is strictly positive by the hypothesis and A.9, independently of whether $z < \zeta$ or $\zeta < z$. (Why?)

Thus with $w = \zeta - z$, recalling the definition of g, we conclude that $f(x, z + w) - f(x, z) > f_z(x, z)w$, when $w \neq 0$, and this establishes the strong convexity of $f(x, z)$. ☐

Remark. If at some $x \in [a, b]$, $f_{zz}(x, z) \equiv 0$ on $[z_1, z_2] \subseteq I$ then $f_z(x, z)$ increases with z, but not strictly, so that $f(x, z)$ is only convex on $[a, b] \times I$.

Example 1. $f(x, z) = \sin^3 x + z^2$ is strongly convex on $\mathbb{R} \times \mathbb{R}$ since

$$f_{zz}(x, z) = 2 > 0.$$

Example 2. $f(x, z) = e^x(\sin^3 x + z^2)$ is also strongly convex on $\mathbb{R} \times \mathbb{R}$ since

$$f_{zz}(x, z) = 2e^x > 0.$$

(In fact the product of a [strongly] convex function by a positive continuous function $p = p(x)$ is again [strongly] convex. See Problem 3.3.)

Example 3. For $r \neq 0$, $f(x, z) = \sqrt{r^2 + z^2}$ is strongly convex on $\mathbb{R} \times \mathbb{R}$, since $\forall z \in \mathbb{R}$:

$$f_z(x, z) = \frac{z}{\sqrt{r^2 + z^2}}, \quad \text{so that}$$

$$f_{zz}(x, z) = \frac{1}{\sqrt{r^2 + z^2}} - \frac{z^2}{(r^2 + r^2)^{3/2}} = \frac{r^2}{(r^2 + z^2)^{3/2}} > 0.$$

Example 4. If $0 < p \in C[a, b]$, and $r \neq 0$, $f(x, z) = p(x)\sqrt{r^2 + z^2}$ is also strongly convex on $[a, b] \times \mathbb{R}$. (See Example 2 above.)

Example 5. $f(z) = -\sqrt{1 - z^2}$ is strongly convex on $I = (-1, 1)$, since

$$f_z(z) = \frac{z}{\sqrt{1 - z^2}} \quad \text{so that for } z \in I,$$

$$f_{zz}(z) = \frac{1}{\sqrt{1 - z^2}} + \frac{z^2}{(1 - z^2)^{3/2}} = \frac{1}{(1 - z^2)^{3/2}} > 0.$$

Example 6. $f(\underline{x}, z) = \underline{x}^3 + e^{\underline{x}}z$ is (only) convex on $\mathbb{R} \times \mathbb{R}$ since $f_{zz} \equiv 0$. Indeed:

$$f(x, z + w) - f(x, z) = e^x w = f_z(x, z)w,$$

and equality holds $\forall\, w \in \mathbb{R}$.

Example 7. $f(\underline{x}, z) = \underline{x}^2 + (\sin \underline{x})z^2$ with $f_{zz}(x, z) = 2(\sin x)$, becomes convex only when $\sin x \geq 0$; e.g., on $[0, \pi] \times \mathbb{R}$; and is strongly convex only when $\sin x > 0$, e.g., on $(0, \pi) \times \mathbb{R}$.

Example 8. Finally, $f(\underline{x}, z) = \underline{x}^2 - z^2$ is never convex, since for $w \neq 0$,

$$f(x, z + w) - f(x, z) - f_z(x, z)w = -(z + w)^2 + z^2 + 2zw$$

$$= -w^2 < 0.$$

(However, $-f(\underline{x}, z) = -x^2 + z^2$ is again strongly convex on $\mathbb{R} \times \mathbb{R}$.)

Of course, the conclusions just obtained are unchanged if z is replaced by y in each occurrence.

When all variables are present in the function $f(x, y, z)$, there are no simplifications such as those just considered. (There is again a second derivative condition which guarantees [strict] convexity, but it is awkward to apply. See Problem 3.5.)

The following general observations (whose proofs are left to Problems 3.2 and 3.3) will be of value:

(3.11) **Fact 1.** *The sum of a [strongly] convex function and one (or more) convex functions is again [strongly] convex.*

Fact 2. *The product of a [strongly] convex function $f(\underline{x}, y, z)$ by a continuous function $[p(x) > 0]\ p(x) \geq 0$ is again [strongly] convex on the same set.*

Fact 3. $f(\underline{x}, y, z) = \alpha(\underline{x}) + \beta(\underline{x})y + \gamma(\underline{x})z$ *is (only) convex for any continuous functions α, β, γ.*

Fact 4. *Each [strongly] convex function $f(\underline{x}, z)$ (or $f(\underline{x}, y)$) is also [strongly] convex when considered as a function $\tilde{f}(\underline{x}, y, z)$ on an appropriate set.*

Example 9. $f(\underline{x}, y, z) = -2(\sin \underline{x})y + z^2$ is the sum of the strongly convex function z^2 (Fact 4) with the convex function $-2(\sin \underline{x})y$ (Fact 3), and hence it is strongly convex on $\mathbb{R} \times \mathbb{R}^2$ (Fact 1).

Similarly, $g(\underline{x}, y, z) = -2(\sin \underline{x})y + z^2 + \underline{x}^2\sqrt{1 + y^2}$, is the sum of the strongly convex function $f(\underline{x}, y, z)$ with the convex function $\underline{x}^2\sqrt{1 + y^2}$ (Fact 2) and so it too is strongly convex on $\mathbb{R} \times \mathbb{R}^2$.

Example 10. $f(y, z) = \sqrt{1 + y^2 + z^2}$ with

$$f_y(y, z) = \frac{y}{\sqrt{1 + y^2 + z^2}} \quad \text{and} \quad f_z(y, z) = \frac{z}{\sqrt{1 + y^2 + z^2}}$$

is more difficult to examine. For its convexity, we require that

$$\sqrt{1 + (y + v)^2 + (z + w)^2} - \sqrt{1 + y^2 + z^2} \geq \frac{yv + zw}{\sqrt{1 + y^2 + z^2}}, \quad (11)$$

or upon introducing the three-dimensional vectors

$$A = (1, y, z) \quad \text{and} \quad B = (1, y + v, z + w),$$

that

$$|B| \geq |A| + \frac{(yv + zw)}{|A|} = \frac{|A|^2 + (yv + zw)}{|A|}$$

$$= \frac{1 + y(y + v) + z(z + w)}{|A|} = \frac{A \cdot B}{|A|},$$

where the dot denotes the scalar product of the vectors.

Since $A \cdot B = |A||B| \cos(A, B) \leq |A||B|$ with equality *iff* A and B are codirected, it is seen that (11) does hold with equality *iff* $(1, y, z)$ and $(1, y + v, z + w)$ are codirected, i.e., *iff* $v = w = 0$. Thus, $f(y, z) = \sqrt{1 + y^2 + z^2}$ is strongly convex on $\mathbb{R}^2$; in fact, it is *strictly* convex on $\mathbb{R}^2$.

Example 11. If $0 < p \in C[a, b]$, then

$$f(\underline{x}, y, z) = p(\underline{x})\sqrt{1 + y^2 + z^2}$$

is strongly convex on $[a, b] \times \mathbb{R}^2$ (Fact 2 and Example 10.)

Example 12. When $b \neq 0$, then $f(y, z) = \sqrt{y^2 + b^2 z^2}$ has derivatives

$$f_y(y, z) = \frac{y}{\sqrt{y^2 + b^2 z^2}} \quad \text{and} \quad f_z(y, z) = \frac{b^2 z}{\sqrt{y^2 + b^2 z^2}},$$

which are discontinuous at the origin. However, on the restricted set $\mathbb{R}^2 \sim \{(0, 0)\}$ this function is again convex but not strongly convex. (See Problem 3.24.)

Example 13. When $\alpha \neq 0$, $f(y, z) = (z + \alpha y)^2$ is only convex in $\mathbb{R}^2$, since (5) holds, but with equality when $(w + \alpha v)^2 = 0$. (See Problem 3.16b.)

(Problems 3.1–3.19)

§3.4. Applications

In this section we show that convexity is present in problems from several diverse fields—at least after suitable formulation—and use previous results to characterize their solutions. Applications, presented in order of increasing difficulty—and/or sophistication, are given which characterize geodesics on a cylinder, a version of the brachistochrone, Newton's profile of minimum drag, an optimal plan of production, and a form of the minimal surface. Other applications in which convexity can be used with profit will be found in Problems 3.20 et seq.

(a) Geodesics on a Cylinder

To find the geodesics on the surface of a right circular cylinder of radius 1 unit, we employ, naturally enough, the cylindrical coordinates (θ, z) shown in Figure 3.1 to denote a typical point. It is obvious that the geodesic joining points $P_1 = (\theta_1, z_1)$, $P_2 = (\theta_1, z_2)$ is simply the vertical segment connecting them. Thus it remains to consider the case where $P_2 = (\theta_2, z_2)$ with $\theta_2 \neq \theta_1$; a little thought shows that by relabeling if necessary, we can suppose that $0 < \theta_2 - \theta_1 \leq \pi$, and consider those curves which admit representation as the graph of a function $z \in \mathscr{D} = \{z \in C^1[\theta_1, \theta_2]: z(\theta_j) = z_j, j = 1, 2\}$.

The spatial coordinates of such a curve are

$$(x(\theta), y(\theta), z(\theta)) = (\cos \theta, \sin \theta, z(\theta)),$$

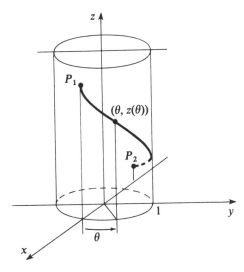

Figure 3.1

so that when $z \in \mathscr{D}$, the resulting curve has the length

$$L(z) = \int_{\theta_1}^{\theta_2} \sqrt{x'(\theta)^2 + y'(\theta)^2 + z'(\theta)^2}\, d\theta = \int_{\theta_1}^{\theta_2} \sqrt{1 + z'(\theta)^2}\, d\theta.$$

With an obvious change in variables, this integrand corresponds to the function of §3.3, Example 3, which is strongly convex. Thus by Corollary 3.8, we conclude that

(3.12) *Among curves which admit representation as the graph of a function* $z \in \mathscr{D}$*, the minimum length is given uniquely for that represented by the function*

$$z_0(\theta) = z_1 + m(\theta - \theta_1) \quad \text{for } m = \frac{z_1 - z_2}{\theta_1 - \theta_2},$$

which describes the circular helix joining the points.

(If the cylinder were "unrolled," this would correspond to the straight line joining the points.) Plants take helical paths when climbing around cylindrical supporting stakes toward the sun [Li].

(b) A Brachistochrone

For our next application, we return to the brachistochrone of §1.2(a). As formulated there, the function $T(y)$ is not of the form covered by Theorem 3.5. (Why not?) However, if we interchange the roles of x and y and consider those curves which admit representation as the graph of a function $y \in \mathscr{D} = \{y \in C^1[0, x_1]: y(0) = 0, \ y(x_1) = y_1\}$ (with x_1 and y_1 both positive) as in Figure 3.2, then in the *new* coordinates, the same analysis as before gives for each such curve the transit time

$$T(y) = \int_0^{x_1} \sqrt{\frac{1 + y'(x)^2}{2gx}}\, dx,$$

which has the strongly convex integrand function of §3.3, Example 4, with

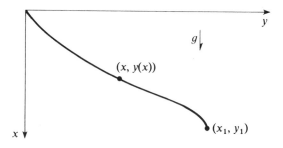

Figure 3.2

$r = 1$ and $p(x) = (2gx)^{-1/2}$ on $(0, x_1]$. Now $p(x)$ is positive and integrable on $[0, x_1]$ and although it is not continuous (at 0), Theorem 3.7 remains valid. (See Problem 3.21.)

Thus we know that among such curves, the minimum transit time would be given uniquely by each $y \in \mathcal{D}$ which makes

$$\frac{y'(x)}{\sqrt{x}\sqrt{1 + y'(x)^2}} \equiv \frac{1}{c} \quad \text{for } some \text{ constant } c.$$

Squaring both sides gives the equation

$$\frac{y'(x)^2}{1 + y'(x)^2} = \frac{x}{c^2}, \quad \text{so} \quad 1 + y'(x)^2 = \frac{c^2}{c^2 - x^2}$$

and

$$y'(x) = \frac{\sqrt{x}}{c}\sqrt{1 + y'(x)^2} = \sqrt{\frac{x}{c^2 - x}} \geq 0. \tag{12}$$

Thus $y'(0) = 0$.

If we introduce the new independent variable θ through the relation $x(\theta) = (c^2/2)(1 - \cos\theta) = c^2 \sin^2(\theta/2)$, then $\theta = 0$ when $x = 0$, and for $\theta < \pi$, θ increases with x. Also, $c^2 - x(\theta) = (c^2/2)(1 + \cos\theta)$. By the chain rule

$$\frac{dy}{d\theta} = y'(x)x'(\theta) = y'(x)\left(\frac{c^2}{2}\sin\theta\right),$$

and from equation (12) we get

$$\frac{dy}{d\theta} = \frac{c^2}{2}\sqrt{\frac{1 - \cos\theta}{1 + \cos\theta}}\sin\theta = \frac{c^2}{2}(1 - \cos\theta).$$

Hence $y(\theta) = (c^2/2)(\theta - \sin\theta) + c_1$, and the requirement $y(0) = 0$ shows that $c_1 = 0$.

Upon replacing the unspecified constant c by $\sqrt{2}c$, we see that the minimum transit time would be given parametrically by a curve of the form

$$\begin{cases} x(\theta) = c^2(1 - \cos\theta), \\ y(\theta) = c^2(\theta - \sin\theta), \end{cases} \quad 0 \leq \theta \leq \theta_1, \tag{13}$$

provided that c^2 and θ_1 can be found to make $x(\theta_1) = x_1$, $y(\theta_1) = y_1$. The curve described by these equations is the cycloid with cusp at $(0, 0)$ which would be traced by a point on the circumference of a disk of radius c^2 as it rolls along the y axis from "below" as shown in Figure 3.3.

For $\theta > 0$, the ratio $y(\theta)/x(\theta) = (\theta - \sin\theta)/(1 - \cos\theta)$ has the limiting value $+\infty$ as $\theta \uparrow 2\pi$, and by L'Hôpital's rule it has the limiting value of 0 as $\theta \downarrow 0$. Its derivative is

$$\frac{(1 - \cos\theta)^2 - \sin\theta(\theta - \sin\theta)}{(1 - \cos\theta)^2} = \frac{2(1 - \cos\theta) - \theta\sin\theta}{(1 - \cos\theta)^2},$$

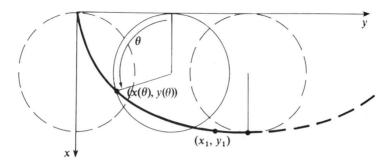

Figure 3.3

which utilizing the half-angle formulae may be rewritten as

$$\frac{\cos \theta/2}{\sin^3 \theta/2}\left(\tan \frac{\theta}{2} - \frac{\theta}{2}\right) \qquad (\theta \neq \pi),$$

and thus is positive for $0 < \theta < 2\pi$. (Why?) $y(\theta)/x(\theta)$ is positive, increases strictly from 0 to $+\infty$ as θ increases from 0 to 2π, and hence from continuity (through the intermediate value theorem of §A.1), assumes each positive value precisely once. In particular, there is a unique $\theta_1 \in (0, 2\pi)$ for which $y(\theta_1)/x(\theta_1) = y_1/x_1$, and for this θ_1, choosing $c^2 = x_1/(1 - \cos \theta_1)$ will guarantee the existence of a (unique) cycloid joining $(0, 0)$ to (x_1, y_1).

Unfortunately, as Figure 3.3 shows, the associated curve can be represented in the form $y = y(x)$ only when $\theta_1 \leq \pi$, i.e., when $y_1/x_1 \leq \pi/2$. Moreover, the associated function $y \in C^1[0, x_1]$ only when $y_1/x_1 < \pi/2$, since the tangent line to the cycloid must be horizontal at the lowest point on the arch. Nevertheless, we do have a nontrivial result:

(3.13) *When $y_1/x_1 < \pi/2$, among all curves representable as the graph of a function $y \in C^1[0, x_1]$ which join $(0, 0)$ to (x_1, y_1), the cycloid provides uniquely the least time of descent.*

Thus we confirm Galileo's belief that the brachistochrone is *not* the straight line and support the classical assertion by Newton and the Bernoullis that it must always be a cycloid.

It is not too difficult to extend our analysis to the case $y_1/x_1 = \pi/2$ (see Problem 3.22*), and it may seem physically implausible to consider curves which fall below their final point or those which have horizontal sections (i.e., those which *cannot* be expressed in the form $y = y(x)$) as candidates for the brachistochrone. However, it is true that the brachistochrone is always the cycloid, but a proof for the general case will be deferred until §8.8.

(c) A Profile of Minimum Drag

One of the first problems to be attacked by a variational approach was that propounded by Newton in his *Principia* (1686) of finding the profile of [the shoulder of] a *projectile* of *revolution* which would offer minimum resistance (or drag) when moved in the direction of its axis at a constant speed w_0 in water. (We can suppose that $w_0 = 1$.)[1]

We adopt the coordinates and geometry shown in Figure 3.4, and postulate with Newton that the resisting pressure at a surface point on the shoulder is proportional to the *square* of the *normal* component of its velocity. A shoulder is obtained by rotating a meridional curve of length l defined by $y = y(x)$ about the y-axis as shown. At a point on the shoulder, let ψ denote the angle between the *normal* to this curve and the positive y-axis. Then we wish to minimize

$$\int_0^l \cos^2 \psi(\sigma) 2\pi x(\sigma) \cos \psi(\sigma)\, d\sigma.$$

Since $\cos \psi(\sigma) = x'(\sigma)$ while $1 + y'(x)^2 = \sec^2 \psi(\sigma)$, we evidently wish to minimize

$$F(y) = \int_a^1 x(1 + y'(x)^2)^{-1}\, dx,$$

on

$$\mathscr{D} = \{y \in C^1[a, 1]: y(a) = h,\ y(1) = 0,\ y(x) \geq 0\},$$

where we suppose that the *positive* constants $a < 1$ and h are given; ($a = 0$ is excluded for reasons which will emerge). Now, if

$$f(x, z) = \frac{x}{1 + z^2}, \quad \text{then } f_z(x, z) = \frac{-2zx}{(1 + z^2)^2},$$

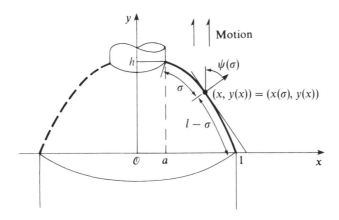

Figure 3.4

[1] See the article "On Newton's Problem of Minimal Resistance," by G. Butazzo and B. Kawohl, in *The Mathematical Intelligencer*, pp. 7–12, Vol. 15, No. 4, 1993. Springer-Verlag, New York.

and for $x > 0$:

$$f_{zz}(x, z) = \frac{2x(3z^2 - 1)}{(1 + z^2)^3} > 0, \quad \text{when } |z| > \frac{1}{\sqrt{3}}.$$

On physical grounds we expect $y' \le 0$ (Why?),[1] and, by Proposition 3.10, $f(\underline{x}, z)$ is strongly convex on $[a, 1] \times (-\infty, -1/\sqrt{3}]$. Hence from Theorem 3.7, we know that if

$$y_0 \in \mathscr{D}' = \{y \in \mathscr{D}: y'(x) \le -1/\sqrt{3}, x \in [a, 1]\}$$

makes

$$f_z(x, y'(x)) = \frac{-2xy'(x)}{(1 + y'(x)^2)^2} = \text{const.} = \frac{2}{c}, \tag{14}$$

say, for a *positive* constant c, then it minimizes F on $\mathscr{D}'$ uniquely. Upon squaring and rewriting, we find that for each $x \in [a, 1]$, $u \overset{\text{def}}{=} 1 + y'^2$ should be a solution to the *quartic* equation $u^4 = c^2x^2(u - 1)$ with $u \ge 1 + (1/\sqrt{3})^2 = 4/3$. From Figure 3.5, we see graphically that this equation has a unique root $u = u(s) \ge 4/3$ when $s = cx \ge 4^2/3^{3/2} = s_0$, say, and that $u(s)$ increases to infinity with s. ($u(s)$ can be determined explicitly by the methods of Cardano–Ferrari, but the result is not simple.[2] Newton's own parametric approach is

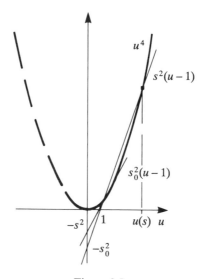

Figure 3.5

taken up in Problem 3.40.) Then, from (14), we get

$$y'(x) = -u^2(cx)/cx, \quad \text{and (since } y(1) = 0),$$

$$y(x) = \int_x^1 \frac{u^2(c\xi)}{c\xi} d\xi = \frac{1}{c} \int_{cx}^c \frac{u^2(s)}{s} ds \qquad (cx \geq s_0), \qquad (14')$$

provided that $c \geq s_0/a$ can be found to make $y(a) = h$. Now $u^2(s)/s = \sqrt{u(s)} - 1$ increases with s, and for fixed x, the first integral in (14') increases continuously to infinity as $c \nearrow +\infty$. Since our restricted minimum-drag problem has at most one solution for given a, it is simpler to choose values of $c \geq c_0 = s_0/a$ and then use (14') to determine associated values of $y(x)$ and $h \overset{\text{def}}{=} y(a)$ by numerical integration. Each value for h exceeding that when $c = c_0$ is achievable. A few of these minimal profiles are presented in Figure 3.6.

Each nontrivial solution thus obtained provides the profile of minimum drag *at least among those in* $\mathscr{D}'$, and this could be used in designing a torpedo or some other missile moving in a medium under *Newton's resistance law*.[1] We cannot claim that our restricted minimum drag profiles remain optimal within a larger class, where, for example, profiles with zero slopes are permitted. See Problems 7.27 and 9.27, and [P] for a more thorough discussion.

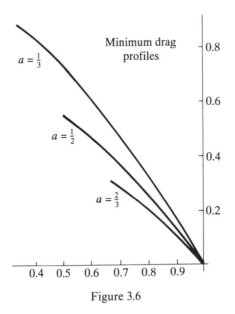

Figure 3.6

[1] Newton himself believed that his results might be applicable in the design of a ship's hull. However, his resistance law is more appropriate to missiles traveling hypersonically in the thin air of our upper atmosphere. See [Fu].

(d) An Economics Problem

All of the classical optimization problems arose in the development of
geometry or physics and are usually concerned with optimizing one of the
fundamental quantities, length, time, or energy under various conditions. For
a change of interest, let's consider a more contemporary problem of production planning (whose statement is taken from [Sm]).

From previous data a manufacturing company with "continuous" inven-
tory has decided that with an initial inventory $\mathcal{I}(0) = \mathcal{I}_0$ and a projected
sales rate $S = S(t)$ over a time interval $[0, T]$, the best production rate at time
t is given by a function $\mathcal{P} = \mathcal{P}(t)$. Assuming that loss during storage occurs
at a rate which is a *fixed* proportion, α, of the associated inventory $\mathcal{I}(t)$ at
time t (perhaps through spoilage), and the rest is sold at the projected rate
$S(t)$, then we should have at time t, the simple differential relation

$$\mathcal{I}'(t) = \mathcal{P}(t) - S(t) - \alpha\mathcal{I}(t)$$
$$(\text{or} \quad \mathcal{P}(t) = \mathcal{I}'(t) + \alpha\mathcal{I}(t) + S(t)). \tag{15}$$

Now suppose that it wishes to maintain the same sales rate $S(t)$ over a
period $[0, T]$ but its actual initial inventory $I(0) = I_0 \neq \mathcal{I}_0$. Then from the
assumed continuity, each projected production rate function $P = P(t)$ results
in an inventory $I(t)$ at time t which differs from $\mathcal{I}(t)$ (at least in a neighborhood of 0). With the same percentage loss, we would have as above,

$$P(t) = I'(t) + \alpha I(t) + S(t). \tag{15'}$$

As a consequence, the company will experience additional operating costs
(perhaps due to handling and storage problems); these costs might be estimated by a function such as

$$C = \int_0^T [\beta^2(I - \mathcal{I})^2(t) + (P - \mathcal{P})^2(t)] \, dt, \tag{16}$$

which takes into account the deviations in both inventory I and associated
production rate P from their "ideal" counterparts. (β is a constant which
adjusts proportions.) This is rather a crude measure of cost, but it possesses
analytical advantages.

Moreover, if we introduce the inventory deviation function $y = I - \mathcal{I}$,
then we see from (15) and (15') that the associated production deviation is
given by

$$P - \mathcal{P} = I' - \mathcal{I}' + \alpha(I - \mathcal{I}) = y' + \alpha y. \tag{16'}$$

Therefore the cost may be regarded as

$$C(y) = \int_0^T [\beta^2 y^2 + (y' + \alpha y)^2](t) \, dt \tag{17}$$

which is to be minimized over $\mathcal{D}^T = \{y \in C^1[0, T]: y(0) = a_0 = I_0 - \mathcal{I}_0\}$.

Now $f(t, y, z) = \beta^2 y^2 + (z + \alpha y)^2$ is strongly convex on $[0, T] \times \mathbb{R}^2$ since the second term is (only) convex. (Recall Example 13 in §3.3.) Here $f_y = 2[\beta^2 y + \alpha(z + \alpha y)]$ and $f_z = 2(z + \alpha y)$. Therefore, by Proposition 3.9(i), C is minimized uniquely by a $y \in \mathscr{D}^T$ for which

$$f_z[y(T)] = 2(y' + \alpha y)(T) = 0 \tag{18}$$

that satisfies the Euler–Lagrange equation

$$\frac{d}{dt}[2(y' + \alpha y)(t)] = 2[(\alpha^2 + \beta^2)y + \alpha y'](t), \qquad 0 < t < T. \tag{18'}$$

From (16') we see that condition (18) simply requires that $P(T) = \mathscr{P}(T)$. Is this reasonable? Why?

If we differentiate in (18') and cancel the $\alpha y'$ terms from each side, then the equation reduces to

$$y'' = (\alpha^2 + \beta^2)y = \gamma^2 y, \quad \text{say,} \tag{19}$$

when we substitute

$$\gamma^2 = \alpha^2 + \beta^2. \tag{19'}$$

The general solution of the differential equation (19) is

$$y_0(t) = c_1 e^{\gamma t} + c_2 e^{-\gamma t}, \quad \text{with} \quad y_0'(t) = \gamma(c_1 e^{\gamma t} - c_2 e^{-\gamma t}), \tag{20}$$

and we must try to find the constants c_1 and c_2 so that $y_0(t)$ satisfies the boundary conditions. We require that

$$y_0(0) = a_0 = c_1 + c_2,$$

and

$$0 = y_0'(T) + \alpha y_0(T)$$
$$= c_1(\gamma + \alpha)e^{\gamma T} + c_2(-\gamma + \alpha)e^{-\gamma T},$$

or that

$$0 = c_1(\gamma + \alpha)e^{2\gamma T} - c_2(\gamma - \alpha).$$

From this last equation, the ratio

$$\rho \stackrel{\text{def}}{=} \frac{\gamma + \alpha}{\gamma - \alpha}e^{2\gamma T} = \frac{c_2}{c_1} \tag{21}$$

is specified, and for this ρ the choices $c_1 = a_0/(1 + \rho)$, $c_2 = a_0\rho/(1 + \rho)$ will satisfy both conditions. This gives the desired conclusion:

(3.14) *Among all inventory functions $I \in C^1[0, T]$ with α, β, and $I(0) = I_0$ prescribed, that given by*

$$I(t) = \mathscr{I}(t) + \frac{(I_0 - \mathscr{I}_0)}{1 + \rho}(e^{\gamma t} + \rho e^{-\gamma t}), \tag{22}$$

with ρ, γ determined by (21) and (19'), respectively, will provide uniquely the minimum cost of operation as assessed by (16). The associated optimal produc-

tion rate is

$$P_0(t) = \mathscr{P}(t) + (1 + \rho)^{-1}(I_0 - \mathscr{I}_0)(\gamma + \alpha)[e^{\gamma t} - e^{\gamma(2T-t)}] \tag{23}$$

Moreover, in this case the minimum cost can easily be computed. Indeed from (17)

$$C(y_0) = \int_0^T [\beta^2 y_0^2 + (y_0' + \alpha y_0)^2] \, dt$$

and we recall that y_0 satisfies (18) and (18′). We see that the integrand is just $[y_0(y_0' + \alpha y_0)]'(t)$, and since $y_0(0) = a_0$, we conclude that

$$C(y_0) = [y_0(y_0' + \alpha y_0)]|_0^T = -\alpha a_0^2 - a_0 y_0'(0),$$

where $y_0'(0) = I'(0) - \mathscr{I}'(0)$ can be obtained by differentiating (22).

Finally recalling that $a_0 = I_0 - \mathscr{I}_0$, we find that

$$C_{\min} = C(y_0) = (I_0 - \mathscr{I}_0)^2 \left[\frac{\gamma(\rho - 1)}{\rho + 1} - \alpha \right], \tag{24}$$

and this expression shows the effects of various choices of α, β, T, and $\mathscr{I}(0)$ on the minimum cost of operation. Observe that it is independent of the sign of the initial inventory deviation.

(e) Minimal Area Problem

Our final example extends the methods of this chapter to a problem in higher dimensions, namely, that of Plateau. In the simplified version formulated in §1.4(b), given a bounded domain D of $\mathbb{R}^2$, and a prescribed smooth boundary function γ, we seek a function $u \in C^1(\bar{D})$ which has these boundary values and minimizes the surface area function

$$S(u) = \iint_D \sqrt{1 + u_x^2 + u_y^2} \, dx \, dy.$$

Introducing $\mathscr{D} = \{u \in C^1(\bar{D}) \text{ with } u|_{\partial D} = \gamma\}$ and $\mathscr{D}_0 = \{v \in C^1(\bar{D}) \text{ with } v|_{\partial D} = 0\}$, we see that this is equivalent to finding a $u \in \mathscr{D}$ for which

$$S(u + v) - S(u) \geq 0, \qquad \forall \, v \in \mathscr{D}_0.$$

Now, the three-dimensional vector inequality used in establishing the strong convexity of $f(y, z) = \sqrt{1 + y^2 + z^2}$ (see Example 10 of §3.3), shows that at each point in D:

$$\sqrt{1 + (u_x + v_x)^2 + (u_y + v_y)^2} - \sqrt{1 + u_x^2 + u_y^2} \geq \frac{u_x v_x + u_y v_y}{\sqrt{1 + u_x^2 + u_y^2}},$$

with equality *iff* $v_x = v_y = 0$. Hence, from the assumed continuity:

$$S(u + v) - S(u) \geq \delta S(u; v) = \iint_D \frac{u_x v_x + u_y v_y}{\sqrt{1 + u_x^2 + u_y^2}} \, dx \, dy$$

(as in §2.4, Example 9) with equality *iff* $v \equiv 0$ (since $v_x = v_y \equiv 0$ in the *domain* $D \Rightarrow v = $ const. $= v|_{\partial D} \equiv 0$). Thus S is strictly convex on $\mathscr{D}$, and again we would seek $u \in \mathscr{D}$ for which $\delta S(u; v)$ vanishes, $\forall\, v \in \mathscr{D}_0$. Such a u would provide the unique minimizing function for S on $\mathscr{D}$. It would, of course, suffice if we could find a u which is even smoother; in particular, if we could find a $u \in \mathscr{D} \cap C^2(D)$ which has these properties.

For $u \in \mathscr{D} \cap C^2(D)$, both

$$U \stackrel{\text{def}}{=} \frac{u_x}{\sqrt{1 + u_x^2 + u_y^2}} \quad \text{and} \quad W \stackrel{\text{def}}{=} \frac{u_y}{\sqrt{1 + u_x^2 + u_y^2}},$$

are in $C^1(D)$ so that the integrand of $\delta S(u; v)$ may be rewritten as $Uv_x + Wv_y = (Uv)_x + (Wv)_y - (U_x + W_y)v$. Now, if we assume that Green's theorem holds for the domain D ([Fl]), then

$$\iint_D [(Uv)_x + (Wv)_y]\, dx\, dy = \int_{\partial D} [(Uv)\, dy - (Wv)\, dx],$$

and for $v \in \mathscr{D}_0$, the line integral vanishes. Thus for $v \in \mathscr{D}_0$,

$$\delta S(u; v) = -\iint_D (U_x + W_y)v\, dx\, dy, \tag{25}$$

and by Proposition 3.3 it is obvious that a minimum area would be given uniquely by each $u \in \mathscr{D} \cap C^2(D)$ which satisfies the *partial differential equation*

$$U_x + W_y = 0 \quad \text{in } D;$$

or upon substitution and simplification, which satisfies the second-order partial differential equation

$$(1 + u_y^2)u_{xx} - 2u_x u_y u_{xy} + (1 + u_x^2)u_{yy} = 0. \tag{26}$$

Equation (26) is called the *minimal surface equation* and it has been studied extensively. Our uniqueness argument shows that this equation cannot have more than one solution u in $\mathscr{D}$, but the existence of a solution depends upon a geometric condition on D:

(3.15) *A domain D is said to be* convex *when it contains the line segment joining each pair of its points.*

A disk is convex while an annulus is not. If the domain D is not convex, it is known that (26) does not always have a solution in the required set $\mathscr{D} \cap C^2(D)$, and we can draw no additional conclusions from the analysis given here. However, it is also known that if D is convex, then (26) has a solution in $\mathscr{D} \cap C^2(D)$ for arbitrary *smooth* γ, which thus describes uniquely the minimal surface; i.e., the surface of minimal area spanning the contour described by the graph of the boundary function γ, among all C^1 surfaces ([Os]). (Actually it does so among all *piecewise* C^1 surfaces, those described by the graph of a piecewise C^1 function $\hat{u}$, which admit internal "roof-shaped" sections. With appropriate definitions, the methods of Chapter 7 can be extended to establish this fact.)

(Problems 3.20–3.26, 3.37)

§3.5. Minimization with Convex Constraints

Convexity may also be of advantage in establishing the minima of functions J that are constrained to the level sets of other functions G (as in the isoperimetric problem). In the formulation suggested by Proposition 2.3, the next result is apparent.

(3.16) **Theorem.** *If D is a domain in $\mathbb{R}^2$, such that for some constants λ_j, $j = 1, 2, \ldots, N$, $f(x, y, z)$ and $\lambda_j g_j(x, y, z)$ are convex on $[a, b] \times D$ [and at least one of these functions is strongly convex on this set], let*

$$\tilde{f} = f + \sum_{j=1}^{N} \lambda_j g_j.$$

Then each solution y_0 of the differential equation

$$\frac{d}{dx} \tilde{f}_z[y(x)] = \tilde{f}_y[y(x)] \quad \text{on } (a, b)$$

minimizes

$$F(y) = \int_a^b f[y(x)]\, dx$$

[uniquely] *on*

$$\mathscr{D} = \{y \in C^1[a, b]: y(a) = y_0(a),\ y(b) = y_0(b);\ (y(x), y'(x)) \in D\}$$

under the constraining relations

$$G_j(y) \overset{\text{def}}{=} \int_a^b g_j[y(x)]\, dx = G_j(y_0), \qquad j = 1, 2, \ldots, N.$$

PROOF. By construction (and 3.11(1)) $\tilde{f}(x, y, z)$ is [strongly] convex on $[a, b] \times D$, so that by Theorem 3.5, y_0 minimizes

$$\tilde{F}(y) = \int_a^b \tilde{f}[y(x)]\, dx = F(y) + \sum_{j=1}^{N} \lambda_j G_j(y)$$

[uniquely] on $\mathscr{D}$. Now apply Proposition 2.3.   □

(3.17) **Remark.** Theorem 3.16 offers a valid approach to minimization in the presence of given isoperimetric constraints as we shall show by example.

However, if we introduce *functions* $\lambda_j = \lambda_j(x)$ in its hypotheses, then as in 2.5

$$\tilde{F}(y) = F(y) + \sum_{j=1}^{N} \int_a^b \lambda_j(x) g_j[y(x)]\, dx,$$

and we conclude that each solution $y_0 \in \mathcal{D}$ of the differential equation for the *new* $\tilde{f}$ minimizes F on $\mathcal{D}$ [uniquely] under the pointwise constraining relations

$$g_j[y(x)] \equiv g_j[y_0(x)], \qquad j = 1, 2, \ldots, N,$$

of Lagrangian form.

Although, in general not even one such $g_j[y_0(x)]$ may be specifiable *a priori* (Why?), the vector-valued version does permit minimization with given Lagrangian constraints. (See Problem 3.35 et seq.)

Corresponding applications involving *inequality* constraints are considered in Problem 3.31 and in §7.4.

Example 1. To minimize

$$F(y) = \int_0^1 (y'(x))^2\, dx$$

on

$$\mathcal{D} = \{y \in C^1[0, 1]: y(0) = 0, y(1) = 0\},$$

when restricted to the set

$$\left\{ y \in C^1[0, 1]: G(y) \stackrel{\text{def}}{=} \int_0^1 y(x)\, dx = 1 \right\},$$

we observe that $f(\underline{x}, y, z) = z^2$ is strongly convex, while $g(\underline{x}, y, z) = y$ is (only) convex, on $\mathbb{R} \times \mathbb{R}^2$. Hence, we set $\tilde{f}(x, y, z) = z^2 + \lambda y$ and try to find λ for which $\lambda g(\underline{x}, y, z)$ remains convex while the differential equation

$$\frac{d}{dx} \tilde{f}_z[y(x)] = \tilde{f}_y[y(x)]$$

has a solution $y_0 \in \mathcal{D}$ for which $G(y_0) = 1$. Now since g is linear in y (and z), $\lambda g(x, y, z) = \lambda y$ is convex for each real λ. Upon substitution for $\tilde{f}$, the differential equation becomes

$$\frac{d}{dx}(2y'(x)) = \lambda \quad \text{or} \quad y''(x) = \frac{\lambda}{2},$$

which has the general solution

$$y(x) = c_1 x + c_2 + \frac{\lambda x^2}{4}.$$

The boundary conditions $y(0) = 0 = c_2$ and $y(1) = 0 = c_1 + \lambda/4$ give

$$y_0(x) = \frac{-\lambda}{4}x(1 - x), \quad \text{which is in } \mathcal{D}.$$

Theorem 3.16 assures us that $y_0(x) = (-\lambda/4)x(1 - x)$ minimizes F on $\mathcal{D}$— even uniquely—under the constraint $G(y) = G(y_0)$. It remains to show that we can choose λ so that $G(y_0) = 1$ (while $\lambda g(x, y, z)$ remains convex).
    Thus we want

$$G(y_0) = 1 = \frac{-\lambda}{4}\int_0^1 x(1 - x)\, dx = \frac{-\lambda}{4}\left(\frac{1}{2} - \frac{1}{3}\right) = \frac{-\lambda}{24}$$

or

$$\lambda = -24,$$

and since $-24g(x, y, z) = -24y$ remains convex, we have found the unique solution to our problem.

(3.18) **Remark.** In this example we can find λ to force y_0 into *any* level set of G we wish, since $\lambda g(x, y, z) = \lambda y$ is always convex for each value of λ. This is *not* the case in general and our approach will work only for a restricted class of level sets of G. (See Problem 3.29.)

The Hanging Cable

Example 2 (The catenary problem). Let's determine the shape which a long inextensible cable (or chain) will assume under its own weight when suspended freely from its end points at equal heights as shown in Figure 3.7. We utilize the coordinate system shown, and invoke Bernoulli's principle that the shape assumed will minimize the potential energy of the system. (See §8.3.)
    We suppose the cable to be of length L and weight per unit length W, and that the supports are separated a distance $H < L$. Then *utilizing the arclength s along the cable as the independent variable*, a shape is specified by a function $y \in \mathcal{Y} = C^1[0, L]$ with $y(0) = y(L) = 0$, which has associated with it the potential energy given within an additive reference constant by the center-of-

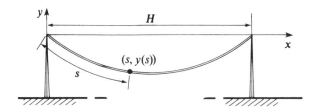

Figure 3.7

mass integral

$$F(y) = W \int_0^L y(s) \, ds.$$

However, in order to span the supports, the function y must satisfy the constraining relation

$$G(y) = \int_0^L \sqrt{1 - y'(s)^2} \, ds = \int_0^L dx(s) = H,$$

where $x(s)$ denotes the horizontal displacement of the point at a distance s along the cable, since then as elementary geometry shows, $x'(s)^2 + y'(s)^2 = 1$. Clearly $|y'(s)| \le 1$ and if $|y'(s_1)| = 1$, then the cable would have a cusp at s_1, since $x'(s_1) = 0$.

Now $f(s, y, z) = Wy$ is (only) convex on $[0, L] \times \mathbb{R}^2$ while $g(s, y, z) = -\sqrt{1 - z^2}$ is by §3.3, Example 5, strongly convex on $[0, L] \times \mathbb{R} \times (-1, 1)$. Thus by 3.11(1), the modified function $\tilde{f}(s, y, z) = Wy - \lambda\sqrt{1 - z^2}$ is strongly convex when $\lambda > 0$. Hence by 3.16, for $\lambda > 0$ we should seek a solution y for the differential equation

$$\frac{d}{ds}\tilde{f}_z[y(s)] = \tilde{f}_y[y(s)] \quad \text{on } (0, L)$$

that is in

$$\mathscr{D} = \{y \in C^1[0, L]: y(0) = y(L) = 0, |y'(s)| < 1, \forall \, s \in (0, L)\}.$$

Upon substitution, the differential equation becomes

$$\frac{d}{ds}\left(\frac{\lambda y'(s)}{\sqrt{1 - y'^2(s)}}\right) = W$$

or

$$\frac{\lambda y'(s)}{\sqrt{1 - y'^2(s)}} = s + c, \tag{27}$$

where we have replaced the unspecified constant λ by $W\lambda$ and introduced a new constant c.

We know that each $y \in \mathscr{D}$ which satisfies this equation for $\lambda > 0$ must be the *unique* shape sought. Hence we can make further simplifying assumptions about y if they do not preclude solution. We could, for example, suppose $y' = \text{const.}$, but it is seen that this could not solve (27). And we can suppose that y is symmetric about $L/2$, which accords with our physical intuition about the shape assumed by the cable. If we set $l = L/2$ it follows that $y'(l) = 0$, so that from (27), $c = -l$; also, we need only determine y on $[0, l]$, where we would expect that $y' \le 0$.

Thus from (27) we should have that

$$y'(s)^2 = \frac{(s - l)^2}{\lambda^2 + (s - l)^2} \quad \text{on } [0, l],$$

and so, with $y(0) = 0$, that

$$y(s) = \int_0^s \frac{(t - l)}{\sqrt{\lambda^2 + (t - l)^2}} dt = \sqrt{\lambda^2 + (t - l)^2}\Big|_0^s$$

or

$$y(s) = \sqrt{\lambda^2 + (l - s)^2} - \sqrt{\lambda^2 + l^2} \quad \text{on } [0, l]. \tag{28}$$

Now we can obviously suppose that $\lambda > 0$; however, we must satisfy the constraining relation

$$\int_0^L \sqrt{1 - y'(s)^2}\, ds = H;$$

or with our symmetry assumption, we require that

$$\int_0^l \sqrt{1 - y'(s)^2}\, ds = \frac{H}{2}.$$

Upon substitution from (28), this becomes

$$\int_0^l \sqrt{1 - \frac{(l - s)^2}{\lambda^2 + (l - s)^2}}\, ds = \int_0^l \frac{\lambda}{\sqrt{\lambda^2 + (l - s)^2}}\, ds = \frac{H}{2}.$$

With the hyperbolic substitution $(l - s) = \lambda \sinh \theta$, we can evaluate the integral and find that

$$h(\alpha) \stackrel{\text{def}}{=} \frac{H}{2l} = \frac{H}{L} = \sinh^{-1}\left(\frac{l}{\lambda}\right)$$

Now, $h(\alpha) \stackrel{\text{def}}{=} (\sinh \alpha)/\alpha$ is continuous and positive on $(0, \infty)$ and has by L'Hôpital's rule as $\alpha \searrow 0$ and $\alpha \nearrow \infty$, the same limits as does $\cosh \alpha$ viz., 1 and ∞, respectively. Thus by the intermediate value theorem (§A.1), h assumes each value on $(1, \infty)$ at least once on $(0, \infty)$. Hence $\exists\, \alpha \in (0, \infty)$ for which $h(\alpha) = L/H$ and for this α, $\lambda = l/\sinh \alpha$ will provide the $y(s)$ sought.

The resulting curve is defined parametrically on $[0, l]$ by

$$y(s) = \sqrt{\lambda^2 + (l - s)^2} - \sqrt{\lambda^2 + l^2},$$
$$x(s) = \int_0^s \sqrt{1 - y'(t)^2}\, dt = \frac{H}{2} - \lambda \sinh^{-1}\left(\frac{l - s}{\lambda}\right), \tag{29}$$

which corresponds to the well-known catenary (Problem 3.30(a)).

(3.19) *Among all curves of length L joining the supports, the catenary of (29) will have (uniquely) the minimum potential energy and should thus represent the shape actually assumed by the cable.*

Remark. This problem is usually formulated with x as the independent variable. However, this results in an energy function which is *not* convex (Problem 3.30(b)).

Optimal Performance

Example 3. (A simple optimal control problem). A rocket of mass m is to be accelerated vertically upward from rest at the earth's surface (assumed stationary) to a height h in time T, by the thrust (mu) of its engine. If we suppose h is so small that both m and g, the gravitational acceleration, remain constant during flight, then we wish to control the thrust to minimize the fuel consumption as measured by, say,

$$F(u) = \int_0^T u^2(t)\, dt, \tag{30}$$

for a given flight time T.

Although T will be permitted to vary later, consider first the problem in which T is fixed. We invoke Newton's second law of motion to infer that at time t, the rocket at height $y = y(t)$ should experience the net acceleration

$$\ddot{y} = u - g, \tag{31}$$

and impose the initial and terminal conditions

$$y(0) = \dot{y}(0) = 0 \quad \text{and} \quad y(T) = h.$$

Since $y(0) = 0$, then $y(T) = \int_0^T \dot{y}(t)\, dt$, so that upon subsequently integrating by parts we obtain

$$y(T) = -(T - t)\dot{y}(t)\Big|_0^T + \int_0^T (T - t)\ddot{y}(t)\, dt.$$

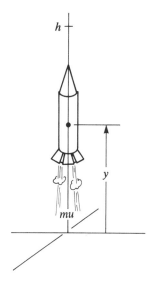

Figure 3.8

From (31) and the remaining boundary conditions, we see that

$$h = y(T) = \int_0^T (T - t)u(t)\, dt - \frac{gT^2}{2}.$$

Hence

$$G(u) \overset{\text{def}}{=} \int_0^T (T - t)u(t)\, dt = h + \frac{gT^2}{2} = k, \quad \text{say}, \tag{32}$$

and we are to minimize F on

$$\mathscr{D} = \{u \in C[0, T], u \geq 0\}$$

subject to the isoperimetric constraint (32).

According to Theorem 3.16, we introduce a constant λ and observe that the modified integrand

$$\tilde{f}(t, u, z) = u^2 + \lambda(T - t)u$$

will be strongly convex for all λ, since the second term is linear in u. Moreover, $\tilde{f}_z \equiv 0$. Thus, a $u_0 \in \mathscr{D}$ which satisfies the equation $\tilde{f}_u[u(t)] = 0 = 2u(t) + \lambda(T - t)$ and meets the constraint (32) will suffice. (See Problem 3.18.)

We require $u_0(t) = -(\lambda/2)(T - t)$, where $\lambda \leq 0$ is to be found to have

$$k = \int_0^T (T - t)u_0(t)\, dt = -\frac{\lambda}{2} \int_0^T (T - t)^2\, dt = -\frac{\lambda T^3}{6},$$

or

$$-\lambda = \frac{6k}{T^3},$$

so that

$$u_0(t) = \frac{3k(T - t)}{T^3}. \tag{33}$$

Observe that from (30) and (32) it follows that

$$F(u_0) = \int_0^T u_0^2(t)\, dt = \frac{9k^2}{T^6} \int_0^T (T - t)^2\, dt = \frac{3k^2}{T^3}$$

$$= \frac{3(2h + gT^2)^2}{4T^3} = 3\left(\frac{h^2}{T^3} + \frac{gh}{T} + \frac{g^2 T}{4}\right);$$

we may now use simple calculus to minimize this expression with respect to T and thus obtain an *optimal* flight time $T_0 = (6h/g)^{1/2}$.

(3.20) **Remark.** We know that (33) provides the unique solution to our problem. However, observe that from (32), the corresponding maximum thrust is

$$u_0(0) = \frac{3k}{T^2} = \left(\frac{3h}{T^2}\right) + 1.5g;$$

when $T = T_0$ as above, $u_0(0) = 2g$ which might not be achievable. A more realistic problem could require $0 \le u \le \beta(< 2g)$. This is a Lagrangian inequality which can also be treated by means of a multiplier *function* as in Proposition 2.5. The resulting solution admits operating at maximal thrust (β) until a switching time τ at which reduction can occur. The details are reserved until Problem 7.23, and a simpler convex problem with Lagrangian inequalities is examined in §7.4. Extensions of this approach are feasible (Problems 3.38 and 3.39) even to problems with discontinuous controls as will be shown in Chapter 10.

(Problems 3.27–3.32)

§3.6. Summary: Minimizing Procedures

In this chapter we have used elementary concepts of convexity to supply the framework for making an educated guess about the solution of a minimization problem. In the presence of *strict* convexity, a guessed solution is the unique solution. However, with each specific problem, it is not essential to establish convexity according to predetermined definitions. Inspection of the results and applications in this chapter suggests the following procedures:

I. To minimize F on $\mathcal{D} \subseteq \mathcal{Y}$ (a linear space):

First. Show that when y, $y + v \in \mathcal{D}$, then $F(y + v) - F(y) \ge I(y; v)$ where $I(y; v)$ is some *new* expression which admits further analysis.

Second. If possible, characterize those v which permit the equality $F(y + v) - F(y) = I(y; v)$. (Ideally, equality at $y \Rightarrow v = 0$.)

Third. Note the restrictions on v which occur when y, $y + v \in \mathcal{D}$, and transform $I(y; v)$ so that conditions (on y) under which it vanishes for *all such* v can be discerned.

Fourth. Show that there is a $y = y_0 \in \mathcal{D}$ which meets these conditions.

If it exists, this y_0 will be a solution (and it may be *the* solution) to the problem. $\qquad\square$

Remarks. To obtain the basic inequality, elementary facts such as $(y'' + v'')^2 - (y'')^2 \ge 2y''v''$ may suffice. It is *not* essential to recognize $I(y; v)$ as $\delta F(y; v)$—or indeed, even to consider this variation.

In transforming $I(y; v)$, we may make further simplifying assumptions about y (which do not exclude y from $\mathcal{D}$). In particular, we may assume that y has as many derivatives as required to integrate an expression such as $\int_a^b y''(x)v''(x)\, dx$ by parts as often as desired. "Natural" boundary conditions for a solution y may arise in this process.

Finally, although we may not be able to guarantee a unique solution y_0 to the problem, it may be possible to use information obtained in the second step to characterize the class of solutions; e.g., $y_0 +$ const.

II. To minimize F on $\mathscr{D} \subseteq \mathscr{Y}$ in the presence of further constraints involving functions $G_j, j = 1, 2, \ldots, n$:

First. Use the device of Lagrangian multipliers λ_j to suggest an augmented function $\tilde{F}$ whose minimization on $\mathscr{D}$ could solve the constrained problem. (See 2.3, 2.5, 3.17 and Problem 3.35.)

Second. Find a $y_0 \in \mathscr{D}$ which minimizes $\tilde{F}$ on $\mathscr{D}$ (possibly under a sign restriction on each λ_j).

Third. Determine the λ_j so that y_0 meets the given constraints.

Fourth. Examine the signs of the λ_j (if necessary) to see that the restrictions in the second step have been met.

If all of these conditions are satisfied then y_0 is a solution to the constrained problem (and it may be *the* solution).   □

III. If a usable basic inequality cannot be obtained for F (or for $\tilde{F}$), it may be possible to reformulate the problem—or consider a restricted version of the problem—(perhaps expressed in other coordinates) in terms of new functions which admit the analysis of I or II. Also, the solution of one minimization problem usually solves some associated problems involving constraints.   □

Success in implementing these programs requires a combination of skill, perseverance, and luck. However, they seem to provide the only possibility of obtaining a solution by methods which can be considered elementary, and successful implementation is possible as the examples and problems of this chapter demonstrate. Alternatives to these procedures require a considerably more sophisticated theoretical framework and are at least as difficult to utilize. (See Chapter 9 and [Ak].)

(Problems 3.33–3.37)

PROBLEMS

3.1.   For which of the following functions f, is $f(\underline{x}, y, z)$ convex on $[a, b] \times \mathbb{R}^2$? For which will $f(\underline{x}, y, z)$ be strongly convex on this set?

    (a) $f(x, y, z) = x + y - z$, $[a, b] = [0, 1]$.

    (b) $f(x, y, z) = x^3 + y^2 + 2z^3$, $[a, b] = [0, 1]$.

    (c) $f(x, y, z) = \sqrt{1 + z^2} + x^2 y^2$, $[a, b] = [0, 1]$.

    (d) $f(x, y, z) = (x \sin x)[y^4 + z^2]$, $[a, b] = [-\pi/2, \pi/2]$.

(e) $f(x, y, z) = -x^4 + e^x y + z^2,$ $\qquad\qquad [a, b] = [-1, 1].$
(f) $f(x, y, z) = yz^2 + \cos x,$ $\qquad\qquad\quad [a, b] = [0, 1].$
(g) $f(x, y, z) = e^y \sec^3 x - z,$ $\qquad\qquad\ [a, b] = [0, \pi/4].$
(h) $f(x, y, z) = -x^2 y + z^4,$ $\qquad\qquad\quad [a, b] = [-1, 1].$
(i) $f(x, y, z) = -xy^2 + z^2,$ $\qquad\qquad\quad [a, b] = [-1, 1].$
(j) $f(x, y, z) = e^x y^4 - xy + 2z^2,$ $\qquad\quad [a, b] = [0, 1].$
(k)* $f(x, y, z) = (1 + \sin x)y^2 + (1 + \cos x)z^2,$ $\ [a, b] = [-8, 8].$
(l) $f(x, y, z) = x^2(cy + z)^2, c = \text{const}.$ $\qquad [a, b] = [-1, 1].$
(m)* $f(x, y, z) = 2z^2 + z|z|.$ Hint: Consider graph.

3.2. Let $f(x, y, z)$ and $g(x, y, z)$ be convex on $S \subseteq \mathbb{R}^3$:
(a) Prove that $f(x, y, z) + g(x, y, z)$ is also convex on S.
(b) Give simple examples of other algebraic combinations of such functions, i.e., the difference, product, and quotient, which are *not* convex.
(c) If $0 \leq p = p(x)$ is continuous, then prove that when defined $p(x)f(x, y, z)$ is convex.
(d) Can the sum in (a) be strongly convex on S when each of its terms is only convex? Justify your conclusion.

3.3. Let $f(x, y, z)$ be strongly convex on $S = [a, b] \times \mathbb{R}^2$, $g(x, y, z)$ be convex on S, and $\alpha, \beta, \gamma \in C[a, b]$.
(a) Prove that $f(x, y, z) + g(x, y, z)$ is strongly convex on S.
(b) If $0 < p \in C[a, b]$, show that $p(x)f(x, y, z)$ is strongly convex on S.
(c) Verify that $l(x, y, z) = \alpha(x) + \beta(x)y + \gamma(x)z$ is (only) convex on S.
(d) If $g \in C^2(\mathbb{R})$ and $g''(z) > 0$, for $z \in \mathbb{R}$, conclude that $f(x, z) = g((\sin x)z)$ is strongly convex on $(0, \pi) \times \mathbb{R}$. Hint: Proposition 3.10. (This result generalizes.)

3.4*. Show that if $f, f_y,$ and f_z are continuous on $[a, b] \times \mathbb{R}^2$, then $f(x, y, z)$ is convex on $[a, b] \times \mathbb{R}^2$ iff

$$f(x, ty_1 + (1 - t)y_2, tz_1 + (1 - t)z_2) \leq tf(x, y_1, z_1) + (1 - t)f(x, y_2, z_2)$$

$$\forall x \in [a, b]; \quad t \in (0, 1); \quad y_j, z_j \in \mathbb{R}, \qquad j = 1, 2.$$

(Hint: See Problem 0.7.)

3.5*. (a) Show that if $f, f_{yy}, f_{yz},$ and f_{zz} are continuous on $[a, b] \times \mathbb{R}^2$, then $f(x, y, z)$ is convex on $[a, b] \times \mathbb{R}^2$ iff the matrix of second derivatives $\begin{bmatrix} f_{yy} & f_{yz} \\ f_{yz} & f_{zz} \end{bmatrix}$ is positive semidefinite on $[a, b] \times \mathbb{R}^2$, i.e.,

$$(u \ \ v) \begin{bmatrix} f_{yy} & f_{yz} \\ f_{yz} & f_{zz} \end{bmatrix} \begin{pmatrix} u \\ v \end{pmatrix} \geq 0, \qquad \forall x \in [a, b]; \quad y, z, u, v \in \mathbb{R}.$$

(Hint: See Problem 0.8.) (Note: This condition is equivalent to requiring that $f_{yy} \geq 0, f_{zz} \geq 0,$ and $\Delta = f_{yy}f_{zz} - (f_{yz})^2 \geq 0$.
(b) Use this approach on Problems 3.1(k) and (l).

3.6–3.15. In these problems, verify that the integrand function is strongly convex (on the appropriate set) and find the unique minimizing function for F
(a) on $\mathscr{D}$. (b) on $\mathscr{D}_1$. (c) on $\mathscr{D}_2$.

3.6. $F(y) = \int_1^2 x^{-1} y'(x)^2 \, dx,$ $\mathcal{D}_1 = \{y \in C^1[1, 2]: y(2) = 3\},$
       $\mathcal{D} = \{y \in C^1[1, 2]: y(1) = 0, y(2) = 3\}.$ $\mathcal{D}_2 = C^1[1, 2].$

3.7. $F(y) = \int_0^1 [2e^x y(x) + y'(x)^2] \, dx,$
       $\mathcal{D} = \{y \in C^1[0, 1]: y(0) = 0, y(1) = 1\}.$ $\mathcal{D}_1 = \{y \in C^1[0, 1]: y(0) = 0\}.$

3.8. $F(y) = \int_5^{10} \sqrt{x}\sqrt{1 + y'(x)^2} \, dx,$ $\mathcal{D}_1 = \{y \in C^1[5, 10]: y(10) = 6\},$
       $\mathcal{D} = \{y \in C^1[5, 10]: y(5) = 4, y(10) = 6\}.$ $\mathcal{D}_2 = C^1[5, 10].$

3.9. $F(y) = \int_1^2 [2y(x)^2 + x^2 y'(x)^2] \, dx,$ $\mathcal{D}_1 = \{C^1[1, 2]: y(1) = 1\},$
       $\mathcal{D} = \{y \in C^1[1, 2]: y(1) = 1, y(2) = 5\}.$ $\mathcal{D}_2 = C^1[1, 2].$
       (Hint: The differential equation has two linearly independent solutions of the
       form x^p, $p \in \mathbb{R}$.)

3.10. $F(y) = \int_0^{\pi/6} [(y'(x) - \cos x)^2 + 4y(x)] \, dx,$
       $\mathcal{D} = \{y \in C^1[0, \pi/6]: y(0) = 0, y(\pi/6) = \frac{1}{2}\}.$
       $\mathcal{D}_1 = \{y \in C^1[0, \pi/6]: y(\pi/6) = \frac{1}{2}\}.$

3.11. $F(y) = \int_1^2 x^{-1} \sqrt{1 + y'(x)^2} \, dx,$
       $\mathcal{D} = \{y \in C^1[1, 2]: y(1) = \sqrt{8}, y(2) = \sqrt{5}\}.$
       $\mathcal{D}_1 = \{y \in C^1[1, 2]: y(1) = \sqrt{8}\}.$

3.12. $F(y) = \int_{-1}^2 e^{y'(x)} \, dx,$
       $\mathcal{D} = \{y \in C^1[-1, 2]: y(-1) = 2, y(2) = 11\}.$

3.13. $F(y) = \int_0^{1/2} [y(x) + \sqrt{1 + y'(x)^2}] \, dx,$
       $\mathcal{D} = \{y \in C^1[0, \frac{1}{2}]: y(0) = -1, y(\frac{1}{2}) = -\sqrt{3/2}\}.$
       $\mathcal{D}_1 = \{y \in C^1[0, \frac{1}{2}]: y(0) = -1\}.$

3.14. $F(y) = \int_0^{\pi/4} y'(x)^4 \sec^3 x \, dx,$
       $\mathcal{D} = \{y \in C^1[0, \pi/4]: y(0) = 0, y(\pi/4) = 1\}.$
       $\mathcal{D}_1 = \{y \in C^1[0, \pi/4]: y(\pi/4) = 1\}.$

3.15. $F(y) = \int_1^8 [y'(x)^4 - 4y(x)] \, dx,$
       $\mathcal{D} = \{y \in C^1[1, 8]: y(1) = 2, y(8) = -37/4\}.$
       $\mathcal{D}_1 = \{y \in C^1[1, 8]: y(1) = 2\}.$

3.16. (a) Demonstrate that the function $F(y) = \int_0^1 y^2(x)y'(x) \, dx$ is convex on

       $$\mathcal{D} = \{y \in C^1[0, 1]: y(0) = 0, y(1) = 1\},$$

       although the integrand function $f(x, y, z) = y^2 z$ is not convex on $[0, 1] \times \mathbb{R}^2$.

       (b)* Prove that the function $f(x, y, z) = (z + 3y)^2$ is (only) convex on $[a, b] \times \mathbb{R}^2$, but $F(y) = \int_a^b (y'(x) + 3y(x))^2 \, dx$ is *strictly* convex on

       $$\mathcal{D} = \{y \in C^1[a, b]: y(a) = a_1, y(b) = b_1\}.$$

       What happens if 3 is replaced by a number α?

3.17. (a) Show that $f(x, y, z) = xz + y$ is convex, but not strongly convex on $[1, 2] \times \mathbb{R}^2$.

       (b) Can you find more than one function which minimizes

       $$F(y) = \int_1^2 [xy'(x) + y(x)] \, dx$$

       on

       $$\mathcal{D} = \{y \in C^1[1, 2]: y(1) = 1, y(2) = 2\}?$$

3.18. Suppose that $f(x, y)$ is [strongly] convex on $[a, b] \times \mathbb{R}$ and set

$$F(y) = \int_a^b f(x, y(x)) \, dx.$$

(a) Show that each $y \in \mathscr{D} = C[a, b]$ which satisfies

$$f_y(x, y(x)) = 0, \qquad \forall\, x \in [a, b]$$

minimizes F [uniquely] on $\mathscr{D}$.
(b) Show that each $y \in \mathscr{D}^* = \{u \in C[a, b]: \int_a^b u(x) \, dx = 0\}$ which satisfies

$$f_y(x, y(x)) = \text{const.}, \qquad \forall\, x \in [a, b]$$

minimizes F [uniquely] on $\mathscr{D}^*$.
(c) Let $f(x, y) = y^2 - g(x)y$, where $g \in C[a, b]$ is a given function. Find the unique minimizing function for F on $\mathscr{D}$ and on $\mathscr{D}^*$.

3.19. (a)   When $b_1 > 0$, explain why the restricted *surface-area-of-revolution* function (see §1.4(a))

$$S(y) = 2\pi \int_1^b x\sqrt{1 + y'(x)^2} \, dx$$

is minimized on $\mathscr{D} = \{y \in C^1[1, b]: y(1) = 0, \, y(b) = b_1\}$ uniquely by a y_0 that makes

$$\frac{xy'(x)}{\sqrt{1 + y'(x)^2}} = \text{const.} = \frac{1}{c}$$

say, on $(1, b)$.
(b)* Show that $cy_0(x) = \cosh^{-1}(cx) - \cosh^{-1} c$ provided that $c > 1$ can be found to make $y_0(b) = b_1$. Graph y_0 and discuss how to guarantee that c exists.

3.20*. Minimum Transit Time of a Boat. (See §1.2(b) and Problem 1.3.) Let

$$T(y) = \int_0^{x_1} [\alpha(x)\sqrt{1 + (\alpha y')^2(x)} - (\alpha^2 r y')] \, dx,$$

where r is a given continuous function, $0 \le r(x) < 1$ on $[0, x_1]$, $\alpha(x) = (1 - r(x)^2)^{-1/2}$, and

$$\mathscr{D} = \{y \in C^1[0, x_1]: y(0) = 0, \, y(x_1) = y_1\} \quad \text{with } x_1 > 0, \, y_1 > 0.$$

(a)   Prove that the integrand function

$$f(\underline{x}, z) = \alpha(\underline{x})\sqrt{1 + (\alpha(\underline{x})z)^2} - \alpha(\underline{x})^2 r(\underline{x})z$$

is strongly convex on $[a, b] \times \mathbb{R}$.
(b)   Show that each $y \in \mathscr{D}$ which makes $(\alpha^3 y')(x)[1 + (\alpha y')^2(x)]^{-1/2} - (\alpha^2 r)(x)$ constant on (a, b) minimizes T uniquely on $\mathscr{D}$.
(c)   Verify that $y(x) = \int_0^x (r + c\alpha^{-2})[(1 - cr)^2 - c^2]^{-1/2}(t) \, dt$ will serve provided that the constant c can be chosen properly.
(d)   Show that the boundary value $y_1 = \int_0^{x_1} r(x) \, dx$ is always achievable with a proper choice of c, and find the minimizing function in this case.
(e)   What happens if $r(x) = r = $ constant?
(f)* For the linear profile $r(x) = (1 - 3x)/2$, $0 \le x \le x_1 = \frac{1}{6}$, show that the admissible choices of c are restricted to $-\frac{4}{3} < c < \frac{2}{3}$, and for this range of

c the integral defining $y(x)$ in part (c) is bounded. (This demonstrates that we cannot always choose c to meet the boundary conditions. Explain why physically.)

3.21. (a) Verify that Theorem 3.7 remains valid for integrands of the form $f(x, z) = p(x)\sqrt{1 + z^2}$, where p is continuous on $(a, b]$, $p(x) > 0$, and $\int_a^b p(x)\,dx < \infty$. (For example, $p(x) = x^{-1/2}$ on $(0, 1]$.)

(b)* More generally, suppose that $f(\underline{x}, y, z)$ is [strongly] convex on $(a, b) \times \mathbb{R}^2$, and $y_0 \in C[a, b]$ is a C^1 solution of the differential equation $(d/dx)f_z[y(x)] = f_y[y(x)]$ on (a, b).

Show that if $[\alpha, \beta] \subseteq (a, b)$ and $v \in C^1(a, b)$ then

$$\int_\alpha^\beta f[(y_0 + v)(x)]\,dx \geq \int_\alpha^\beta f[y_0(x)]\,dx + v(x)f_z[y_0(x)]\Big|_\alpha^\beta.$$

Thus when $\max|f_z[y_0(x)]| \leq M < +\infty$, conclude that y_0 minimizes $F(y) = \int_a^b f[y(x)]\,dx$ [uniquely] on $\mathcal{D}^* \subseteq \mathcal{Y} = C[a, b] \cap C^1(a, b)$, where

$$\mathcal{D}^* = \{y \in \mathcal{Y}: y(a) = y_0(a),\ y(b) = y_0(b);$$
$$F(y) \text{ exists as an } \textit{improper} \text{ Riemann integral}\}.$$

(This extension of Theorem 3.5 to improper integral functions F also permits consideration of functions y_0 whose derivative "blows up" at the end points.)

(c)* Make similar extensions of Theorem 3.7 and 3.9.

(d)* Suppose $f_z[y_0(x)]$ is bounded near $x = a$, but only $(b - x)f_z[y_0(x)]$ is bounded near $x = b$. Show that we can reach the same conclusion as in part (b) on $\mathcal{D}' = \{y \in \mathcal{D}^*: y'(x) \to b_1'$ as $x \nearrow b\}$. Hint: Use the mean value theorem on v near b.

3.22*. A Brachistochrone. (See §3.4(b).)

(a) Show that the time of transit along the cycloid joining the origin to the end point (x_1, y_1) for $y_1/x_1 < \pi/2$ is given by $T_{\min} = \sqrt{(x_1/2g)}[\theta_1/\sin(\theta_1/2)]$ where θ_1 is the parameter angle of the end point. Hint: Use equations (12), (13).

(b) For the case $x_1 = y_1 = 1$, compute θ_1 and $T_{\min}$. Compare with the answers obtained for the straight line and the quarter circle in Problem 1.2.

(c)* Use the results of Problem 3.21 to extend the analysis in §3.4(b) to the case when $y_1/x_1 = \pi/2$.

(d)** Can you use the methods of this chapter to establish the minimality of the cycloid when $y_1/x_1 > \pi/2$ for *some* class of curves?

3.23. A Seismic Wave Problem. (See Problem 1.8(b) and Figure 1.12(b).)

(a) Show that the integrand function $f(\underline{x}, z) = \sqrt{1 + z^2}/x$ is strongly convex on $[a, b] \times \mathbb{R}$ when $a > 0$.

(b) Conclude that each $y \in \mathcal{D} = \{y \in C^1[a, b]: y(a) = 0,\ y(b) = b_1 > 0\}$ which satisfies for some positive constant r, the equation $y'/\sqrt{1 + y'(x)^2} = x/r$ on (a, b) will minimize the time-of-travel integral $T(y) = p\int_a^b[\sqrt{1 + y'(x)^2}/x]\,dx$ on $\mathcal{D}$ uniquely.

(c) Show that the associated path is along a circular arc of radius r with center on the y axis.

(d) Use the parametrization $x = r \sin \theta$ to evaluate the time-of-travel integral $T(y) = p \int_a^b ds/x$ along this path, in terms of α, and β, the initial and final values of θ, when $0 < \alpha < \beta < \pi/2$, as in Figure 1.12(b).

(This last result affords an experimental determination of the unknown physical constants σ and p, by measuring the actual time of travel required for seismic waves from explosions on the earth's surface to reach a point in a mineshaft some distance away. Thus far, however, we must demand that the resulting geometry be compatible with our assumption that $0 < \alpha < \beta < \pi/2$.)

3.24*. Geodesics on a Cone: I.

Each point on the surface S of a right circular cone of apex angle 2α has the cartesian coordinates $(x, y, z) = (r \cos \theta, r \sin \theta, ar)$, where $a = \cot \alpha$, and r, θ are the polar coordinates shown in Figure 3.9.

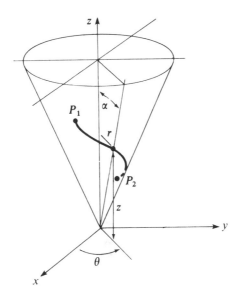

Figure 3.9

(a) If a curve on S can be represented by $r = r(\theta)$, with $r \in \mathcal{Y} = C^1[\theta_1, \theta_2]$, show that its *arc length* is $L(r) = \int_{\theta_1}^{\theta_2} \sqrt{r(\theta)^2 + b^2 r'(\theta)^2} \, d\theta$, where $b^2 = 1 + a^2$.

(b) Consider the convexity of $f(y, z) = \sqrt{y^2 + b^2 z^2}$, and explain why it would be appropriate to consider minimizing L on

$$\mathcal{D}^* = \{r \in \mathcal{Y}: r(\theta_j) = r_j, j = 1, 2; r(\theta)^2 + r'(\theta)^2 > 0\}.$$

(c)* Show that a function $r \in \mathcal{D}^*$, which for $R = b \log r$ satisfies the equation

$$\frac{d}{d\theta} \frac{bR'(\theta)}{\sqrt{1 + R'(\theta)^2}} = \frac{1}{\sqrt{1 + R'(\theta)^2}} \quad \text{on } (\theta_1, \theta_2),$$

should minimize L on $\mathcal{D}^*$.

(d) Make the substitution $R'(\theta) = \tan \varphi(\theta)$ in (c), and conclude that the resulting equation is satisfied when $\varphi'(\theta) = 1/b = \sin \alpha$, or when

$$r(\theta) = c_1 \sec(b^{-1}\theta + c)$$

for constants c, c_1.

(e) Suppose that $\theta_1 = 0$, and $0 < \beta \overset{\text{def}}{=} b^{-1}\theta_2 < \pi$. Prove that an $r \in \mathscr{D}^*$ of the form in (d) can be found if $c_1 = r_1 \cos c$, where

$$\tan c = \frac{\cos \beta - (r_1/r_2)}{\sin \beta}$$

and argue that this *is* possible.

(f)* Show that even though $f(y, z)$ is *not* strongly convex, L is strictly convex on $\mathscr{D}^*$, so that the minimizing function found in (e) is unique. Hint: Prove that if $r, r + v \in \mathscr{D}^*$, then $L(r + v) - L(r) \geq \delta L(r; v)$ with equality *iff* $vr' = v'r$, or $v = \text{const. } r$.

(g) Conclude that when $r_1 = r_2$, the circular arc $r(\theta) = \text{const.}$ is *not* the geodesic as might have been conjectured.

Geodesics on a Cone: II.

Consider the right circular cone shown in Figure 3.9. To find geodesic curves of the form $\theta = \theta(r)$ joining points (r_1, θ_1) and (r_2, θ_2), we assume without loss of generality that $r_2 > r_1 > 0$, $\theta_1 = 0$, and $0 \leq \theta_2 \leq \pi$.

(h) Suppose that $\theta \in C^1[r_1, r_2]$, derive the length function $L(\theta)$. Is it convex?

(i) When $\theta_2 \neq 0$, prove that $L(\theta)$ is minimized uniquely on

$$\mathscr{D} = \{\theta \in C^1[r_1, r_2]: \theta(0) = 0, \theta(r_2) = \theta_2\},$$

by $\theta = b \sec^{-1}(r/c) - b \sec^{-1}(r_1/c)$, provided that the constant c can be chosen to make $\theta(r_2) = \theta_2$.

(j) What happens when $\theta_2 = 0$?

3.25*. Beam Deflection. When a cantilevered beam of length L is subjected to a distributed load (force per unit length) $p(x)$ as shown in Figure 3.10(a), it will deflect to a new position which can be described by a function $y \in C^2[0, L]$. According to the theory of linear elasticity, the potential energy is approximated by

$$U(y) = \int_0^L \left[\tfrac{1}{2}\mu y''(x)^2 - p(x)y(x)\right] dx,$$

where μ is a positive constant (called the flexural rigidity) determined by the beam cross section and material, and the shape assumed by the deflected beam will minimize U on

$$\mathscr{D} = \{y \in C^2[0, L]: y(0) = y'(0) = 0\}.$$

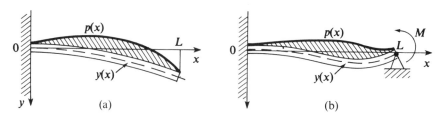

Figure 3.10

(a) Prove that each $y \in \mathscr{D} \cap C^4[0, L]$ which satisfies the differential equation $\mu y^{IV}(x) = p(x)$ and the "natural" boundary conditions $y''(L) = 0$, $y'''(L) = 0$ is the unique minimizing function for U on $\mathscr{D}$. (The physical meaning of the natural boundary conditions is that both the bending moment and shear force are zero at the free end of the beam.) Hint: Show that U is strictly convex on $\mathscr{D}$ and integrate the $y''v''$ term of $\delta U(y; v)$ by parts *twice*.

(b) Solve the differential equation from part (a) when $p(x) = w = \text{const.}$, selecting the constants of integration so that the solution is in $\mathscr{D}$ and satisfies the given natural boundary conditions. (This would be the case for deflection of a beam under its own weight.) What is the maximum deflection and where does it occur?

(c) If at $x = L$, the beam is pinned and has a concentrated moment M applied as shown in Figure 3.10(b), the potential energy is approximated by

$$U^*(y) = \int_0^L \left[\tfrac{1}{2}\mu y''(x)^2 - p(x)y(x)\right] dx + My'(L),$$

and the shape assumed by the deflected beam will minimize U^* on

$$\mathscr{D}^* = \{y \in C^2[0, L]: y(0) = y'(0) = y(L) = 0\}.$$

Find a differential equation and a natural boundary conditon at $x = L$ which are sufficient to determine the unique minimizing function for U^* on $\mathscr{D}^*$.

(d) Solve the differential equation obtained in part (c) when $p(x) \equiv 0$, choosing the constants of integration so that the solution is in $\mathscr{D}^*$. What is the maximum deflection and where does it occur?

(e) Define [strong] convexity for the function $f(\underline{x}, y, z, r)$ on say $[a, b] \times \mathbb{R}^3$.

(f) Use your definition in (e) to conclude that the integrands of U and of U^* are strongly convex.

(g) Use your definition in (e) to characterize the [strict] convexity of the integral function of Problem 2.9 on $\mathscr{D} = \{y \in \mathscr{Y}: y(a) = a_1, \ y(b) = b_1\}$. Hint: If $[v(x)v'(x)]v''(x) \equiv 0$ on (a, b), then $v''(x) \equiv 0$.

3.26.   Dirichlet's Integral. Let

$$J(u) = \frac{1}{2} \int_D (u_x^2 + u_y^2) \, dA$$

and

$$\mathscr{D} = \{u \in C^1(\bar{D}): u|_{\partial D} = \gamma\},$$

where D is a Green's domain the x, y plane and γ is a given smooth function on ∂D.

(a)   Show that each $u \in \mathscr{D}$ which satisfies Laplace's equation, $u_{xx} + u_{yy} = 0$, in D is the unique minimizing function for J on $\mathscr{D}$.

(b)   Find the minimizing function for J on $\mathscr{D}$ when D is the annulus

$$\{(x, y): 1 < x^2 + y^2 < 4\}$$

and γ is given by

$$\gamma = \begin{cases} 0, & \text{on } x^2 + y^2 = 1, \\ \ln 2, & \text{on } x^2 + y^2 = 4. \end{cases}$$

(Hint: Look for a solution to Laplace's equation of the form $u(x, y) = \bar{u}(x^2 + y^2)$ and find an ordinary differential equation satisfied by $\bar{u}$.)

(c)* Define [strong] convexity for a function $f(\underline{X}, y, Z)$ on $\bar{D} \times \mathbb{R}^{d+1}$, where D is a Green's domain in $\mathbb{R}^d$, and show that when $d = 2$, the integrand function in (a) is strongly convex on $\bar{D} \times \mathbb{R}^3$. (See §6.9.)

(d)* Find an analogue of Theorem 3.5 for the integral function

$$F(u) = \int_D f(X, u(X), \nabla u(X))\, dX = \int_D f[u(X)]\, dX$$

on

$$\mathcal{D} = \{u \in C^1(\bar{D}): u|_{\partial D} = \gamma\}.$$

Hint: Use the divergence theorem of vector calculus.

3.27. Find the unique minimizing function for

$$F(y) = \int_0^1 xy(x)\, dx$$

on

$$\mathcal{D} = \{y \in C^1[0, 1]: y(0) = y(1) = 0\},$$

when restricted to the set where

$$G(y) \overset{\text{def}}{=} \int_0^1 y'(x)^2\, dx = 1.$$

3.28. Show that $y_0(x) = -1 + \sqrt{2 - (x - 1)^2}$ is the unique minimizing function for

$$F(y) = \int_0^1 \sqrt{1 + y'(x)^2}\, dx$$

on

$$\mathcal{D} = \{y \in C^1[0, 1]: y(0) = 0,\ y(1) = \sqrt{2} - 1\},$$

when constrained to the set where

$$G(y) \overset{\text{def}}{=} \int_0^1 y(x)\, dx = \frac{\pi}{4} - \frac{1}{2}.$$

3.29. Is there a value of λ which permits Theorem 3.16 to be used to minimize

$$F(y) = \int_0^\pi y'(x)^2\, dx$$

on

$$\mathcal{D} = \{y \in C^1[0, \pi]: y(0) = y(\pi) = 0\},$$

when further constrained to the set where

$$\int_0^\pi y^2(x)\, dx = 1?$$

3.30. Catenary Problem. (See §3.5, Example 2.)

(a) Verify equation (29) and eliminate the parameter s to obtain the equation

$$y = \lambda \cosh\left(\frac{x - h}{\lambda}\right) - \sqrt{\lambda^2 + l^2},$$

for $0 \leq x \leq 2h = H$. (This is a more common representation for the catenary joining the given points.)

(b) Formulate the problem using x as the independent variable and conclude that this results in an energy function U which is not convex on $\mathscr{D} = \{y \in C^1[0, H]: y(0) = y(H) = 0\}$. (Hint: Use $v = -y$ to show that $U(y + v) - U(y)$ is not always greater than or equal to $\delta U(y; v)$ when y, $y + v \in \mathscr{D}$.)

(c)* Use the arc length s, as a parameter to reformulate the problem of finding the minimal surface of revolution (as in §1.4(a)) among all curves of *fixed length* L joining the required points. (Take $a = 0$ and $a_1 = 1 \le b_1$.)

(d) Conclude that the problem in (c) is identical to that of a hanging cable for an appropriate W, and hence there can be at most one minimizing surface. (See §3.5.)

3.31. Determine the (unique) function $y \in C[0, T]$ which maximizes

$$U(y) = \int_0^T e^{-\beta t} \log(1 + y(t)) \, dt$$

subject to the constraint $L(y) = \int_0^T e^{-\alpha t} y(t) \, dt \le l$, where α, β, and l are positive constants. Hint: Problem 3.18, with 2.4, 3.17. (This may be given the interpretation of finding that consumption rate function y which maximizes a measure of utility (or satisfaction) U subject to a savings-investment constraint $L(y) \le l$. See [Sm], p. 80. y is positive when l is sufficiently large relative to α, β, and T.)

3.32*. Dido's Problem.

Convexity may be used to provide partial substantiation of Dido's conjecture from Problem 1.5, in the reformulation suggested by Figure 3.11.

Verify her conjecture to the following extent:

(a) If $b > l/\pi$, prove that the function representing a circular arc (uniquely) maximizes

$$A(y) \equiv \int_{-b}^{b} y(x) \, dx$$

on

$$\mathscr{D} = \{y \in C^1[-b, b]; y(b) = y(-b) = 0\},$$

when further constrained to the l level set of $L(y) \equiv \int_{-b}^{b} \sqrt{1 + y'(x)^2} \, dx$.

(b) If $b = l/\pi$, show that the function representing the semicircle accomplishes the same purpose for a suitably chosen $\mathscr{D}*$ (see Problem 3.21).

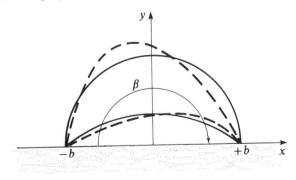

Figure 3.11

(c)* In parts (a) and (b), compute the maximal area as a function of β, the angle subtended by the arc; show that this function increases with β on $(0, \pi]$.

(d) Why does this not answer the problem completely? Can you extend the analysis to do so?

3.33. Let I be an interval in $\mathbb{R}$ and D be a domain of $\mathbb{R}^{2d}$. For $x \in \mathbb{R}$, and $Y = (y_1, \ldots, y_d)$, $Z = (z_1, \ldots, z_d) \in \mathbb{R}^d$ a function $f(\underline{x}, Y, Z)$ is said to be [strongly] convex on $I \times D$ if f, $f_Y = (f_{y_1}, \ldots, f_{y_d})$, and $f_Z = (f_{z_1}, \ldots, f_{z_d})$ are defined and continuous on this set[1] and satisfy the inequality

$$f(x, Y + V, Z + W) - f(x, Y, Z) \geq f_Y(x, Y, Z) \cdot V + f_Z(x, Y, Z) \cdot W,$$

$$\forall \, (x, Y, Z), (x, Y + V, Z + W) \in I \times D$$

[with equality at (x, Y, Z) only if $V = 0$ or $W = 0$].

(a) Show that if $f(\underline{x}, Y, Z)$ is [strongly] convex on $[a, b] \times \mathbb{R}^{2d}$, then

$$F(Y) = \int_a^b f[Y(x)] \, dx = \int_a^b f(x, Y(x), Y'(x)) \, dx$$

is [strictly] convex on

$$\mathscr{D} = \{ Y \in (C^1[a, b])^{2d} : Y(a) = A, \, Y(b) = B \},$$

where A, $B \in \mathbb{R}^d$ are given.

(b) If $f(\underline{x}, Y, Z)$ is strongly convex on $[a, b] \times \mathbb{R}^{2d}$, then prove that each $Y \in \mathscr{D}$ which satisfies the vector differential equation $(d/dx) f_Z[Y(x)] = f_Y[Y(x)]$ (i.e.; $(d/dx) f_{z_j}[Y(x)] = f_{y_j}[Y(x)]$, $j = 1, 2, \ldots, d$) on (a, b) is the unique minimizing function for F on $\mathscr{D}$.

3.34. Use the results in Problem 3.33 to formulate and prove analogous vector valued versions of: (a) Theorem 3.7, (b) Corollary 3.8, and (c) Proposition 3.9 in §3.2.

3.35. (a) Formulate and prove a vector valued version of Theorem 3.16 in §3.5.

(b) Modify Theorem 3.16 to cover the case of a single Lagrangian constraint, and prove your version. Hint: Proposition 2.5 with 3.17.

(c) Formulate a vector valued version of the modified problem in (b) that covers both isoperimetric and Lagrangian constraints.

3.36. (a) Show that $y_0(x) = \sinh x = (e^x - e^{-x})/2$ minimizes $\int_0^1 (y_1(x)^2 + y_1'(x)^2) \, dx$ uniquely for $y_1 \in C^1[0, 1]$ when $y_1(0) = 0$, $y_1(1) = \sinh 1$.

(b) Apply 3.16 as extended above to the problem of minimizing

$$F(Y) = \int_0^1 [y_1'(x)^2 + y_2'(x)^2] \, dx$$

on

$$\mathscr{D} = \{ Y = (y_1, y_2) \in (C^1[0, 1])^2 : Y(0) = (0, 1), \, Y(1) = (\sinh 1, \cosh 1) \}$$

under the constraint: $g[Y(x)] \stackrel{\text{def}}{=} y_2'(x) - y_1(x) \equiv 0$ on $(0, 1)$.

(c) Compare (a) and (b) to make the "correct' choice for λ. Can you find a means of choosing this λ without using the result of (a)?

---

[1] That is, their components are continuous on this set.

3.37. Minimal Area with Free Boundary. (See §3.4(e) and §6.9.)
  (a) When $u \in C^1(\bar{D})$ and $v \in C^1(\bar{D})$, verify that *formally*

$$\delta S(u; v) = -\int_D (U_x + W_y)v \, dA + \int_{\partial D} \frac{(\nabla u \cdot N)v \, d\sigma}{\sqrt{1 + u_x^2 + u_y^2}},$$

  where N is the *outward* normal to ∂D and $d\sigma$ is the element of arc length
  along ∂D. Hint: Green's theorem in its divergence form.
  (b) Conclude that if $U_x + W_y \equiv 0$ in D and $\nabla u \cdot N = 0$ on a *subarc* $\tilde{K} \subseteq \partial D$,
  then u provides uniquely the minimal surface area among those competing
  functions which agree with it on K, that part of the boundary *complementary* to $\tilde{K}$.

3.38*. An Optimal Control Problem with Damping.
  If the motion of the rocket discussed in Example 3 of §3.5 is opposed by its
  speed through the atmosphere, then the equation of motion becomes approximately

$$\ddot{y} = u - g - \alpha\dot{y}, \quad \text{for a constant } \alpha > 0.$$

  (a) Rewrite the last equation as $(d/dt)(e^{\alpha t}\dot{y}) = e^{\alpha t}(u - g)$ and integrate it as in
  the text under the same conditions to obtain the new isoperimetric condition

$$G(u) \stackrel{\text{def}}{=} \int_0^T (1 - e^{-\alpha(T-t)})(u(t) - g) \, dt = \alpha h.$$

  (b) Show that the function $u_0(t) = \lambda_0(1 - e^{-\alpha(T-t)})$ will minimize $F(u) = \int_0^T u^2(t)$ on $C[0, T]$ uniquely, subject to the constraint in (a) for an appropriate choice of λ_0.
  (c) For $\alpha = 1$, find λ_0 and determine $F(u_0)$.

3.39. A heavy rocket-propelled sled is accelerated from rest over a straight *horizontal* track of length l under velocity-proportional air resistance as in the previous problem.
  (a) To optimize the fuel consumption over time T show that we might consider minimizing

$$F(v) = \int_0^T (\dot{v} + \alpha v)^2 \, dt$$

  on

$$\mathcal{D} = \{v \in C^1[0, T]: v(0) = 0\}$$

  under the isoperimetric condition $G(v) \stackrel{\text{def}}{=} \int_0^T v(t) \, dt = l$, where $v(t)$ is the
  velocity at time t and α is a positive constant.
  (b) Find a minimizing velocity function v_0. Is it unique? Hint: $f(y, z) = (z + \alpha y)^2$ is convex! (Why?)
  (c) When $\alpha = 1$ and $l = h$, compare $F(v_0)$ with $F(u_0)$ in part (c) of the previous
  problem under zero gravity conditions. Should they be the same? Explain.

3.40*. (Newton's parametrization.) In (14), let $t = -y'$ so that

$$cx = \frac{(1 + t^2)^2}{t} = \frac{1}{t} + 2t + t^3 = cx(t) \quad \text{say, for } c > 0.$$

  (a) Show that for $t > 1/\sqrt{3}$, $dx/dt > 0$, so that t can be used as an independent
  variable when $t \geq 1/\sqrt{3}$.

(b) Conclude that

$$c\frac{dy}{dt} = \frac{1}{t} - 2t - 3t^3$$

so that *if we replace c by 4/c*, we get the parametric equations

$$x = \frac{c\,(1+t^2)^2}{4}\,\frac{1}{t}; \qquad y = \frac{c}{4}\left(\log t - t^2 - \frac{3t^4}{4} + \frac{7}{4}\right) + y_1 \qquad (t \geq 1/\sqrt{3}),$$

where at $t = 1$: $x = c$ and $y = y_1$. Plot some points on this curve when $c = y_1 = 4$ using a graphing calculator, if necessary, and compare with the curves graphed in Figure 3.6.

(c) In general, we need to choose c and y_1 to force our curve to pass through given points. For simplicity, suppose that at $t = 1$, $x = c = a < 1$ and $y = y_1 = h$. (This choice is further motivated in Problem 7.27). Then to have $y = 0$ when $x = 1$, prove that there is a unique $T > 1$ at which

$$x(T) = 1 = \frac{a\,(1+T^2)^2}{4}\,\frac{1}{T}$$

and

$$H(T) \stackrel{\text{def}}{=} T(\tfrac{3}{4}T^4 + T^2 - \log T - \tfrac{7}{4})/(1+T^2)^2 = h.$$

Hint: Show that for $T > 1$, $H'(T) > 0$ and that H assumes each positive value. (Recall the argument in §3.4(b).)

3.41. (Newton's minimum drag problem.) If we permit $a = 0$ in the problem of §3.4(c), then we wish to minimize the drag integral

$$F(y) = \int_0^1 \frac{x\,dx}{1 + y'(x)^2}$$

on

$$\mathcal{D}^* = \{y \in C^1[0, 1]: y(0) = h, \; y(1) = 0, \; y'(x) \leq 0\},$$

and we are now seeking the profile of minimum drag for an entire body of revolution, not just a shoulder.

(a) Show that if we remove the last restriction from $\mathcal{D}^*$ and admit zig-zag profiles y with large slopes, then we could obtain arbitrarily small values for $F(y)$!

(b) When $h = 1$, compare the drag values of the profiles for a cone C, a hemisphere H, and a truncated cone T in which $y(x) = h$ when $x \leq \tfrac{1}{2}$.

(c) When $h = 2$, compare the drag values of the profiles for a cone C, a paraboloid of revolution P, and a truncated cone T in which $y(x) = h$ when $x \leq \tfrac{1}{2}$.

(d) When $h = \tfrac{1}{3}$, repeat part (c) and conjecture about the superiority of truncated cones or other flattened objects. In particular, using $m = h/(1 - a)$ can you find the "best" truncated cone T_0 for given h?

(e) Show that with initial flattening permitted, we would need to minimize the Bolza-type function $G(y, a) = a^2/2 + F(y)$ on $\mathcal{D} \times [0, 1]$, where F and $\mathcal{D}$ are as in §3.4(c), and we require $y' \leq 0$. Can our previous results from convexity be used in attacking this problem? How?

CHAPTER 4

The Lemmas of Lagrange and du Bois-Reymond

In most of the examples in Chapter 3, we examined a real-valued function F defined on a domain of functions $\mathcal{D}$. We obtained for F an integral condition in the form $I(y; v) = 0$, $\forall v$ in an auxiliary domain $\mathcal{D}_0$, which is *sufficient* to guarantee that each $y \in \mathcal{D}$ that satisfies it must minimize F on $\mathcal{D}$.

By inspection (after reformulation if necessary) we were able to guess a restricted class of y which could meet this condition and, in most cases, find a particular $y \in \mathcal{D}$ which would do so. This, of course, leaves open the possibility that other minimizing functions might exist. In the presence of strict convexity, we have seen that this cannot occur (Proposition 3.3). Without strict convexity, we may have an alternative possibility. (See Problem 3.17.) To explore this fully we should determine conditions *necessary* for a minimizing y and this will be carried out in the next chapter.

However, there are already related necessary conditions of both mathematical interest and importance which we can consider here. For instance, in several examples we observed that the choice $h(x) = \text{const.}$ would make $\int_a^b h(x)v'(x)\,dx = 0$, whenever $v(a) = v(b) = 0$. Thus the constancy of h is *sufficient* to guarantee the vanishing of all such integrals. But is it *necessary*? i.e., will the vanishing of these integrals guarantee the constancy of h? YES.

(4.1) **Lemma** (du Bois-Reymond). *If $h \in C[a, b]$ and $\int_a^b h(x)v'(x)\,dx = 0$,*

$$\forall\, v \in \mathcal{D}_0 = \{v \in C^1[a, b]: v(a) = v(b) = 0\},$$

then $h = \text{const. on } [a, b]$.

PROOF. For a constant c, the function

$$v(x) \stackrel{\text{def}}{=} \int_a^x (h(t) - c)\,dt$$

is in $C^1[a, b]$ (by A.8, the fundamental theorem of calculus) with derivative $v'(x) = h(x) - c$ on (a, b), and it satisfies the condition $v(a) = 0$. It will be in $\mathcal{D}_0$ if in addition $v(b) = 0$; i.e., if

$$\int_a^b (h(t) - c)\, dt = 0 \quad \text{or} \quad c = \frac{1}{b - a} \int_a^b h(t)\, dt.$$

Thus for *this c and v*, we have from the hypothesis that

$$0 \le \int_a^b (h(x) - c)^2\, dx = \int_a^b (h(x) - c)v'(x)\, dx$$

$$= \int_a^b h(x)v'(x)\, dx - cv(x)\Big|_a^b = 0.$$

Hence, from A.9 it follows that on $[a, b]$, the continuous integrand $(h(x) - c)^2 \equiv 0$ or $h(x) = c = \text{const.}$ as asserted. $\qquad\square$

The next result should be compared with Theorem 3.5.

(4.2) **Proposition.** *If $g, h \in C[a, b]$ and $\int_a^b [g(x)v(x) + h(x)v'(x)]\, dx = 0$,*

$$\forall\, v \in \mathcal{D}_0 = \{v \in C^1[a, b]\colon v(a) = v(b) = 0\},$$

then $h \in C^1[a, b]$ and $h' = g$.

PROOF. Let $G(x) = \int_a^x g(t)\, dt$ for $x \in [a, b]$. Then $G \in C^1[a, b]$ and $G' = g$ by A.8. Hence integrating the first term of the integral by parts gives

$$\int_a^b [g(x)v(x) + h(x)v'(x)]\, dx = \int_a^b [h(x) - G(x)]v'(x)\, dx + G(x)v(x)\Big|_a^b,$$

so that

$$\int_a^b [h(x) - G(x)]v'(x)\, dx = 0, \qquad \forall\, v \in \mathcal{D}_0,$$

and by the preceding lemma 4.1,

$$h(x) - G(x) = \text{const.} = c, \quad \text{say, on } [a, b].$$

But then

$$h = G + c \in C^1[a, b] \quad \text{and} \quad h' = G' = g \quad \text{as asserted.} \qquad\square$$

Setting $h \equiv 0$ in this proposition gives the

(4.3) **Corollary.** *If $g \in C[a, b]$ and $\int_a^b g(x)v(x)\, dx = 0$.*

$$\forall\, v \in \mathcal{D}_0 = \{v \in C^1[a, b]\colon v(a) = v(b) = 0\},$$

then $g \equiv 0$ on $[a, b]$. $\qquad\square$

This result admits generalization:

(4.4) **Lemma** (Lagrange). *If $g \in C[a, b]$ and for some $m = 0, 1, 2, \ldots$*

$$\int_a^b g(x)v(x)\, dx = 0,$$

$$\forall \, v \in \mathscr{D}_0 = \{v \in C^m[a, b]: v^{(k)}(a) = v^{(k)}(b) = 0, k = 0, 1, 2, \ldots, m\},$$

then $g \equiv 0$ on $[a, b]$. (Here $C^0[a, b] \equiv C[a, b]$.)

PROOF. Suppose $g(c) > 0$ for some $c \in (a, b)$. Then from the hypothesized continuity of g, c is contained in an interval $[\alpha, \beta] \subseteq (a, b)$ in which $|g(x) - g(c)| \leq g(c)/2$ or $g(x) \geq g(c)/2 > 0$. On the other hand, the function

$$v(x) \overset{\text{def}}{=} \begin{cases} [(x - \alpha)(\beta - x)]^{m+1} & x \in [\alpha, \beta], \\ 0, & x \notin [\alpha, \beta], \end{cases}$$

is in $C^m(\mathbb{R})$ (Why?), and nonnegative. It follows that on $[a, b]$ the product gv is continuous, nonnegative, and not identically zero. Thus from A.9, $0 < \int_a^b g(x)v(x)\, dx$, contradicting the hypothesis.

Similarly, the supposition that $g(c) < 0$ or $-g(c) > 0$ leads to a contradiction and we conclude that $g(c) = 0$, $\forall \, c \in (a, b)$. But since g is continuous, it must also vanish at the end points of the interval; i.e., $g \equiv 0$ on $[a, b]$. $\square$

Lemma 4.1 of du Bois-Reymond also generalizes but with more difficulty:

(4.5) **Proposition.** *If $h \in C[a, b]$ and for some $m = 1, 2, \ldots$*

$$\int_a^b h(x)v^{(m)}(x)\, dx = 0, \qquad \forall \, v \in \mathscr{D}_0,$$

where

$$\mathscr{D}_0 = \{v \in C^m[a, b]: v^{(k)}(a) = v^{(k)}(b) = 0, k = 0, 1, 2, \ldots, m - 1\},$$

then on $[a, b]$, h is a polynomial of degree $< m$.

PROOF*. By a translation we may assume that $a = 0$. The function

$$H(x) \overset{\text{def}}{=} \int_0^x dt_1 \int_0^{t_1} dt_2 \cdots \int_0^{t_{m-1}} h(t)\, dt, \tag{1}$$

is in $C^m[0, b]$ with derivative

$$H^{(m)}(x) = h(x),$$

as is shown by repeated application of the fundamental theorem of calculus. Moreover, $H(0) = H'(0) = \cdots = H^{(m-1)}(0) = 0$, since each successive differentiation eliminates one integral.

Similarly, if q is a polynomial of degree less than m, then $P(x) \overset{\text{def}}{=} x^m q(x)$ vanishes at $x = 0$, together with $P^{(j)}(x)$ for $j < m$, while $p(x) \overset{\text{def}}{=} P^{(m)}(x)$ is another polynomial of degree less than m.

Let

$$v(x) = H(x) - P(x) \tag{2}$$

so that

$$v^{(m)}(x) = h(x) - p(x).$$

We must next show that with the proper choice of q we can make $v^{(k)}(b) = 0$ for $k = 0, 1, 2, \ldots, m - 1$, and this is possible. (See Problem 4.6*.) Assuming that this choice has been made, the resulting $v \in \mathscr{D}_0$, and it follows from repeated partial integrations that

$$\int_0^b p(x)v^{(m)}(x)\, dx = -\int_0^b p'(x)v^{(m-1)}(x)\, dx$$

$$= \cdots (-1)^m \int_0^b p^{(m)}(x)v(x)\, dx = 0,$$

since the boundary terms vanish. Thus, finally, from the hypothesis and construction:

$$0 \le \int_0^b (h(x) - p(x))^2\, dx = \int_0^b (h(x) - p(x))v^{(m)}(x)\, dx$$

$$= \int_0^b h(x)v^{(m)}(x)\, dx = 0,$$

so that $h(x) = p(x)$ on $[0, b]$.   □

It is straightforward to obtain the vector valued analogue of 4.2:

(4.6) **Proposition.** *If $d = 2, 3, \ldots$ and for $G, H \in (C[a, b])^d$,*

$$\int_a^b [G(x) \cdot V(x) + H(x) \cdot V'(x)]\, dx = 0,$$

$$\forall\, V \in \mathscr{D}_0 = \{V \in \mathscr{Y} \colon V(a) = V(b) = 0\},$$

where $\mathscr{Y} = (C^1[a, b])^d$, then $H \in \mathscr{Y}$ and $H' = G$ (i.e., $h_j'(x) = g_j(x), j = 1, 2, \ldots, d$).

PROOF. If we restrict attention to those $V = (v, 0, 0, 0, \ldots, 0) \in \mathscr{Y}$, then the integral condition resuces to

$$\int_a^b [g_1(x)v(x) + h_1(x)v'(x)]\, dx = 0, \quad \forall\, v \in C^1[a, b] \quad \text{with } v(a) = v(b) = 0.$$

Hence from 4.2, it follows that

$$h_1 \in C^1[a, b] \quad \text{and} \quad h_1' = g_1.$$

We can obviously apply the same technique to each component and conclude that $h_j' = g_j, j = 1, 2, \ldots, m$.   □

(4.7) **Corollary.** *When $H \in (C[a, b])^d$ and*

$$\int_a^b H(x) \cdot V'(x) \, dx = 0, \qquad \forall \, V \in \mathscr{D}_0$$

as above, then $H(x) = \text{const.} = C \in \mathbb{R}^d$.

PROOF. Set $G \equiv \mathcal{O}$ in the proposition.   □

A multidimensional version of Lemma 4.4 is presented in Problem 4.5.

PROBLEMS

4.1. Carry out the steps in the proof of Proposition 4.2 in the special case that $g(x) = \sin x$.

4.2. If $h \in C[a, b]$ and

$$\int_a^b h(x)u(x) \, dx = 0, \qquad \forall \, u \in \mathscr{D}_0 = \left\{ u \in C[a, b]: \int_a^b u(x) \, dx = 0 \right\},$$

show that $h(x) = \text{const.}$ on $[a, b]$. Hint: Consider $v(x) = \int_a^x u(t) \, dt$.

4.3. (a) Suppose that $h \in C^1[a, b]$ and

$$\int_a^b h(x)v'(x) \, dx = 0, \qquad \forall \, v \in \mathscr{D}_0 = \{v \in C^1[a, b]:$$
$$v(a) = v'(a) = v(b) = v'(b) = 0\}.$$

Use integration by parts and the proof of Lemma 4.4 to conclude that $h = \text{const.}$ on $[a, b]$. (Do *not* use Lemma 4.1.)

(b) If $g \in C[a, b]$ and

$$\int_a^b [g(x)v(x) + h(x)v'(x)] \, dx = 0, \qquad \forall \, v \in \mathscr{D}_0,$$

conclude that $h' = g$ on $[a, b]$, *without* invoking 4.2.

4.4. Formulate and prove a vector valued analogue of the result in Problem 4.2.

4.5. (a) Prove the following multidimensional version of Corollary 4.3 for a bounded domain D of $\mathbb{R}^d$: If $u \in C(\bar{D})$ and $\int_D uv \, dX = 0$, $\forall \, v \in \mathscr{D}_0 = \{v \in C^1(\bar{D}): v|_{\partial D} = 0\}$, then $u \equiv 0$ in D. (Existence of the integrals with respect to dX, the d-dimensional element of volume, is to be assumed.) Hint: For each closed box $B = \{X \in D: a_j \le x_j \le b_j; j = 1, 2, \ldots, d\}$, consider functions $v(X) = v_1(x_1)v_2(x_2)\ldots v_d(x_d)$, where each v_j is a function of the type used in the proof of Lemma 4.4 (for $m = 0$).

(b) Formulate and prove a corresponding extension of Lagrange's Lemma, 4.4.

4.6*. In proving Proposition 4.5 we needed to show that there is a polynomial q of degree less than m such that $P(x) = x^m q(x)$ has at $b \ne 0$, prescribed derivatives $P^{(k)}(b) = p_k$, $k = 0, 1, 2, \ldots$. (Indeed, we want $p_k = H^{(k)}(b)$ in (2).)

(a) Let $q_k = q^{(k)}(b)$, $k = 0, 1, 2, \ldots, m - 1$. By repeated differentiation show that

$$p_0 = b^m q_0,$$

$$p_1 = b^m q_1 + mb^{m-1} q_0,$$

$$p_2 = b^m q_2 + 2mb^{m-1} q_1 + m(m - 1)b^{m-2} q_0,$$

etc.,

and that conversely these equations can be solved *uniquely* for the q_k. Thus the q_k are known *implicitly* (and recursively).

(b) With the values of q_k from part (a), show that

$$q(x) = \sum_{k=0}^{m-1} \frac{q_k}{k!}(x - b)^k$$

is a polynomial for which $q^{(k)}(b) = q_k$, $k = 0, 1, \ldots$.

(c) Conclude that $P(x) = x^m q(x)$ will meet the required conditions.

4.7. Carry out the steps of the proof of Proposition 4.5 when $m = 2$.

CHAPTER 5

Local Extrema in Normed Linear Spaces

In $\mathbb{R}^d$, it is possible to give conditions which are *necessary* in order that a function f have a *local* extremal value on a subset D, expressed in terms of the vanishing of its gradient ∇f (§0.5). In this chapter, we shall obtain analogous variational conditions which are necessary to characterize *local* extremal values of a function J on a subset $\mathcal{D}$ of a linear space $\mathcal{Y}$ supplied with a norm which assigns a "length" to each $y \in \mathcal{Y}$.

In §5.1 we characterize norms and in the next two sections use them to forge the analytical tools of convergence, compactness, and continuity, presuming that the reader is familiar with these concepts in $\mathbb{R}^d$. After some obvious terminology concerning (local) extremal points (§5.4) we come to the heart of the chapter: the observation that at such a point, the Gâteaux variation of a function must vanish in each admissible direction (§5.5). In the next section, the norm is used to extend the development of differentiation from the Gâteaux variations of Chapter 2 to that of the derivative in the sense of Fréchet. In the concluding section, we introduce the method of Lagrangian multipliers to characterize the local extremal points of one function when it is restricted (constrained) to the level sets of others.

§5.1. Norms for Linear Spaces

Analysis in $\mathbb{R}^d$ is described most easily through inequalities between the lengths of its vectors. Similarly, in the real linear space $\mathcal{Y}$, we shall assume that we can assign to each $y \in \mathcal{Y}$ a nonnegative number, denoted $\|y\|$, which exhibits the following properties:

$$\|y\| \geq 0, \qquad \forall \; y \in \mathcal{Y} \quad \text{with equality } \textit{iff } y = \mathcal{O}; \tag{1a}$$

$$\|cy\| = |c| \, \|y\|, \qquad \forall \, c \in \mathbb{R}, \quad y \in \mathcal{Y}; \tag{1b}$$

$$\|y + \tilde{y}\| \leq \|y\| + \|\tilde{y}\|, \qquad \forall \; y, \tilde{y} \in \mathcal{Y}. \tag{1c}$$

Thus $\|\cdot\|$ is simply a real valued function on $\mathcal{Y}$ which by (1a) is *positive definite*, by (1b) is *positive homogeneous*, and by (1c) satisfies the *triangle inequality*. Each function with these properties is called a *norm* for $\mathcal{Y}$. There may be more than one norm for a linear space, although in a specific example, one may be more natural or more useful than another. Every norm also satisfies the so-called reverse triangle inequality

$$\left| \|y\| - \|\tilde{y}\| \right| \leq \|y - \tilde{y}\|, \qquad \forall \; y, \tilde{y} \in \mathcal{Y}. \tag{1d}$$

(Problem 5.1.)

Example 1. For $\mathcal{Y} = \mathbb{R}^d$ with $y = (y_1, y_2, \ldots, y_d)$, the choice $\|y\| = |y| = (\sum_{j=1}^{d} y_j^2)^{1/2}$ defines a norm, called the Euclidean norm, but the verification of the triangle inequality (1c) is not trivial (Problem 5.2).

The choice $\|y\|_1 = \sum_{j=1}^{d} |y_j|$ also defines a norm and now all of the properties are easily verified. In particular, for (1c):

$$\|y + \tilde{y}\|_1 = \sum_{j=1}^{d} |y_j + \tilde{y}_j| \leq \sum_{j=1}^{d} (|y_j| + |\tilde{y}_j|),$$

or

$$\|y + \tilde{y}\|_1 \leq \sum_{j=1}^{d} |y_j| + \sum_{j=1}^{d} |\tilde{y}_j| = \|y\|_1 + \|\tilde{y}\|_1.$$

Still another norm is the *maximum* norm $\|y\|_M = \max_{j=1,2,\ldots,d} |y_j|$. However, the simpler choice, $\|y\| = |y_1|$ does *not* yield a norm for $\mathbb{R}^d$ if $d \geq 2$ because it is not positive definite. Indeed, the nonzero vector $(0, 1, 1, \ldots, 1)$ would have a zero "norm" with this assignment.

Example 2. For $\mathcal{Y} = C[a, b]$, it is useful to think of the values $y(x)$ as the "components" of the "vector" $y \in \mathcal{Y}$. Then the choice $\|y\|_M = \max |y(x)| = \max_{x \in [a,b]} |y(x)|$ determines a norm, the so-called *maximum* norm. That it is even defined and finite is not obvious and it requires the knowledge that:

(i) $y \in C[a, b] \Rightarrow |y| \in C[a, b]$;
(ii) $[a, b]$ is compact;
(iii) a continuous function ($|y|$) on a compact interval ($[a, b]$) assumes a maximum value ($\|y\|_M$).

(i) is a consequence of the (reverse) triangle inequality in $\mathbb{R}^1$, while (ii) is established in §A.0; (iii) will be established in Proposition 5.3.

Accepting its definition, it is straightforward to verify that $\|\cdot\|_M$ is

(a) positive definite: Since $0 \le |y(x)| \le \|y\|_M$, then

$$\|y\|_M = 0 \Rightarrow |y(x)| = 0 \quad \text{or} \quad y(x) = 0, \qquad \forall\, x \in [a, b];$$

i.e., $y = \mathcal{O}$;

(b) positive homogeneous:

$$\|cy\|_M = \max |cy(x)| = \max |c|\,|y(x)| = |c| \max |y(x)| = |c|\,\|y\|_M;$$

and

(c) satisfies the triangle inequality:

$$|(y + \tilde{y})(x)| = |y(x) + \tilde{y}(x)| \le |y(x)| + |\tilde{y}(x)|$$

$$\le \|y\|_M + \|\tilde{y}\|_M, \qquad x \in [a, b];$$

thus $\|y + \tilde{y}\|_M = \max |(y + \tilde{y})(x)| \le \|y\|_M + \|\tilde{y}\|_M$.

Another choice of norm for $C[a, b]$ is $\|y\|_1 = \int_a^b |y(x)|\, dx$. (See **Problem 5.3**.) Observe that $\|y\|_1 \le (b - a)\|y\|_M$. Are there other norms for this space?

Example 3. For $\mathcal{Y} = C^1[a, b]$, analogous norms are defined by

$$\|y\|_M = \max_{x \in [a,b]} (|y(x)| + |y'(x)|)$$

and

$$\|y\|_1 = \int_a^b (|y(x)| + |y'(x)|)\, dx.$$

To establish the triangle inequality for $\|\cdot\|_M$, observe that for $x \in [a, b]$:

$$|(y + \tilde{y})(x)| + |(y + \tilde{y})'(x)| \le |y(x)| + |\tilde{y}(x)| + |y'(x)| + |\tilde{y}'(x)|$$

$$= (|y(x)| + |y'(x)|) + (|\tilde{y}(x)| + |\tilde{y}'(x)|)$$

$$\le \|y\|_M + \|\tilde{y}\|_M.$$

Now maximizing over x yields the inequality

$$\|y + \tilde{y}\|_M = \max\{|y + \tilde{y})(x)| + |(y + \tilde{y})'(x)|\} \le \|y\|_M + \|\tilde{y}\|_M,$$

as desired.

Observe that for $x \in [a, b]$: $|y(x)| \le \|y\|_M$. Thus $\|y\|_M = 0 \Rightarrow y = \mathcal{O}$; i.e., $\|\cdot\|_M$ is positive definite, as is $\|\cdot\|_1$, since $\|y\|_1 \le (b - a)\|y\|_M$. Can you devise corresponding norms for $C^2[a, b]$? For $C^m[a, b]$?

(5.0) **Remark.** Since for each $m = 2, 3, \ldots,$ $C^m[a, b] \subseteq C^1[a, b] \subseteq C[a, b]$, then each norm for $C[a, b]$ from the previous example will serve also as a norm for $C^1[a, b]$ (or for $C^m[a, b]$). However, these norms do not take cognizance of the differential properties of the functions and supply control only over their continuity.

Example 4. If $\|\cdot\|$ is a norm for the linear space $\mathcal{Y}$, then $\|y\| + |t|$ is a norm for the linear space $\mathcal{Y} \times \mathbb{R}$ consisting of the pairs (y, t) with componentwise addition and scalar multiplication. (See Problem 5.8.)

Example 5. $\mathcal{Y} = C(a, b]$ is a linear space which has no obvious choice for a norm. Now, $\max|y(x)|$ can easily be infinite as can $\int_a^b |y(x)|\, dx$; the function $y(x) = 1/(x - a)$, $x \ne a$ is in $\mathcal{Y}$ and realizes both of these possibilities. In fact, this space cannot be normed usefully.

Example 6. $\mathcal{Y} = (C[a, b])^d$, the space of d-dimensional vector functions with components in $C[a, b]$, is a linear space with a norm for $Y = (y_1, y_2, \ldots, y_d)$ (i.e., $Y(x) = (y_1(x), y_2(x), \ldots, y_d(x))$, $x \in [a, b]$), given by

$$\|Y\|_M = \max|Y(x)|,$$

or by

$$\|Y\| = \sum_{j=1}^{d} \max|y_j(x)|,$$

or by

$$\|Y\| = \sum_{j=1}^{d} \int_a^b |y_j(x)|\, dx,$$

or by

$$\|Y\|_1 = \int_a^b \left(\sum_{j=1}^{d} y_j^2(x) \right)^{1/2} dx = \int_a^b |Y(x)|\, dx$$

The verification is left to Problem 5.6

Example 7. $\mathcal{Y} = (C^1[a, b])^d$, the space of d-dimensional vector functions with components in $C^1[a, b]$, is a linear space with a norm

$$\|Y\|_M = \max(|Y(x)| + |Y'(x)|),$$

or

$$\|Y\| = \sum_{j=1}^{d} \max(|y_j(x)| + |y_j'(x)|),$$

Can you think of other norms for this space?

As these examples show, the discovery of a suitable norm for a given linear space is not immediate, is seldom trivial, and may not be possible. Fortunately, the spaces of interest to us in this text do have standard norms, and these have been given in the examples above.

(Problems 5.1–5.8)

§5.2. Normed Linear Spaces: Convergence and Compactness

When a norm $\|\cdot\|$ for a real linear space $\mathcal{Y}$ has been assigned, it is straightforward to define an associated (topological) structure on $\mathcal{Y}$ which will permit analysis in the *normed linear space* $(\mathcal{Y}, \|\cdot\|)$.

First, we define the *"distance"* between vectors y and $\tilde{y}$ by $\|y - \tilde{y}\|$. The triangle inequality shows that for any three vectors x, y, $z \in \mathcal{Y}$: $\|x - z\| \leq \|x - y\| + \|y - z\|$ and this has the familiar geometrical interpretation.

Next, we introduce the concept of *convergence* by declaring that if $y_n \in \mathcal{Y}$, $n = 1, 2, \ldots$, then the sequence $\{y_n\}_1^{\infty}$ has the *limit* $y_0 \in \mathcal{Y}$ (denoted $\lim_{n \to \infty} y_n = y_0$, or $y_n \to y_0$ as $n \to \infty$) *iff* $\|y_n - y_0\| \to 0$ as $n \to \infty$. A given sequence need not have a limit, but if it does its limit is unique.

[Indeed, were y_0 and $\tilde{y}_0$ both limits of the same sequence $\{y_n\}$ from $\mathcal{Y}$, then by the triangle inequality for each n, $0 \leq \|y_0 - \tilde{y}_0\| \leq \|y_0 - y_n\| + \|y_n - \tilde{y}_0\|$. The right side can be made as small as we wish by choosing n sufficiently large; thus $\|y_0 - \tilde{y}_0\| = 0$; but by the positive definiteness of the norm, this means that $y_0 - \tilde{y}_0 = \mathcal{O}$ or $y_0 = \tilde{y}_0$.]

Alternatively, we can introduce the (open) spherical δ-neighborhood of y_0 for each $\delta > 0$ defined by $S_\delta(y_0) = \{y \in \mathcal{Y}: \|y - y_0\| < \delta\}$ and note that for a sequence $\{y_n\}$: $\lim_{n \to \infty} y_n = y_0$ *iff* the y_n are eventually in each $S_\delta(y_0)$; i.e., for each $\delta > 0$, $\exists\, N_\delta$ such that $n \geq N_\delta \Rightarrow y_n \in S_\delta(y_0)$.

Using these concepts, we can identify the *compact* subsets of $\mathcal{Y}$ as those sets K with the property that each sequence $\{y_n\}$ from K contains a convergent *subsequence* $\{y_{n_j}\}$ with limit $y_0 \in K$; i.e., for integers $n_1 < n_2 < \cdots < n_j < n_{j+1} < \cdots$, $\lim_{j \to \infty} y_{n_j} = y_0 \in K$. (Equivalently, K has the property that each covering of K by a collection of open spheres can be accomplished by using only a *finite* number of these spheres. See [F1].)

For $\mathcal{Y} = C[a, b]$, with the maximum norm (Example 2 of §5.1), the δ-sphere of a function $y_0 \in \mathcal{Y}$ is the set of functions y whose graphs are *uniformly* within δ of the graph of y_0. (See Figure 5.1.) The associated convergence of a sequence is accordingly referred to as *uniform* convergence.

Now although $C[a, b]$ is a linear space,

$$\mathcal{D} = \{y \in C[a, b]: y(a) = 0, y(b) = 1\}$$

is *not*, since the sum of two of its functions cannot satisfy the boundary

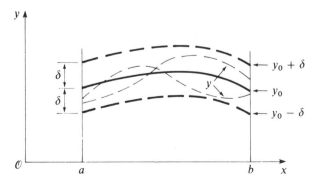

Figure 5.1

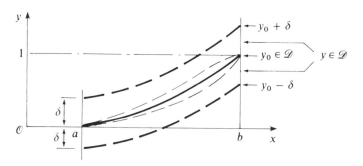

Figure 5.2

condition at b. $\mathscr{D}$ is simply a subset of $\mathscr{Y} = C[a, b]$, and as such automatically inherits the norm-distance properties from $\mathscr{Y}$. Notice also that with the maximum norm, $\mathscr{D}$ does *not* contain a full δ-neighborhood of any of its elements. (See Figure 5.2.) $\mathscr{D}$ is not compact, for reasons that are deferred until the discussion of continuity in the next section.

For $\mathscr{Y} = \mathbb{R}^d$ with the standard Euclidean norm (Example 1 of §5.1), the δ-spheres are the open d-dimensional spheres which in $\mathbb{R}^3$ are ordinary spheres, in $\mathbb{R}^2$, disks, and in $\mathbb{R}^1$, open intervals. With the maximum norm $\|y\|_M = \max_{j=1,2,\ldots,d} |y_j|$, the δ-spheres are d-dimensional "cubes." With either norm, the convergence of a sequence of vectors from $\mathbb{R}^d$ has the same geometrical interpretation, and the *compact* subsets are precisely those which are *bounded* (contained inside some δ-sphere) and *closed* (contain the limits of each of their convergent sequences). In particular, each "box" of the form $B = \{y \in \mathbb{R}^d : a_j \le y_j \le b_j; j = 1, 2, \ldots, d\}$ is compact (§A.0).

(Problems 5.9–5.12)

§5.3. Continuity

If $\mathscr{D}$ is a subset of $\mathscr{Y}$, then we can consider it as the domain of various kinds of functions. For example, a $\mathscr{Y}$-valued function $\mathscr{F}: \mathscr{D} \to \mathscr{Y}$ could be defined by requiring $\mathscr{F}(y) = 0$, $\forall\, y \in \mathscr{D}$. For our purposes, the most important functions are those which are *real* valued, i.e., those of the form $J: \mathscr{D} \to \mathbb{R}$, of which we have already encountered many examples in the previous chapters.

When $\mathscr{Y}$ is supplied with a norm $\|\cdot\|$, we simply adopt the standard $\varepsilon - \delta$ definition for the continuity of a real valued function. (See §0.3.)

(5.1) **Definition.** In a normed linear space $(\mathscr{Y}, \|\cdot\|)$, if $\mathscr{D} \subseteq \mathscr{Y}$, a function $J: \mathscr{D} \to \mathbb{R}$ is said to be *continuous* at $y_0 \in \mathscr{D}$ *iff* for each $\varepsilon > 0$, $\exists$ a $\delta > 0$ such that

$$|J(y) - J(y_0)| < \varepsilon, \qquad \forall\, y \in \mathscr{D} \quad \text{with } \|y - y_0\| < \delta.$$

We observe that the definition of continuity of J at y_0 is a mathematical statement of the intuitive requirement that the smallness of $\|y - y_0\|$ control that of $|J(y) - J(y_0)|$.

Equivalently, J is continuous at $y_0 \in \mathscr{D}$ iff for each sequence $\{y_n\}$ from $\mathscr{D}$, $\lim_{n \to \infty} y_n = y_0 \Rightarrow \lim_{n \to \infty} J(y_n) = J(y_0)$. The proof of this fact is left to Problem 5.13.

We also say that J is *continuous on* $\mathscr{D}$ iff J is continuous at each point $y_0 \in \mathscr{D}$. Observe that if $J: \mathscr{Y} \to \mathbb{R}$ is continuous, and $\mathscr{D} \subseteq \mathscr{Y}$, then the restriction $J|_{\mathscr{D}}: \mathscr{D} \to \mathbb{R}$ is also continuous with respect to the *same* norm.

Example 1. In any normed linear space $(\mathscr{Y}, \|\cdot\|)$, the norm function $J(y) = \|y\|$ is always continuous on $\mathscr{Y}$ and hence on any subset $\mathscr{D}$ of $\mathscr{Y}$. Indeed, from the reverse triangle inequality (1d), $|J(y) - J(y_0)| = |\|y\| - \|y_0\|| \leq \|y - y_0\|$, and hence making $\|y - y_0\|$ small $(< \delta)$ makes $|J(y) - J(y_0)|$ at least as small. (In fact, $\|\cdot\|$ is *uniformly* continuous on $\mathscr{Y}$. See Lemma 5.2 below.)

Example 2. For $\mathscr{Y} = C[a, b]$ with the maximum norm $\|\cdot\|_M$ of §5.1 Example 2, the function

$$J(y) = \int_a^b [\sin^3 x + y^2(x)] \, dx$$

is defined for each $y \in \mathscr{Y}$. To establish its continuity at y_0 we must estimate

$$|J(y) - J(y_0)| = \left| \int_a^b [y(x)^2 - y_0(x)^2] \, dx \right|$$

$$\leq \int_a^b |y(x)^2 - y_0(x)^2| \, dx$$

$$= \int_a^b |y(x) - y_0(x)| \, |y(x) + y_0(x)| \, dx$$

$$\leq \|y - y_0\|_M \|y + y_0\|_M \int_a^b dx,$$

or

$$|J(y) - J(y_0)| \leq (b - a)\|y + y_0\|_M \|y - y_0\|_M.$$

Now, if $\|y - y_0\|_M < 1$, then $\|y\|_M < 1 + \|y_0\|_M$ (Why?) and so $\|y + y_0\|_M \leq \|y\|_M + \|y_0\|_M < 1 + 2\|y_0\|_M$. Thus when $\|y - y_0\|_M < 1$:

$$|J(y) - J(y_0)| < (b - a)(1 + 2\|y_0\|_M)\|y - y_0\|_M$$

$$= A_0 \|y - y_0\|_M,$$

say, for the constant $A_0 = (b - a)(1 + 2\|y_0\|_M)$.

This last estimate shows that for each $\varepsilon > 0$, we can make $|J(y) - J(y_0)| < \varepsilon$ provided that we take $\|y - y_0\| < \delta \overset{\text{def}}{=} \min(1, \varepsilon/A_0)$. Hence J is continuous at each $y_0 \in \mathscr{Y}$, and so J is continuous on $\mathscr{Y}$.

Example 3. With a given $\alpha \in C[a, b]$, the function of §2.2, Example 2; viz., $J(y) = \int_a^b \alpha(x)\sqrt{1 + y'(x)^2}\, dx$ is defined $\forall\, y \in \mathcal{Y} = C^1[a, b]$. Direct examination of its continuity with respect to the maximum norm $\|\ \|_M$ is facilitated by the following *uniform* estimate for $f(z) = \sqrt{1 + z^2}$:

$$|f(z) - f(z_0)| \le |z - z_0|, \qquad z, z_0 \in \mathbb{R}$$

[This is an immediate consequence of the mean value theorem and the fact that

$$|f'(z)| = \frac{|z|}{\sqrt{1 + z^2}} \le 1, \qquad z \in \mathbb{R}.]$$

Thus for $y, y_0 \in \mathcal{Y}$ we have $\forall\, x \in [a, b]$ the *uniform* estimate

$$|f(y'(x)) - f(y_0'(x))| \le |y'(x) - y_0'(x)| \le \|y - y_0\|_M.$$

Hence

$$|J(y) - J(y_0)| \le \int_a^b |\alpha(x)|\, |f(y'(x)) - f(y_0'(x))|\, dx$$

$$\le \|y - y_0\|_M \int_a^b |\alpha(x)|\, dx = A\, \|y - y_0\|_M,$$

say, and the (uniform) continuity of J on $\mathcal{Y}$ should now be evident.

To obtain uniform estimates of the type used in this last example, we shall make frequent appeal to the following technical

(5.2) Lemma. *If K is a compact set in a normed linear space $(\mathcal{Y}, \|\cdot\|)$, then a continuous function $F: K \to \mathbb{R}$ is uniformly continuous on K; i.e., given $\varepsilon > 0$, $\exists\, \delta > 0$ such that $y, \tilde{y} \in K$ and $\|y - \tilde{y}\| < \delta \Rightarrow |F(y) - F(\tilde{y})| < \varepsilon.$*

PROOF. We shall establish the contrapositive implication. Suppose the lemma does not hold. Then, for some $\varepsilon_0 > 0$, and each $n = 1, 2, \ldots$, $\exists$ points $y_n, \tilde{y}_n \in K$ with $\|y_n - \tilde{y}_n\| < 1/n$ for which $|F(y_n) - F(\tilde{y}_n)| \ge \varepsilon_0$. However, since K is compact, there is a *subsequence* $y_{n_j} \to y_0 \in K$ as $j \to \infty$, and since for each $j = 1, 2, \ldots$:

$$\|\tilde{y}_{n_j} - y_0\| \le \|y_{n_j} - y_0\| + \|y_{n_j} - \tilde{y}_{n_j}\| \le \|y_{n_j} - y_0\| + \frac{1}{n_j},$$

it follows that $\tilde{y}_{n_j} \to y_0$ as $j \to \infty$. But, $\forall\, j$:

$$0 < \varepsilon_0 \le |F(y_{n_j}) - F(\tilde{y}_{n_j})| \le |F(y_{n_j}) - F(y_0)| + |F(\tilde{y}_{n_j}) - F(y_0)|,$$

and hence F cannot be continuous at y_0 since continuity would demand that *both* terms on the right $\to 0$ as $j \to \infty$. Thus we have shown that if F is not uniformly continuous on K, then there is at least one point y_0 at which it is not continuous. $\qquad \square$

Example 4. When $f \in C([a, b] \times \mathbb{R}^2)$, the function

$$F(y) = \int_a^b f[y(x)] \, dx = \int_a^b f(x, y(x), y'(x)) \, dx$$

is defined $\forall \, y \in \mathcal{Y} = C^1[a, b]$. To establish its continuity with respect to the maximum norm $\|\cdot\|_M$ of Example 3 in §5.1, we can use Lemma 5.2 as follows:

f is uniformly continuous on each *compact* box $[a, b] \times [-c, c]^2$ when $c > 0$. Thus for a *fixed* $y_0 \in \mathcal{Y}$, when $y \in S_1(y_0)$ we have $\forall \, x \in [a, b]$ that both

$$|y(x)|, |y'(x)| \le |y(x)| + |y'(x)| \le \|y\|_M < 1 + \|y_0\|_M = c_0, \quad \text{say}.$$

Then with $c = c_0$, it follows from the aforementioned uniform continuity that given $\varepsilon > 0$, $\exists$ a $\delta \in (0, 1)$ such that $\|y - y_0\|_M < \delta (< 1) \Rightarrow$

$$|f[y(x)] - f[y_0(x)]| = |f(x, y(x), y'(x)) - f(x, y_0(x), y_0'(x))| < \varepsilon,$$

$$\forall \, x \in [a, b].$$

This *uniform* estimate gives

$$|F(y) - F(y_0)| \le \int_a^b |f[y(x)] - f[y_0(x)]| \, dx \le \varepsilon(b - a),$$

when

$$\|y - y_0\| < \delta < 1,$$

and the continuity of F at the arbitrary point $y_0 \in \mathcal{Y}$ is established. Observe that δ depends on c_0 (and so on y_0) and will in general decrease as c_0 increases.

Example 5. For $\mathcal{Y} = C[a, b]$ with the maximum norm $\|y\|_M = \max |y(x)|$ of §5.1, Example 2, the evaluation function $L(y) = y(a)$ is defined $\forall \, y \in \mathcal{Y}$ and L is even *linear* in that for $y, \tilde{y} \in \mathcal{Y}$ and $c, \tilde{c} \in \mathbb{R}$, we have

$$L(cy + \tilde{c}\tilde{y}) = cL(y) + \tilde{c}L(\tilde{y}),$$

(since $(cy + \tilde{c}\tilde{y})(a) = cy(a) + \tilde{c}\tilde{y}(a)$). Observe that $L(\mathcal{O}) = 0$.

L is also continuous: For a linear function L it suffices to establish continuity at $\mathcal{O}$ since

$$L(y) - L(y_0) = L(y - y_0),$$

so that $|L(y) - L(y_0)|$ is controllably small as $v = y - y_0 \to \mathcal{O}$ iff $L(v) \to 0 = L(\mathcal{O})$ as $v \to \mathcal{O}$. Here, clearly $|L(v)| = |v(a)| \le \|v\|_M$ so that $v \to \mathcal{O} \Rightarrow L(v) \to 0$.

However, if instead we use the *integral* norm $\|y\|_1 = \int_a^b |y(x)| \, dx$ on this *same* space, as in §5.1, Example 2, then this *same* function $L(y) = y(a)$ remains linear, but it is *not* continuous anywhere. From our observation, it suffices to eliminate continuity at $\mathcal{O}$, and for this it suffices to exhibit a sequence $y_n \in C[a, b]$ for which $\|y_n\|_1 \to 0$ as $n \to \infty$, but $L(y_n) = y_n(a) \ge 1$. The functions $y_n(x)$ shown in Figure 5.3 have this property, since geometrically

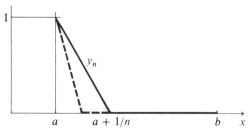

Figure 5.3

for $n > (b - a)^{-1}$,

$$\|y_n\|_1 = \left(\frac{1}{2}\right)(1)\left(\frac{1}{n}\right) = \frac{1}{2n} \to 0 \quad \text{as } n \to \infty,$$

while $y_n(a) = 1, \forall n$.

We are of course interested in characterizing the maximum and minimum values of such real valued functions J and the points at which they occur. Unfortunately, as the simple example $y(x) = x, \ x \in (-1, 1) \subseteq \mathbb{R}$, shows, a continuous function on a subset of a linear space may have neither a maximum value nor a minimum value *on this set—unless the set is compact.*

(5.3) Proposition. *A continuous real valued function J on a compact subset K of a normed linear space $(\mathcal{Y}, \|\cdot\|)$ assumes both maximum and minimum values at points in K. In particular, these values are finite.*

PROOF. To establish that J assumes a maximum value, observe that we can at least find a sequence $\{y_n\}$ from K for which the corresponding sequence $\{J(y_n)\}$ of real numbers increases to this "maximum" (which may be infinite). However, since K is compact, we can extract from this sequence a convergent *sub*sequence $\{y_{n_j}\}$ with limit y_0 in K. From the continuity of J it follows that $J(y_0) = \lim_{j \to \infty} J(y_{n_j})$. But $J(y_{n_j})$ must have the same limit as the original sequence $J(y_n)$. (Why?) Hence this common limit is $J(y_0)$, i.e., the maximum value is assumed. The proof for the assumption of the minimum value is similar.   □

A consequence already utilized in §5.1, Example 2, is that every real valued continuous function on the compact interval $[a,b]$ is bounded and it assumes a maximum value.

However, as attractive as this solution to the problem of establishing the existence of maxima and minima may appear, it will be of little help to us because most of the sets of interest to us are too "large" to be compact. One application of the proposition will be in forestalling attempts to establish compactness!

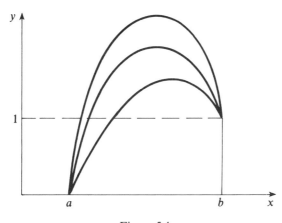

Figure 5.4

For example, $C[0, 1]$ is *not* compact with the maximum norm since the continuous function $J(y) = y(1)$ is unbounded on $C[0, 1]$. To see this, we consider the functions $y_n(x) = nx$, $x \in [0, 1]$ for which $J(y_n) = n \to +\infty$.

More generally, in a nontrivial normed linear space $(\mathcal{Y}, \|\cdot\|)$, $\mathcal{Y}$ itself is *never* compact (Problem 5.17). In particular, there is no clever assignment of a norm to $C[a, b]$ that makes this space compact.

Similarly, $\mathcal{D} = \{y \in C^1[a, b] \colon y(a) = 0, y(b) = 1\}$, which is one of the sets of concern in Chapter 2, is not compact with respect to the maximum norm $\|\cdot\|_M$ or the integral norm $\|\cdot\|_1$ of §5.1, Example 3. For as we have seen in Example 1, the norm function $J(y) = \|y\|$ is always continuous on $\mathcal{Y}$ and hence on any subset $\mathcal{D}$ of $\mathcal{Y}$. But $J_M(y) = \|y\|_M$ is unbounded on the set $\mathcal{D}$ above, as is shown by the sequence of parabolic functions sketched in Figure 5.4, which have maximum values as large as desired while remaining in $\mathcal{D}$. Clearly $\|y\|_M \geq \max |y(x)|$ and so $J_M(y)$ will be unbounded on this sequence. It is also plausible graphically that the integral norm function $J_1(y) = \|y\|_1$ ($\geq \int_0^1 |y(x)|\, dx$) is unbounded on this sequence, since the area under the curve can be made as large as we wish. In particular, the apparently reasonable problem of finding the maximum of the function $J(y) = \int_0^1 |y(x)|\, dx$ on $\mathcal{D}$ has no solution. (In fact, the problem of minimizing J on $\mathcal{D}$ has no solution. See Problem 5.19.)

Thus, the results of Chapter 3 in which we actually obtained minimum values for rather complicated functions on sets such as $\mathcal{D}$ become even more remarkable. (As we saw, an underlying convexity was responsible for our success in these cases.)

We shall proceed with the theoretically unattractive task of seeking maxima and minima of functions which need *not* have them, as the above example shows. Note that to maximize J it suffices to minimize $-J$, and when the underlying set $\mathcal{D}$ is not compact it is probably of little value to establish the continuity of a function J which we wish to minimize on $\mathscr{D}$. (However, continuity of related functions will be required in §5.6 and §5.7.)

Thus in general, neither continuity nor compactness alone can provide us with useful information. As suggested by the analogous situation in $\mathbb{R}^d$, we must consider necessary conditions expressed in terms of an appropriately defined concept of differentiation. We introduce first some obvious terminology.

(Problems 5.13–5.19)

§5.4. (Local) Extremal Points

When $\mathscr{Y}$ is a linear space and J is a real valued function on $\mathscr{D} \subseteq \mathscr{Y}$, then a point $y_0 \in \mathscr{D}$ at which J assumes its maximum value or its minimum value is called an *extremal point* for J on $\mathscr{D}$. This assignment is global in nature and may be made without consideration of a norm. However, the presence of a norm permits an analogous description of the *local* behavior of J at a point y_0:

(5.4) **Definition.** In a normed linear space $(\mathscr{Y}, \|\cdot\|)$ a point $y_0 \in \mathscr{D} \subseteq \mathscr{Y}$ is said to be a *local extremal point* for J on $\mathscr{D}$ if for some $r > 0$, y_0 is an extremal point for J on $\mathscr{D}_r(y_0)$, where $\mathscr{D}_r(y_0) = \{y \in \mathscr{D}: \|y - y_0\| < r\}$; i.e., either

$J(y) \leq J(y_0), \quad \forall \, y \in \mathscr{D}_r(y_0) \quad (y_0 \text{ is a } local \; maximum \; point \text{ for } J \text{ on } \mathscr{D}),$ or

$J(y) \geq J(y_0), \quad \forall \, y \in \mathscr{D}_r(y_0) \quad (y_0 \text{ is a } local \; minimum \; point \text{ for } J \text{ on } \mathscr{D}).$

Of course, each extremal point is automatically a local extremal point whatever norm is used. However, y_0 may be a local extremal point with respect to one norm but *not* with respect to another. (See Problem 5.20)

Now, the Gâteaux variations of Chapter 2 may also be formed without consideration of a norm and when *nonvanishing*, they preclude local extremal behavior with respect to *any* norm.

Suppose, for example, that at a point y_0, the function J has a *positive* variation $\delta J(y_0; v)$ in the direction $v \in \mathscr{Y}$. From Definition 2.6, it follows that $\forall \, \varepsilon$ sufficiently small, the ratio $[J(y_0 + \varepsilon v) - J(y_0)]/\varepsilon$ is also positive, so that $J(y_0 + \varepsilon v) - J(y_0)$ has the sign of ε. Hence,

$$\delta J(y_0; v) > 0 \Rightarrow J(y_0 - \varepsilon v) < J(y_0) < J(y_0 + \varepsilon v), \qquad \forall \text{ small } \varepsilon > 0, \quad (2)$$

and we say that at y_0, J *increases strictly* in the direction v (and *decreases strictly* in the opposite direction $-v$). When $\delta J(y_0; v) < 0$, then $\delta J(y_0; -v) > 0$ (Why?) so that the preceding inequalities and assertions are reversed. In either case, since as $\varepsilon \searrow 0$, $\|(y_0 \pm \varepsilon v) - y_0\| = \varepsilon \|v\| \searrow 0$, the points $y_0 \pm \varepsilon v$ in (2) are eventually in each norm neighbourhood of y_0. Thus local extremal behavior of J at y_0 is not possible in the direction v.

The function J of §2.4, Example 2, has at $y_0(x) = x^2$, the variation

$$\delta J(y_0; v) = 2 \int_a^b y_0(x)v(x)\, dx, \qquad \forall\, v \in \mathcal{Y} = C[a, b],$$

which is clearly nonvanishing in the direction $v(x) = e^x$. Thus, in this direction, J cannot exhibit local extremal behavior at y_0, and as a consequence, y_0 cannot be a local extremal point for J on $\mathcal{Y}$ regardless of the norm employed for $\mathcal{Y}$.

§5.5. Necessary Conditions: Admissible Direction

In minimizing a real valued function J over $\mathcal{D} \subseteq \mathcal{Y}$, where $(\mathcal{Y}, \|\cdot\|)$ is a normed linear space, it is natural to consider for each $y \in \mathcal{D}$ those directions $v \in \mathcal{Y}$ in which the *restricted* function $J|_{\mathcal{D}}$ admits variation at y; i.e., we wish to distinguish those directions $v \in \mathcal{Y}$ for which:

(i) $y + \varepsilon v \in \mathcal{D}$, $\forall$ sufficiently small ε; and
(ii) $\delta J(y; v)$ exists.

Such directions will be termed *admissible* at y for $\mathcal{D}$, or $\mathcal{D}$-admissible at y (for J). Observe that if v is $\mathcal{D}$-admissible at y, then so is each scalar multiple cv for $c \in \mathbb{R}$; $\mathcal{O}$ is always admissible.

To see their significance, suppose that y_0 is a (local) minimum point for J on $\mathcal{D}$. Then J cannot decrease strictly in any direction v admissible at y_0 for $\mathcal{D}$, and hence from (2) (et seq.) $\delta J(y_0; v) = 0$ in such distinguished directions. The same result is obtained when y_0 is a (local) maximum point on $\mathcal{D}$. Thus we have established the following:

(5.5) **Proposition.** *In a normed linear space $(\mathcal{Y}, \|\cdot\|)$, if $y_0 \in \mathcal{D} \subseteq \mathcal{Y}$ is a (local) extremal point for a real valued function J on $\mathcal{D}$, then*

$$\delta J(y_0; v) = 0, \ \forall \ \text{directions } v \ \text{which are } \mathcal{D}\text{-admissible at } y_0. \qquad \square$$

Our hope is that there will be "enough" admissible directions so that the condition $\delta J(y_0; v) = 0$ can determine y_0. Observe, though, that with this condition alone we cannot distinguish between a local maximum and a local minimum point—or between a local minimum point and a global minimum point. Moreover, as in $\mathbb{R}^d$, we must admit the possibility of *stationary points* (such as saddle points) which satisfy this condition but may be neither local maximum points nor local minimum points.

Clearly then, in many senses this condition is *necessary but not sufficient* for, say, a (local) minimum point. However, its analysis forms a large part of the classical calculus of variations. We shall give a geometrical interpretation for it in the next section and return to its classical treatment in the next chapter.

(Note that Proposition 5.5 admits a local extremal point $y_0 \in \mathcal{D}$ which has no nonzero $\mathcal{D}$-admissible directions, and such points must also be considered as candidates for local extrema.)

Example 1. To characterize local extrema for the function (of §2.4, Example 2)

$$J(y) = \int_a^b [\sin^3 x + y(x)^2]\, dx$$

on the domain

$$\mathcal{D} = \{y \in C[a, b]: y(a) = a_1,\, y(b) = b_1\},$$

(where $a_1, b_1 \in \mathbb{R}$ are given), we know that $\delta J(y; v)$ is defined $\forall\, y, v \in C[a, b]$. However, the only $\mathcal{D}$-admissible directions at $y \in \mathcal{D}$ are those for which $y + \varepsilon v \in \mathcal{D}$ for sufficiently small $\varepsilon \neq 0$. Thus we require that

$$a_1 = (y + \varepsilon v)(a) = y(a) + \varepsilon v(a) = a_1 + \varepsilon v(a),$$

$$b_1 = (y + \varepsilon v)(b) = y(b) + \varepsilon v(b) = b_1 + \varepsilon v(b),$$

so that $v(a) = v(b) = 0$. Hence v is $\mathcal{D}$-admissible at $y \in \mathcal{D}$ iff

$$v \in \mathcal{D}_0 = \{v \in C[a, b]: v(a) = v(b) = 0\}.$$

Here the class $\mathcal{D}_0$ is not y-dependent, and by Proposition 5.5 the condition necessary for a (local) extremal point $y_0 \in \mathcal{D}$ is that

$$\delta J(y_0; v) = 2 \int_a^b y_0(x)v(x)\, dx = 0, \qquad \forall\, v \in \mathcal{D}_0.$$

This condition is surely fulfilled when $y_0 \equiv 0$, but this function is not in $\mathcal{D}$ (unless $a_1 = b_1 = 0$). And as Lemma 4.4 of Lagrange shows there is no other possibility; i.e., when $a_1 \neq 0$ or $b_1 \neq 0$ there are "too many" admissible directions to permit any function $y_0 \in \mathcal{D}$ to satisfy all of the conditions $\delta J(y_0; v) = 0$ necessary for a (local) extremum. Thus no such local extremum exists.

If on the other hand, we attempt to minimize J over

$$\mathcal{D}^* = \left\{y \in C[0, 1]: \int_0^1 xy(x)^{4/3}\, dx = 1\right\},$$

then it is difficult to characterize the $\mathcal{D}^*$-admissible directions. (This problem will be considered again in Example 1 of §5.7.)

Example 2*. Let's characterize local minima for the function J of §2.4, Example 3,

$$J(y) = \int_a^b \alpha(x)\sqrt{1 + y'(x)^2}\, dx,$$

for a given $\alpha \in C[a, b]$, on the domain

$$\mathcal{D} = \{y \in C^1[a, b]: y(a) = a_1,\, y(b) = b_1\},$$

with given $a_1, b_1 \in \mathbb{R}$. We again know that $\delta J(y; v)$ is defined $\forall$ y, $v \in \mathcal{Y} = C^1[a, b]$, and as in the previous example conclude that v is $\mathcal{D}$-admissible (at y) iff

$$v \in \mathcal{D}_0 = \{v \in C^1[a, b]: v(a) = v(b) = 0\}.$$

Hence from Proposition 5.5, the necessary condition that $y \in \mathcal{D}$ be a (local) extremum is that

$$\delta J(y; v) = \int_a^b \frac{\alpha(x)y'(x)v'(x)}{\sqrt{1 + y'(x)^2}} \, dx = 0, \qquad \forall\, v \in \mathcal{D}_0.$$

Thus from Lemma 4.1 of du Bois-Reymond, we know that this necessary condition is satisfied only by a $y \in \mathcal{D}$ for which the continuous function

$$\frac{\alpha(x)y'(x)}{\sqrt{1 + y'(x)^2}} = c = \text{const.}, \tag{3}$$

or after obvious algebra, for which

$$y'^2 = \frac{c^2}{\alpha^2 - c^2}. \tag{4}$$

We shall suppose that α does not vanish identically on any subinterval of $[a, b]$. (See Problem 5.21.) If α vanishes at a single point, then from (3), $c = 0$ and so $y = \text{const.}$ which would require $y(a) = a_1 = y(b) = b_1$. Thus unless $a_1 = b_1$, solution to (3) is possible only when $\alpha^2(x) > c^2 > 0$, and for a continuous α this requires that either $\alpha(x) > |c| > 0$ on $[a, b]$ or $\alpha(x) < -|c| < 0$ on $[a, b]$. It suffices to consider the first alternative where $\alpha > 0$ since the second is reduced to this case with the replacement of J by $-J$. Now, when $\alpha > 0$ on $[a, b]$, the integrand function is strongly convex (§3.3, Example 4) and we know from Theorem 3.7 that each $y \in \mathcal{D}$ which satisfies (3) must supply uniquely the *minimum* value for J on $\mathcal{D}$. We see that (3) is both necessary and sufficient for a minimum; that there can be at most one such minimum point; and that there are no other (local) extremal points for J on $\mathcal{D}$. However, as yet, we have no assurance that there is a $y \in \mathcal{D}$ which satisfies (3).

Supposing that $a_1 < b_1$, and that $\alpha(x) \geq \alpha_0 > 0$ on $[a, b]$, we require from (4) that

$$y'(x) = [(\alpha(x)/c)^2 - 1]^{-1/2},$$

or upon incorporating the requirement $y(a) = a_1$, that

$$y(x) = a_1 + \int_a^x [(\alpha(t)/c)^2 - 1]^{-1/2} \, dt, \qquad x \in [a, b].$$

It remains to show that $c \in (0, \alpha_0)$ can be chosen to satisfy the other boundary condition $y(b) = b_1$; i.e., to make

$$h(c) \stackrel{\text{def}}{=} \int_a^b [(\alpha(x)/c)^2 - 1]^{-1/2} \, dx = b_1 - a_1. \tag{5}$$

When $\alpha(x) = \text{const.} = \alpha_0$ on $[a, b]$, then $h(c) = [(\alpha_0/c)^2 - 1]^{-1/2}(b - a)$, is continuous and strictly increasing on $(0, \alpha_0)$ with the limiting values 0 $(= \lim_{c\downarrow 0} h(c))$ and $+\infty$ $(= \lim_{c\uparrow \alpha_0} h(c))$. Hence in this case, there is precisely one c for which (5) is satisfied and precisely one $y \in \mathcal{D}$ which satisfies (3).

When α is not constant, it is more difficult to analyze $h(c)$. (See Problem 5.32*.) However, it is important to realize that there may be *no* solution. For example, when $a = 1$, $b = 2$, and $\alpha(x) = x$ on $[1, 2]$, then we may take $\alpha_0 = 1$, and for $c \in (0, 1)$:

$$\frac{\alpha^2(x)}{c^2} - 1 = \frac{x^2}{c^2} - 1 \geq x^2 - 1;$$

hence

$$h(c) = \int_1^2 \left(\frac{\alpha^2(x)}{c^2} - 1\right)^{-1/2} dx \leq \int_1^2 (x^2 - 1)^{-1/2} dx$$

$$\leq \int_1^2 (x^2 - 1)^{-1/2} x\, dx = (x^2 - 1)^{1/2}\Big|_1^2 = \sqrt{3}.$$

Thus when $b_1 - a_1 > \sqrt{3}$, there is no $y \in \mathcal{D}$ which satisfies the necessary condition $\delta J(y, v) = 0$, $\forall$ $\mathcal{D}$-admissible directions $v \in \mathcal{D}_0$; in this case, there are *no* local extremal points for J on $\mathcal{D}$.

Example 3. For $\mathcal{Y} = \mathbb{R}^2$ with the standard Euclidean norm $|\cdot|$, and

$$\mathcal{D} = \{y \in \mathcal{Y} : |y| = 1\},$$

there are no $\mathcal{D}$-admissible directions $v \neq \mathcal{O}$ at any $y \in \mathcal{D}$ for any function J, because if $y \in \mathcal{D}$ and $\mathcal{O} \neq v \in \mathcal{Y}$, then $y + \varepsilon v \notin \mathcal{D}$ except at most for one value of ε, as the simple geometry of Figure 5.5(a) shows.

On the other hand, at each point in the square $\mathcal{D}_1 = \{y \in \mathcal{Y} : \|y\|_M = 1\}$ of Figure 5.5(b) there is always one *possible* nonzero $\mathcal{D}$-admissible direction, except at the corner points where again there are none.

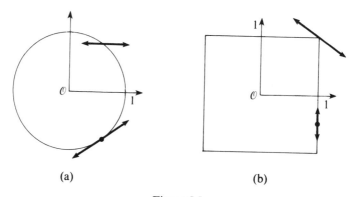

(a)   (b)

Figure 5.5

Example 4*. The brachistochrone function of §1.2(a), viz.,

$$T(y) = \frac{1}{\sqrt{2g}} \int_0^{x_1} \frac{\sqrt{1 + y'(x)^2}}{\sqrt{y(x)}} \, dx$$

is not defined $\forall \, y \in \mathcal{Y} = C^1[0, x_1]$ because of the denominator $\sqrt{y(x)}$. It is defined on

$$\mathcal{D}^* = \left\{ y \in \mathcal{Y} : y \geq 0 \text{ and } \int_0^{x_1} y(x)^{-1/2} \, dx < +\infty \right\},$$

which contains, for example, all of those functions $y \in \mathcal{Y}$ for which $y(x) \geq m > 0$ on $[0, x_1]$.

However, the only functions which could be considered as solutions for the brachistochrone problem are those which satisfy the boundary conditions $y(0) = 0$ and $y(x_1) = y_1$ for some given $y_1 > 0$. Among such functions $y(x) = ax^2$ for $a > 0$, on $[0, x_1]$ is *not* in $\mathcal{D} = \{ y \in \mathcal{D}^* : y(0) = 0, \, y(x_1) = y_1 \}$, since

$$\int_0^{x_1} a^{-1/2} x^{-1} \, dx = +\infty.$$

And even for one such as $y(x) = x^{3/2}$, which *is* in $\mathcal{D}$, it may be that only a restricted class of directions $v \in \mathcal{Y}$ can be $\mathcal{D}$-admissible at y, since in order for $T(y + \varepsilon v)$ to be defined we must have $y + \varepsilon v \geq 0$ for small ε both positive and negative. For example, $v(x) = \sin(\pi x/x_1)$ (which does meet the boundary conditions required to be $\mathcal{D}$-admissible) is *not* an admissible direction for this y since for each $\varepsilon > 0$, $x^{3/2} - \varepsilon \sin(\pi x/x_1)$ is negative when x is sufficiently near (but not equal to) 0. (Why?) Consideration of admissible directions is *essential* to this problem and *cannot* be avoided.

If $y \in \mathcal{D}$, then each $v \in \mathcal{Y}$ with $v(0) = v(x_1) = 0$ and $|v(x)| \leq y(x)$, $\forall \, x \in [0, x_1]$ is a *possible* $\mathcal{D}$-admissible direction at y (as is, of course, any scalar multiple of such a direction). Indeed for such v and $|\varepsilon| \leq \frac{1}{2}$:

$$y(x) + \varepsilon v(x) \geq y(x) - |\varepsilon| |v(x)| \geq y(x) - \tfrac{1}{2} y(x) = \tfrac{1}{2} y(x) \geq 0$$

and

$$\int_0^{x_1} (y + \varepsilon v)^{-1/2}(x) \, dx \leq \sqrt{2} \int_0^{x_1} y(x)^{-1/2} \, dx < +\infty.$$

Finally, formally differentiating $T(y + \varepsilon v)$ with respect to ε under the integral sign as in §2.4, Example 4, with

$$f(x, y, z) = \frac{1}{\sqrt{2g}} \sqrt{\frac{1 + z^2}{y}},$$

so that

$$f_y(x, y, z) = \frac{-1}{2\sqrt{2g}} \sqrt{\frac{1 + z^2}{y^{3/2}}},$$

and

$$f_z(x, y, z) = \frac{z}{\sqrt{2g} \sqrt{y} \sqrt{1 + z^2}},$$

we have

$$\delta T(y; v) = \frac{1}{\sqrt{2g}} \int_0^{x_1} \left[-\frac{1}{2} \frac{\sqrt{1 + y'(x)^2}}{y(x)^{3/2}} v(x) + \frac{y'(x)}{\sqrt{y(x)}\sqrt{1 + y'(x)^2}} v'(x) \right] dx,$$
(6)

which is finite in general only if $\int_0^{x_1} y(x)^{-3/2} dx$ is finite.

Thus we see that if

$$\mathcal{D}_1 = \{ y \in \mathcal{Y} : y(0) = 0, y(x_1) = y_1, y(x) \geq 0 \text{ on } [0, x_1],$$

$$\text{and } \int_0^{x_1} y(x)^{-3/2} dx < +\infty \},$$

then for each $y \in \mathcal{D}_1$, $\delta T(y; v)$ is defined by (6) $\forall$ directions $v \in \mathcal{Y}$ at y for which $v(0) = v(x_1) = 0$ and $|v(x)| \leq y(x)$ (or scalar multiples of these directions). We shall return to this problem in §6.2(c), Example 4.

In summary, as these examples show, there may be "too many" nontrivial admissible directions v to allow any $y_0 \in \mathcal{D}$ to fulfill the necessary condition $\delta J(y_0; v) = 0$; or there may be just enough to permit this condition to determine y_0, or there may be many, but not readily usable—or there may be just one—or even none. Nevertheless, when present, they provide the most obvious approach to attacking problems in optimization, and should always be considered before investigating alternatives such as the method of Lagrangian multipliers to be introduced in §5.7. Finally, as with the brachistochrone function of Example 4*, they may be essential to the problem.

(Problems 5.21–5.32)

§5.6*. Affine Approximation: The Fréchet Derivative

As we have seen, the Gâteaux variation in a normed linear space is analogous to the directional derivative in $\mathbb{R}^d$. In particular, without further information, we cannot expect to use these variations to provide a good approximation to a function which has them—except, of course. in each separate direction. For this purpose in $\mathbb{R}^d$, we required that the function satisfy the stronger requirement of differentiability (§0.10), and we shall simply lift the definition employed there, together with the associated terminology, to our normed linear space $(\mathcal{Y}, \|\cdot\|)$.

In $\mathbb{R}^d$ with the Euclidean norm $|\cdot|$, a real valued function f is said to be differentiable at $y_0 \in \mathbb{R}^d$ provided that it is defined in a sphere $S(y_0)$ and there

$$f(y) = f(y_0) + l(y - y_0) + |y - y_0| \mathfrak{z}(y - y_0),$$

where $\mathfrak{z}(y - y_0)$ is a function with zero limit as $y - y_0 \to \mathcal{O}$, and l is the *continuous* *linear* function defined on $\mathbb{R}^d$ by $l(v) = \nabla f(y_0) \cdot v$. See Problem 5.15.

Now, for $(\mathscr{Y}, \|\cdot\|)$ a linear function $L: \mathscr{Y} \to \mathbb{R}$ need *not* be continuous (§5.3, Example 5), and we must require this continuity. Accordingly we make the following:

(5.6) **Definition.** In a normed linear space $(\mathscr{Y}, \|\cdot\|)$, a real valued function J is said to be *differentiable* (in the sense of Fréchet) at $y_0 \in \mathscr{Y}$ provided that J is defined in a sphere $S(y_0)$ and there exists a continuous linear function $L: \mathscr{Y} \to \mathbb{R}$ for which

$$J(y) = J(y_0) + L(y - y_0) + \|y - y_0\| \mathfrak{z}(y - y_0), \tag{7}$$

where $\mathfrak{z}(y - y_0)$ is a real valued function (defined when $y - y_0 \neq \mathcal{O}$ by this equation) which has zero limit as $y - y_0 \to \mathcal{O}$ or as $\|y - y_0\| \to 0$.

(5.7) **Proposition.** *If J is (Fréchet) differentiable at y_0, then J has the Gâteaux variation $\delta J(y_0; v) = L(v)$ in each direction $v \in \mathscr{Y}$.*

PROOF. For $v \in \mathscr{Y}$ and $\varepsilon \neq 0$ sufficiently small, set $y = y_0 + \varepsilon v$ in (7). Then

$$J(y_0 + \varepsilon v) - J(y_0) = L(\varepsilon v) + \|\varepsilon v\| \mathfrak{z}(\varepsilon v).$$

Thus using the assumed linearity of L and the homogeneous property of the norm (1b) we have that

$$\frac{J(y_0 + \varepsilon v) - J(y_0)}{\varepsilon} = L(v) + \frac{|\varepsilon|}{\varepsilon} \|v\| \mathfrak{z}(\varepsilon v).$$

Now as $\varepsilon \to 0$, $\varepsilon v = y - y_0 \to \mathcal{O}$ and hence $\mathfrak{z}(\varepsilon v) \to 0$, while $|\varepsilon|/\varepsilon = \pm 1$. Thus

$$\delta J(y_0; v) = \lim_{\varepsilon \to 0} \frac{J(y_0 + \varepsilon v) - J(y_0)}{\varepsilon} = L(v)$$

as asserted. $\qquad\square$

It follows that the linear function L of the definition is uniquely determined. It is denoted $J'(y_0)$ and called the *Fréchet derivative* of J at y_0.

Observe that differentiability implies continuity as we should wish:

(5.8) **Proposition.** *In a normed linear space $(\mathscr{Y}, \|\cdot\|)$ if a real valued function J is differentiable at $y_0 \in \mathscr{Y}$, then it is continuous at y_0.*

PROOF. From Definition 5.6,

$$|J(y) - J(y_0)| \leq |L(y - y_0)| + \|y - y_0\| |\mathfrak{z}(y - y_0)|.$$

Now as $y \to y_0$, from the linearity and continuity of L,

$$|L(y - y_0)| = |L(y) - L(y_0)| \to 0;$$

also $\|y - y_0\| \to 0$, and $\mathfrak{z}(y - y_0) \to 0$ from its definition. Thus as $y \to y_0$,

$$|J(y) - J(y_0)| \to 0,$$

and this establishes the continuity. $\square$

As in $\mathbb{R}^d$, the converses of these propositions need not hold. Continuous functions are seldom differentiable. Moreover, if J admits the Gâteaux variation $\delta J(y_0; v)$ in each direction $v \in \mathcal{Y}$, the resulting function of v may be neither linear nor continuous—and even these properties may not suffice for differentiability. Some additional conditions are required.

Proposition 5.7 provides the key for establishing the differentiability of a suitably defined function J at a point y_0 in a normed linear space $(\mathcal{Y}, \|\cdot\|)$.

First: Check that $\delta J(y_0; v)$ exists, $\forall\, v \in \mathcal{Y}$.

Next: Prove that $L(v) \stackrel{\text{def}}{=} \delta J(y_0; v)$ is linear and continuous in v.

Last: Show that for $y \neq y_0$,

$$\mathfrak{z}(y - y_0) \stackrel{\text{def}}{=} \frac{J(y) - J(y_0) - \delta J(y_0; y - y_0)}{\|y - y_0\|} \to 0 \quad \text{as } \|y - y_0\| \to 0.$$

This program may either be applied directly to a specific function (Example 3 below and Problem 5.22), or it may be used to suggest alternate conditions which would imply differentiability as in the following:

(5.9) Theorem. *In a normed linear space $(\mathcal{Y}, \|\cdot\|)$, if a real valued function J has at each $y \in S_r(y_0)$ Gâteaux variations $\delta J(y; v)$, $\forall\, v \in \mathcal{Y}$ and*

(a) $\delta J(y_0; v)$ *is linear and continuous in v;*
(b) *as $y \to y_0$, $|\delta J(y; u) - \delta J(y_0, u)| \to 0$ uniformly for $u \in B = \{u \in \mathcal{Y}: \|u\| = 1\}$;*

then J is differentiable at y_0.

PROOF*. From condition (a) we may express $\delta J(y_0; u) = L(u)$ for a *linear* function $L: \mathcal{Y} \to \mathbb{R}$. Each $y \in S_r(y_0) \sim \{y_0\}$ may be expressed (uniquely) as $y = y_0 + tu$ for $t = \|y - y_0\| < r$ and $u \in B$. (Why?)

Moreover, for each *fixed* $u \in B$, $f(t) \stackrel{\text{def}}{=} J(y_0 + tu)$ is differentiable on $(-r, r)$ since at $t_1 \in (-r, r)$, with $\varepsilon = t - t_1 \neq 0$ and $y_1 = y_0 + t_1 u$, we have

$$y_0 + tu = y_1 + \varepsilon u,$$

so that

$$\frac{f(t) - f(t_1)}{t - t_1} = \frac{J(y_1 + \varepsilon u) - J(y_1)}{\varepsilon} \to \delta J(y_1; u), \quad \text{as } t \to t_1.$$

Thus $f'(t_1) = \delta J(y_1; u)$.

Also as $t \searrow 0$, $f(t) = J(y_0 + tu) = J(y) \to J(y_0)$. (Why?) Hence we have by the law of the mean (§A.1):

$$J(y) - J(y_0) = J(y_0 + tu) - J(y_0) = f(t) - f(0)$$
$$= f'(t_1)t, \quad \text{for some } t_1 \in (-t, t),$$
$$= \delta J(y_1; u)t \quad \text{as above.}$$

Observe also that $\|y_1 - y_0\| = |t_1| < |t| = \|y - y_0\|$. Hence

$$J(y) - J(y_0) - L(y - y_0) = J(y) - J(y_0) - tL(u)$$
$$= [\delta J(y_1; u) - \delta J(y_0; u)]t,$$

so that $y \to y_0 \Rightarrow y_1 \to y_0$ and

$$\left| \frac{J(y) - J(y_0) - L(y - y_0)}{\|y - y_0\|} \right| = |\delta J(y_1; u) - \delta J(y_0; u)| \to 0,$$

if we utilize the *uniformity* expressed in (b) to make the last assertion. Thus

$$\mathfrak{z}(y - y_0) \overset{\text{def}}{=} \frac{J(y) - J(y_0) - L(y - y_0)}{\|y - y_0\|}, \quad y \neq y_0,$$

has the requisite zero limiting value to satisfy Definition 5.6 for the differentiability of J at y_0. □

Remark. This theorem is the most usable for our purposes. Other sufficient conditions are known, but all involve some additional uniformity such as that in condition (b). Without this uniformity, it is only possible to characterize the behavior as $y = y_0 + tu \to y_0$ for *fixed* u. On the other hand, part of condition (a) *is* superfluous. The *linearity* of $\delta J(y_0; v)$ is a consequence of condition (b). See [V].

Conditions (a) and (b) also imply a weak continuity of δJ at y_0 in the sense of the following:

(5.10) **Definition.** In a normed linear space $(\mathcal{Y}, \|\cdot\|)$, the Gâteaux variations $\delta J(y; v)$ of a real valued function J are said to be *weakly continuous* at $y_0 \in \mathcal{Y}$ provided that for each $v \in \mathcal{Y}$: $\delta J(y; v) \to \delta J(y_0; v)$ as $y \to y_0$. [See Problem 5.34.]

Example 1. The function of §5.3, Example 2, viz.,

$$J(y) = \int_a^b [\sin^3 x + y(x)^2] \, dx,$$

is defined if $y \in \mathcal{D} = C[a, b]$; using the maximum norm $\|\cdot\|_M$ of §5.1, Example 2, we know that J is continuous at each $y_0 \in \mathcal{Y}$. Moreover, from (8) in §2.4 we know that $\forall \, y, v \in \mathcal{Y}$,

$$\delta J(y; v) = 2 \int_a^b y(x)v(x) \, dx$$

and the linearity in v is apparent (Why?) Thus to establish the continuity in v of $\delta J(y_0; v)$ it suffices to establish continuity at $v = \mathcal{O}$ [§5.3, Example 5]. But since $\delta J(y_0; \mathcal{O}) = 0$,

$$|\delta J(y_0; v) - \delta J(y_0; \mathcal{O})| = |\delta J(y_0; v)|$$

$$= 2\left|\int_a^b y_0(x)v(x)\,dx\right|$$

$$\leq 2\int_a^b |y_0(x)|\,|v(x)|\,dx$$

$$\leq 2\|y_0\|_M \|v\|_M (b - a), \quad \text{(Why?)},$$

and for fixed $y_0 \in \mathcal{Y}$ this last term $\to 0$ as $v \to \mathcal{O}$. Thus J as above satisfies condition (a) of Theorem 5.9.

For condition (b) we suppose $u \in \mathcal{Y}$ with $\|u\|_M = 1$, and estimate similarly to obtain

$$|\delta J(y; u) - \delta J(y_0; u)| = 2\left|\int_a^b [y(x) - y_0(x)]u(x)\,dx\right|$$

$$\leq 2\|y - y_0\|_M \|u\|_M (b - a)$$

$$= 2\|y - y_0\|_M (b - a). \tag{8}$$

We observe that the last term $\to 0$ as $y \to y_0$ and is *independent of u*. Hence the left side of (8) $\to 0$ uniformly in u when $\|u\|_M = 1$ as required.

It follows that J is differentiable at each $y_0 \in \mathcal{Y}$.

Example 2. Similarly, the function J of §5.3, Example 3, viz.,

$$J(y) = \int_a^b \alpha(x)\sqrt{1 + y'(x)^2}\,dx,$$

for given $\alpha \in C[a, b]$, is defined if $y \in \mathcal{Y} = C^1[a, b]$, and it is continuous in the maximum norm $\|\cdot\|_M$ [§5.1, Example 3]. We also know from Example 2 of §5.5 that $\forall\, y, v \in \mathcal{Y}$:

$$\delta J(y; v) = \int_a^b \frac{\alpha(x)y'(x)v'(x)}{\sqrt{1 + y'(x)^2}}\,dx;$$

again the linearity in v is evident, and it will lead to continuity at $v = \mathcal{O}$ essentially as in the previous example. Indeed:

$$|\delta J(y_0; v)| \leq \int_a^b \frac{|\alpha(x)|\,|y_0'(x)|\,|v'(x)|}{\sqrt{1 + y_0'(x)^2}}\,dx$$

$$\leq \int_a^b |\alpha(x)|\,|v'(x)|\,dx, \quad \left(\text{since } \frac{|z|}{\sqrt{1 + z^2}} \leq 1, \forall\, z \in \mathbb{R}\right),$$

$$\leq \|\alpha\|_M \int_a^b |v'(x)|\,dx$$

$$\leq A\|v\|_M, \quad \text{say.}$$

To establish condition (b) (of 5.9) we observe that $f(z) = z/\sqrt{1 + z^2}$ is uniformly continuous on $\mathbb{R}$ since $|f'(z)| = (1 + z^2)^{-3/2} \le 1$, $\forall z \in \mathbb{R}$ and so (by the law of the mean) $|f(z) - f(z_0)| \le |z - z_0|$, $\forall z, z_0 \in \mathbb{R}$. Now we make the following estimate at $y_0 \in \mathscr{Y}$:

$$|\delta J(y; u) - \delta J(y_0; u)| \le \int_a^b |\alpha(x)| |f(y'(x)) - f(y_0'(x))| |u'(x)| \, dx$$

$$\le \|u\|_M \int_a^b |\alpha(x)| |y'(x) - y_0'(x)| \, dx \le A \|u\|_M \gamma,$$

when $\|y - y_0\|_M < \gamma$. Thus, if $\|u\|_M = 1$,

$$|\delta J(y; u) - \delta J(y_0; u)| < \varepsilon,$$

when $\|y - y_0\|_M < \gamma = \varepsilon/A$, and since ε can be made as small as we please condition (b) is satisfied; the differentiability at an arbitrary $y_0 \in \mathscr{Y}$ follows from Theorem 5.9.

Example 3. The length function of §1.1(a), viz.,

$$L(Y) = \int_0^1 |Y'(t)| \, dt \quad \text{for } Y \in \mathscr{Y} \equiv (C^1[0, 1])^d$$

has at each $Y \in \mathscr{D}^* = \{Y \in \mathscr{Y} : |Y'(t)| \ne 0\}$, the Gâteaux variations

$$\delta L(Y; V) = \int_0^1 \frac{Y'(t)}{|Y'(t)|} \cdot V'(t) \, dt, \qquad \forall V \in \mathscr{Y},$$

which are linear and continuous in V by standard estimates.

L is differentiable at each $Y_0 \in \mathscr{D}^*$ in the maximum norm $\|Y\|_M$, since

$$0 \le |L(Y) - L(Y_0) - \delta L(Y_0; Y - Y_0)|$$

$$= \left| \int_0^1 \left[|Y'| - |Y_0'| - \frac{Y_0'}{|Y_0'|} \cdot (Y' - Y_0') \right](t) \, dt \right|$$

$$= \left| \int_0^1 \frac{[|Y'||Y_0'| - Y' \cdot Y_0'](t)}{|Y_0'(t)|} \, dt \right|$$

$$\le \int_0^1 \frac{|Y' - Y_0'|^2}{|Y_0'|}(t) \, dt$$

$$\le A \|Y - Y_0\|_M^2 \int_0^1 \frac{1}{|Y_0'(t)|} \, dt$$

$$= (\|Y - Y_0\|_M)(A_0 \|Y - Y_0\|_M), \quad \text{say,}$$

so that

$$\left| \frac{L(Y) - L(Y_0) - \delta L(Y_0; Y - Y_0)}{\|Y - Y_0\|_M} \right| \le A_0 \|Y - Y_0\|_M \to 0 \quad \text{as } Y \to Y_0.$$

Here, we have utilized the vector inequality

$$|A||B| - A \cdot B \le |A - B|^2 \quad \text{for } A, B \in \mathbb{R}^d,$$

which is a consequence of the fact that $2|A||B| \le |A|^2 + |B|^2$.

Remarkably, it is still more difficult to establish the weak continuity of $\delta L(Y; V)$ at $Y_0 \in \mathscr{D}^*$ which depends upon the previous inequality in another form. (See Problem 5.35.)

(5.11) **Proposition.** *When $f = f(x, y, z)$, f_y, and $f_z \in C([a, b] \times \mathbb{R}^2)$, then*

$$F(y) = \int_a^b f(x, y(x), y'(x))\, dx$$

is differentiable and has weakly continuous variations at each $y_0 \in \mathscr{Y} = C^1[a, b]$ in the maximum norm $\|y\|_M$.

PROOF. By (10) of §2.4, the Gâteaux variations

$$\delta F(y; v) = \int_a^b (f_y[y(x)]v(x) + f_z[y(x)]v'(x))\, dx$$

are obviously linear in v and continuous at each y_0 (Problem 5.36). To show that they satisfy condition (b) at y_0, note that

$$|\delta F(y; v) - \delta F(y_0; v)| \le \int_a^b |f_y[y(x)] - f_y[y_0(x)]|\,|v(x)|\, dx$$

$$+ \int_a^b |f_z[y(x)] - f_z[y_0(x)]|\,|v'(x)|\, dx$$

by standard estimates.

Now

$$f_y[y(x)] - f_y[y_0(x)] = f_y(x, y(x), y'(x)) - f_y(x, y_0(x), y_0'(x)),$$

and since f_y is uniformly continuous on each box $[a, b] \times [-c, c]^2$ (Lemma 5.2), it follows that $|f_y(x, y, z) - f_y(x, y_0, z_0)| < \varepsilon$ if $|y|, |y_0|, |z|, |z_0| \le c$ and $|y - y_0| + |z - z_0| < r = r(\varepsilon)$.

Thus for the given $y_0 \in \mathscr{Y}$, we can choose c so large that $\|y - y_0\|_M \le 1 \Rightarrow |y(x)|, |y_0(x)|, |y'(x)|, |y_0'(x)| \le c, \forall\, x \in [a, b]$, and hence for a given $\varepsilon > 0$, conclude that $\exists\, r > 0$ such that

$$\|y - y_0\|_M < r \le 1 \Rightarrow |f_y[y(x)] - f_y[y_0(x)]| \le \varepsilon, \quad \forall\, x \in [a, b].$$

Similarly, for perhaps a smaller r, we can have that

$$|f_z[y(x)] - f_z[y_0(x)]|$$

$$= |f_z(x, y(x), y'(x)) - f_z(x, y_0(x), y_0'(x))| \le \varepsilon, \qquad \forall x \in [a, b].$$

Thus

$$|\delta F(y; v) - \delta F(y_0; v)| \le \varepsilon \int_a^b (|v(x)| + |v'(x)|)\, dx$$

$$= \varepsilon V, \quad \text{say, when } \|y - y_0\|_M \le r \le 1,$$

$$[\le \varepsilon(b - a) \quad \text{when } \|v\|_M \le 1];$$

since ε is arbitrary, it follows that

$$\delta F(y; v) \to \delta F(y_0; v) \quad \text{as } y \to y_0. \quad [\text{uniformly when } \|v\|_M \le 1],$$

and this establishes both the weak continuity and condition (b). Differentiability follows from Theorem 5.9.

(When $f = f(x, y)$ alone, the variations are weakly continuous in the stronger maximum norm $\|y\|_M = \max|y(x)|$ of Remark 5.0.) (See Problem 5.24.)   □

Tangency

By 5.6 and 5.7, a function J which is differentiable at a point y_0 in a normed linear space $(\mathscr{Y}, \|\cdot\|)$, with Frechét derivative $J'(y_0)$, admits a good approximation near y_0 by the affine function

$$T(y) = J(y_0) + J'(y_0)(y - y_0), \tag{9}$$

which is defined $\forall\, y \in \mathscr{Y}$. The approximation is "good" in the sense that for y near y_0 and $y \ne y_0$,

$$\frac{J(y) - T(y)}{\|y - y_0\|} = \mathfrak{z}(y - y_0) \to 0 \quad \text{as } y \to y_0.$$

As in Euclidean space (§0.10) the graph of T (a "hyperplane" in $\mathscr{Y} \times \mathbb{R}$) may be said to be tangent to that of J at the point $(y_0, J(y_0)) = (y_0, T(y_0))$. Moreover, comparison with Definition 3.1, shows that a convex differentiable function is one whose graph lies "above" each of these tangent hyperplanes in $\mathscr{Y} \times \mathbb{R}$.

However, there is also an intrinsic sense in which T can provide tangency for J at y_0 in $\mathscr{Y}$ itself: namely, between the respective level sets of these functions.

Now the level set of T through y_0,

$$T_{y_0} \overset{\text{def}}{=} \{y \in \mathscr{Y}: T(y) = T(y_0)\} = \{y \in \mathscr{Y}: J'(y_0)(y - y_0) = \mathcal{O}\},$$

is by definition a hyperplane, and if we introduce the corresponding level set of J through y_0, viz.,

$$J_{y_0} = \{y \in \mathcal{Y}: J(y) = J(y_0)\},$$

we see that for $y \in J_{y_0}$, $y \neq y_0$,

$$J'(y_0)\left(\frac{y - y_0}{\|y - y_0\|}\right) = \frac{T(y) - J(y)}{\|y - y_0\|} = -\mathfrak{z}(y - y_0),$$

so that if a sequence $y_n \in J_{y_0}$, $(y_n \neq y_0)$ $n = 1, 2, \ldots$, provides *unit* directions $\tau_n = (y_n - y_0)/(\|y_n - y_0\|)$ with a limit direction, τ, as $y_n \to y_0$, it follows from the assumed continuity of $J'(y_0)$ that

$$J'(y_0)\tau = \lim_{n \to \infty} J'(y_0)\tau_n = \lim_{n \to \infty} \mathfrak{z}(y_n - y_0) = 0.$$

By (9), $T(y_0 + \tau) = J(y_0) = T(y_0)$, and so any possible limit direction τ must furnish a point $y_0 + \tau$ in the hyperplane T_{y_0}. Conversely, each τ such that $y_0 + \tau \in T_{y_0}$ will make $J'(y_0) = 0$. (Why?) Accordingly we make the following

(5.12) **Definition.** In a normed linear space $(\mathcal{Y}, \|\cdot\|)$, if a real valued function J is differentiable at $y_0 \in \mathcal{Y}$, then we introduce

$$T(y) = J(y_0) + J'(y_0)(y - y_0),$$

and say that the level set T_{y_0} is tangent to the level set J_{y_0} at y_0. Moreover, each nonzero direction $\tau \in \mathcal{Y}$ for which $\delta J(y_0; \tau) = J'(y_0)\tau = 0$ is called a *tangential direction* to J_{y_0} at y_0.[1]

When $\mathcal{Y} = \mathbb{R}^3$ has the standard Euclidean norm, then a function J differentiable at Y_0 has the Gâteaux variation

$$\delta J(Y_0; V) = \nabla J(Y_0) \cdot V, \qquad \forall\, V \in \mathbb{R}^3.$$

In this case, the linear function

$$L(\tau) = J'(Y_0)\tau = \nabla J(Y_0) \cdot \tau,$$

and the tangent directions τ are precisely those which are orthogonal to the gradient vector $\nabla J(Y_0)$. If $\nabla J(Y_0) \neq \mathcal{O}$, then it is perpendicular to the plane T_{Y_0} through Y_0 determined by these tangent vectors, and hence $\nabla J(Y_0)$ is normal to the level surface J_{Y_0} through this point. Thus our definition of tangency admits this well-known interpretation in $\mathbb{R}^3$.

(Problems 5.33–5.36)

---

[1] Although this definition provides suggestive terminology, it avoids the deeper question of whether each such tangential direction τ is *geometrically* tangent in that it is the limit of a sequence $\{\tau_n\}$, of the type described above. This does hold under more stringent requirements on J. See §A.7 and Liusternik's Theorem in [I–T].

§5.7. Extrema with Constraints: Lagrangian Multipliers

In §5.5, we saw that in finding (local) extrema of a real valued function J on a subset $\mathcal{D}$ in a normed linear space $(\mathcal{Y}, \|\cdot\|)$, there may be enough $\mathcal{D}$-admissible directions for J at a typical point $y_0 \in \mathcal{D}$ to provide a usable characterization of possible local extremal points. However, we also saw that even in $\mathbb{R}^2$, a domain as simple as a circle has no nontrivial admissible directions for any function J.

Observe that this domain is itself a level set of another function. For example, the unit circle $\{y \in \mathbb{R}^2 : |y| = 1\}$ is the one-level set of the function $G(y) = |y|$ which is differentiable at each point of that level set. In this section we shall develop the method of Lagrangian multipliers for characterizing the local extrema of a function J in a normed linear space when restricted to one or more level sets of other such functions. In this context, the level sets involved are called constraints, and the equations defining the sets are referred to as constraining relations.

We have already encountered many examples of constraining relations in the previous sections. For instance, a set of the form

$$\mathcal{D} = \{y \in C[a, b]: y(a) = a_1, y(b) = b_1\}$$

is the intersection of the a_1-level set of the function $G_1(y) = y(a)$ with that of the b_1-level set of the function $G_2(y) = y(b)$. The set

$$\mathcal{D} = \left\{y \in C[a, b]: \int_a^b y(x)\, dx = 1\right\}$$

is the one-level set of the function $G(y) = \int_a^b y(x)\, dx$. In fact, many sets $\mathcal{D}$ considered previously can be described in terms of level sets of appropriately defined functions.

To motivate the ensuing development, we consider first the problem of characterizing a (local) extremal point y_0 of a real valued function J in a normed linear space $(\mathcal{Y}, \|\cdot\|)$ when constrained to a level set of a real valued function G. Thus we should have that when y is sufficiently near y_0 and $G(y) = G(y_0)$, either $J(y) \geq J(y_0)$, $\forall$ such y, or $J(y) \leq J(y_0)$, $\forall$ such y.

This possibility is eliminated if there exists a direction v and scalars $\bar{r}$, $\underline{r} \in \mathbb{R}$, as small as we wish, such that upon setting $\bar{y} = y_0 + \bar{r}v$ and $\underline{y} = y_0 + \underline{r}v$, we have

$$J(\underline{y}) < J(y_0) < J(\bar{y}) \quad \text{while } G(\underline{y}) = G(y_0) = G(\bar{y}). \tag{10}$$

However, it is more useful to consider a pair of directions v, w for which $\exists$ pairs of scalars $(\bar{r}, \bar{s})$ and $(\underline{r}, \underline{s})$ as small as we please such that these same requirements (10) hold for

$$\bar{y} = y_0 + \bar{r}v + \bar{s}w,$$

$$\underline{y} = y_0 + \underline{r}v + \underline{s}w.$$

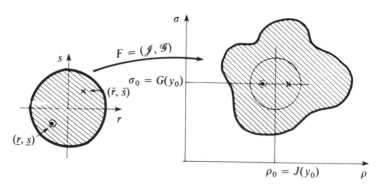

Figure 5.6

We shall now assume that both J and G are defined in a neighborhood of y_0, and consider for *fixed* directions v, w the auxiliary functions

$$\rho = \mathscr{J}(r, s) = J(y_0 + rv + sw),$$

$$\sigma = \mathscr{G}(r, s) = G(y_0 + rv + sw),$$

which are defined in some two-dimensional neighborhood of the origin in $\mathbb{R}^2$. (Why?) The pair of these functions F maps this neighborhood into a (ρ, σ) set in $\mathbb{R}^2$ which contains the point

$$(\rho_0, \sigma_0) = (\mathscr{J}(0, 0), \mathscr{G}(0, 0)) = (J(y_0), G(y_0)).$$

If it also contains a full neighborhood of (ρ_0, σ_0), then there are preimage points $(\bar{r}, \bar{s})$ and $(\underline{r}, \underline{s})$ and associated $\bar{y}$, $\underline{y}$ for which the conditions (10) are met. This is readily seen in Figure 5.6.

Finally, to have $(\bar{r}, \bar{s})$, $(\underline{r}, \underline{s})$ as near $(0, 0)$ as we please we would require that each small neighborhood of $(0, 0)$ map onto a set which contains a full neighborhood of (ρ_0, σ_0). All of this is assured if the mapping $F \equiv (\mathscr{J}, \mathscr{G})$ has an inverse defined in a neighborhood of (ρ_0, σ_0) which is continuous at (ρ_0, σ_0).[1]

The simplest conditions which provide this continuous local inverse are well known, and form the content of the inverse function theorem which we state without proof. (See [Ed].)

(5.13) **Theorem.** *For $X_0 \in \mathbb{R}^d$ and $\tau > 0$, if a vector valued function F: $S_\tau(X_0) \to \mathbb{R}^d$ has continuous first partial derivatives in each component with nonvanishing Jacobian determinant at X_0, then F provides a continuously invertible mapping between a neighborhood of X_0 and a region containing a full neighborhood of $F(X_0)$.* □

[1] A weaker open mapping result which suffices is established in §A.4.

For $F = (f_1, f_2, \ldots, f_d)$ we require in this theorem that the matrix with (continuous) elements $\partial f_i/\partial x_j$, $i, j = 1, 2, \ldots, d$, arranged in natural order, have a nonzero determinant when evaluated at X_0. If F defines a *linear* transformation of $\mathbb{R}^d$ into itself, then this becomes the familiar condition for invertibility of the matrix representing the transformation.

Now, with $y = y_0 + rv + sw$, the partial derivative

$$\mathcal{J}_r(r, s) = \frac{\partial}{\partial r} J(y_0 + rv + sw) = \lim_{\varepsilon \to 0} \frac{1}{\varepsilon}[J(y_0 + (r + \varepsilon)v + sw) - J(y)]$$

$$= \lim_{\varepsilon \to 0} \frac{J(y + \varepsilon v) - J(y)}{\varepsilon} = \delta J(y; v),$$

by Definition 2.6 *provided that this variation exists*. Similarly,

$$\mathcal{J}_s(r, s) = \delta J(y, w); \qquad \mathcal{G}_r(r, s) = \delta G(y; v); \qquad \mathcal{G}_s(r, s) = \delta G(y; w),$$

provided that these variations exist.

Evaluating at $(r, s) = (0, 0)$ where $y = y_0$, gives us the following:

(5.14) **Proposition.** *In a normed linear space* $(\mathcal{Y}, \|\cdot\|)$, *if real valued functions J and G are defined in a neighborhood of y_0 and have there in any pair of (fixed) directions v, w, Gâteaux variations which are continuous in this neighborhood and satisfy the Jacobian condition*

$$\begin{vmatrix} \delta J(y_0; v) & \delta J(y_0; w) \\ \delta G(y_0; v) & \delta G(y_0; w) \end{vmatrix} \neq 0,$$

then J cannot have a local extremal point at y_0 (even) when constrained to G_{y_0}, the level set of G through y_0.

Remark. The hypotheses of Proposition 5.14 also imply that G cannot have a local extremal point at y_0 (even) when constrained to J_{y_0}, the level set of J through y_0.

PROOF. Since the nonvanishing determinant of the hypothesis is precisely the Jacobian determinant $\partial(\mathcal{J}, \mathcal{G})/\partial(r, s)$ evaluated at $r = s = 0$, we can apply the inverse function theorem (5.13) to the vector valued function $F = (\mathcal{J}, \mathcal{G})$ provided that it has continuous partial derivatives in a neighborhood of $X_0 = (0, 0)$.

It suffices to establish the continuity of, say,

$$\mathcal{J}_r(r, s) = \delta J(y_0 + rv + sw; v),$$

for *fixed* v, w, in a neighborhood of $(0, 0)$.

But if r_1, s_1 are such that $y_1 = y_0 + r_1 v + s_1 w$ is in $S_r(y_0)$, the neighborhood given by the hypothesis, then $y = y_0 + rv + sw$ is within any given τ_1 of y_1 if $|r - r_1|, |s - s_1| < \tau_1/[2(\|v\| + \|w\|)]$ since $\|y - y_1\| \leq |r - r_1| \|v\| + |s - s_1| \|w\|$; (Why?) And by the continuity of $\delta J(y; v)$ at y_1 we know that given $\varepsilon_1 > 0$, $\exists\, \tau_1 > 0$ such that $|\delta J(y; v) - \delta J(y_1; v)| < \varepsilon_1$ when $\|y - y_1\| < \tau_1$.   □

With this preparation, it is easy to give conditions necessary for a local extremal point in the presence of a constraint. We first recall Definition 5.10:

Definition. In a normed linear space $(\mathcal{Y}, \|\cdot\|)$ the Gâteaux variations $\delta J(y; v)$ of a real valued function J are said to be *weakly continuous* at $y_0 \in \mathcal{Y}$ provided that for each $v \in \mathcal{Y}$: $\delta J(y; v) \to \delta J(y_0; v)$ as $y \to y_0$.

(5.15) **Theorem** (Lagrange). *In a normed linear space* $(\mathcal{Y}, \|\cdot\|)$, *if real valued functions* J *and* G *are defined in a neighborhood of* y_0, *a local extremal point for* J *constrained to* G_{y_0}, *and have there weakly continuous Gâteaux variations, then either*

(a) $\delta G(y_0; w) \equiv 0,\ \forall\, w \in \mathcal{Y}$; *or*
(b) *there exists a constant* $\lambda \in \mathbb{R}$ *such that* $\delta J(y_0; v) = \lambda \delta G(y_0; v),\ \forall\, v \in \mathcal{Y}$.

PROOF. If (a) does not hold, then $\exists\, w \in \mathcal{Y}$ for which $\delta G(y_0; w) \neq 0$. With *this* w and any $v \in \mathcal{Y}$, by Proposition 5.14 we must have that the determinant

$$\begin{vmatrix} \delta J(y_0; v) & \delta J(y_0; w) \\ \delta G(y_0; v) & \delta G(y_0; w) \end{vmatrix} = 0.$$

Hence with $\lambda \stackrel{\text{def}}{=} \delta J(y_0; w)/\delta G(y_0; w)$, it follows that $\delta J(y_0; v) = \lambda \delta G(y_0; v)$, $\forall\, v \in \mathcal{Y}$ as was to be proven.   □

The parameter λ which appears in the conclusion of the theorem is called a Lagrangian multiplier, and in application the theorem is usually referred to as the *Method of Lagrangian Multipliers*. It is easier to apply the method than it is to understand it! However, the following geometrical interpretation provides some insight.

Utilizing the terminology of directional derivatives appropriate to $\mathbb{R}^d$, the Lagrange condition $\delta J(y_0; v) = \lambda \delta G(y_0; v)$ says simply that the directional derivatives of J are proportional to those of G at y_0. If we suppose that J and G are differentiable at y_0 as in the last section, then in the directions τ tangent to the level set G_{y_0} at y_0, we know that $\delta G(y_0; \tau) = 0$.

Hence $\delta J(y_0; \tau) = 0$ in these tangential directions, and this is what we should expect for the constrained extremum point. Moreover, unless $\lambda = 0$ (in which case $\delta J(y_0; \cdot) \equiv 0$), $\delta J(y_0; v)$ is zero only in those directions of tangency, i.e., the level set J_{y_0} of J (unconstrained to G_{y_0}) through y_0 has precisely the same directions of tangency and nontangency as does G_{y_0}. Thus in general, Lagrange's condition means that the level sets of J and G through y_0 share the same tangent hyperplane at y_0, or meet tangentially at y_0 as illustrated for $\mathbb{R}^3$ in Figure 5.7.

Recalling the linearity of the Gâteaux variation established in §2.4 and replacing λ by $-\lambda$ we can also write condition (b) in the form $\delta(J + \lambda G)(y_0; \cdot) \equiv 0$, which suggests consideration of the augmented function $J + \lambda G$ *without constraints*, as in §2.3.

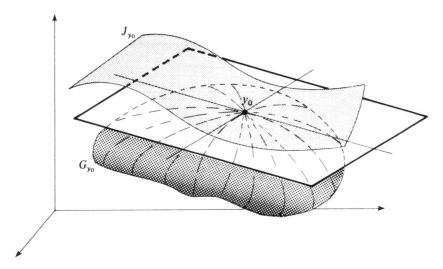

Figure 5.7

Example 1. We return now to the second problem of §5.5, Example 1, that of minimizing

$$J(y) = \int_0^1 [\sin^3 x + y(x)^2] \, dx$$

over

$$\mathcal{D} = \left\{ y \in \mathcal{Y} = C[0, 1]: \int_0^1 x[y(x)]^{4/3} \, dx = 1 \right\},$$

or that of minimizing J subject to the constraining relation

$$G(y) \equiv \int_0^1 x[y(x)]^{4/3} \, dx = 1.$$

We know from §2.4, Example 2, that $\delta J(y; v)$ exists for all y, $v \in \mathcal{Y}$ and is given by

$$\delta J(y; v) = 2 \int_0^1 y(x)v(x) \, dx.$$

Similarly, simple computation shows that $\delta G(y; v)$ exists for y, $v \in \mathcal{Y}$ and is given by

$$\delta G(y; v) = \frac{4}{3} \int_0^1 x[y(x)]^{1/3} v(x) \, dx.$$

In the maximum norm, both $\delta J(y; v)$ and $\delta G(y; v)$ are weakly continuous by Proposition 5.11. Thus by Theorem 5.15, a point $y_0 \in \mathcal{Y}$ which minimizes $J|_{G_{y_0}}$ must satisfy either

(a) $\delta G(y_0; w) = \frac{4}{3} \int_0^1 x y_0(x)^{1/3} w(x) \, dx \equiv 0$, $\forall \, w \in \mathcal{Y}$, [and this condition cannot hold since by Lemma 4.4 it would imply that $x y_0(x)^{1/3} \equiv 0$ or $y_0(x) \equiv 0$ on $[0, 1]$, while $y_0 = \mathcal{O}$ is not in $\mathcal{D}$]; or

(b) $\exists \lambda \in \mathbb{R}$ such that

$$\delta(J + \lambda G)(y_0; v) = \int_0^1 \left[2y_0(x) + \frac{4}{3}\lambda x y_0(x)^{1/3} \right] v(x) \, dx \equiv 0, \qquad \forall \, v \in \mathcal{Y}.$$

(11)

By Lemma 4.4, condition (11) implies that $2y_0(x) + \frac{4}{3}\lambda x y_0(x)^{1/3} \equiv 0$ on $[0, 1]$, and so by continuity either $y_0 \equiv 0$ (which we reject as before) or $y_0(x) = \pm(-\frac{2}{3}\lambda x)^{3/2}$. The constant $-\lambda > 0$ must be chosen to make $\int_0^1 x y_0(x)^{4/3} = 1$ which requires the value $\lambda = -3$. Thus the only possible minimizing (or maximizing) functions for $J|_{G_{y_0}}$ are $y_0(x) = \pm(2x)^{3/2}$. $(J - 3G$ is not convex (Why?) and we cannot use Theorem 3.16 to further characterize y_0.)

It is straightforward, albeit rather technical, to extend the method of Lagrangian multipliers so that it is applicable to a problem involving any finite number of constraints.

(5.16) **Theorem.** *In a normed linear space $(\mathcal{Y}, \|\cdot\|)$, let real valued functions J, $G_1, G_2, \ldots, G_N$ be defined in a neighborhood of y_0, a local extremal point for J constrained to $G_{y_0} \overset{\text{def}}{=} \{y \in \mathcal{Y}: G_i(y) = G_i(y_0), i = 1, 2, \ldots, N\}$, and have there weakly continuous Gâteaux variations.*

Then either:

(a)
$$\begin{vmatrix} \delta G_1(y_0; v_1) & \delta G_1(y_0; v_2) & \cdots & \delta G_1(y_0; v_N) \\ \delta G_2(y_0; v_1) & \delta G_2(y_0; v_2) & \cdots & \delta G_2(y_0; v_N) \\ \vdots & & & \vdots \\ \delta G_N(y_0; v_1) & \delta G_N(y_0; v_2) & \cdots & \delta G_N(y_0; v_N) \end{vmatrix} = 0,$$

$\forall \, v_j \in \mathcal{Y}, j = 1, 2, \ldots, N;$

or

(b) *there exist constants $\lambda_i \in \mathbb{R}, i = 1, 2, \ldots, N$ for which*

$$\delta J(y_0; v) = \sum_{i=1}^N \lambda_i \delta G_i(y_0; v), \qquad \forall \, v \in \mathcal{Y}.$$

PROOF. If condition (a) does *not* hold for one set of directions $v_1, v_2, \ldots, v_N \in \mathcal{Y}$, suppose $\exists$ *one* direction $v \in \mathcal{Y}$ for which the $(N + 1) \times (N + 1)$ determinant

$$\begin{vmatrix} \delta J(y_0; v) & \delta J(y_0; v_1) & \cdots & \delta J(y_0; v_N) \\ \hline \delta G_1(y_0; v) & & & \\ \vdots & & \delta G_i(y_0; v_j) & \\ & & i, j = 1, 2, \ldots, N & \\ \delta G_N(y_0; v) & & & \end{vmatrix}$$

(12)

(having the determinant of condition (a) in its lower right corner) is non-vanishing. Then the inverse function theorem in $\mathbb{R}^{N+1}$ can be used as before to find $(N + 1)$-tuples of scalars $(\bar{r}, \bar{s}_1, \bar{s}_2, \ldots, \bar{s}_N)$ and $(\underline{r}, \underline{s}_1, \underline{s}_2, \ldots, \underline{s}_N)$ as near $(0, 0, \ldots, 0)$ as we wish for which the points

$$\bar{y} = y_0 + \bar{r}v + \sum_{j=1}^{N} \bar{s}_j v_j,$$

$$\underline{y} = y_0 + \underline{r}v + \sum_{j=1}^{N} \underline{s}_j v_j,$$

satisfy the conditions

$$J(\underline{y}) < J(y_0) < J(\bar{y}),$$

and

$$G_i(\underline{y}) = G_i(y_0) = G_i(\bar{y}); \qquad i = 1, 2, \ldots, N.$$

We thereby *exclude* a local extremum for J constrained to G_{y_0}, contradicting the hypothesis.

Thus for the specific set of directions $v_1, v_2, \ldots, v_N$ the determinant (12) must vanish for each $v \in \mathcal{Y}$, and if we expand it by minors of the *first column*, we have upon dividing by the *cofactor* of $\delta J(y_0; v)$ (see [N]) an equation equivalent to condition (b), viz.,

$$\delta J(y_0; v) - \sum_{i=1}^{N} \lambda_i \delta G_i(y_0; v) = 0, \qquad \forall\, v \in \mathcal{Y}$$

where for each $i = 1, 2, \ldots, N$, the constant

$$\lambda_i = -\frac{\text{cofactor of } \delta G_i(y_0; v)}{\text{cofactor of } \delta J(y_0; v)},$$

is defined since the denominator is precisely the nonvanishing determinant

$$\left| \begin{matrix} \delta G_i(y_0; v_j) \\ i, j = 1, 2, \ldots, N \end{matrix} \right|. \qquad \qquad \square$$

(3.6) **Remarks.** Condition (a) holds if the constraining relations are locally *linearly dependent* in that there exist constants μ_i, $i = 1, 2, \ldots, N$ not all zero, for which $\sum_{i=1}^{N} \mu_i G_i(y) = 0$, $\forall\, y$ near y_0. Indeed, from the linearity of the Gâteaux variation (see §2.4) it would follow that $\sum_{i=1}^{N} \mu_i \delta G_i(y_0; v) = 0$ for each direction $v \in \mathcal{Y}$. Thus for each set of directions $v_1, v_2, \ldots, v_N \in \mathcal{Y}$, the rows of the determinant of condition (a) are linearly *dependent* and so it must vanish.

Conversely, if condition (a) is satisfied for any set of directions $v_1, v_2, \ldots, v_N \in \mathcal{Y}$, then in general the rows (and columns) of the determinant are linearly dependent. Indeed, upon expanding it by the minors of the *first column* as in proof of Theorem 5.16, we would have that $\sum_{i=1}^{N} \mu_i \delta G_i(y_0; v_j)$ must always vanish for $j = 1, 2, \ldots, N$, since this represents the expansion of a determinant having two identical columns. Thus the rows of the determinant are linearly dependent (unless $\mu_i = 0$ for $i = 1, 2, \ldots, N$).

Similarly, Lagrange's condition (b) implies that the variations $\delta J(y_0; v)$, $\delta G_1(y_0; v), \ldots, \delta G_N(y_0; v)$ are linearly dependent for each $v \in \mathcal{Y}$. Utilizing the geometric language of $\mathbb{R}^d$, we see that when all functions $J, G_i, i = 1, 2, \ldots, N$ are differentiable at y_0 as in §5.6 and $\tau \in \mathcal{Y}$ is a direction simultaneously tangent to each level set $G_{iy_0} i = 1, 2, \ldots, N$, then it must also be tangential to J_{y_0}, the level set of J *unconstrained* by the G_i at y_0, and thus $\delta J(y_0; \tau) = 0$ for all such directions as we should expect.

However, when $N > 1$, it is possible that $\delta J(y_0; v) = 0$ for directions v which are *not* tangential to any of the level sets G_{iy_0} at y_0. Thus we *cannot* assert that Lagrange's condition implies common tangency of the level sets at y_0, as was the case for $N = 1$.

Observe that upon replacement of λ_i by $-\lambda_i$ condition (b) can also be restated in the form $\delta(J + \sum_{i=1}^N \lambda_i G_i)(y_0; \cdot) \equiv 0$ which suggests consideration of the augmented function $J + \sum_{i=1}^N \lambda_i G_i$ without constraints, again as in §2.3.

(5.17) **Remark.** If $y_0 \in \mathcal{D} \subseteq \mathcal{Y}$, and the $\mathcal{D}$-admissible directions for J at y_0 form a linear *subspace* $\mathcal{Y}_0 \subseteq \mathcal{Y}$ (so that $v, w \in \mathcal{Y}_0 \Rightarrow rv + sw \in \mathcal{Y}_0 \ \forall$ scalars $r, s \in \mathbb{R}$), then it is seen that all arguments used in proving Theorem 5.16 remain valid when the weak continuity and directions $v, v_j, j = 1, 2, \ldots, N$ are restricted to $\mathcal{Y}_0$. Hence the conclusions of the theorem hold *for these restricted directions* and provide the conditions characterizing y_0 as a local extremal point of $J|_{\mathcal{D}}$ when further constrained to G_{y_0}. This observation will lead to a more efficient but admittedly hybrid approach to certain problems involving multiple constraints:

Those constraints on J which determine a domain $\mathcal{D}$ having a linear subspace $\mathcal{Y}_0$ of $\mathcal{D}$-admissible directions usable in the sense of §5.5, may be taken into account simply by restricting the supply of directions used when applying the method of Lagrangian multipliers to the remaining constraint(s).

Example 2*. To find the possible (local) extremal points for

$$F(y) = \int_{-1}^0 y'(x)^3 \, dx,$$

on

$$\mathcal{D} = \left\{ y \in \mathcal{Y} = C^1[-1, 0]: y(-1) = 0, \ y(0) = \frac{2}{3} \right\}$$

under the constraining relation

$$G(y) \stackrel{\text{def}}{=} \int_{-1}^0 xy'(x) \, dx = -\frac{4}{15}, \tag{13}$$

we may either characterize $\mathcal{D}$ by means of the two additional constraining relations $G_1(y) \stackrel{\text{def}}{=} y(-1) = 0, G_2(y) \stackrel{\text{def}}{=} y(0) = \frac{2}{3}$ and apply Theorem 5.16 with

$N = 3$; or, we may utilize Remark 5.17, since clearly the $\mathscr{D}$-admissible directions for F at any point $y \in \mathscr{D}$ are precisely those in

$$\mathscr{Y}_0 = \{v \in C^1[-1, 0]: v(-1) = v(0) = 0\},$$

which forms a *subspace* of $\mathscr{Y}$.

We now invoke Theorem 5.16 for these restricted directions. Here

$$\delta F(y; v) = \int_{-1}^{0} 3y'(x)^2 v'(x)\, dx,$$

$$\delta G(y; v) = \int_{-1}^{0} x v'(x)\, dx,$$

and these variations are weakly continuous everywhere by Proposition 5.11.

Thus at a local extremal point, y, either:

(a) $\delta G(y; v) = \int_{-1}^{0} x v'(x)\, dx = 0$, $\forall\, v \in \mathscr{Y}_0$ [which would imply by Lemma 4.1 that the continuous function $h(x) = x$ is constant on $[-1, 0]$ and this is false]; or
(b) $\exists\, \lambda$ such that $\delta(F + \lambda G)(y; v) = \int_{-1}^{0} [3y'(x)^2 + \lambda x]v'(x)\, dx = 0$, $\forall\, v \in \mathscr{Y}_0$.

Again by Lemma 4.1, we conclude that $3y'(x)^2 + \lambda x$ is constant on $[-1, 0]$, or, upon replacing λ by -3λ, we have for an appropriate c that $y'(x)^2 = c + \lambda x\,(\geq 0$ on $(-1, 0))$. Thus on $(-1, 0)$, either $y'(x) = -\sqrt{c + \lambda x}$ which cannot satisfy (13) (Why?); or[1]

$$y'(x) = \sqrt{c + \lambda x}. \tag{14}$$

Similarly, the possibility that $\lambda = 0$ in (14) requires that $y'(x) = \sqrt{c}$, and to satisfy (13) we must take $\sqrt{c} = \frac{8}{15}$, but then $y(x) = \frac{8}{15}x + $ const. cannot be in $\mathscr{D}$. (Why?).

When $\lambda \neq 0$, integration of (14) gives for some constant c_1:

$$y(x) = \frac{2}{3\lambda}(c + \lambda x)^\alpha + c_1, \tag{15}$$

where $\alpha = \frac{3}{2}$. Now

$$y(0) = \frac{2}{3} \Rightarrow c_1 = \frac{2}{3} - \frac{2c^\alpha}{3\lambda},$$

while

$$y(-1) = 0 \Rightarrow c_1 = -\frac{2}{3\lambda}(c - \lambda)^\alpha. \tag{16}$$

Upon subtracting and simplifying, we obtain

$$\lambda = c^\alpha - (c - \lambda)^\alpha. \tag{17}$$

---

[1] y' cannot change sign at a point $x_0 \in (-1, 0)$ since $y'(x_0)^2 = c + \lambda x_0$ cannot be zero.

Next, to satisfy the constraining relation (13) we require that

$$-\frac{4}{15} = G(y) = \int_{-1}^{0} x\sqrt{c + \lambda x}\, dx$$

$$= \frac{2}{3\lambda}\left[x(c + \lambda x)^{\alpha}\Big|_{-1}^{0} - \int_{-1}^{0}(c + \lambda x)^{3/2}\, dx \right]$$

$$= \frac{2}{3\lambda}\left[(c - \lambda)^{\alpha} - \frac{2}{5\lambda}(c^{5/2} - (c - \lambda)^{5/2}) \right],$$

or

$$\lambda^2 = \frac{-5\lambda}{2}(c - \lambda)^{\alpha} + c^{\alpha+1} - (c - \lambda)^{\alpha+1}. \tag{18}$$

We need to solve the nonlinear system, (17) and (18), for $c \geq 0$ and $\lambda \leq c$ with $\lambda \neq 0$. *By inspection*, $\lambda = c = 1$ constitutes one such solution, and hence from (15), (16):

$$y_0(x) = \tfrac{2}{3}(x + 1)^{3/2} \text{ provides a possible local extremal}$$
$$\text{point in } \mathscr{D} \text{ under (13).}$$

[In fact, there is no other admissible solution to this system.[1] To establish this, we use (17) to replace $(c - \lambda)^{\alpha}$ in (18). We get

$$\lambda^2 = \frac{5\lambda}{2}(\lambda - c^{\alpha}) + c^{\alpha+1} + (c - \lambda)(\lambda - c^{\alpha}),$$

or

$$(\tfrac{1}{2})\lambda^2 = (\tfrac{3}{2})\lambda c^{\alpha} - \lambda c, \quad \text{and since } \lambda \neq 0:$$

$$\lambda = 3c^{\alpha} - 2c, \quad \text{so that with } \alpha = \tfrac{3}{2},$$

$$c - \lambda = 3(c - c^{\alpha}) = 3c(1 - \sqrt{c}) \geq 0. \tag{19}$$

Thus $0 \leq c \leq 1$; but $c = 1$ in (19) leads to the case $\lambda = c = 1$ already considered, while $c = 0$, gives $\lambda = c = 0 = y'(x)$ which violates the constraining relation (13).

Upon substitution of (19) into (17), we obtain for $0 < c < 1$:

$$3c^{\alpha} - 2c = c^{\alpha} - (3c)^{\alpha}(1 - \sqrt{c})^{\alpha} \quad \text{or} \quad 2c(\sqrt{c} - 1) = -(3c)^{\alpha}(1 - \sqrt{c})^{\alpha},$$

and with $\alpha = \tfrac{3}{2}$:

$$2 = 3^{\alpha}\sqrt{c}(1 - \sqrt{c})^{1/2},$$

so that

$$4 = 27c(1 - \sqrt{c}),$$

or finally

$$c - \tfrac{4}{27} = c^{3/2}.$$

---

[1] This also follows from convexity. See Problem 5.41(e).

With $\sqrt{c} = t$, this last equation becomes $t^3 - t^2 + \frac{4}{27} = 0$ which factors into $(t - \frac{2}{3})^2(t + \frac{1}{3}) = 0$. The only feasible solution is $\sqrt{c} = t = \frac{2}{3}$ or $c = \frac{4}{9}$. But then from (19), $\lambda = 3(\frac{8}{27}) - (\frac{8}{9}) = 0$, a possibility already excluded.]

We conclude that $y_0(x) = \frac{2}{3}(x + 1)^{3/2}$ is the only possible local extremal point for our problem, but we still do not know if it *is* a local extremal point. However, now convexity can be used to show that y_0 *cannot* be a local maximum point with respect to the maximum norm $\|\cdot\|_M$ of §5.1, Example 3. (See Problem 5.41*.)

The foregoing analysis involves technical complications which, unfortunately, typify the difficulties encountered when applying the method of Lagrangian multipliers. However, had we attempted a solution using the additional constraining functions, G_1 and G_2, we would have been burdened at the outset with two more unknowns, namely the Lagrangian multipliers λ_1, λ_2 associated with these functions.

(Problems 5.37–5.41)

PROBLEMS

5.1. Reverse Triangle Inequality. Show that if $(\mathcal{Y}, \|\cdot\|)$ is a normed linear space, then

$$\left| \|y\| - \|\tilde{y}\| \right| \leq \|y - \tilde{y}\|, \qquad \forall\, y, \tilde{y} \in \mathcal{Y}.$$

5.2. Let $\mathcal{Y} = \mathbb{R}^d$ with the Euclidean norm

$$\|y\| = \left(\sum_{j=1}^{d} y_j^2 \right)^{1/2}, \quad \text{for } y = (y_1, y_2, \ldots, y_d).$$

(a) Cauchy Inequality. Verify that

$$\left| \sum_{j=1}^{d} y_j \tilde{y}_j \right| \leq \|y\| \cdot \|\tilde{y}\|, \qquad \forall\, y, \tilde{y} \in \mathcal{Y}.$$

(Hint: See Problem 0.1.)
(b) Use the result of part (a) to establish the triangle inequality (1c).
(c) Show also for this norm that

$$\|y + \tilde{y}\|^2 + \|y - \tilde{y}\|^2 = 2(\|y\|^2 + \|\tilde{y}\|^2), \qquad \forall\, y, \tilde{y} \in \mathbb{R}^d.$$

Parts (a) and (c) do *not* hold for all norms on $\mathcal{Y}$.
(d) Verify that $\|y\|_M = \max_{j=1, 2, \ldots, d} |y_j|$ is a norm for $\mathbb{R}^d$, which does *not* have the properties in (a) and (c) above when $d \geq 2$.

5.3. Let $\mathcal{Y} = C[a, b]$.
(a) Verify that $\|y\|_1 = \int_a^b |y(x)|\, dx$ defines a norm for $\mathcal{Y}$.
(b) Does $\|y\| = |\int_a^b y(x)\, dx|$ define a norm for $\mathcal{Y}$?

5.4. Show that $\|y\| = \max_{x \in [a,b]} |y'(x)|$ defines a norm for the linear space $\mathcal{Y}_0 = \{y \in C^1[a, b] : \int_a^b y(x)\, dx = 0\}$, but does not define a norm for $\mathcal{Y} = C^1[a, b]$.

5.5.   (a) Verify that $\|y\| = |y(a)| + \max_{x \in [a,b]} |y'(x)|$ defines a norm for $\mathscr{Y} = C^1[a, b]$.

(b) Show that $\max_{x \in [a,b]} |y(x)| \leq (1 + b - a)\|y\|, \forall y \in \mathscr{Y}$. Hint:

$$y(x) = y(a) + \int_a^x y'(t)\, dt.$$

5.6.   (a) Verify that each of the functions, $\|Y\|_M$ and $\|Y\|_1$, given in §5.1, Example 6, gives a norm for $\mathscr{Y} = (C[a, b])^d$.

(b) Show that the remaining functions in the example also give norms for $\mathscr{Y}$.

5.7.   Suppose that both $\|\cdot\|_1$ and $\|\cdot\|_2$ are norms for the linear space $\mathscr{Y}$.

(a) Show that $\|y\| = \|y\|_1 + \|y\|_2$ defines a norm on $\mathscr{Y}$.

(b) Does $\|y\| = \|y\|_1 \cdot \|y\|_2$ also define a norm for $\mathscr{Y}$?

5.8.   (a) Verify the assertion of §5.1, Example 4.

(b) When $(\mathscr{Y}_j, \|\cdot\|_j)$ are each normed linear spaces for $j = 1, 2$, find a corresponding norm for the linear space $\mathscr{Y}_1 \times \mathscr{Y}_2$.

5.9.   With $\mathscr{Y} = C[0, 1]$ and $\{y_n\} = \{(x/2)^n\}$,

(a) Show that $y_n \to \mathcal{O}$ as $n \to \infty$, using $\|y\|_1 = \int_0^1 |y(x)|\, dx$.

(b) Show also that $y_n \to \mathcal{O}$ as $n \to \infty$, using $\|y\|_M = \max_{x \in [0, 1]} |y(x)|$.

5.10. Let $\mathscr{Y} = C[0, 1]$ and $\{y_n\} = \{x^n\}$; i.e., $y_n(x) = x^n$, $n = 1, 2, \ldots$.

(a) Establish that $y_n \to \mathcal{O}$ as $n \to \infty$, using $\|\cdot\|_1$, but

(b) $y_n \nrightarrow \mathcal{O}$ using $\|\cdot\|_M$, where $\|\cdot\|_1$ and $\|\cdot\|_M$ are as in Problem 5.9.

(Note: This shows that a sequence from $\mathscr{Y}$ may converge to $y_0 \in \mathscr{Y}$ with respect to one norm, but not with respect to another.)

5.11. Let $(\mathscr{Y}, \|\cdot\|)$ be a normed linear space, and $\{y_n\}$, $\{\tilde{y}_n\}$ be sequences from $\mathscr{Y}$. Show that if $y_n \to y_0$ and $\tilde{y}_n \to \tilde{y}_0$ as $n \to \infty$, then $(y_n + \tilde{y}_n) \to (y_0 + \tilde{y}_0)$ as $n \to \infty$.

5.12. Suppose that $(\mathscr{Y}, \|\cdot\|)$ is a normed linear space, and let $\{y_n\}$ be a sequence from $\mathscr{Y}$.

(a) Show that if $y_n \to y_0$ as $n \to \infty$, then $\|y_n\| \to \|y_0\|$ as $n \to \infty$.

(b) Give an example to illustrate that the converse of (a) is false.

5.13. Use Definition 5.1 to prove that in a normed linear space $(\mathscr{Y}, \|\cdot\|)$, a real valued function J is continuous at $y_0 \in \mathscr{Y}$ iff for each sequence $\{y_n\}$ from $\mathscr{Y}$,

$$\lim_{n \to \infty} y_n = y_0 \Rightarrow \lim_{n \to \infty} J(y_n) = J(y_0).$$

5.14. Let $\mathscr{Y} = C[a, b]$ and use Definition 5.1 to establish that $J(y) = \int_a^b (\sin x) y(x)\, dx$ is continuous on $\mathscr{Y}$ using:

(a) $\|y\|_M = \max_{x \in [a,b]} |y(x)|$.

(b) $\|y\|_1 = \int_a^b |y(x)|\, dx$.

Make a similar analysis for $F(y) = \int_a^b \sin(y(x))\, dx$. Hint: Use a mean value inequality.

5.15. Let $(\mathscr{Y}, \|\cdot\|)$ be a normed linear space and L be a real valued linear function on $\mathscr{Y}$ (i.e., $L(cy + \tilde{c}\tilde{y}) = cL(y) + \tilde{c}L(\tilde{y})$, $\forall y, \tilde{y} \in \mathscr{Y}$ and $\forall c, \tilde{c} \in \mathbb{R}$). Prove that L is continuous on $\mathscr{Y}$ iff there exists a constant A such that $|L(y)| \leq A\|y\|$, $\forall y \in \mathscr{Y}$.

5.16. Suppose that $\|\cdot\|_1$ and $\|\cdot\|_2$ are both norms for the linear space $\mathcal{Y}$ and there is a constant A such that $\|y\|_1 \le A\|y\|_2, \forall\, y \in \mathcal{Y}$.
 (a) Show that if $y_n \to y_0$ as $n \to \infty$ using $\|\cdot\|_2$, then also $y_n \to y_0$ using $\|\cdot\|_1$.
 (b) Prove that if a real valued function J on $\mathcal{Y}$ is continuous with respect to $\|\cdot\|_1$, then it is also continuous with respect to $\|\cdot\|_2$.

5.17. Let $(\mathcal{Y}, \|\cdot\|)$ be a normed linear space.
 (a) Show that if K is a compact subset of $\mathcal{Y}$, then K is bounded, i.e., there is a constant k such that $\|y\| \le k, \forall\, y \in K$.
 (b) Conclude that if $\mathcal{Y} \neq \{\mathcal{O}\}$, then $\mathcal{Y}$ itself cannot be compact.
 Let $\mathcal{Y} = C[a, b]$ and $K = \{y \in \mathcal{Y}: \int_a^b y(x)\, dx = 1\}$. Is K compact if we use:
 (c) $\|y\|_1 = \int_a^b |y(x)|\, dx$?
 (d) $\|y\|_M = \max_{x \in [a,b]} |y(x)|$?

5.18. Let $(\mathcal{Y}, \|\cdot\|)$ be a normed linear space and J, G, be real valued functions on $\mathcal{Y}$ which are continuous at $y_0 \in \mathcal{Y}$. Prove that for $c \in \mathbb{R}$, the following functions are also continuous at y_0:
 (a) cJ;   (b) $J + G$;   (c) JG.
 Hint for JG: $ab - a_0 b_0 = (a - a_0)(b - b_0) + (a - a_0)b_0 + a_0(b - b_0)$.

5.19. Verify that $J(y) = \int_0^1 |y(x)|\, dx$ does not achieve a minimum value on

$$\mathcal{D} = \{y \in C[0, 1]: y(0) = 0, y(1) = 1\},$$

although J is bounded below (i.e., $J(y) \ge 0$) on $\mathcal{D}$. Does Proposition 5.3 cover this?

5.20*. Let $\mathcal{Y} = C[0, 1]$ and $J(y) = 2y(0)^3 - 3y(0)^2$.
 (a) Prove that $y_0(x) \equiv 1$ is a local minimum point for J on $\mathcal{Y}$ using $\|y\|_M = \max_{x \in [0, 1]} |y(x)|$. (Hint: Show that $y \in S_1(y_0) \Rightarrow J(y) \ge -1 = J(y_0)$. Consider minimizing the cubic polynomial $p(t) = 2t^3 - 3t^2$ on $\mathbb{R}$.)
 (b) Prove that $y_0(x) \equiv 1$ is *not* a local minimum point for J on $\mathcal{Y}$ using $\|y\|_1 = \int_0^1 |y(x)|\, dx$. (Hint: Consider the *continuous* function

$$y_\varepsilon(x) = \begin{cases} -1 + 2x/\varepsilon, & 0 \le x \le \varepsilon, \\ 1, & \varepsilon < x \le 1, \end{cases}$$

for each fixed $\varepsilon > 0$ and show that $\|y_\varepsilon - y_0\|_1$ can be made as small as we please by choosing ε small, while $J(y_\varepsilon) = -5 < J(y_0), \forall\, \varepsilon > 0$.)

5.21. For Example 2 of §5.5, discuss what happens if α vanishes identically on a subinterval of $[a, b]$.

5.22. Let $\mathcal{Y} = C[a, b], J(y) = \int_a^b [\sin^3 x + y(x)^2]\, dx$, and $\mathcal{D} = \{y \in \mathcal{Y}: \int_a^b y(x)\, dx = 1\}$.
 (a) What are the $\mathcal{D}$-admissible directions for J?
 (b) Find all possible (local) extremal points for J on $\mathcal{D}$. (See Problem 4.2.)
 (c)* Prove directly that J is differentiable at each $y_0 \in \mathcal{Y}$. (See §5.6, Example 3.)

5.23. Let $\mathcal{Y} = C^1[a, b]$, $\mathcal{D} = \{y \in \mathcal{Y}: y(a) = a_1, y(b) = b_1\}$, and $J(y) = \int_a^b f(x, y'(x))\, dx$, where $f(x, z)$ and $f_z(x, z)$ are continuous on $[a, b] \times \mathbb{R}$.
 (a) What are the $\mathcal{D}$-admissible directions for J?
 (b) Show that if y is a (local) extremal point for J on $\mathcal{D}$, then $f_{y'}(x) \overset{\text{def}}{=} f_z(x, y'(x)) = \text{const.}$ on $[a, b]$.

5.24.   Let $\mathcal{Y} = C[a, b]$, $\mathcal{D} = \{y \in \mathcal{Y}: y(a) = a_1, y(b) = b_1\}$, and $J(y) = \int_a^b f(x, y(x))\, dx$, where $f(x, y)$ and $f_y(x, y)$ are continuous on $[a, b] \times \mathbb{R}$.
   (a)   What are the $\mathcal{D}$-admissible directions for J on $\mathcal{D}$?
   (b)   Show that if $y \in \mathcal{D}$ is a (local) extremal point for J on $\mathcal{D}$, then $f_y(x) \overset{\text{def}}{=} f_y(x, y(x)) = 0$ on $[a, b]$.
   (c)* Prove that the variations $\delta J(y; v)$ are weakly continuous in the maximum norm. Hint: See the proof of Proposition 5.11.
   (d)   Conclude that if $\alpha \in C[a, b]$, and $J(y) \overset{\text{def}}{=} \int_a^b \alpha(x)e^{y(x)}\, dx$, then J cannot have a (local) extremum on $\mathcal{D}$ for any values of a_1, b_1, unless $\alpha = 0$.

In Problems 5.25–5.31 find all possible (local) extremal points for J (a) on $\mathcal{D}$; (b) on $\mathcal{D}_1$.

5.25. $J(y) = y'(0)^2 - y'(0)^3$,
   (a) $\mathcal{D} = C^1[0, 1]$.
   (b) $\mathcal{D}_1 = \{y \in C^1[0, 1]: y'(0) = y(1) = 0\}$.

5.26. $J(y) = \int_0^1 \cos y(x)\, dx$,
   (a) $\mathcal{D} = C[0, 1]$.
   (b) $\mathcal{D}_1 = \{y \in C[0, 1]: y(0) = y(1) = \pi\}$.

5.27. $J(y) = \int_1^2 x^{-1} y'(x)^3\, dx$,
   $\mathcal{D} = \{y \in C^1[1, 2]: y(1) = 1, y(2) = 8\}$.

5.28. $J(y) = \int_0^1 [y(x)^2 + 2xy(x)]\, dx$,
   $\mathcal{D} = \{y \in C[0, 1]: y(0) = 0, y(1) = -1\}$.

5.29. $J(y) = \int_0^{\pi/6} (\sec^2 x) y'(x)^3\, dx$,
   $\mathcal{D} = \{y \in C^1[0, 1]: y(0) = 1, y(\pi/6) = \frac{3}{2}\}$.

5.30. $J(y) = \int_0^1 [y(x)^3 + e^x y(x)]\, dx$,
   (a) $\mathcal{D} = C[0, 1]$.
   (b) $\mathcal{D}_1 = \{y \in C[0, 1]: y(0) = y(1) = 0\}$.

5.31. $J(y) = \int_1^2 [xy'(x) - e^{y'(x)}]\, dx$,
   $\mathcal{D} = \{y \in C^1[1, 2]: y(1) = -1, y(2) = 2(\ln 2 - 1)\}$.

5.32*. In Example 2 of §5.5, let $\alpha \in C[a, b]$ with $\alpha \geq \alpha_0 > 0$ on $[a, b]$.
   (a)   Show that there exists a $\delta_0 > 0$ such that if $0 < b_1 - a_1 < \delta_0$, then there is precisely one $c \in (0, \alpha_0)$ for which (5) is satisfied (and hence precisely one $y \in \mathcal{D}$ which satisfies (3)).
   (b)   What happens if $a_1 = b_1$?

5.33.   Suppose that $(\mathcal{Y}, \|\cdot\|)$ is a normed linear space for which $L: \mathcal{Y} \to \mathbb{R}$ is continuous and linear (i.e., $L(cy + \tilde{c}\tilde{y}) = cL(y) + \tilde{c}L(\tilde{y})$, $\forall y, \tilde{y} \in \mathcal{Y}$, and $c, \tilde{c} \in \mathbb{R}$). Show that L is Fréchet differentiable at each $y_0 \in \mathcal{Y}$:
   (a)   by using Definition 5.6; and
   (b)   by using Theorem 5.9.
   (c)   If $L \not\equiv 0$, prove that $\exists v_1 \in \mathcal{Y}$ with $L(v_1) = 1$, and thus $L(\tau) = 0$ when $\tau = y - L(y)v_1$, if $y \in \mathcal{Y}$.
   (d)   In Definition 5.12, take $L = J'(y_0)$ and conclude that "most" directions are tangential.

5.34. If $(\mathcal{Y}, \|\cdot\|)$ is a normed linear space and $J: \mathcal{Y} \to \mathbb{R}$ has at each $y \in \mathcal{Y}$ Gâteaux variations which satisfy conditions (a) and (b) of Theorem 5.9, verify that $\delta J(y; v)$ is weakly continuous at y_0. Hint: Each $v \in \mathcal{Y}$ may be expressed as $v = \|v\| v_1$, with $\|v_1\| = 1$.

5.35*. In Example 3 of §5.6*, use the vector inequality of Problem 0.2, viz.,

$$\left| \frac{A}{|A|} - \frac{B}{|B|} \right| \le \frac{\sqrt{2}}{\sqrt{|A| \cdot |B|}} |A - B|, \qquad 0 \ne A, B, \in \mathbb{R}^n,$$

to establish the weak continuity of $\delta L(Y_0; V)$ at each $Y_0 \in \mathcal{Y}$.

5.36. Establish the linearity and continuity in v of the Gâteaux variations $\delta F(y; v)$ utilized in the first part of the proof of Proposition 5.11.

In Problems 5.37–5.39, use the method of Lagrangian multipliers to determine all *possible* (local) extremal points for J on $\mathcal{D}$.

5.37. $J(y) = \int_0^1 x^2 y(x)\, dx$,
       $\mathcal{D} = \{y \in C[0, 1]: \int_0^1 y(x)^5\, dx = 1\}$.

5.38*. $J(y) = \int_0^1 y(x)^3\, dx$,
       $\mathcal{D} = \{y \in C[0, 1]: \int_0^1 y(x)\, dx = \frac{2}{3}, \int_0^1 xy(x)\, dx = \frac{2}{5}\}$.

5.39. $J(y) = \int_0^1 y'(x)^{4/3}\, dx$,
       $\mathcal{D} = \{y \in C^1[0, 1]: y(0) = -5/4, y(1) = 5, \int_0^1 xy'(x)\, dx = 5\}$.

5.40*. Suppose $f = f(x, y, z)$ and its partial derivatives f_y and f_z are continuous only on $(a, b] \times D$ where D is a domain in $\mathbb{R}^2$. The *improper* integral $F(y) = \int_a^b f(x, y(x), y'(x))\, dx$ may still be finite for some functions y. For given values a_1, b_1, let

$$\mathcal{D}^* = \{y \in C[a, b]: y(a) = a_1, y(b) = b_1, \text{ with } y' \in C(a, b] \text{ and } F(y) \text{ finite.}\}$$

(a) Show that if $y \in \mathcal{D}^*$, then each v in

$$\mathcal{D}_0^* = \{v \in C^1[a, b]: v(b) = 0 \text{ and } v(x) \equiv 0 \text{ in a } neighborhood \text{ of } a\}$$

is $\mathcal{D}^*$-admissible at y and

$$\delta F(y; v_0) = \int_{x_0}^b [f_y(x)v_0(x) + f_{y'}(x)v_0'(x)]\, dx$$

when $v_0 \in \mathcal{D}_0^*$ and $v_0(x) \equiv 0$ on $[a, x_0]$. (See §1.5 for the notation.) This relaxation of conditions near an end point will be required for a careful analysis of the brachistochrone where

$$f(x, y, z) = \frac{\sqrt{1 + z^2}}{\sqrt{2gy}}$$

(See Example 4* in §5.5, and Problems 6.14*, 6.15*.)
(b) Formulate and prove a vector valued analogue of this result.

5.41*. (a) For Example 2 of §5.7, prove that for $\lambda \in \mathbb{R}$, the function $\tilde{f}(\underline{x}, z) = z^3 + \lambda \underline{x} z$ is strongly convex on $[-1, 0] \times [0, \infty)$.
       (b) Conclude that when $y \in \mathcal{D}$ and $y'(x) \ge 0$ on $[-1, 0]$ then for an appropriate λ, $\tilde{F}(y) > \tilde{F}(y_0)$, when $y_0(x) = \frac{2}{3}(x + 1)^{3/2}$ and $y \ne y_0$.

(c)   Draw a sketch to show that in each $\|\cdot\|_M$ neighborhood of y_0, $\exists\, y \in \mathscr{D}$ with $F(y) > F(y_0)$ and $G(y) = \int_{-1}^{0} xy(x)\, dx = G(y_0)$.

(d)   Can convexity be used to prove that y_0 is a local minimum point for this problem? Explain.

(e)   Use part (b) to conclude that system (17), (18) has at most one solution λ, c for which $\lambda x + c \geq 0$ on $[-1, 0]$. Hint: Each solution pair (λ, c) gives a $y_0 \in \mathscr{D}^* = \{ y \in \mathscr{D} : y' \geq 0 \}$ that minimizes $\tilde{F} = F - 3\lambda G$ on $\mathscr{D}^*$ uniquely!

(f)* Redo the problem of this Example when $-\frac{1}{3}$ replaces $-\frac{4}{15}$ in (13).

5.42. When D is a bounded domain in $\mathbb{R}^d$ (for $d \geq 2$) with a smooth boundary, verify *formally*, that $\|u\|_M \equiv \max_{X \in \bar{D}}(|u(X)| + |\nabla u(X)|)$ defines a norm for $\mathscr{Y} = C^1(\bar{D})$. ($\bar{D}$ is compact.) See §6.9.

5.43. Find all possible functions that maximize $J(y) = y^2(1)$ on $\mathscr{D} = \{ y \in C^1[0, 1] : y(0) = 0 \}$ under the constraint $G(y) \overset{\text{def}}{=} \int_0^1 y'(x)^2\, dx = 1$.

CHAPTER 6

The Euler–Lagrange Equations

Jakob Bernoulli's solution of 1696 to his brother Johann's problem of the brachistochrone (§1.2) marked the introduction of variational considerations. However, it was not until the work of Euler (c. 1742) and Lagrange (1755) that the systematic theory now known as the calculus of variations emerged. Initially, it was restricted to finding conditions which were *necessary* in order that an integral function

$$F(y) = \int_a^b f(x, y(x), y'(x)) \, dx = \int_a^b f[y(x)] \, dx$$

should have a (local) extremum on a set

$$\mathcal{D} \subseteq \{y \in C^1[a, b] : y(a) = a_1; y(b) = b_1\}.$$

For specified a_1, b_1 this is a *fixed end point* problem. However, it was already of interest to Jakob Bernoulli to seek (local) extrema for a larger set

$$\mathcal{D}^b \subseteq \{y \in C^1[a, b] : y(a) = a_1\},$$

in describing a modified brachistochrone for which it is desired to descend over a given horizontal distance $(b - a)$ in minimum time, without specifying the vertical distance to be covered (Figure 6.1(a)). This type of problem is said to have *one free end point*.

There are also problems with *two free end points* where local extrema on arbitrary subsets of $C^1[a, b]$ are desired.

A related problem with variable end point conditions is that of characterizing the brachistochrone joining fixed *curves* called *transversals* (Figure 6.1(b)) which would require minimizing the integral with variable limits,

$$F(y; x_1, x_2) = \int_{x_1}^{x_2} f(x, y(x), y'(x)) \, dx = \int_{x_1}^{x_2} f[y(x)] \, dx$$

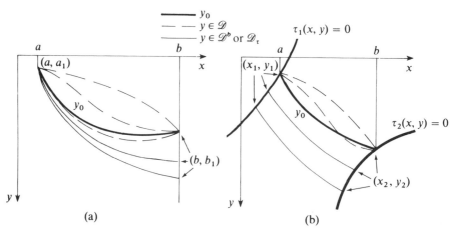

Figure 6.1

over a set

$$\mathcal{D}_\tau \subseteq \{y \in C^1[x_1, x_2]: \tau_j(x_j, y(x_j)) = 0; j = 1, 2\},$$

where $[x_1, x_2] \subseteq \mathbb{R}$, and the τ_j are given functions.

All of these problems admit a common variational approach: If $y_0 \in C^1[a, b]$ is, say, a (local) minimizing function for one of these problems, then with appropriate selection of a_1, b_1 and $\mathcal{D}$, we may suppose that $y_0 \in \mathcal{D}$, and, as required, that $\mathcal{D} \subseteq \mathcal{D}^b$, or $\mathcal{D} \subseteq \mathcal{D}_\tau$. (This possibility is illustrated in Figure 6.1.) In each case, y_0 is a (local) minimizing function for F on $\mathcal{D}$, the *fixed* end point problem considered initially. Consequently, from Proposition 5.5,

$$\delta F(y_0; v) = 0, \qquad \forall\, v \in \mathcal{D}_0 = \{v \in C^1[a, b]: v(a) = v(b) = 0\},$$

which are $\mathcal{D}$-admissible for F at y_0.

When f is sufficiently differentiable, there are enough such directions to infer that on (a, b), y_0 is a solution of the first and second equations of Euler–Lagrange. These equations, whose C^1 solutions are by definition stationary functions for f, are the subject of the initial sections of this chapter. The additional freedom of working in $\mathcal{D}^b$ or in $\mathcal{D}_\tau$ permits variation in other "directions" specifically related to the end point freedom, and this will give rise to the corresponding natural boundary conditions of §6.4 which the extremal function should satisfy. Problems involving isoperimetric constraints are considered in §6.5 through the device of Lagrangian multipliers and this approach is extended to cover Lagrangian constraints of a simple form.

In the concluding sections, we examine various extensions of these methods: first, to integrals involving derivatives higher than the first (§6.6), next, to integrals involving C^1 vector valued functions (§6.7), and finally (in §6.9) to integrals over higher-dimensional space. Invariance of stationarity with respect to change in coordinates is examined in §6.8.

Many of these results were obtained first by Lagrange (1738–1813) who began his investigations in the subject at age sixteen[1]; however, his successors have added mathematical rigor to the original discoveries.

In this chapter, only those conditions necessary for a local extremum are considered, and although the methods developed are applied to significant problems of classical interest (including that of the brachistochrone), the final disposition of such problems must await the discussion of sufficiency in Chapter 9. It should be noted, however, that the initial investigators in these fields, often regarded a function which satisfied the necessary conditions as the extremal function sought, and the practice continues today in elementary treatments of the subject.

Throughout this chapter, we shall supply the space $C^1[a, b]$ with the maximum norm $\|y\|_M = \max(|y(x)| + |y'(x)|)$ of §5.1, Example 3, and its vector valued counterpart $(C^1[a, b])^d$ with the corresponding norm

$$\|Y\|_M = \max(|Y(x)| + |Y'(x)|).$$

Other norms will be introduced as needed. However, for many of our considerations the particular norm in use is not significant.

§6.1. The First Equation: Stationary Functions

For simplicity, suppose initially that the function $f = f(x, y, z)$, together with its derivatives f_y and f_z, is continuous on $[a, b] \times \mathbb{R}^2$.

Then for each $y \in \mathscr{Y} = C^1[a, b]$:

$$F(y) = \int_a^b f(x, y(x), y'(x)) \, dx = \int_a^b f[y(x)] \, dx$$

is defined. From Example 4 of §2.4, F has in each direction v the Gâteaux variation

$$\delta F(y; v) = \int_a^b [f_y(x)v(x) + f_{y'}(x)v'(x)] \, dx, \tag{1}$$

where for the given $y \in \mathscr{Y}$, we use the compressed notation from §1.5:

$$f_y(x) \stackrel{\text{def}}{=} f_y[y(x)] \quad \text{and} \quad f_{y'}(x) \stackrel{\text{def}}{=} f_z[y(x)]. \tag{2}$$

(6.1) **Proposition.** *If* $y \in \mathscr{Y}$ *makes* $\delta F(y; v) = 0$, $\forall v$ *in* $\mathscr{D}_0 = \{v \in \mathscr{Y}: v(a) = v(b) = 0\}$, *then* $f_{y'}$ *is* C^1, *and*

$$\frac{d}{dx} f_{y'}(x) = f_y(x), \qquad x \in (a, b), \tag{3}$$

[1] The same age at which Euler obtained his master's degree!

so that

$$\delta F(y; v) = f_{y'}(x)v(x)\Big|_a^b, \qquad \forall\, v \in \mathcal{Y}. \tag{3'}$$

PROOF. The first assertions are a restatement of Proposition 4.2 for the continuous functions $g(x) = f_y(x)$ and $h(x) = f_{y'}(x)$. But then (3) permits the integrand of (1) to be recognized as $(d/dx)\,[f_{y'}(x)v(x)]$ and thus integrated to produce (3'). $\qquad\square$

(3) is the *first* differential equation of Euler (who obtained it heuristically in 1736 by varying the vertices of an imagined polygonal solution curve) and Lagrange (who obtained it in 1755 (incorrectly) by integrating the *second* term of (1) by parts. Why would this not be permissible?) The correct derivation as above was given (in 1879) by P. du Bois-Reymond. As we have seen, (3) must be satisfied along each curve which could provide a local extremal for F on $\mathcal{D}$ (or on $\mathcal{D}^b$ or $\mathcal{D}_\tau$) as in the introduction. Moreover, (3) is precisely the equation (8) already obtained in §3.2.

(6.2) **Definition.** Each C^1 function y which satisfies the differential equation (3) (i.e., for which $(d/dx)f_z(x, y(x), y'(x)) = f_y(x, y(x), y'(x))$) on some interval will be called a *stationary function* for f (of x, y, y').

(An old and rather entrenched tradition calls such functions *extremal* functions or simply extremals, although they may provide neither a local maximum nor a local minimum for the problem.) Observe that we do not require that a stationary function satisfy any particular boundary conditions, although in each problem, we might be interested only in those which meet given boundary conditions.

Now, as in §5.5, certain functions f with their derivatives f_y and f_z are defined only for a restricted class of functions y, (e.g., $y \geq 0$) so that variation of F at y can be performed only for a reduced class of v (e.g., those for which $|v(x)| \leq |y(x)|$). As the preceding discussion shows, when y is stationary and meets the restrictions, then $\delta F(y; v) = 0$, $\forall\, v \in \mathcal{D}_0$ for which the variation at y is defined. However, there may also be *nonstationary* functions η which make $\delta F(\eta; v) = 0$ for the reduced class of v, and these may provide the true extremals. (See Problem 6.13.)

(Problem 6.1)

§6.2. Special Cases of the First Equation

Although *every* C^1 function y is stationary for $f(x, y, z) = z$ or yz, in general, it is difficult to find *any* solutions for the first equation (3). However, when one or more of the variables of f is not present explicitly, then we can at least obtain a first integral of the differential equation. We shall analyze three such cases in this section.

(a) When $f = f(z)$

Then $f_y \equiv 0$ so that (3) becomes $(d/dx)f_{y'}(x) = 0$ or $f_{y'}(x) = $ const. Thus $f_z(y'(x)) = $ const. and the stationary functions y have derivatives y' which lie in the level sets of $f_{z'}$. In particular, the *linear functions*, for which $y' = $ const., *must be stationary*.

Example 1. In characterizing the geodesics on a right circular cylinder of unit radius, we were led in §3.4(a) to minimize

$$L(y) = \int_0^{\theta_2} \sqrt{1 + [y'(\theta)]^2} \, d\theta$$

on

$$\mathcal{D} = \{y \in C^1[0, \theta_2]: y(0) = y_1; y(\theta_2) = y_2\}.$$

Here

$$f = f(z) = \sqrt{1 + z^2}, \qquad f_z(z) = \frac{z}{\sqrt{1 + z^2}};$$

hence, a necessary condition that a given $y \in \mathcal{D}$ minimize L on $\mathcal{D}$ is that y be stationary, or that

$$\frac{y'}{\sqrt{1 + y'^2}} = \text{const.} \quad \text{or} \quad (y')^2 = \text{const.},$$

so that $y' = $ const. In this case, the only stationary functions are the linear functions $y(\theta) = c_1\theta + c_2$ corresponding to the circular helices on the cylinder. (From this analysis alone, however, we cannot say that a helix provides the minimum sought. We would need, in addition, an argument such as that used in §3.4(a).)

(b) When $f = f(x, z)$

Then again $f_y(x, z) = 0$ so that the stationarity condition (3) is

$$f_z(x, y'(x)) = \text{const.}$$

Example 2. To characterize those smooth geodesics on a sphere of radius $\mathbb{R}$ which can be parametrized by functions $\theta = y(\varphi)$ (see §1.1(b)), we should examine

$$L(y) = R \int_0^{\varphi_1} \sqrt{1 + (y'(\varphi) \sin \varphi)^2} \, d\varphi$$

on

$$\mathcal{D}^1 = \{y \in C^1[0, \varphi_1]: y(\varphi_1) = 0\}.$$

Here

$$f = f(\varphi, z) = R\sqrt{1 + z^2 \sin^2 \varphi},$$

so that

$$f_z(\varphi, z) = \frac{Rz \sin^2 \varphi}{\sqrt{1 + z^2 \sin^2 \varphi}}.$$

Thus the stationary functions y are those for which

$$\frac{Ry'(\varphi) \sin^2 \varphi}{\sqrt{1 + y'(\varphi)^2 \sin^2 \varphi}} = \text{const.} = 0 \quad (\text{at } \varphi = 0);$$

i.e., the stationary functions in $\mathcal{D}^1$ are those for which $y'(\varphi) = 0$ which correspond to the great circles. Again, the fact that such a function minimizes L requires separate analysis, as in §1.1(b).

(c) When $f = f(y, z)$

Then with the abbreviation $f(x) = f(y(x), y'(x))$ it follows from the chain rule that *when y is C^2*:

$$\frac{d}{dx} f(x) = \frac{d}{dx} f(y(x), y'(x)) = f_y(x)y'(x) + f_{y'}(x)y''(x).$$

Upon substitution and cancellation we see that

$$\frac{d}{dx}[f(x) - y'(x)f_{y'}(x)] = \frac{d}{dx} f(x) - y''(x)f_{y'}(x) - y'(x)\frac{d}{dx}f_{y'}(x)$$

$$= -y'(x)\left[\frac{d}{dx}f_{y'}(x) - f_y(x)\right],$$

and when y is stationary, the right side vanishes by (3). Thus on each interval of stationarity of y:

$$f(x) - y'(x)f_{y'}(x) = \text{const.} \tag{4}$$

Conversely, if (4) holds on an interval in which y' does not vanish, then y is stationary. (Why?) In this case stationarity is characterized by (4) which is a first integral of (3).[1] The additional smoothness requirement that y be C^2 can be removed if y is assumed to be a local extremal function. See the next section.

Example 3. For the function $f(y, z) = y^2(1 - z)^2$ where $f_z(y, z) = 2y^2(z - 1)$, the (C^2) stationary functions $y = y(x)$ satisfy (4). Thus on each interval, for some constant c

$$y^2(1 - y')^2 - y'[2y^2(y' - 1)] = c,$$

---

[1] However, this integral is usually nonlinear in y' while the original Euler–Lagrange equation is sometimes linear.

so that upon simplification,

$$y^2(1 - y'^2) = c, \quad \text{or} \quad y^2 y'^2 = y^2 - c.$$

With the substitution $u = y^2$ (so that $u' = 2yy'$), we obtain the new equation $u'^2 = 4(u - c)$ which has by inspection, the *singular* solution $\tilde{u}_0(x) \equiv c$. For $u > c$, we get $(\sqrt{u - c})' = \pm 1$ and hence, the *general* solution

$$y^2(x) = u(x) = (x + c_1)^2 + c. \tag{5}$$

The constants c and c_1 may be found so that y meets given boundary conditions. For example, the conditions $y(-1) = 0$ and $y(1) = 1$ produce constants $c = -(\frac{3}{4})^2$ and $c_1 = \frac{1}{4}$; but the resulting function, viz.,

$$y_0(x) = \sqrt{(x + \tfrac{1}{4})^2 - (\tfrac{3}{4})^2} = \sqrt{(x + 1)(x - \tfrac{1}{2})}$$

is defined only for $x \geq \frac{1}{2}$, or $x \leq -1$, and is C^2 only for $x > \frac{1}{2}$ or $x < -1$. (Why?) Moreover, the singular solution $\tilde{y}_0(x) \equiv \sqrt{c}$, cannot satisfy these boundary conditions. For this f, there are *no* stationary functions in

$$\mathscr{D} = \{y \in C^2[-1, 1]: y(-1) = 0, y(1) = 1\}.$$

On the other hand, y_0 is stationary for f on $[1, 2]$ and

$$y_0 \in \mathscr{D}_1 = \{y \in C^1[1, 2]: y(1) = 1, y(2) = 3/\sqrt{2}\}.$$

We shall return to this problem in §7.3, Example 2.

Example 4*. For the brachistochrone problem as formulated in §5.5, Example 4, we must minimize

$$T(y) = \frac{1}{\sqrt{2g}} \int_0^{x_1} \frac{\sqrt{1 + y'(x)^2}}{\sqrt{y(x)}} \, dx$$

on

$$\mathscr{D} = \left\{ 0 \leq y \in C^1[0, x_1]: y(0) = 0, y(x_1) = y_1; \int_0^{x_1} (y(x))^{-3/2} \, dx < +\infty \right\}.$$

Here

$$f = f(y, z) = \frac{\sqrt{1 + z^2}}{\sqrt{y}}$$

(within a constant factor) and

$$f_z(y, z) = \frac{z}{\sqrt{y}\sqrt{1 + z^2}}.$$

The (C^2) stationary functions $y = y(x)$ satisfy (4) so that

$$\frac{\sqrt{1 + y'^2}}{\sqrt{y}} - y'\left(\frac{y'}{\sqrt{y}\sqrt{1 + y'^2}}\right) = \text{const.},$$

or

$$\frac{1}{\sqrt{y}\sqrt{1 + y'^2}} = \text{const.} = \frac{1}{c}, \quad \text{say;}$$

squaring gives $y(1 + y'^2) = c^2$, or

$$\sqrt{\frac{y}{c^2 - y}}\, y' = 1. \tag{6}$$

With the introduction of the dependent variable $\theta = \theta(x)$ such that

$$y = c^2 \sin^2 \frac{\theta}{2} = \frac{c^2}{2}(1 - \cos \theta), \quad \text{for } 0 \le \theta < 2\pi,$$

then

$$c^2 - y = c^2 \cos^2 \frac{\theta}{2} \quad \text{and} \quad y' = c^2 \sin \frac{\theta}{2} \cos \frac{\theta}{2} \theta'$$

By substituting these expressions into equation (6) we obtain

$$c^2 \left(\sin^2 \frac{\theta}{2} \right) \theta' = 1 \quad \text{or} \quad \frac{c^2}{2}(1 - \cos \theta)\theta' = 1.$$

Integrating gives $(c^2/2)(\theta - \sin \theta) = x - c_1$, for a constant c_1. Replacing $c^2/2$ by c^2, we get the parametric equations

$$\begin{cases} x = c^2(\theta - \sin \theta) + c_1, \\ y = c^2(1 - \cos \theta), \end{cases} \quad 0 \le \theta \le \theta_1, \tag{6'}$$

and we see that the only stationary functions are those which determine cycloids. In order that $y \in \mathscr{D}$, we need $y(0) = x(0) = 0$, which implies that $c_1 = 0$. From the corresponding analysis performed in §3.4(b), we know that unique constants c^2 and $\theta_1 < 2\pi$ can be found to make the resulting cycloid with cusp at the origin pass through the given point (x_1, y_1). Since $x'(\theta) = c^2(1 - \cos \theta) > 0$ on $(0, 2\pi)$ it is seen that the x equation can be solved (implicitly) for $\theta = \theta(x)$, with as many derivatives as desired. The composite $y(x) = y(\theta(x))$ is C^2 (at least) on $(0, 2\pi)$, and except at the bottom of the cycloidal arch, $y'(x) \ne 0$. Thus from (4), this $y(x)$ is a stationary function for the problem. However, at the origin $y'(\theta(0)) = +\infty$, so that this function is *not* in $\mathscr{D}$. Although it may represent the brachistochrone sought, we do not yet have it within the framework of the analysis employed.

There is also a more subtle point to consider before regarding this stationary function as a candidate for representing the brachistochrone; it arises from the proof of Lemma 4.1 as follows:

Our analysis that a minimizing function for T on $\mathscr{D}$ must be a stationary function, utilized the fact that $\delta T(y; v) = 0$ for a *particular* $v \in \mathscr{D}_0$. However, as we know from the discussion in §5.5, Example 4, at a given $y \in \mathscr{D}$, the only $v \in \mathscr{D}_0$ which are definitely $\mathscr{D}$-admissible for variation are those for which $|v(x)| \le y(x)$ on $[0, x_1]$ (or their scalar multiples). Unless the particular v used to establish stationarity meets this condition, the analysis is not conclusive, and the true brachistochrone in this class may be provided by a nonstationary function.

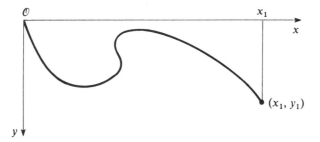

Figure 6.2

Finally, in deciding to consider only a special class of curves, i.e., those with graphs representable by functions of the form $y = y(x)$ or $x = x(y)$, we have excluded a general curve of the form shown in Figure 6.2 which might be the "true" mathematical brachistochrone.

We shall return to this elusive problem, which so far remains just outside the methods being used to analyze it, in §6.4, in §8.8, and in Chapter 9. (See, however, Problem 6.15* for a reformulation which circumvents some of the above difficulties.)

(Problems 6.2–6.16)

§6.3. The Second Equation

When $f = f(x, y, z)$ is C^1 and y is a C^1 solution of the first equation (3) on $[a, b]$, then integration gives

$$f_{y'}(x) = \int_a^x f_y(t) \, dt + \text{const.} \tag{7}$$

When y is C^2, then with the abbreviations

$$f(x) \stackrel{\text{def}}{=} f(x, y(x), y'(x)) \quad \text{and} \quad f_x(x) \stackrel{\text{def}}{=} f_x[y(x)],$$

we have

$$\frac{d}{dx} f(x) = f_x(x) + f_y(x)y'(x) + f_{y'}(x)y''(x), \quad \text{(Why?)}$$

$$= f_x(x) + \frac{d}{dx}(y'(x)f_{y'}(x)), \quad \text{by (3)}.$$

Thus

$$\frac{d}{dx}[f(x) - y'(x)f_{y'}(x)] = f_x(x),$$

or

$$f(x) - y'(x)f_{y'}(x) = \int_a^x f_x(t) \, dt + c_0, \quad \text{for a constant } c_0.$$

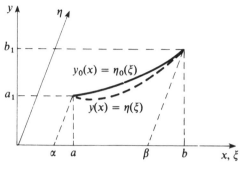

Figure 6.3

This equation resembles (7), the integral form of the first, and, moreover, it does not exhibit explicitly the C^2 requirement on y used in its derivation. Hence we can hope to obtain it directly. This is indeed possible (for extremal functions) but it is surprisingly complicated to do so in view of the simplicity of the underlying strategy: viz., to conduct the original variational operations in terms of coordinate axes which are skewed slightly with respect to the original x, y axes as in Figure 6.3.

Let

$$F(y) = \int_a^b f(x, y(x), y'(x)) \, dx$$

and

$$\mathcal{D} = \{ y \in C^1[a, b] : y(a) = a_1, y(b) = b_1 \}.$$

(6.3) Proposition. *If $f \in C^1([a, b] \times \mathbb{R}^2)$ and $y_0 \in \mathcal{D}$ is a local extremal function for F on $\mathcal{D}$, then on $[a, b]$, y_0 satisfies the second Euler–Lagrange equation*

$$f(x) - y'(x)f_{y'}(x) = \int_a^x f_x(t) \, dt + c_0 \tag{8}$$

for some constant c_0.

PROOF*. For the hypothesized y_0, $\exists \, c > 0$ so small that the affine transformation

$$x = \xi + c\eta,$$

$$y = \eta$$

permits the associated extremal curve to be represented as the graph of a function $\eta_0 \in \widetilde{\mathcal{D}} = \{ \eta \in C^1[\alpha, \beta] : \eta(\alpha) = a_1, \eta(\beta) = b_1 \text{ and } 1 + c\eta'(\xi) > 0 \}$. Here, $\alpha = a - ca_1$ and $\beta = b - cb_1$.

A "nearby" curve represented by $\eta \in \widetilde{\mathcal{D}}$ should also admit the representation $y \in \mathcal{D}$.[1] Thus if $y = y(x)$ corresponds to $\eta = \eta(\xi)$, we should have that $\eta(\xi) \equiv y(\xi + c\eta(\xi))$ so that by the chain rule, $\eta'(\xi) = y'(\xi + c\eta(\xi))(1 + c\eta'(\xi))$

---

[1] Since $1 + c\eta'(\xi) > 0$, we can take $y(x) = \eta(\xi)$ for the *unique* ξ such that $\xi + c\eta(\xi) = x$.

and the derivatives are related through the equations

$$y'(x) = \frac{\eta'(\xi)}{1 + c\eta'(\xi)} \quad \text{or} \quad \frac{1}{1 + c\eta'(\xi)} = 1 - cy'(x). \tag{9}$$

Under the substitution $x = \xi + c\eta(\xi)$, the integral for $F(y)$ transforms as follows: (Problem 6.35(a))

$$F(y) = \int_a^b f(x, y(x), y'(x)) \, dx = \int_\alpha^\beta \tilde{f}(\xi, \eta(\xi), \eta'(\xi)) \, d\xi = \tilde{F}(\eta), \quad \text{say,}$$

where

$$\tilde{f}(\xi, \eta, \zeta) \overset{\text{def}}{=} f\left(\xi + c\eta, \eta, \frac{\zeta}{1 + c\zeta}\right)(1 + c\zeta). \tag{10}$$

Since y_0 is by assumption locally extremal for F on $\mathcal{D}$, it follows that η_0 is locally extremal for $\tilde{F}$ on $\tilde{\mathcal{D}}$. Hence η_0 will be a solution to the *first* equation for $\tilde{f}$ in the integral form of (7):

$$\tilde{f}_\zeta(\xi, \eta(\xi), \eta'(\xi)) = \int_\alpha^\xi \tilde{f}_\eta(\tau, \eta(\tau), \eta'(\tau)) \, d\tau + c_1, \tag{11}$$

for some constant c_1.

But from (10):

$$\tilde{f}_\eta(\xi, \eta, \zeta) = f_x\left(\xi + c\eta, \eta, \frac{\zeta}{1 + c\zeta}\right)(1 + c\zeta)c + f_y\left(\xi + c\eta, \eta, \frac{\zeta}{1 + c\zeta}\right)(1 + c\zeta),$$

while

$$\tilde{f}_\zeta(\xi, \eta, \zeta) = cf\left(\xi + c\eta, \eta, \frac{\zeta}{1 + c\zeta}\right) + f_z\left(\xi + c\eta, \eta, \frac{\zeta}{1 + c\zeta}\right)(1 + c\zeta)^{-1}.$$

Using (9) to return to the original variables, (11) becomes (Problem 6.35(b))

$$cf(x, y(x), y'(x)) + f_z(x, y(x), y'(x))(1 - cy'(x))$$

$$= \int_a^x cf_x(t, y(t), y'(t)) \, dt + \int_a^x f_y(t, y(t), y'(t)) \, dt + c_1,$$

or with the usual abbreviations,

$$c\left[f(x) - y'(x)f_{y'}(x) - \int_a^x f_x(t) \, dt\right] = -\left[f_{y'}(x) - \int_a^x f_y(t) \, dt - c_1\right]. \tag{11'}$$

Finally, upon substituting (7) and subsequently dividing by c, we obtain (8) as desired. □

Observe that when $f = f(y, z)$ alone as in case (c) discussed in the last section, then a local extremal function $y \in \mathcal{D}$, must also satisfy the equation $(d/dx)(f(x) - y'(x)f_{y'}(x)) = 0$ *without* additional smoothness assumptions.

Remark. When y is only stationary, this proof does not yield the second equation unless y is C^2. (See Problem 6.35(c), (d).) However, if f_z is C^1 then y is C^2 when f_{zz} is nonvanishing. (Theorem 7.14)

§6.4. Variable End Point Problems: Natural Boundary Conditions

To find conditions necessary to minimize

$$F(y) = \int_a^b f(x, y(x), y'(x))\, dx$$

(locally) on

$$\mathscr{D}^a = \{y \in C^1[a, b]: y(b) = b_1\},$$

where the value of y at a is unspecified, we know that at a local extremal point $y \in \mathscr{D}^a$, we should have

$$\delta F(y; v) = 0, \qquad \forall\, v \in C^1[a, b]$$

which are $\mathscr{D}^a$-admissible at y (Proposition 5.5).

When f, f_y, and f_z are continuous on $[a, b] \times \mathbb{R}^2$, then also each $v \in \mathscr{D}_0^a = \{v \in C^1[a, b]: v(b) = 0\}$ ($\supseteq \mathscr{D}_0 = \{v \in C^1[a, b]: v(a) = v(b) = 0\}$) is $\mathscr{D}$-admissible; but by Proposition 6.1, $\delta F(y; v) = 0$, $\forall\, v \in \mathscr{D}_0$ implies that y is stationary on (a, b), and from (3'),

$$0 = \delta F(y; v) = f_{y'}(x)v(x)\Big|_a^b = -f_{y'}(a)v(a), \qquad \forall\, v \in \mathscr{D}_0^a.$$

Since $v(x) = b - x$ gives a $v \in \mathscr{D}_0^a$ for which $v(a) \neq 0$, y *must be a stationary function that satisfies the following "natural" boundary condition at the free end:*

$$f_{y'}(a) = 0 \qquad (\text{or } f_z[y(a)] = 0). \tag{12a}$$

Similarly, if y minimizes F on $\mathscr{D}^b = \{y \in C^1[a, b]: y(a) = a_1\}$ (locally) then y must be a stationary function which satisfies the natural boundary condition

$$f_{y'}(b) = 0 \qquad (\text{or } f_z[y(b)] = 0). \tag{12b}$$

Finally, if y minimizes F on $\mathscr{Y} = C^1[a, b]$ (locally) then $\delta F(y; v) = 0$, $\forall\, v \in \mathscr{Y} \supseteq \mathscr{D}_0^a \supseteq \mathscr{D}_0$. Thus y must be a stationary function which satisfies the natural boundary conditions (12a) and, from symmetry, (12b)). (All of these conditions were utilized in Proposition 3.9.)

Application: Jakob Bernoulli's Brachistochrone

In 1696, Jakob Bernoulli publicly challenged his younger brother Johann to find the solutions to several problems in optimization including that of the brachistochrone which covered a given horizontal distance, x_1 (thereby initiating a long, bitter, and pointless rivalry between two representatives of the best minds of their era). Since the time-of-descent function is the same as in §1.2 and §6.2(c); viz.,

$$T(y) = \frac{1}{\sqrt{2g}} \int_0^{x_1} \frac{\sqrt{1 + y'(x)^2}}{\sqrt{y(x)}} \, dx,$$

we should attempt to minimize T on

$$\mathscr{D}_1 = \left\{ 0 \leq y \in C^1[0, x_1] : y(0) = 0, \int_0^{x_1} (y(x))^{-3/2} \, dx < +\infty \right\}.$$

From our general analysis, we know that a minimizing function $y \in \mathscr{D}_1$ should be a stationary function which satisfies in addition the natural boundary condition (12b)

$$0 = f_{y'}(x_1) = \frac{y'(x_1)}{\sqrt{y(x_1)}\sqrt{1 + y'(x_1)^2}}.$$

As in §6.2(c), Example 4, y must represent a cycloid which satisfies the natural boundary condition $y'(x_1) = 0$. Since this requirement can be met only at the lowest point on the cycloidal arch, which corresponds to $\theta = \theta_1 = \pi$, it follows from equations (6') that we require $y(x_1) = (2/\pi)x_1$, and this will be obtained (uniquely) for the cycloid represented by (6') with $c^2 = x_1/\pi$. We still have the difficulty that $y'(0) = +\infty$. Moreover, from this analysis alone, we cannot conclude (nor could either of the Bernoullis)[1] that this cycloid does in fact provide the brachistochrone sought. We know simply that it is the only curve which supports the variational requirements for a minimum. [See, however, the relevant comments in §6.2(c).] But, this cycloid meets the condition that $x_1/y_1 \leq \pi/2$, and so from the analysis in §3.4(b) we can safely say that it provides (uniquely) the least time of descent among all curves joining the origin to (x_1, y_1) which can *also* be represented as the graph of a function in $C^1[0, y_1]$. In particular, it cannot give a (local) *maximum* value for T.

Transversal Conditions*

To obtain the natural boundary conditions associated with more general end point constraints provided by transversals such as that illustrated in Figure 6.4, it is more convenient to use Lagrangian multipliers. Here we suppose the integral

$$J(y, t) = \int_a^t f(x, y(x), y'(x)) \, dx = \int_a^t f[y(x)] \, dx \tag{13}$$

is to be minimized over

$$\mathscr{D}_\tau = \{ y \in C^1[a, t] : y(a) = a_1; \tau(t, y(t)) = 0 \}.$$

---

[1] But see Carathéodory's article in the historical references.

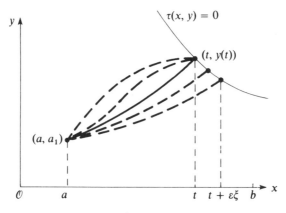

Figure 6.4

Assume that f and the constraining function τ are C^1 on domains large enough to admit all functions of interest and that $\nabla \tau \neq \mathcal{O}$. If $y \in C^1[a, t]$ minimizes J, then varying y by functions v in $\mathcal{D}_0 = \{v \in C^1[a, t] : v(a) = v(t) = 0\}$ shows as usual that y is a stationary function and thus is a solution of (3), the Euler–Lagrange equation $(d/dx) f_{y'}(x) = f_y(x)$ on (a, t).

For the proper natural boundary condition at the right end, we must admit more general variations. To provide a convenient framework, we suppose that the functions y are defined by extension on a fixed large interval $[a, b]$, and introduce the linear space

$$\mathcal{Y} = C^1[a, b] \times \mathbb{R},$$

with the norm $\|(y, t)\| = \|y\|_M + |t|$. (See Problem 5.8.)

A general variation for J in this space in the "direction" (v, ξ) is obtained by differentiating $J(y + \varepsilon v, t + \varepsilon \xi)$ with respect to ε and setting $\varepsilon = 0$. By Leibniz' rule (A.14), we get that with the usual abbreviations:

$$\delta J(y, t; v, \xi) = f(t)\xi + \int_a^t (f_y(x)v(x) + f_{y'}(x)v'(x))\, dx,$$

and for a *stationary function* y, the integrand is the derivative of $f_{y'}(x)v(x)$ so that

$$\delta J(y, t; v, \xi) = f(t)\xi + f_{y'}(x)v(x)\Big|_a^t. \tag{14}$$

The right end point constraint may be expressed as the zero level set of the function

$$G(y, t) = \tau(t, y(t)) = \tau[y(t)],$$

so that upon differentiating $\tau(t + \varepsilon\xi, (y + \varepsilon v)(t + \varepsilon\xi))$ with respect to ε and evaluating at $\varepsilon = 0$, we obtain

$$\delta G(y, t; v, \xi) = \tau_x[y(t)]\xi + \tau_y[y(t)](y'(t)\xi + v(t)). \tag{14'}$$

(It may be shown that variations (14) and (14′) are weakly continuous.) Now, let y be a local extremum point for J of (13). According to Theorem 5.15, unless $\delta G(y, t; \cdot, \cdot) \equiv 0$ (which would require the vanishing of both $\tau_x[y(t)]$ and $\tau_y[y(t)]$; Why?), then $\exists\ \lambda \in \mathbb{R}$ such that

$$\delta(J + \lambda G)(y, t; \cdot, \cdot) \equiv 0.$$

Hence restricting attention to those $v \in \mathscr{D}_0$ as before which vanish at a and t, we have from (14) and (14′) that

$$\{f(t) + \lambda(\tau_x[y(t)] + \tau_y[y(t)]y'(t))\}\xi = 0, \qquad \forall\ \xi \text{ sufficiently small.}$$

Similarly, if we consider variations $(v, 0)$ for which $\xi = v(a) = 0$, then

$$\{f_{y'}(t) + \lambda\tau_y[y(t)]\}v(t) = 0, \qquad \forall\ v(t) \text{ sufficiently small.}$$

Dividing these last equations by ζ, $v(t)$, respectively, and eliminating λ between them shows that *a local extremal point y for J on $\mathscr{D}_\tau$ is a stationary function on (a, t) which meets the transversal condition*

$$f(t)\tau_y[y(t)] = f_{y'}(t)\{\tau_x[y(t)] + \tau_y[y(t)]y'(t)\} \tag{15}$$

(15) is the desired natural boundary condition. Note that when $\tau(x, y) = b - x$ so that $\tau_y \equiv 0$, then (15) reduces to $f_{y'}(b) = 0$ as obtained earlier.

Similarly, when $\tau(x, y) = y - b_1$ for given b_1, the terminal value t of x is unspecified, and at (t, b_1) an optimal solution should meet the transversal condition:

$$f(t) - y'(t)f_{y'}(t) = 0. \tag{15′}$$

In economics, this would be called a *free-horizon* problem. If the terminal value b_1 is also unspecified, then in addition to (15′), an optimal solution must meet the free-end condition $f_{y'}(t) = 0$, at its terminal point $(t, y(t))$. Why?

If both end points lie on curves of this type, as in Figure 6.1(b), then a local extremal function will be stationary on an interval for which it satisfies (15) at the right endpoint and the corresponding condition at the left.

Some other types of constraints amenable to the use of Lagrangian multipliers will be treated in §6.7 in connection with vector valued extremals.

Example 1. If the brachistochrone joining the origin to a "lower" curve which is the zero level set of the function τ as in Figure 6.5, can be represented as the graph of a function $y \in C^1[0, t]$ for some $t > 0$, then we should expect that y is stationary for the time-of-descent function $f(y, z) = \sqrt{1 + z^2}/\sqrt{y}$ (as in §6.2(c)) and satisfies the obvious boundary condition $y(0) = 0$ together with the natural boundary condition (15).

According to the analysis in §6.2(c), y must represent a cycloid joining the origin to a point (t, y), at which

$$\frac{\sqrt{1 + y'(t)^2}}{\sqrt{y(t)}}\tau_y[y(t)] = \frac{y'(t)}{\sqrt{y(t)}\sqrt{1 + y'(t)^2}}(\tau_x[y(t)] + \tau_y[y(t)]y'(t)),$$

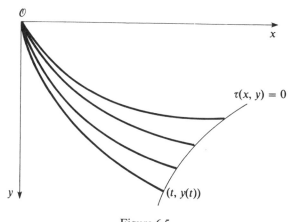

Figure 6.5

or, after obvious algebra, at which

$$\tau_y[y(t)] = \tau_x[y(t)]y'(t).$$

Thus, the point(s) (t, y) are those which permit a cycloid joining (them) to the origin to meet the τ curve *orthogonally*. There may be only one such point, or there may be many. (See Problem 6.21.) Observe that this result generalizes that obtained for Jakob Bernoulli's brachistochrone from this section in which we found that the cycloid in question should at (t, y) have a horizontal tangent ($y'(t) = 0$).

(Problems 6.17–6.22)

§6.5. Integral Constraints: Lagrangian Multipliers

As we observed in the preceding section, end point constraints do not affect the stationarity of the possible extremal functions for integrals such as

$$F(y) = \int_a^b f(x, y(x), y'(x))\, dx, = \int_a^b f[y(x)]\, dx$$

but do control the boundary conditions which the extremal function should satisfy.

However, frequently present are other constraints which operate over the entire interval $[a, b]$. When each of these can also be expressed in integral form say by requiring that a function

$$G(y) = \int_a^b g(x, y(x), y'(x))\, dx = \int_a^b g[y(x)]\, dx,$$

assume a prescribed value, then we can employ the method of Lagrangian multipliers (cf. §5.7) since, in general, the linearity and weak continuity of the variations δF and δG is assured by Proposition 5.11.

(6.4) **Theorem.** *Suppose that $f = f(x, y, z)$ and $g_i = g_i(x, y, z)$, $i = 1, 2, \ldots, N$, together with their y and z partial derivatives, are continuous on $[a, b] \times \mathbb{R}^2$. Let y_0 be a local extremal function for*

$$F(y) = \int_a^b f[y(x)] \, dx$$

on

$$\mathcal{D} = \{y \in C^1[a, b]: y(a) = a_1; y(b) = b_1\},$$

when further constrained to the set

$$G_{y_0} = \left\{y \in C^1[a, b]: G_i(y) \overset{\text{def}}{=} \int_a^b g_i[y(x)] \, dx = G_i(y_0), i = 1, 2, \ldots, N\right\}.$$

Then either:

(a) *the $N \times N$ determinant*

$$\left| \begin{matrix} \delta G_i(y_0; v_j) \\ i, j = 1, \ldots, N \end{matrix} \right| = 0,$$

whenever $v_j \in \mathcal{D}_0 = \{v \in C^1[a, b]: v(a) = v(b) = 0\}, j = 1, 2, \ldots, N$;

or

(b) *$\exists \, \lambda_i \in \mathbb{R}, i = 1, 2, \ldots, N$ that make y_0 stationary for the modified function $\tilde{f} = f + \sum_{i=1}^n \lambda_i g_i$; i.e., y_0 is a solution of the equation*

$$\frac{d}{dx} \tilde{f}_{y'}(x) = \tilde{f}_y(x) \qquad \text{on } (a, b).$$

PROOF. As noted, the hypotheses on f and the g_i assure that the variations $\delta F(y; v)$, $\delta G_i(y; v)$ are linear in v and weakly continuous for all v in the subspace $\mathcal{D}_0$. Hence from Theorem 5.16 (and subsequent remarks), either condition (a) holds, $\forall \, v_j \in \mathcal{D}_0$ or $\exists \, \lambda_i \in \mathbb{R}$, $i = 1, 2, \ldots, N$, for which $\delta \tilde{F}(y_0; v) = 0$, $\forall \, v \in \mathcal{D}_0$, where

$$\tilde{F}(y) = F(y) + \sum_{i=1}^n \lambda_i G_i(y) = \int_a^b \tilde{f}[y(x)] \, dx,$$

with $\tilde{f}$ defined as in condition (b). Hence by Proposition 6.1, y_0 is stationary for $\tilde{f}$. □

(6.5) **Remark.** As in the general theory, the hope is that the Lagrangian multipliers λ_i can be determined so that the stationary function $y_0 \in \mathcal{D}$ will provide *prescribed* values for $G_i(y_0)$, $i = 1, 2, \ldots, N$.

(6.6) **Remark.** If in the theorem, $\mathcal{D}$ is replaced by $\mathcal{D}^b = \{y \in C^1[a, b]: y(a) = a_1\}$ (as in §6.4) then $\mathcal{D}_0$ is replaced by the *subspace* $\mathcal{D}_0^b = \{v \in C^1[a, b]: v(a) = 0\}$ and it is seen that in addition to condition (b), y_0 must also satisfy the natural boundary condition $\tilde{f}_{y'}(b) = 0$. Similarly, if $\mathscr{D}$ is replaced by $\mathscr{Y} = C^1[a, b]$, then y_0 must in addition satisfy $\tilde{f}_{y'}(a) = \tilde{f}_{y'}(b) = 0$. Finally, if y_0 is required to meet boundary conditions such as $y(a) = a_1$, $\tau(t, y(t)) = 0$ for some function τ, at $t \in [a, b]$, where $\nabla\tau \neq \mathcal{O}$, and

$$F(y, t) = \int_a^t f(x, y(x), y'(x))\, dx,$$

as in §6.4, and where

$$G_i(y, t) = \int_a^t g_i(x, y(x), y'(x))\, dx$$

is prescribed at y_0 for $i = 1, 2, \ldots, N$, then, in general, y_0 will be stationary for $\tilde{f}$ on (a, t) and satisfy the transversal condition

$$\tilde{f}(t)\tau_y[y(t)] = \tilde{f}_{y'}(t)\{\tau_x[y(t)] + \tau_y[y(t)]y'(t)\}. \tag{16}$$

(Problems 6.23–6.24)

§6.6. Integrals Involving Higher Derivatives

It is straightforward to obtain results analogous to those in the preceding sections of this chapter that characterize the local extremals for the function

$$F(y) \stackrel{\text{def}}{=} \int_a^b f(x, y(x), y'(x), y''(x))\, dx = \int_a^b f[y(x)]\, dx$$

for given $f \in C^1([a, b] \times \mathbb{R}^3)$ on domains $\mathscr{D} \subseteq C^2[a, b]$ defined by prescribing (possibly) $y(a)$, $y'(a)$, $y(b)$, or $y'(b)$. Indeed, if $y, v \in C^2[a, b]$, then

$$\delta F(y; v) \stackrel{\text{def}}{=} \frac{\partial}{\partial\varepsilon} J(y + \varepsilon v)\Big|_{\varepsilon=0},$$

or from A.13

$$\delta F(y; v) = \int_a^b (f_y(x)v(x) + f_{y'}(x)v'(x) + f_{y''}(x)v''(x))\, dx, \tag{17}$$

(see Problem 2.9); where $f_{y''}(x) \stackrel{\text{def}}{=} f_r(x, y(x), y'(x), y''(x))$, when $f = f(x, y, z, r)$, with corresponding extensions for $f_y(x)$ and $f_{y'}(x)$.

For definiteness, suppose that

$$\mathscr{D} = \{y \in C^2[a, b] : y(a) = a_1, y(b) = b_1, y'(a) = a_1'\},$$

where a_1, b_1, and a_1' are given real numbers, and let

$$\mathscr{D}_0 = \{v \in C^2[a, b] : v(a) = v(b) = v'(a) = 0\}.$$

Then if $y \in \mathscr{D}$ is locally extremal for F on $\mathscr{D}$, from Proposition 5.5 it follows that $\delta F(y; v) = 0$, $\forall \, v \in \mathscr{D}_0$. Introducing the C^1 functions

$$g(x) = \int_a^x f_y(t) \, dt$$

and $\qquad\qquad\qquad\qquad\qquad\qquad\qquad\qquad\qquad\qquad (18)$

$$h(x) = \int_x^b [f_{y'}(t) - g(t)] \, dt,$$

we have upon integrating (17) by parts *twice* in succession that for $v \in \mathscr{D}_0$:

$$\delta F(y; v) = \int_a^b [(f_{y'}(x) - g(x))v'(x) + f_{y''}(x)v''(x)] \, dx + g(x)v(x) \Big|_a^b$$

$$= \int_a^b [h(x) + f_{y''}(x)]v''(x) \, dx - h(x)v'(x) \Big|_a^b.$$

Here, the definition of h assures its vanishing at $x = b$, and hence the vanishing of the boundary term $h(x)v'(x)|_a^b$ for $v \in \mathscr{D}_0$. However, since $v \in \mathscr{D}_0 \Rightarrow v(a) = v(b) = 0$, the definition of g is less critical and $g(x) \overset{\text{def}}{=} \int_a^x f_y(t) \, dt + \text{const.}$ would also suffice. In any case, as a necessary condition that y be locally extremal, we have that

$$\delta F(y; v) = \int_a^b [h(x) + f_{y''}(x)]v''(x) \, dx = 0, \qquad \forall \, v \in \mathscr{D}_0, \qquad (19)$$

and hence by Proposition 4.5, that for some constants c and c_1:

$$h(x) + f_{y''}(x) = c_1 x + c, \qquad\qquad\qquad\qquad\qquad (20)$$

or by (18),

$$f_{y''}(x) = -\int_x^b [f_y(t) - g(t)] \, dt + c_1 x + c, \qquad \forall \, x \in [a, b].$$

Thus for $x \in (a, b): f_{y''}(x) \in C^1$ and

$$\frac{d}{dx} f_{y''}(x) = f_y(x) - g(x) + c_1;$$

similarly, the combination $(d/dx)f_{y''}(x) - f_{y'}(x)$ is C^1 and

$$\frac{d}{dx}\left[\frac{d}{dx} f_{y''}(x) - f_{y'}(x) \right] = -f_y(x). \qquad\qquad\qquad (21)$$

(6.7) **Definition.** (21) is the appropriate Euler–Lagrange equation for a C^1 function f (of x, y, y' and y''). Those C^2 functions y which satisfy (21) on some interval will be termed *stationary functions for f*.

As might be expected in this case, the additional freedom in the derivative $y'(b)$ gives rise to a corresponding natural boundary condition as in §6.4. To discover it we use the Euler–Lagrange equation (21) to replace $f_y(x)$ in (17) and then integrate by parts to get

$$\delta F(y; v) = \left[\left(f_{y'}(x) - \frac{d}{dx} f_{y''}(x) \right) v(x) \right] \Big|_a^b$$

$$+ \int_a^b \left[\left(\frac{d}{dx} f_{y''}(x) \right) v'(x) + f_{y''}(x) v''(x) \right] dx$$

so that

$$\delta F(y; v) = \left[\left(f_{y'}(x) - \frac{d}{dx} f_{y''}(x) \right) v(x) \right] \Big|_a^b + \left[f_{y''}(x) v'(x) \right] \Big|_a^b. \qquad (21')$$

In our case, when $v \in \mathcal{D}_0$, the first term on the right in (21') vanishes and the second reduces to $f_{y''}(b) v'(b)$. Hence from (19), we see that the appropriate natural boundary condition is

$$f_{y''}(b) = 0; \quad (\text{or } f_r(b, y(b), y'(b), y''(b)) = 0). \qquad (22)$$

Other boundary conditions are considered in Problems 6.25–6.28, together with the derivation of the Euler–Lagrange equation for functions f involving derivatives higher than the second. A corresponding "second" equation is obtained in Problem 6.34.

Application: Buckling of a Column under Compressive Loading

It is a fact of experience that work is required to bend a long straight thin elastic rod of uniform cross section and material, and that more work is required to bend it further (although not necessarily in proportion). As a result of experiments, Daniel Bernoulli (the son of Johann) concluded (c. 1738) that the work required per unit length is proportional to the square of the resulting mean curvature $\bar{k}$ of that length.[1] Hence the work required to bend an entire rod of initial length l into a form whose center line is described by a function $y(x)$, $0 \le x \le l$, as in Figure 6.6(a), is given by

$$W_B = \mu \int_0^l k^2(x) \, ds(x),$$

where the constant μ is determined by the material and the cross-sectional shape.[2] Also,

$$s(x) \stackrel{\text{def}}{=} \int_0^x \sqrt{1 + y'(t)^2} \, dt, \qquad 0 \le x \le l,$$

is the arc length of the center line between 0 and x, while $k(x)$, the local curvature at x, is from calculus, given by $k(x) = |y''(x)|/(1 + y'(x)^2)^{3/2}$, for

---

[1] See [Ra], page 256, for a simple explanation.

[2] $\mu = EI/2$, where E is Young's modulus between stress and strain for the bar, and I is the moment of inertia of the cross-section.

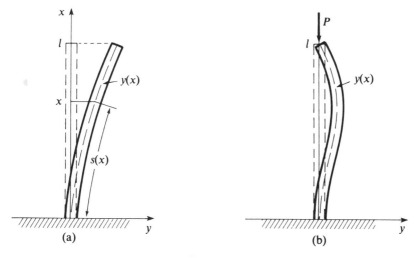

Figure 6.6

each x at which y'' is defined. This work may also be regarded as the potential energy of strain stored in the rod as it is bent from an initial unstressed configuration (supposed straight). Bernoulli conjectured that when the rod is bent by external forces, it will assume a shape which minimizes the potential energy. We have already utilized this principle in the analysis of the catenary problem of §3.5, and what we have thus far would suffice to describe a situation in which other types of strain energy (work) can be considered negligible. (See Problem 6.29.)

However, if the bending of a column is produced through *buckling* under a longitudinal compressive force of magnitude P applied to the end as in Figure 6.6(b), then work is also done in compressing the bar. If we regard the bent bar as an elastic spring of the "new" length

$$\int_0^l ds(x) = \int_0^l \sqrt{1 + y'(x)^2}\, dx,$$

then the compressive strain energy *in the absence of bending* may be determined from the work done by P in restoring the bar to its original length l; viz., from

$$W_C = P\left(\int_0^l \sqrt{1 + y'(x)^2}\, dx - l\right).$$

Disregarding further interaction effects, the total potential energy of the buckled column is $W_B - W_C$, where the negative sign accounts for the fact that upon buckling, the strain energy of compression is released to be transformed (partially) into that of bending, (For a more rigorous derivation which permits direct interaction, see [Se], while Euler's original solution of 1744 is discussed in [Fu].)

If the bar is clamped at its lower end as in Figure 6.6(b), then $y(0) = y'(0) = 0$, while if we suppose that the upper end remains essentially fixed so that $y(l) = 0$, we would wish to minimize the *potential energy*

$$U(y) \equiv \int_0^l \left[\mu \frac{y''(x)^2}{(1 + y'(x)^2)^{5/2}} - P(\sqrt{1 + y'(x)^2} - 1) \right] dx \qquad (23)$$

on

$$\mathscr{D} = \{ y \in C^2[0, l] : y(0) = y(l) = y'(0) = 0 \}.$$

This is the situation discussed in the text (with $\mathscr{D}_0 = \mathscr{D}$) for the function:

$$f(x, y, z, r) = \mu \frac{r^2}{(1 + z^2)^{5/2}} - P((1 + z^2)^{1/2} - 1). \qquad (24)$$

Since $f_y \equiv 0$, it follows from (21) that a stationary function, y, for f, makes $f_{y''}(x)$ continuously differentiable and satisfies the equation

$$\frac{d}{dx} f_{y''}(x) - f_{y'}(x) = \text{const.} = \bar{c}, \quad \text{say}, \qquad (24')$$

or with (24), it makes $y''/(1 + y'^2)^{5/2}$ (*and hence y''*) continuously differentiable, and

$$\left(2\mu \frac{y''}{(1 + y'^2)^{5/2}} \right)' + 5\mu(y'')^2 \frac{y'}{(1 + y'^2)^{7/2}} + P \frac{y'}{(1 + y'^2)^{1/2}} = \bar{c}.$$

After differentiating and simplifying, we find that

$$\frac{y'''}{(1 + y'^2)^{5/2}} + \frac{y'}{2\sqrt{1 + y'^2}} \left\{ \frac{P}{\mu} - \frac{5y''^2}{(1 + y'^2)^3} \right\} = \frac{\bar{c}}{2\mu} = c, \quad \text{say}. \qquad (25)$$

The natural boundary condition (22) associated with the unspecified slope at $x = l$ is (from (24)) given by

$$\frac{2\mu y''(l)}{(1 + y'(l)^2)^{5/2}} = 0, \quad \text{or} \quad y''(l) = 0. \qquad (26)$$

Thus it would be necessary to solve the third-order nonlinear differential equation (25) on $(0, l)$ where c is to be determined, if possible, to satisfy the boundary conditions,

$$y(0) = y'(0) = y(l) \equiv y''(l) = 0. \qquad (27)$$

To carry this out is a highly nontrivial task and requires numerical approximation methods. (However, one more integration is possible. See Problem 6.34.)

To simplify the analysis, we shall make the assumption that buckling occurs with a beam geometry for which $|y''|$ is small, in particular, one for which $\max |y''| \ll 1$.[1] Then for $y'(0) = 0$, it follows that $\max |y'| \ll 1$ also (Why?) and (25) may be approximated by the following *linear* equation with

---

[1] It is sufficient to assume that $|y'| \ll 1$. See Problem 6.34.

constant coefficients:

$$y''' + \omega^2 y' = c, \quad \text{where } \omega^2 = P/2\mu, \tag{28}$$

subject to the homogeneous boundary conditions (27). Integrating and using the conditions $y(l) = y''(l) = 0$ gives the second-order equation

$$y'' + \omega^2 y = c(x - l), \tag{29}$$

whose general solution is known; (see, for example, [B–diP]). It is given by

$$y(x) = A \cos \omega x + B \sin \omega x + Cx + D, \tag{30}$$

for constants A, B, C, and D to be found to satisfy (29) and the remaining boundary conditions, $y(0) = y'(0) = 0$.

From these last two conditions, we must have

$$0 = A + D \quad \text{or} \quad A = -D,$$

and

$$0 = B\omega + C \quad \text{or} \quad B = -C/\omega.$$

Differentiating (30) twice gives

$$y''(x) + \omega^2 y(x) = \omega(Cx + D),$$

and the right side agrees with that of (29) if and only if

$$C\omega^2 = c \quad \text{and} \quad D\omega^2 = -cl.$$

Thus the solution is given by a constant multiple (c) of

$$y(x) = \frac{1}{\omega^2} \cos \omega x - \frac{1}{\omega^3} \sin \omega x + \frac{1}{\omega^2}(x - l), \tag{31}$$

where it remains to select $\omega^2 = P/2\mu$ to make $y(l) = 0$; i.e., to make $\omega l \cos \omega l - \sin \omega l = 0$, or

$$\omega l = \tan \omega l. \tag{32}$$

Now (32) has an infinite set of solutions $\omega_n l \in (n\pi, (n + 1/2)\pi)$, $n = 1, 2, \ldots$, as is evident graphically from Figure 6.7. The least of these, $\omega_1 l$, determines the load at which buckling "first" occurs; viz.,

$$P_1 = 2\mu\omega_1^2 \simeq \frac{\mu}{2}\left(\frac{3\pi}{l}\right)^2.$$

With $\omega = \omega_n$, (31) defines a sequence of stationary mode functions, y_n in $\mathcal{D}$, each of which satisfies the natural boundary condition (26) under the additional linearizing assumption that

$$\max |y''| \ll 1.$$

From (31), with $\omega = \omega_n$ (so that $\omega_n l > n\pi$), it follows that

$$|y_n''(x)|^2 = l^2 \left[\frac{\sin \omega_n x}{\omega_n l} - \cos \omega_n x \right]^2 \quad (= l^2, \text{ when } \omega_n x = n\pi).$$

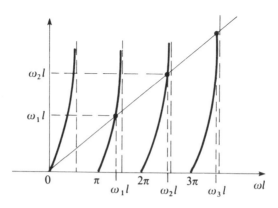

Figure 6.7

Thus the smallness of y'' for the actual deflection curve $y(x) = cy_n(x)$ is possible *iff* c itself is small, which means that the maximum deflection must itself be small. Such linear analysis is usually termed *small deflection theory*, and the approximations are often made in (23), the integral expression defining U itself. (However, if the estimate max $|y'|^2 \ll 1$ is used there uncritically, the term involving P disappears. See Problem 6.30.)

Our linearization has resulted in another difficulty: Since each multiple of y_1 (or y_n) is another stationary function which meets the required boundary conditions, it is not evident in what sense the potential energy U could be minimized by such functions. What must be realized is that once buckling has occurred with the critical load P_1 in the mode described by y_1, then further bending can occur in this mode *without additional load* until the nonlinear effects excluded by our analysis become prominent. In particular, the assumptions of small deflection theory may be violated, even though they are valid at the instant when buckling first occurs.

Another anomaly requires explanation; namely, whether buckling can occur only at the critical loads $P_n = 2\mu\omega_n^2$. If, for example, the column is encased in a more rigid structure before loading, loaded by P without buckling, and then uncased, it is in unstable equilibrium at the critical loads P_n and buckling in the associated mode y_n can be induced, with the buckled bar in static equilibrium. However, with the load P_2, say, the column cannot buckle in a mode y_n for $n > 2$, since more energy would be required than can be sustained by P_2. On the other hand, with this loading (or by any $P > P_1$) buckling in the mode y_1 could not retain the static equilibrium of the bar (at least as described by small deflection theory). Thus with moderate loading P, buckling may be prevented by supporting the bar only at the points of maximum deflection of the lower mode shapes.

(Problems 6.25–6.34)

§6.7. Vector Valued Stationary Functions

If we wish to examine whether the curve shown in Figure 6.2 could represent the (mathematical) brachistochrone, or seek geodesics on a general surface as in §1.1(c), then we must utilize integrals depending upon vector valued functions. Fortunately, it is straightforward to generalize the analysis in the preceding sections of this chapter to obtain necessary conditions which characterize a vector valued extremal function in any finite number (d) of dimensions. We recall from §5.1, Example 7, that for the linear space

$$\mathscr{Y} = (C^1[a, b])^d$$

of elements

$$Y = (y_1, y_2, \ldots, y_d),$$

having derivatives

$$Y' = (y_1', y_2', \ldots, y_d'),$$

a suitable norm is given by

$$\|Y\| = \max(|Y(x)| + |Y'(x)|)$$

Thus when $f \in C^1([a, b] \times \mathbb{R}^{2d})$, to characterize a function $Y_0 \in \mathscr{Y}$ which provides a local extremum for the function

$$F(Y) \stackrel{\text{def}}{=} \int_a^b f(x, Y(x), Y'(x)) \, dx = \int_a^b f[Y(x)] \, dx$$

on

$$\mathscr{D} = \{Y \in (C^1[a, b])^d \colon Y(a) = A; \ Y(b) = B\},$$

where $A, B \in \mathbb{R}^d$ are prescribed, we should introduce vector directions in

$$\mathscr{D}_0 = \{V \in (C^1[a, b])^d \colon V(a) = V(b) = \mathcal{O}\},$$

consider $F(Y_0 + \varepsilon V)$ for $V \in \mathscr{D}_0$ and sufficiently small $\varepsilon \in \mathbb{R}$, and require that $\delta F(Y_0; V) = 0, \ \forall \ V \in \mathscr{D}_0$.

As in Example 8 of §2.4,

$$\delta F(Y; V) \stackrel{\text{def}}{=} \frac{\partial}{\partial \varepsilon} F(Y + \varepsilon V) \Big|_{\varepsilon = 0}$$

or

$$\delta F(Y; V) = \int_a^b [f_Y(x) \cdot V(x) + f_{Y'}(x) \cdot V'(x)] \, dx \tag{33}$$

by Leibniz' rule for differentiating under the integral (A.14), where, as explained in §1.5, $f_Y(x)$ is the *vector valued* function with components $f_{y_j}(x) \stackrel{\text{def}}{=} f_{y_j}[Y(x)], j = 1, 2, \ldots, d$; and $f_{Y'}(x)$ is the *vector valued* function with components $f_{z_j}(x) \stackrel{\text{def}}{=} f_{z_j}[Y(x)], j = 1, 2, \ldots, d$. (Here we regard $f \stackrel{\text{def}}{=} f(x, Y, Z) = f(x, y_1, y_2, \ldots, y_d, z_1, z_2, \ldots, z_d)$.) The dot denotes the ordinary scalar product in $\mathbb{R}^d$, and is used for convenience in notation.

Now, by assumption, both $f_Y(x)$ and $f_{Y'}(x)$ have continuous components; hence from (33), if $\delta F(Y_0; V) = 0, \forall\, V \in \mathscr{D}_0$, then Proposition 4.6 is applicable and by generalization we obtain the following result:

(6.8) Theorem. *If $f = f(x, Y, Z) \in C^1([a, b] \times D)$ for a domain D of $\mathbb{R}^{2d}$ and $Y_0 \in \mathscr{Y} = (C^1[a, b])^d$ is a (local) extremal function for*

$$F(Y) = \int_a^b f[Y(x)]\, dx$$

on

$$\mathscr{D} = \{Y \in \mathscr{Y} \colon Y(a) = Y_0(a);\ Y(b) = Y_0(b);\ (Y(x),\ Y'(x)) \in D\},$$

then on (a, b), Y_0 satisfies the equation(s)

$$\frac{d}{dx} f_{Y'}(x) = f_Y(x); \quad \left[or \quad \frac{d}{dx} f_{y'_j}(x) = f_{y_j}(x),\, j = 1, 2, \ldots, d \right]. \tag{34}$$

PROOF*. When $D = \mathbb{R}^{2d}$, the result is an immediate consequence of the preceding arguments, which also apply to a general *subset* D if all directions $V \in \mathscr{D}_0 = \{V \in \mathscr{Y} \colon V(a) = V(b) = \mathcal{O}\}$ are $\mathscr{D}$-admissible at Y_0. (Why?) But this is true for each (*open*) domain D, by an appeal to the compactness of $[a, b]$, which we shall only outline: For each $x \in [a, b]$, the point $(Y_0(x), Y_0'(x))$ is the center of a spherical neighborhood $\subseteq D$, of *positive maximal* radius $r(x) \le 1$. Moreover, the continuity of both Y_0 and Y_0' on $[a, b]$ will guarantee that $r \in C[a, b]$; it follows from Proposition 5.3 that r assumes a *minimum value* at some x_0 so that $r(x) \ge r(x_0) = r_0 > 0$. Thus, when $\varepsilon < r_0$ and $V \in \mathscr{D}_0$ with $\|V\| \le 1$ then $Y_0 + \varepsilon V \in \mathscr{D}$, and we conclude that all such V *together with their scalar multiples* are $\mathscr{D}$-admissible at Y_0. □

(6.9) Definition. Equations (34) constitute the vector valued Euler–Lagrange equations; their C^1 solutions, Y, are called the *stationary functions* for f (of x, Y, Y').

In general, Y is stationary for f on (a, b) iff $\delta F(Y; V) = 0, \forall\, V \in \mathscr{D}_0$, where F is the associated integral function of Theorem 6.8. Indeed, such stationarity permits the use of (34) in (33) to obtain

$$\delta F(Y; V) = f_{Y'}(x) \cdot V(x) \Big|_a^b = 0, \quad \text{if} \quad V(a) = V(b) = \mathcal{O},$$

while the converse assertion was established in proving Theorem 6.8.

(Problem 6.40)

There is also an analogous *second equation* for local extremal Y (cf. §6.3); viz.,

$$f(x) - Y'(x) \cdot f_{Y'}(x) = \int_a^x f_x(t)\, dt + c, \quad x \in (a, b), \tag{35}$$

where $f(x) = f[Y(x)]$ and $f_x(x) = f_x[Y(x)]$. However here, the second equation is *scalar* and cannot characterize stationarity of Y as in the one-dimensional case (Problem 6.36).

As in §6.4, if, say, $y_j(b)$ is left unspecified for *some* value of $j = 1, 2, \ldots, d$, there results the associated natural boundary condition $f_{y_j'}(b) = 0$. (See Problem 6.37.)

To consider more general boundary conditions and/or constraints of the form $G(Y) = $ const., we may employ the method of Lagrangian multipliers developed in §5.7. From Theorem 5.15 and Remark 5.17, in general with a single constraint, we should expect to characterize each local extremal function Y_0 of $F|_{\mathscr{D}}$ when further constrained to G_{Y_0}, the level set of G through Y_0, by a $\lambda \in \mathbb{R}$ for which

$$\delta(F + \lambda G)(Y_0; V) = 0, \qquad \forall\, V \in \mathscr{D}_0,$$

(since in this case $\mathscr{D}_0 = \mathscr{Y}_0$ is a subspace).

In particular, if the constraining function G is itself defined by an integral in the form

$$G(Y) = \int_a^b g(x, Y(x), Y'(x))\, dx,$$

then Y_0 will be stationary for the *modified* function $f + \lambda g$, and so should satisfy the corresponding Euler–Lagrange equation(s):

$$\frac{d}{dx}(f + \lambda g)_{Y'}(x) = (f + \lambda g)_Y(x) \tag{36}$$

and an analogous *second* equation (35). Multiple constraining functions defined by integrals are amenable to a similar extension of Theorem 5.16 (and Remark 5.17).

In applications, another symbol (usually "t") may be used to represent the independent variable in the above formulas thereby freeing x to represent one component of the vector valued Y. For example, in problems involving planar curves, the use of $Y = (x, y)$ is both more suggestive and less cumbersome than $Y = (y_1, y_2)$.

Application 1: The Isoperimetric Problem

For the original isoperimetric problem as formulated in §1.3, we are led to consider the area function of Example 7 of §2.4

$$A(Y) \stackrel{\text{def}}{=} \int_0^1 x(t)y'(t)\, dt$$

which by an easy extension of Proposition 5.11, has the weakly continuous variations in the direction $V = (u, v)$ given by

$$\delta A(Y; V) = \int_0^1 [x(t)v'(t) + y'(t)u(t)] \, dt.$$

We wish to maximize A on

$$\mathscr{D} = \{Y \in \mathscr{Y} = (C^1[0, 1])^2 : Y(0) = Y(1)\},$$

subject to the isoperimetric condition

$$L(Y) = \int_0^1 |Y'(t)| \, dt = l \quad \text{(given)}.$$

Without loss of generality we can suppose that the domain D whose area is to be maximized has the origin $\mathcal{O}$ in its boundary, and require that the curves represented by Y originate and end there, so that $Y(0) = Y(1) = \mathcal{O}$; we may also take $\mathscr{D}_0 = \mathscr{D} = \mathscr{Y}_0$. [Note, however, that we have not excluded the possibility that the origin is a *corner* point for the curve.]

Formally (from Example 8 of §2.4),

$$\delta L(Y; V) = \int_0^1 \frac{Y'(t)}{|Y'(t)|} \cdot V'(t) \, dt,$$

so that in order to apply Proposition 5.5, we must further restrict attention to

$$\mathscr{D}^* = \{Y \in \mathscr{D} : |Y'| \neq 0 \text{ on } [0, 1]\}.$$

Curves defined by $Y \in \mathscr{D}^*$ are said to be *smooth*, and a typical curve in $\mathscr{D}^*$ is sketched in Figure 6.8.

$\delta L(Y; V)$ is also weakly continuous on $\mathscr{D}^*$. (This is not immediate: see Problem 5.35.)

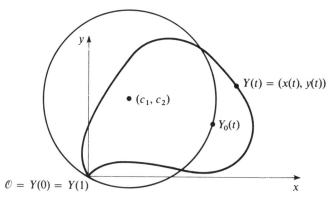

Figure 6.8

Now, if for some $Y \in \mathscr{D}^*$, $\delta L(Y; V) = 0$, $\forall V \in \mathscr{D}_0$, then from Corollary 4.7 it would follow that the unit vector $Y'/|Y'| = $ const. so that Y' has a constant direction, but such functions are not in $\mathscr{D}^*$. (The function $Y(t) \equiv (0, 0)$ is in $\mathscr{D}$ but it is *not* in $\mathscr{D}^*$.)

Hence from Theorem 5.15 and Remark 5.17, if $Y_0 \in \mathscr{D}^*$ maximizes A (locally) on $\mathscr{D}^*$ when restricted to the l level set of L, then $\exists \lambda \in \mathbb{R}$ such that $\delta(A + \lambda L)(Y_0; V) = 0$, $\forall V \in \mathscr{D}_0$; Y_0 is stationary for the function $xy' + \lambda |Y'|$ and satisfies the associated Euler–Lagrange equation(s) (34):[1]

$$\frac{d}{dt}\left(\frac{\lambda x'}{|Y'|}\right) = y' \quad \text{and} \quad \frac{d}{dt}\left(x + \frac{\lambda y'}{|Y'|}\right) = 0.$$

Thus $y - (\lambda x'/|Y'|) = c_2$ and $x + (\lambda y'/|Y'|) = c_1$, for constants c_1 and c_2. Hence $c_2 y' + c_1 x' = yy' + xx'$, or upon integrating:

$$x^2 + y^2 - 2c_1 x - 2c_2 y = \text{const.} = 0,$$

when evaluated at $t = 0$; this gives $(x - c_1)^2 + (y - c_2)^2 = c_1^2 + c_2^2$ and the curve represented by Y_0 is seen to lie on a circle through the origin. However, closure requires that the circle be traversed completely at least once, while maximility in A for the given length l of traversal, could be associated only with a *single* traversal. On geometric grounds, $c_1^2 + c_2^2 = l^2/4\pi^2$ (Figure 6.8).

A could not not have a minimum value on $\mathscr{D}^*$ (or on $\mathscr{D}$). (See Problem 6.38.) Thus we support but still have not proven Pappus' conjecture that the maximal curve is the circle. (See §8.8 and §9.5.) Observe that the circle does *not* exhibit a corner point at the origin.

(Problems 6.36–6.39)

Lagrangian Constraints*

The method of Lagrangian multipliers may also be adapted to the case of constraints of the form

$$g[Y(x)] = g(x, Y(x), Y'(x)) \equiv 0, \qquad \forall x \in [a, b],$$

where $g \in C^1(D)$ for a suitable domain D of $\mathbb{R}^{2d+1}$. We shall consider only the simple case of a single constraint $g(Y(x)) \equiv 0$, $\forall x \in [a, b]$ which is required for the discussion of Hamiltonian mechanics in §8.6. The general case will be treated in §11.3.

(6.10) **Theorem** (Lagrange). *For $f = f(x, Y, Z)$ and $f_{z_j} \in C^1([a, b] \times \mathbb{R}^{2d})$, $j = 1, 2, \ldots, d$, suppose Y_0 is C^2 and it minimizes*

$$F(Y) = \int_a^b f(x, Y(x), Y'(x)) \, dx$$

---

[1] For this problem the second equation (35) is satisfied trivially.

locally on

$$\mathcal{D} = \{Y \in \mathcal{Y} = (C^1[a, b])^d : Y(a) = Y_0(a), Y(b) = Y_0(b)\},$$

when subject to the constraint

$$g(Y(x)) \equiv 0, \qquad \forall \, x \in [a, b],$$

where $g = g(Y)$ is a C^2 function for which

$$\nabla g(Y_0(x)) \neq \mathcal{O} \qquad (d > 1).$$

Then $\exists \, \lambda \in C[a, b]$ such that Y_0 is stationary for the modified function $f + \lambda g$.

Remark. If $g_{y_j}(Y_0(x)) \neq 0$ for some j, then by relabelling, if necessary, we can arrange that $Y = (y_j, \overline{Y})$. The method of proof involves the local elimination of this distinguished variable, and to simplify the presentation, we assume initially that g has a form which readily permits this.

PROOF*. Suppose that $g(Y) = y - \psi(\overline{Y})$ for some C^2 function ψ. Then $g(Y) = 0$ iff $y = \psi(\overline{Y})$, and, in particular, $y_0(x) \equiv \psi(\overline{Y}_0(x))$ if we represent $Y_0(x) = (y_0(x), \overline{Y}_0(x))$.

Introduce $\overline{\mathcal{D}} = \{\overline{Y} \in C^1([a, b])^{d-1} : \overline{Y}(a) = \overline{Y}_0(a); \overline{Y}(b) = \overline{Y}_0(b)\}$ and observe that for each $\overline{Y} \in \overline{\mathcal{D}}$, we may define $y(x) = \psi(\overline{Y}(x))$, to obtain $Y = (\psi(\overline{Y}), \overline{Y}) \in \mathcal{D}$ since $y(a) = \psi(\overline{Y}(a)) = \psi(\overline{Y}_0(a)) = y_0(a)$ (and similarly $y(b) = y_0(b)$). Moreover, by construction, $g(Y(x)) = y(x) - \psi(\overline{Y}(x)) \equiv 0$; i.e., these Y *automatically* satisfy the constraining equation. We also have $y'(x) = \nabla \psi(\overline{Y}(x)) \cdot \overline{Y}'(x)$ expressed in terms of the gradient of ψ.

For $\overline{Y} \in \overline{\mathcal{D}}$, we may consider the *unconstrained* function

$$\overline{F}(\overline{Y}) \stackrel{\text{def}}{=} F(\psi(\overline{Y}), \overline{Y}) = \int_a^b \overline{f}[\overline{Y}(x)] \, dx,$$

where

$$\overline{f}(x, \overline{Y}, \overline{Z}) \stackrel{\text{def}}{=} f(x; y, \overline{Y}; z, \overline{Z}), \text{ with } y = \psi(\overline{Y}) \text{ and } z = \nabla \psi(\overline{Y}) \cdot \overline{Z}.$$

From the chain rule, it follows that in abbreviated form:

$$\overline{f}_{\overline{Z}} = f_z \nabla \psi + f_{\overline{Z}},$$

while

$$\overline{f}_{\overline{Y}} = f_y \nabla \psi + f_{\overline{Y}} + f_z H,$$

where $H(\overline{Y}, \overline{Z}) \stackrel{\text{def}}{=} (\nabla \psi(\overline{Y}) \cdot \overline{Z})_{\overline{Y}}$.

Now $\overline{Y}_0$ minimizes $\overline{F}$ (locally) on $\overline{\mathcal{D}}$ (Why?), and hence as in 6.8, it is a solution of the first equation in the form

$$\frac{d}{dx} \overline{f}_{\overline{Z}}[\overline{Y}(x)] = \overline{f}_{\overline{Y}}[\overline{Y}(x)].$$

Upon substitution of the preceding equations and subsequent simplification it can be seen that $Y_0 = (\psi(\overline{Y}_0), \overline{Y}_0)$ is a solution of the equation

$$\frac{d}{dx}[f_{y'}(x)\nabla\psi(\overline{Y}(x)) + f_{\overline{Y}'}(x)]$$

$$= f_y(x)\nabla\psi(\overline{Y}(x)) + f_{\overline{Y}}(x) + f_{y'}(x)\frac{d}{dx}\nabla\psi(\overline{Y}(x)). \tag{37}$$

(In obtaining the last term, we have utilized the identities

$$\frac{\partial}{\partial y_i}(\nabla\psi(\overline{Y}(x))\cdot\overline{Y}'(x)) = \sum_{j=2}^{d}\psi_{y_iy_j}(\overline{Y}(x))y_j'(x) = \frac{d}{dx}\psi_{y_i}(\overline{Y}(x)),$$

for $i = 2, 3, \ldots, d$, which hold since ψ is C^2.)

By hypothesis, $f_z[Y_0(x)]$ is C^1. (Why?) Hence when $Y = Y_0$, each term in the bracketed expression of (37) is C^1 so that Y_0 also satisfies the equation

$$\frac{d}{dx}f_{\overline{Y}'}(x) - f_{\overline{Y}}(x) = -\left[\frac{d}{dx}f_{y'}(x) - f_y(x)\right]\nabla\psi(\overline{Y}(x)).$$

Finally, since $g_{y'} \equiv 0$ and $g_{\overline{Y}'} \equiv 0$, while $g_{\overline{Y}} \equiv -g_y\nabla\psi$; (here, $g_y \equiv 1$), then for each $\lambda \in C[a, b]$,

$$\frac{d}{dx}(f + \lambda g)_{\overline{Y}'}(x) - (f + \lambda g)_{\overline{Y}}(x) = \frac{d}{dx}f_{\overline{Y}'}(x) - f_{\overline{Y}}(x) - \lambda(x)g_{\overline{Y}}(x)$$

$$= -\left[\frac{d}{dx}f_{y'}(x) - f_y(x) - \lambda(x)g_y(x)\right]\nabla\psi(\overline{Y}(x)). \tag{37'}$$

Now, since $g_y \neq 0$ (here, $g_y \equiv 1$), for $x \in [a, b]$, we may define

$$\lambda(x) = \frac{(d/dx)f_{y'}(x) - f_y(x)}{g_y(x)}\bigg|_{Y(x)=Y_0(x)}, \tag{37''}$$

and see that this λ in $C[a, b]$ forces the vanishing of the bracketed terms on the *right side* of the last equation which in turn makes the *left side* vanish as well. Upon combining these assertions we have for *this* λ that Y_0 is a solution of the equations

$$\frac{d}{dx}(f + \lambda g)_{Y'}(x) = (f + \lambda g)_Y(x),$$

and so it is stationary for the modified function $f + \lambda g$ under the simplifying assumption $g(Y) = y - \psi(\overline{Y})$.

In any case, the hypotheses require the nonvanishing of some $g_{y_j}(Y_0(\xi))$ for each $\xi \in [a, b]$. Thus by implicit function theory ([Ed]), for each such ξ there is a *locally determined* C^2 function ψ for which $g(Y) = 0$ iff $y = y_j = \psi(\overline{Y})$ for all those Y near $Y_0(\xi)$ in $\mathbb{R}^d$. Differentiating the resulting local

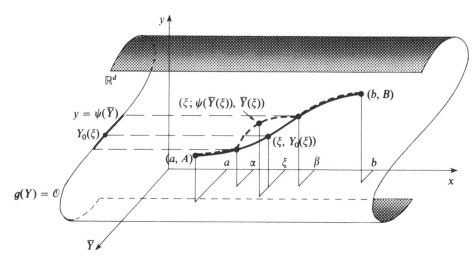

Figure 6.9

identity $g(\psi(\overline{Y}), \overline{Y}) \equiv 0$ shows that $g_{\overline{Y}}(\psi(\overline{Y}), \overline{Y}) \equiv -g_y(\psi(\overline{Y}), \overline{Y})\nabla\psi(\overline{Y})$, or in abbreviated form: $g_{\overline{Y}} = -g_y\nabla\psi$.

We now consider those $\overline{Y} \in \mathscr{D}$ near $\overline{Y}_0$ which differ from $\overline{Y}_0$ only in a *small interval* $[\alpha, \beta]$ containing ξ, and suppose that $y(x) = \psi(\overline{Y}(x))$ is defined in $[\alpha, \beta]$. Then we can set $y(x) = y_0(x)$ *outside* $[\alpha, \beta]$ to obtain a $Y = (y, \overline{Y}) \in \mathscr{D}$ as before. See Figure 6.9. ($A = Y_0(a)$ and $B = Y_0(b)$.)

For such Y, with $\overline{f}$ defined as above, we have

$$\int_\alpha^\beta \overline{f}[\overline{Y}(x)]\, dx - \int_\alpha^\beta \overline{f}[\overline{Y}_0(x)]\, dx = F(Y) - F(Y_0);$$

thus $\int_\alpha^\beta \overline{f}[\overline{Y}(x)]\, dx$ is minimized (locally) at $\overline{Y}_0|_{[\alpha,\beta]}$. Each such $\overline{Y}$ yields a $\overline{V} \equiv \overline{Y} - \overline{Y}_0$, vanishing at α and β *together with its derivative,* $\overline{V}'$, which is admissible for variation (on $[\alpha, \beta]$). $\overline{f}_{\overline{z}}[\overline{Y}_0(x)]$ is C^1 (Why?), so that *on* (α, β), $\overline{Y}_0$ is stationary for $\overline{f}$. (See Problem 4.3.)

Each step of the preceding argument is now usable and in particular, we can again use the λ as given by equation (37″) to make $Y_0|_{[\alpha,\beta]}$ stationary for $f + \lambda g$ on (α, β). But, the point ξ determining $[\alpha, \beta]$ is arbitrary, and from (37′) we see that $\lambda(\xi)$ is given by (37″) *independently of which* $y = y_j$ *we take* as long as $g_{y_j}[Y_0(\xi)] \neq 0$. Thus λ is a well-defined function in $C[a, b]$ and Y_0 is stationary for $f + \lambda g$ on (a, b).   □

(6.11) **Remark.** It is straightforward to extend this method of proof to $N < d$ constraining functions $g_j, j = 1, 2, \ldots, N$ of the *same* form, provided that a suitable $N \times N$ Jacobian determinant of the constraining functions is non-vanishing along the stationary trajectory. The conclusion is that there exist

N functions $\lambda_j \in C[a, b]$ such that Y_0 is stationary for the modified function

$$f + \sum_{j=1}^{N} \lambda_j g_j.$$

When $N = d$, then Y_0 will in general be the only function satisfying the constraining equations.

Application 2: Geodesics on a Surface

In $\mathbb{R}^3$, the surface of an ellipsoid is one of many that can readily be described as the zero-level set of a smooth function g. If we assume that there is a geodesic curve (Y_0) joining points A and B on this surface, and that along this curve $\nabla g \neq \mathcal{O}$, then we may use the previous theorem to characterize this geodesic when both Y_0 and g are C^2.

As in §1.1, suppose that a general curve joining A and B is the range of $Y \in (C^1[a, b])^3$. Its length is, of course,

$$L(Y) = \int_a^b |Y'(t)|\, dt,$$

and by assumption, Y_0 minimizes L among such curves which lie on the surface, i.e., for which $g(Y(t)) \equiv 0, t \in [a, b]$.

It follows that $\exists\, \lambda \in C[a, b]$ for which Y_0 is stationary for $f + \lambda g$ where $f(t, Y, Z) = |Z|$; thus from (34):

$$\frac{d}{dt}[(f + \lambda g)_{Y'}(t)] = (f + \lambda g)_Y(t),$$

or, since $f_Y \equiv g_{Y'} \equiv \mathcal{O}$,

$$\frac{d}{dt}(f_{Y'}(t)) = (\lambda g_Y)(t),$$

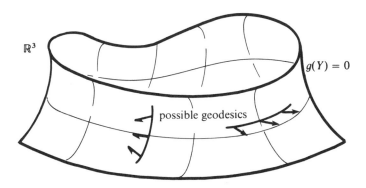

Figure 6.10

or, after substitution,

$$\frac{d}{dt}\left(\frac{Y'}{|Y'|}(t)\right) = \lambda(t)\nabla g(Y(t)).$$

Along Y_0 we may use the arc length s as the parameter. Then $|Y_0'(s)| \equiv 1$, and for a new λ, the above equation becomes

$$Y_0''(s) = \lambda(s)\nabla g(Y_0(s)),$$

which shows that in general the *principal normal* to a geodesic on a surface is in the direction of the (nonvanishing) gradient and so is *normal to the surface* at each point.

Observe that we have not established the existence of geodesics for a general surface, but we have obtained valuable insight as to the manner in which such geodesics should lie on the surface. (See Figure 6.10.)

§6.8*. Invariance of Stationarity

For certain physical applications, and especially for Hamilton's principle (Chapter 8), it is important to know that the property of being a stationary function does not change with the coordinate system used to describe it. As indicated in Problem 6.35*, this need *not* be the case if the function is only C^1. However, when the function is C^2, so that both the first and second equations are satisfied (§6.7), then invariance can be established under the following general transformation:

We suppose that a portion of $\mathbb{R}^{d+1}$ containing the *graph* of a function $Y \in (C^1[a, b])^d$ is mapped into $\mathbb{R}^{d+1}$ under the invertible transformation

$$\begin{aligned}\xi &= \tilde\varphi(x, Y), & x &= \varphi(\xi, H), \\ H &= \tilde\Psi(x, Y), & Y &= \Psi(\xi, H),\end{aligned} \tag{38}$$

which carries the graph of Y onto the graph of $H \in (C^1[a, \beta])^d$ as shown in Figure 6.11. The functions φ, Ψ and $\tilde\varphi$, $\tilde\Psi$ are to be C^2. We further suppose that, $\varphi(\xi, H(\xi))$ *increases strictly* from a to b with a positive derivative, and, conversely, that $\tilde\varphi(x, Y(x))$ *increases strictly* from α to β, say.

All of these conditions are realized in the case of the simple skew transformation considered in §6.3 and the proof of the next theorem is accomplished by an appropriate extension of the formula obtained in Problem 6.35(d). The reader should examine this formula and consider the simpler transformation at each step of the following argument, which is straightforward but complicated by the generality.

Since the point $(x, Y(x))$ is transformed by (38) into $(\xi, H(\xi))$ we must have $\Psi(\xi, H(\xi)) \equiv Y(\varphi(\xi, H(\xi)))$, so that upon differentiation,

$$(\Psi_\xi + \Psi_H H')(\xi) = Y'(x)u(\xi), \tag{39}$$

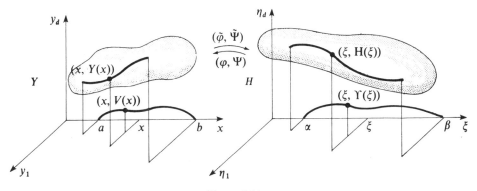

Figure 6.11

where

$$u(\xi) = \frac{d}{d\xi} \varphi(\xi, H(\xi)) = (\varphi_\xi + \varphi_H \cdot H')(\xi) > 0, \tag{39'}$$

in view of our assumptions.

Thus we have the following correspondence between triples:

$$(x, Y(x), Y'(x)) = \left(\varphi(\xi, H(\xi)), \Psi(\xi, H(\xi)), \frac{(\Psi_\xi + \Psi_H H')(\xi)}{u(\xi)} \right); \tag{40}$$

(Ψ_H is the Jacobian matrix having elements $\partial\psi_i/\partial\eta_j$, $i, j = 1, 2, \ldots, d$, arranged in natural order with its *rows* indexed by i).

Now, when $f \in C^1([a, b] \times \mathbb{R}^3)$, then under the transformation $x = \varphi(\xi, H(\xi))$, [so that (formally) $dx = u(\xi) \, d\xi$], we have

$$F(Y) = \int_a^b f[Y(x)] \, dx = \int_\alpha^\beta \tilde{f}[H(\xi)] \, d\xi = \tilde{F}(H),$$

say, if we define

$$\tilde{f}(\xi, H, Z) = f\left(\varphi(\xi, H), \Psi(\xi, H), \frac{\Psi_\xi(\xi, H) + \Psi_H(\xi, H)Z}{u} \right) u, \tag{41}$$

where the function

$$u = u(\xi, H, Z) = \varphi_\xi(\xi, H) + \varphi_H(\xi, H) \cdot Z, \tag{42}$$

is that used to produce (39').

The desired invariance of stationarity is expressed in the

(6.12) **Theorem.** *If* $Y \in (C^2[a, b])^d$ *and* $\delta F(Y; V) = 0$, $\forall V$ *in*

$$\mathcal{D}_0 = \{V \in (C^1[a, b])^d : V(a) = V(b) = \mathcal{O}\},$$

then under the transformation (38), H *is* C^2 *and* $\delta\tilde{F}(H; \Upsilon) = 0$, $\forall \Upsilon$ *in* $\tilde{\mathcal{D}}_0 = \{\Upsilon \in (C^1[\alpha, \beta])^d : \Upsilon(\alpha) = \Upsilon(\beta) = \mathcal{O}\}$.

Remarks. Each $V \in \mathcal{D}_0$ provides $\Upsilon \in \tilde{\mathcal{D}}_0$ defined by

$$\Upsilon(\xi) = V(\varphi(\xi, H(\xi))). \tag{43}$$

Now, the *graphs* of V and Υ need *not* correspond under (38), and may lie outside the domain(s) of the transformation; however, we shall prove that for *this* Υ:

$$\delta\tilde{F}(H; \Upsilon) = \delta F(Y; \Psi_H V) + \delta_2 F(Y; \varphi_H \cdot V), \tag{44}$$

where

$$\delta_2 F(Y; w) \overset{\text{def}}{=} \int_a^b [f_x(x)w(x) + (f - Y' \cdot f_{Y'})(x)w'(x)]\, dx. \tag{45}$$

Here, $\delta_2 F(Y; w) = (f - Y' \cdot f_{Y'})(x)w(x)|_a^b$, since this Y is C^2 and stationary from the hypothesis so that it satisfies the *second* equation, (35), viz., $(d/dx)(f - Y' \cdot f_{Y'})(x) = f_x(x)$.[1]

In (44), φ_H and Ψ_H must be understood as being retransformed into functions of x; i.e., φ_H is actually $\varphi_H(\tilde{\varphi}(x, Y(x)), \tilde{\Psi}(x, Y(x)))$, and Ψ_H is actually $\Psi_H(\tilde{\varphi}(x, Y(x)), \tilde{\Psi}(x, Y(x)))$.

Then $\Psi_H V$ is a vector function in $\mathcal{D}_0$ as is easily verified by matrix multiplication, and the *first* term of (44) vanishes by hypothesis. Similarly, $w(x) \overset{\text{def}}{=} (\varphi_H \cdot V)(x)$ vanishes at a and b since $V \in \mathcal{D}_0$; thus by (45) et seq. the *second* term vanishes as well, and we see that for this particular Υ: $\delta\tilde{F}(H; \Upsilon) = 0$.

Conversely, each $\Upsilon \in \tilde{\mathcal{D}}_0$ arises from that $V \in \mathcal{D}_0$ given by $V(x) \overset{\text{def}}{=} \Upsilon(\tilde{\varphi}(x, Y(x)))$. (Why?) Hence we conclude that $\delta\tilde{F}(H; \Upsilon) = 0, \forall \Upsilon \in \tilde{\mathcal{D}}_0$.

To avoid complications, we shall carry out the derivation of (44) only in the case $d = 1$, being careful to preserve the order which would be essential for the corresponding matrices of the higher-dimensional version. Thus H, Z, Υ, Ψ, Y, Z, V reduce to η, ζ, v, ψ, y, z, v, respectively. (However, see Chapter 9 where analogous vector valued computations are carried out.)

PROOF*. From (43), $v(\xi) = v(\varphi(\xi, \eta(\xi)))$ so that $v'(\xi) = v'(x)u(\xi)$. Then

$$\delta\tilde{F}(\eta; v) = \int_\alpha^\beta (\tilde{f}_\eta[\eta(\xi)]v(\xi) + \tilde{f}_\zeta[\eta(\xi)]v'(\xi))\, d\xi, \tag{46}$$

but from (41) and the chain rule with the usual abbreviations,

$$\tilde{f}_\eta[\eta(\xi)] = f(x)u_\eta + f_y(x)\psi_\eta u + f_{y'}(x)(\psi_{\xi\eta} + \psi_{\eta\eta}\eta')$$
$$- f_{y'}(x)\left[\frac{\psi_\xi + \psi_\eta\eta'}{u}\right]u_\eta + f_x(x)\varphi_\eta u,$$

---

[1] δ_2 represents the effect in the x-direction of a variation γ defined in a different coordinate system. We utilized this effect to derive the second equation in §6.3.

and

$$\tilde{f}_\zeta[\eta(\xi)] = f(x)\varphi_n + f_{y'}(x)\psi_n - f_{y'}(x)\left[\frac{\psi_\xi + \psi_n\eta'}{u}\right]\varphi_n,$$

(since from (42), $u_\zeta(\xi, \eta, \zeta) = \varphi_n(\xi, \eta)$). The bracketed term in each case is seen to be simply $y'(x)$ by equation (39), and hence, under the transformation $x = \varphi(\xi, \eta(\xi))$, $dx = u(\xi)\,d\xi$, the integrals in (46), may be recognized as arising from the integrals:

$$\int_a^b \left\{ f_y(x)\psi_n v(x) + f_{y'}(x)\left[\psi_n v'(x) + \left(\frac{\psi_{\xi\eta} + \psi_{\eta\eta}\eta'}{u}\right)v(x)\right]\right\} dx$$

$$+ \int_a^b \left\{ f_x(x)\varphi_n v(x) + (f - f_{y'}y')(x)\left[\varphi_n v'(x) + \frac{u_\eta}{u}v(x)\right]\right\} dx,$$

where we have restored $v(\xi)$ to $v(x)$ and so $v'(\xi)$ to $v'(x)u(\xi)$.

To complete the proof, it is necessary only to observe that by the chain rule (formally)

$$\frac{d}{dx} = \frac{1}{u(\xi)}\frac{d}{d\xi},$$

so that

$$\frac{d}{dx}\varphi_n = \frac{1}{u(\xi)}\frac{d}{d\xi}\varphi_n = \frac{1}{u(\xi)}(\varphi_{\xi\eta} + \varphi_{\eta\eta}\eta') = \frac{u_\eta(\xi)}{u(\xi)} \quad \text{by (42)};$$

and similarly,

$$\frac{d}{dx}\psi_n = \frac{1}{u(\xi)}\frac{d}{d\xi}\psi_n = \frac{\psi_{\xi\eta} + \psi_{\eta\eta}\eta'}{u(\xi)}.$$

Then we recognize each bracketed term in the last integrals as the derivative of a product, and we obtain $\delta\tilde{F}(\eta; v) = \delta F(y; \psi_n v) + \delta_2 F(y; \varphi_n v)$ in view of (45).

Observe that under the transformation (38) both Y and H are assumed to be C^1, and from (39) and (40) it would follow that

$$Y''(x) = \frac{1}{u(\xi)}\frac{d}{d\xi}Y'(x) = \frac{1}{u(\xi)}\frac{d}{d\xi}\frac{(\Psi_\xi + \Psi_H H')(\xi)}{u(\xi)}$$

is defined and continuous when H is C^2. By the corresponding argument we could conclude that conversely when Y is C^2, then H is C^2 also. $\square$

§6.9. Multidimensional Integrals

As we have seen in §1.4(c) and §3.4(e), it is frequently required to optimize functions F defined by an integral over a domain D of $\mathbb{R}^d$ where $d > 1$. When the structure of D and its boundary ∂D are sufficiently regular, it is straightforward to obtain formal analogues for most of the one-dimensional results obtained hitherto in this chapter. (Without this regularity, however, far more sophisticated tools are required to handle the delicate questions concerning behavior at the boundary. See [G–T].)

A typical point in $\mathbb{R}^d$ will be denoted by $X = (x_1, x_2, \ldots, x_d)$ in Cartesian coordinates, and the d-dimensional element of integration by dX. D is assumed to be a *bounded Green's domain* in $\mathbb{R}^d$—i.e., one for which the boundary ∂D consists of $(d-1)$-dimensional surfaces on which integration is possible such that Green's theorem holds in the *divergence* form

$$\int_D (\nabla \cdot U)\, dX = \int_{\delta D} (U \cdot N)\, d\sigma, \qquad \forall\, U \in (C^1(\overline{D}))^d. \tag{47}$$

Here $C^1(\overline{D})$ denotes the set of real valued functions $u \in C(\overline{D})$ which in D have first partial derivatives admitting continuous extensions to $\overline{D}$; $U = (u_1, u_2, \ldots, u_d)$ is a d-tuple of such functions with the *divergence*

$$\nabla \cdot U \overset{\text{def}}{=} \sum_{j=1}^d (u_j)_{x_j};$$

N is the *outward* pointing unit normal vector on ∂D (which is defined except at a set negligible with respect to the surface integration), and $d\sigma$ denotes the element of integration on ∂D (Figure 6.12(a)). (See [Ed].)

For example, in $\mathbb{R}^3$, all of the above hold when D is the interior of a rectangular box with faces parallel to the coordinate planes (Figure 6.12(b)). Then $dX = dx_1\, dx_2\, dx_3$ and the integration over D can be expressed through iterated integrals. ∂D consists of the six faces of the box which meet only at the edges. Finally on each face, $d\sigma = dx_1\, dx_2$, or $dx_2\, dx_3$, or $dx_1\, dx_3$ so that the surface integrals can also be expressed through iterated integrals, and

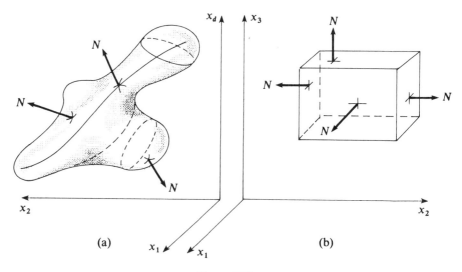

(a)   (b)

Figure 6.12

Green's theorem can be verified by partial integrations. However, all of the above also hold when D is the interior of an ellipsoid in $\mathbb{R}^3$, but in this case, it is difficult to express dX, N and $d\sigma$ in forms convenient enough to verify Green's theorem easily. A typical $u \in C^1(\bar{D})$ has in D, the gradient $\nabla u \overset{\text{def}}{=} (u_{x_1}, u_{x_2}, \ldots, u_{x_d})$, and hence

$$\|u\|_M = \max_X \left(|u| + |\nabla u|\right)(X)$$

will supply a norm for this space. (See Problem 5.42.)

Given $f \in C^1(\bar{D} \times \mathbb{R} \times \mathbb{R}^d)$ and $\gamma \in C(\partial D)$, we shall be interested in finding conditions necessary that $u \in C^1(\bar{D})$ be a local extremal function for

$$F(u) = \int_D f(X, u(X), \nabla u(X)) \, dX = \int_D f[u(X)] \, dX$$

on

$$\mathscr{D} \overset{\text{def}}{=} \{u \in C^1(\bar{D}): u|_{\partial D} = \gamma\}.$$

We introduce the *admissible* directions (§5.5)

$$\mathscr{D}_0 = \{v \in C^1(\bar{D}): v|_{\partial D} \equiv 0\},$$

and observe that if u is a local extremal point for F on $\mathscr{D}$, then

$$\delta F(u; v) = \frac{\partial}{\partial \varepsilon} F(u + \varepsilon v)\bigg|_{\varepsilon=0} = 0, \qquad \forall \, v \in \mathscr{D}_0.$$

The differentiability assumptions on f permit this variation to be calculated by Leibniz' formula as follows:

$$\delta F(u; v) = \int_D [f_u(X)v(X) + f_{\nabla u}(X) \cdot \nabla v(X)] \, dX, \tag{48}$$

where $f_u(X) = f_u(X, u(X), \nabla u(X))$ and $f_{\nabla u}(X)$ is the *vector-valued* function with components

$$f_{z_j}(X) = f_{z_j}(X, u(X), \nabla u(X)), \qquad j = 1, 2, \ldots, d;$$

(where $f = f(X, u, Z) = f(X, u, z_1, z_2, \ldots, z_d)$).

Next, we suppose that $f \in C^2(\bar{D} \times \mathbb{R} \times \mathbb{R}^d)$, and that $u \in \mathscr{D} \cap C^2(D)$ (as in §3.4(e)) so that we can integrate the *second* term of (48) by parts using Green's theorem (47), for $U(X) = v(X)f_{\nabla u}(X)$, as follows:

$$\int_D [f_{\nabla u}(X) \cdot \nabla v(X)] \, dX = \int_D \nabla \cdot (v(X)f_{\nabla u}(X)) \, dX - \int_D [v(X)\nabla \cdot f_{\nabla u}(X)] \, dX$$

$$= \int_{\partial D} v(X)f_{\nabla u}(X) \cdot N(X) \, d\sigma - \int_D v(X)\nabla \cdot f_{\nabla u}(X) \, dX$$

$$= -\int_D v(X)(\nabla \cdot f_{\nabla u}(X)) \, dX, \qquad \forall \, v \in \mathscr{D}_0, \tag{49}$$

since the boundary integrand vanishes when $v \in \mathscr{D}_0$.

Thus, finally, if $u \in \mathscr{D} \cap C^2(D)$, then $\forall\, v \in \mathscr{D}_0$

$$\delta F(u; v) = \int_D [f_u(X) - \nabla \cdot f_{\nabla u}(X)]v(X)\, dX. \tag{50}$$

In particular, if u in $\mathscr{D} \cap C^2(D)$ is a local extremal for F on $\mathscr{D}$ then the integral in (50) vanishes $\forall\, v \in \mathscr{D}_0$, and we wish to conclude from this that the bracketed term in the integral vanishes identically in D. This can be accomplished by a natural extension of the proof of Lemma 4.4 of Lagrange. (See Problem 4.5.) Thus, finally, we obtain the desired analogue of the fundamental result of Euler–Lagrange.

(6.13) **Theorem.** *Let D be a Green's domain of $\mathbb{R}^d$ and suppose that functions $f \in C^2(\bar{D} \times \mathbb{R} \times \mathbb{R}^d)$ and $\gamma \in C(\partial D)$ are given. Then in order that $u_0 \in C^2(D)$ be a (local) extremal for*

$$F(u) = \int_D f(X, u(X), \nabla u(X))\, dX = \int_D f[u(X)]\, dX$$

on

$$\mathscr{D} = \{u \in C^1(\bar{D}): u|_{\partial D} = \gamma\}$$

it is necessary that u_0 be stationary for f in D; i.e., that u_0 satisfy the equation

$$\nabla \cdot f_{\nabla u}(X) = f_u(X), \qquad \forall\, X \in D. \tag{51}$$

$\square$

Application: Minimal Area Problem

For the minimal area problem of Plateau discussed in §1.4(c) and §3.4(e), we take $d = 2$. Then $f(X, u, Z) = (1 + z_1^2 + z_2^2)^{1/2}$ has continuous derivatives of all orders; $f_u = 0$ and $f_{z_j} = z_j(1 + z_1^2 + z_2^2)^{-1/2}, j = 1, 2$. Denoting X by (x, y), so that $\nabla u = (u_x, u_y)$, the Euler–Lagrange equation (51) for this problem is

$$\nabla \cdot \left(\frac{\nabla u}{(1 + u_x^2 + u_y^2)^{1/2}} \right) = 0, \tag{52}$$

which agrees with Equation 26 of §3.4. Thus in order that $u_0 \in C^2(D)$ have a graph with a local extremal surface area among all such functions with the same continuous boundary values, it is necessary *and* sufficient that u_0 satisfy the *minimal* surface equation ((26) of §3.4) or its equivalent. As we have noted in §3.4(e), this equation has a solution with arbitrarily prescribed continuous boundary values *iff* the domain D is convex [Os].

Multidimensional problems arise naturally when Hamilton's principle is applied to obtain the equations governing the motions of elastic bodies. We shall reserve further discussion until §8.9.

(Problem 6.41)

Natural Boundary Conditions

Multidimensional problems permit considerable flexibility in the specification of the boundary conditions, and boundary point freedom from specification gives rise to natural boundary conditions which a local extremal function must satisfy.

(6.14) **Corollary.** *If as in Theorem 6.13, $u_0 \in C^2(D)$ is a local extremal function for F on*

$$\tilde{\mathscr{D}} = \{u \in C^1(\overline{D}): u(X) = \gamma(X), X \in K\},$$

where K is a compact subset of ∂D, then

(i) *in D, u_0 satisfies (51);*
(ii) *at each boundary point $X_0 \notin K$ having a neighborhood of ∂D in which N is continuous: $f_{\nabla u}(X_0) \cdot N(X_0) = 0$.*

PROOF*. Each $v \in \tilde{\mathscr{D}}_0 = \{v \in C^1(\overline{D}): v(X) = 0, X \in K\}$ is $\tilde{\mathscr{D}}$-admissible (Why?), and since $\tilde{\mathscr{D}}_0 \supseteq \mathscr{D}_0$, we conclude as before that u_0 satisfies (51).

Then, when $v \in \tilde{\mathscr{D}}_0$, it follows from (48) using Green's theorem as in (49), that for $u = u_0$:

$$0 = \delta F(u; v) = \int_D [f_u(X) - (\nabla \cdot f_{\nabla u})(X)] v(X) \, dX$$

$$+ \int_{\partial D} v(X)(f_{\nabla u}(X) \cdot N(X)) \, d\sigma$$

$$= \int_{\tilde{K}} v(X) f_{\nabla u}(X) \cdot N(X) \, d\sigma,$$

since (i) holds and v vanishes on K. ($\tilde{K} = \partial D \sim K$.)

Now, each $X_0 \notin K$ is a positive distance from the *compact* set K. [Otherwise we could find a sequence of points in K with limit point $X_0 \notin K$ (see §5.2).] Thus we can construct functions $v \in C^1(\overline{D})$, which are *positive* at X_0 and vanish outside neighborhoods of X_0 so small that $v \in \tilde{\mathscr{D}}_0$. By hypothesis $f_{\nabla u}(X)$ is continuous at X_0; hence if also N is continuous at X_0, we can use a version of the standard local arguments to conclude that $f_{\nabla u}(X_0) \cdot N(X_0) = 0$. (See Problem 4.5.) □

Consider the surface area function of the previous example when the boundary values are specified *only* on a compact subarc K of the boundary. Then a minimizing function u should satisfy the minimal surface equation (52) in the domain D and have the prescribed values on K; however, on $\tilde{K}$, we expect in general that

$$\frac{\nabla u \cdot N}{(1 + u_x^2 + u_y^2)^{1/2}} = 0 \quad \text{or} \quad \nabla u \cdot N = 0;$$

i.e., that the derivative of u in the direction *normal* to the boundary curve should vanish. Conversely, we may use convexity as in §3.4(e), to show that such u will in fact minimize the surface area function uniquely under these conditions (Problem 3.37).

PROBLEMS

6.1. Give the first Euler–Lagrange equation for f when:
 (a) $f(x, y, z) = \sin z$.
 (b) $f(x, y, z) = x^3 z^3$.
 (c) $f(x, y, z) = \sqrt{1 + z^2}/x, (x \neq 0)$.
 (d) $f(x, y, z) = y^2 - z^2$.
 (e) $f(x, y, z) = 2xy - y^2 + 3zy^2$.

6.2. Find the stationary functions for f which belong to $\mathcal{D}$ if:
 (a) $f(x, y, z) = \sin z$, and
   $\mathcal{D} = \{y \in C^1[0, 1]: y(0) = -5, y(1) = 2\}$.
 (b) $f(x, y, z) \sqrt{1 + z^2}/x$, and
   $\mathcal{D} = \{y \in C^1[0, 3]: y(0) = 0; y(3) = 3\}$.
 (c) $f(x, y, z) = y^2 - z^2$, and
   $\mathcal{D} = \{y \in C^1[0, \pi/2]: y(0) = 0, y(\pi/2) = 1\}$.
 (d) $f(x, y, z) = y^2 - z^2$, and
   $\mathcal{D} = \{y \in C^1[0, \pi]: y(0) = y(\pi) = 0\}$.
 (e) $f(x, y, z) = 2xy - y^2 + 3zy^2$, and
   $\mathcal{D} = \{y \in C^1[0, 1]: y(0) = 0, y(1) = 1\}$.
 (f) $f(y, z) = \sqrt{1 + z^2}/y, y > 0$
   $\mathcal{D} = \{y \in C^1[0, 2]: y(0) = y(2) = 1, y(x) > 0\}$.

In Problems 6.3–6.12 find the possible local extremal functions for F on $\mathcal{D}$.

6.3 $F(y) = \int_0^\pi [y(x)^2 - y'(x)^2] \, dx$, and
  $\mathcal{D} = \{y \in C^1[0, \pi]: y(0) = y(\pi) = 0\}$.

6.4. $F(y) = \int_1^2 x^3 y'(x)^2 \, dx$, and
  $\mathcal{D} = \{y \in C^1[1, 2]: y(1) = 5, y(2) = 2\}$.

6.5. $F(y) = \int_1^2 y'(x)^3/x^2 \, dx$, and
  $\mathcal{D} = \{y \in C^1[1, 2]: y(1) = 1, y(2) = 7\}$.

6.6. $F(y) = \int_{1/2}^1 \sqrt{1 + y'(x)^2}/x \, dx$, and
  $\mathcal{D} = \{y \in C^1[\frac{1}{2}, 1]: y(\frac{1}{2}) = -\sqrt{3}/2, y(1) = 0\}$.

6.7. $F(y) = \int_0^1 [y(x)^3 + 3x^2 y'(x)] \, dx$, and
 (a) $\mathcal{D} = \{y \in C^1[0, 1]: y(0) = 0, y(1) = 1\}$.
 (b) $\mathcal{D} = \{y \in C^1[0, 1]: y(0) = 0, y(1) = 2\}$.

6.8. $F(y) = \int_0^1 [1 + y(x)^2]/y'(x)^2 \, dx$, and
  $\mathcal{D} = \{y \in C^1[0, 1]: y(0) = 0, y(1) = \sinh 1;$ with $y'(x) \neq 0$ on $(0, 1)\}$.

6.9. $F(y) = \int_0^1 [2xy(x) - y'(x)^2 + 3y'(x)y(x)^2] \, dx$, and
  $\mathcal{D} = \{y \in C^1[0, 1]: y(0) = 0, y(1) = -1\}$.

6.10. $F(y) = \int_0^1 [2xy(x)^3 + e^x \sin y(x) + 3x^2 y(x)^2 y'(x) + y'(x)e^x \cos y(x)] \, dx$, and
   (a) $\mathscr{D} = \{y \in C^1[0, 1]: y(0) = 0, \, y(1) = 1\}$.
   (b) $\mathscr{D} = \{y \in C^1[0, 1]: y(0) = \pi, \, y(1) = \sqrt{8}\}$.

6.11. $F(y) = \int_0^\pi [4y'(x)^2 + 2y(x)y'(x) - y(x)^2] \, dx$, and
   $\mathscr{D} = \{y \in C^1[0, \pi]: y(0) = 2; \, y(\pi) = 0\}$.

6.12. $F(y) = \int_0^1 [y'(x)^2 - 6x^2 y(x)] \, dx$, and
   $\mathscr{D} = \{y \in C^1[0, 1]: y(0) = \frac{3}{4}, \, y(1) = \frac{7}{4}\}$.

6.13. Let $F(y) = \int_0^1 \sqrt{y(x) - x} \, dx$ and
   $\mathscr{D} = \{y \in C[0, 1]: y(0) = 0, \, y(1) = 1, \, y(x) \geq x \text{ on } [0, 1]\}$.
   (a) Show that there are no stationary functions for $f(x, y, z) = \sqrt{y - x}$.
   (b) Show directly that $y_0(x) = x$ minimizes F on $\mathscr{D}$.
   (c) Verify that $\delta F(y_0; v) = 0$ for all v which are $\mathscr{D}$-admissible for the function
   $y_0(x) = x$. (Are there any ?)
   (d) What does this example demonstrate?

6.14*. With the same definitions as in Problem 5.40*, duplicate the analysis in §6.1 to
   prove that if $\delta F(y; v) = 0$, $\forall \, v \in \mathscr{D}_0^*$ then $f_{y'}(x) + g(x) = c_0$ on $(a, b]$, where
   $g(x) = \int_x^b f_y(t) \, dt$ when $x \in (a, b]$. Conclude that y is stationary for f on $(a, b]$.
   (Hint: On *each* interval $[x_0, b]$, apply the result of Problem 4.3.)

6.15*. For the brachistochrone problem in Example 4* of §6.2, use Problem 6.14* to
   show that the *first* equation $(d/dx)f_{y'}(x) = f_y(x)$ can be integrated as it stands
   upon multiplication by

$$f_{y'}(x) = \frac{y'(x)}{\sqrt{y(x)}\sqrt{1 + y'(x)^2}}.$$

Conclude that with an *appropriate* $\mathscr{D}^*$, the only possible minimizing function
for T on $\mathscr{D}^*$ is that representing the cycloid given parametrically by equation
(6). Why is this an improvement?

6.16*. Minimal Surface of Revolution. To find a smooth planar curve represented by
   $y \in C^1[0, b]$, joining points $(0, 1)$ and (b, b_1) with $b_1 > 0$, which when rotated
   about the x-axis will have the smallest possible surface area of revolution, we
   should minimize the surface area integral

$$S(y) = 2\pi \int_0^b y(x)\sqrt{1 + y'(x)^2} \, dx$$

on

$$\mathscr{D} = \{y \in C^1[0, b]: y(0) = 1, \, y(b) = b_1; \, y(x) \geq 0\}$$

Prove that if such a curve exists, it must be of the form

$$y(x) = c_1 \cosh\left[\frac{(x - c_2)}{c_1}\right],$$

where c_1 and c_2 are constants. (Note: Depending on the location of the points,
there may be one, two, or no curves of this type which meet the given boundary
conditions. This problem will be dealt with again in Chapters 7 and 9.)

In Problems 6.17–6.20, find all possible (local) extremal functions for F on $\mathcal{D}$.

6.17. $F(y) = \int_0^{\pi/2} [y(x)^2 - y'(x)^2] \, dx$:
  (a) $\mathcal{D} = \{y \in C^1[0, \pi/2]: y(0) = 0\}$.
  (b) $\mathcal{D} = \{y \in C^1[0, \pi/2]: y(0) = 1\}$.

6.18. $F(y) = \int_0^1 [(y'(x) - x)^2 + 2xy(x)] \, dx$
  $\mathcal{D} = \{y \in C^1[0, 1]: y(0) = 1\}$.

6.19. $F(y) = \int_0^1 \cos y'(x) \, dx$
  $\mathcal{D} = \{y \in C^1[0, 1]: y(0) = 0\}$.

6.20. $F(y) = \int_0^1 [xy(x) - y'(x)^2] \, dx$:
  (a) $\mathcal{D} = \{y \in C^1[0, 1]: y(0) = 1\}$.
  (b) $\mathcal{D} = \{y \in C^1[0, 1]: y(1) = 1\}$.

6.21. Consider the problem of finding a smooth curve of the form $y(x)$ which will provide the shortest distance from the origin to the parabola given by $y = x^2 - 1$.
  (a) What are the stationary functions for this problem?
  (b) Show that there are precisely two points (t, y) and $(\tilde{t}, \tilde{y})$ on the parabola which satisfy the transversal condition (15).
  (c) Find the associated curves which represent the possible extremals.
  (d) Use a direct argument to show that a minimum is actually achieved for each of the curves found in part (c).
  (e) What happens if the parabola is replaced by the circle $x^2 + y^2 = 1$?

6.22. Brachistochrone. (See §6.4, Example 1.) A brachistochrone joining the origin to the straight line $y = 1 - x$ is sought. Show that there is precisely one point (t, y) on the line which satisfies the transversal condition (15) and find the associated cycloid which might be the brachistochrone in question.

6.23. (a) Use the method of Lagrangian multipliers to find all possible (local) extremal functions for
$$F(y) = \int_0^1 y(x)^2 \, dx$$
  on
$$\mathcal{D} = \{y \in C^1[0, 1]: y(0) = y(1) = 0\},$$
  when further constrained to $\{y \in C^1[0, 1]: \int_0^1 y'(x)^2 \, dx = 1\}$.
  (b) Can the convexity methods of Chapter 3 be used to conclude that a minimum is achieved in part (a)? Explain.

6.24. (a) Use the method of Lagrangian multipliers to find all possible (local) extremal functions for
$$F(y) = \int_0^\pi [2(\sin x)y(x) + y'(x)^2] \, dx$$
  on
$$\mathcal{D} = \{y \in C^1[0, \pi]: y(0) = y(\pi) = 0\}$$
  when further constrained to $\{y \in C^1[0, \pi]: \int_0^\pi y(x) \, dx = 1\}$.
  (b) Can the convexity methods of Chapter 3 be used to conclude that a minimum is achieved in part (a)?

6.25. Let $f \in C^1([a, b] \times \mathbb{R}^3)$. Show that the natural boundary conditions associated with minimizing

$$F(y) = \int_a^b f(x, y(x), y'(x), y''(x)) \, dx$$

locally on:
(i) $\mathscr{D} = \{y \in C^2[a, b]: y(a) = a_1, y(b) = b_1\}$ are $f_{y''}(a) = f_{y''}(b) = 0$.
(ii) $\mathscr{D} = \{y \in C^2[a, b]: y(a) = a_1, y'(a) = a'\}$ are $f_{y''}(b) = 0$ and

$$\left. \frac{d}{dx} f_{y''}(x) \right|_{x=b} = f_{y'}(b).$$

6.26. Let $f \in C^1([a, b] \times \mathbb{R}^3)$. Find the natural boundary conditions associated with minimizing

$$F(y) = \int_a^b f(x, y(x), y'(x), y''(x)) \, dx$$

on each of the following sets:
(i) $\mathscr{D} = \{y \in C^2[a, b]: y'(a) = a', y'(b) = b'\}$.
(ii) $\mathscr{D} = \{y \in C^2[a, b]: y(a) = a_1, y'(b) = b'\}$.

6.27. Let $f \in C^1([a, b] \times \mathbb{R}^4)$:
(a) Show that an Euler–Lagrange equation for a function which minimizes

$$F(y) = \int_a^b f(x, y(x), y'(x), y''(x), y'''(x)) \, dx$$

on

$$\mathscr{D} = \{y \in C^3[a, b]: y(a) = a_1, y'(a) = a', y''(a) = a'',$$
$$y(b) = b_1, y'(b) = b', y''(b) = b''\},$$

is

$$f_y(x) - \frac{d}{dx} f_{y'}(x) + \frac{d^2}{dx^2} f_{y''}(x) - \frac{d^3}{dx^3} f_{y'''}(x) = 0,$$

where we employ the usual abbreviations, and *assume sufficient differentiability*.
(b)* *Without* the additional differentiability, what form will this equation take?

6.28*. Let $f \in C^1([a, b] \times \mathbb{R}^{n+1})$. Show that an Euler–Lagrange equation for a function which minimizes

$$F(y) = \int_a^b f(x, y(x), y'(x), y''(x), \ldots, y^{(n)}(x)) \, dx$$

locally on

$$\mathscr{D} = \{y \in C^n[a, b]: y(a) = a_1, y'(a) = a^1, \ldots, y^{(n-1)}(a) = a^{(n-1)},$$
$$y(b) = b_1, y'(b) = b^1, \ldots, y^{(n-1)}(b) = b^{(n-1)}\}$$

is

$$f_y(x) - \frac{d}{dx} f_{y'}(x) + \frac{d^2}{dx^2} f_{y''}(x) - \cdots + (-1)^n \frac{d^n}{dx^n} f_{y^{(n)}}(x) = 0,$$

where we employ the usual abbreviations, *and assume sufficient differentiability*.

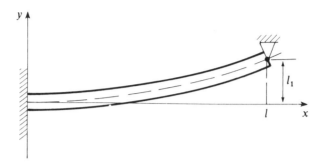

Figure 6.13

6.29. A thin elastic rod of initial length l clamped at one end and pinned at the other is deflected as shown in Figure 6.13 from its straight unstressed state. If the center line of the rod is described by a smooth function $y(x)$, $0 \le x \le l$, then the associated potential energy is given by

$$U(y) = \mu \int_0^l \frac{y''(x)^2}{[1 + y'(x)^2]^{5/2}} \, dx,$$

where μ is a constant. The physically imposed boundary conditions are $y(0) = y'(0) = 0$, and $y(l) = l_1$.
   (a) Assuming that the shape of the rod minimizes the potential energy, find a third-order differential equation satisfied by $y(x)$.
   (b) What is the natural boundary condition at $x = l$?
   (c) Find a suitable linearized version of the differential equation from part (a) by supposing that both $|y'(x)|$ and $|y''(x)|$ are very small on $[0, l]$.
   (d) Solve the linear equation found in part (c), choosing the constants to satisfy the boundary conditions.

6.30. Buckling of a Column. (See §6.6.) In small deflection theory, approximations are often made in the potential energy, rather than in the differential equation.
   (a)   Show that the approximation of (23) by

$$\tilde{U}(y) = \int_0^l \left[\mu y''(x)^2 - \frac{P}{2} y'(x)^2 \right] dx$$

leads to the same linear differential equation (28).
   (b)* Explain how the approximation in part (a) was obtained. Hint: Consider the effect of the complementary factor $(\sqrt{1 + (y')^2} + 1)$.
   (c)   Use (29) to determine the buckling loads when the end at $x = 0$ is pinned so that $y''(0) = 0$ instead of $y'(0) = 0$.

In Problems 6.31–6.33, find all possible (local) extremal functions for F on $\mathscr{D}$.

6.31. $F(y) = \int_0^{\pi/2} [y''(x)^2 - y(x)^2] \, dx$,
   $\mathscr{D} = \{y \in C^2[0, \pi/2]: y(0) = 1, \, y'(0) = y(\pi/2) = 0, \, y'(\pi/2) = -1\}$.

6.32. $F(y) = \int_0^\pi [y''(x)^2 - y'(x)^2] \, dx$,
   $\mathscr{D} = \{y \in C^2[0, \pi]: y(0) = 0, \, y'(0) = 2, \, y(\pi) = \pi, \, y'(\pi) = 0\}$.

6.33. $F(y) = \int_1^2 [x^3 y''(x)^2 - 12xy(x)] \, dx$,
$\mathscr{D} = \{ y \in C^2[1, 2]: y(1) = \frac{3}{2}, y'(1) = 3, y(2) = \frac{14}{3} + \ln 2, y'(2) = \frac{9}{2} \}$.

6.34. (a) Let $f \in C^1([a, b] \times \mathbb{R}^3)$. Show that if $y \in C^3[a, b]$ is a solution of the Euler
–Lagrange equation (21), then y must also satisfy a *second* equation

$$f(x) - y'(x) \left[f_{y'}(x) - \frac{d}{dx} f_{y''}(x) \right] - y''(x) f_{y''}(x) = \int_a^x f_x(t) \, dt + c,$$

for some constant c, where we employ the usual abbreviations.

(b) Use this to obtain a second-order equation replacing (25) and show that
assuming $|y'| \ll 1$ in this equation gives us an integral of (28). Hint: see
Problem 6.30(b).

6.35*. The Second Equation (Proposition 6.3).
(a) Verify the transformation of the integral $F(y)$ into $\tilde{F}(\eta)$ under the substitution $x = \xi + c\eta(\xi)$, where $\tilde{f}$ is defined by (10).
(b) Verify the retransformation of (11) into (11'), supposing that y_0 is an
extremal point for F on $\mathscr{D}$.
(c) With $y \in \mathscr{D}$, $\eta \in \tilde{\mathscr{D}}$ and c as above, take v in $\mathscr{D}_0 = \{ v \in C^1[a, b]: v(a) = v(b) = 0 \}$, and define v by $\quad v(\xi) = v(\xi + c\eta(\xi))$, so that $v'(\xi) = v'(x)(1 + c\eta'(\xi))$. Show that $v \in \tilde{\mathscr{D}}_0 = \{ v \in C^1[\alpha, \beta]: v(\alpha) = v(\beta) = 0 \}$.
(d)* Prove that for this v, v,

$$\delta\tilde{F}(\eta; v) = \int_\alpha^\beta (\tilde{f}_\eta[\eta(\xi)]v(\xi) + \tilde{f}_{\eta'}[\eta(\xi)]v'(\xi)) \, d\xi$$

$$= \delta F(y; v) + \delta_2 F(y; cv),$$

where

$$\delta_2 F(y; v) \overset{\text{def}}{=} \int_a^b [f_x(x)v(x) + (f - y'f_{y'})(x)v'(x)] \, dx.$$

(e)* Use the formula in (d) to discuss the invariance of C^1 stationarity of y
under this skew transformation.

6.36. The Second Equation for Vector Valued Functions. Assume that

$$f \in C^1([a, b] \times \mathbb{R}^{2d})$$

for $d > 1$.
(a) Derive the second equation (35) satisfied by a function $Y \in (C^2[a, b])^d$
which is stationary for f.
(b)* Obtain the second equation when $Y \in \mathscr{Y} = (C^1[a, b])^d$ supplies a local
extremal point for

$$F(Y) = \int_a^b f(Y(x)] \, dx$$

on

$$\mathscr{D} = \{ Y \in \mathscr{Y}: Y(a) = A; Y(b) = B \}.$$

Hint: Use the skew transformation

$$\begin{cases} \xi = x + c\eta, \\ H = Y \end{cases} \quad \text{for } H = (\eta_1, \eta_2, \eta_3, \ldots, \eta_d), \text{ and } a \text{ small } c$$

to duplicate the proof of Proposition 6.3.

6.37. Let $f \in C^1([a, b] \times \mathbb{R}^{2d})$. Show that the natural boundary condition associated with minimizing

$$F(Y) = \int_a^b f(x, Y(x), Y'(x)) \, dx$$

locally on

$$\mathcal{D}_1 = \{Y \in C^1([a, b])^d \colon Y(a) = A, \ y_i(b) = b_i, \ i = 2, 3, \ldots, d\}$$

(where $y_1(b)$ remains unspecified), is $f_{y_1'}(b) = 0$.

6.38. **Isoperimetric Problem.** (See §6.7.) Show that A cannot achieve a (positive) minimum value on $\mathcal{D}$ or $\mathcal{D}^*$. [Reason geometrically.]

6.39. Prove that in $\mathbb{R}^d$, a geodesic curve which can be parametrized by $Y \in (C^1[0, 1])^d$ with $Y'(t) \neq \mathcal{O}$, is a straight line segment. Hint: See §1.1(a), and recall that the unit vector $Y'/|Y'|$ is tangent to the curve.

6.40. For each of the following functions f defined on a subset of $\mathbb{R}^{2d+1}$, write the differential equations whose solutions Y will be stationary for f. Also, give an example of an integral function F on a set $\mathcal{D}$, which could have such Y as local extrema.
(a) $f(x, Y, Z) = x^2 + |Y|^2 + 3z_1, (d = 2)$.
(b) $f(x, Y, Z) = x|Z|, (d \geq 2)$.
(c) $f(x, Y, Z) = y_1|Z| - (\sin z_1)y_2, (d = 3)$.
(d) $f(x, Y, Z) = |Z|/\sqrt{y_1}, (d = 2)$.
(Do *not* attempt to solve the differential equations.)

6.41. In (a)–(c), find a partial differential equation which is satisfied by $u \in \mathcal{D} \cap C^2(D)$, if u is a (local) extremal function for F on $\mathcal{D} = \{u \in C^1(\overline{D}) \colon u|_{\partial D} = \gamma\}$, where D is a Green's domain in the $x-y$ plane and γ is a given continuous function on ∂D. (Do *not* attempt to solve the equation.)
(a) $F(u) = \int_D [\frac{1}{2}(u_x^2 + u_y^2) - (x^2 + y^2)u] \, dA$.
(b) $F(u) = \int_D [\frac{1}{2}(u_x^2 + u_y^2) + \frac{1}{3}u^3] \, dA$.
(c) $F(u) = \int_D [u_x^3 - u_y^4 + u^2] \, dA$.

6.42. (a) Show that a function which maximizes

$$F(y) = \int_0^l y(s)\sqrt{1 - y'(s)^2} \, ds$$

on

$$\mathcal{D}^* = \{y \in C^1[0, l] \colon y(0) = y(l) = 0; \ y(s) \geq 0\},$$

must satisfy the equation $y/\sqrt{1 - y'^2} = r = \text{const}$.
(b) Solve the equation in (a) for y' and make the substitution $y(s) = r \sin \theta(s)$ to conclude that $\theta'(s) = 1/r$, or $\theta(s) = s/r + c$.
(c) Show that also $y \in \mathcal{D}^*$, when $y(s) = r \sin(s/r)$, where $r = l/\pi$. Use $x(s) = \int_0^s \sqrt{1 - y'(t)^2} \, dt$ to eliminate the parameter s and conclude that a maximizing function must describe a semicircle of radius r.

6.43. The Zenodoros Problem (in its nonisoperimetric formulation from Problem 1.9(b)).

(a) Show that a function y_0 which maximizes

$$F(y) = \int_0^T y(t)[1 - (yy')(t)^2)]^{1/2} \, dt$$

on

$$\mathscr{D}^* = \{ y \in C^1[0, T] : y(0) = y(T) = 0; \ y(t) \geq 0, |yy'|(t) < 1 \}$$

must satisfy the equation $\sqrt{(1 - (yy')^2)}/y - \text{const.} = c > 0$.

(b) Conclude that when $y_0(0) = 0$, then $y_0(t) = \sqrt{2t - c^2 t^2}$, so that when $y_0 \in \mathscr{D}^*$, $y_0(t) = \sqrt{2t(1 - tT^{-1})}$.

6.44. Over some future time interval $[0, T]$, a strip-mining company intends to remove all of the iron ore from a region that contains an estimated Q tons. As it is extracted, they will sell it for processing at a net price per ton of $p(y, y') = P - \alpha y - \beta y'$ for positive constants P, α, and β, where $y(t)$ is the total tonnage sold by time t. (This pricing model allows the cost of mining to increase with the extent of the mined region and speed of production).

(a) If the company wishes to control its rate of production $y'(t)$, to maximize its total profit as represented by

$$F(y) = \int_0^T p(y, y')y' \, dt, \quad \text{when } y(0) = 0, \text{ and } y(T) = Q,$$

how might it proceed? (T is unspecified, but we need $y \geq 0$.)

(b) If future money is discounted continuously at a constant rate r, then we can assess the *present* value of profits from this mining operation by introducing a factor of e^{-rt} in the integrand of (a). How will this affect optimal mine operation?

PART TWO

ADVANCED TOPICS

Paris, 1900

AN ADDRESS

"*As long as a branch of science offers an abundance of problems, so long is it alive: a lack of problems foreshadows extinction or the cessation of independent development. Moreover, a mathematical problem should be difficult in order to entice us, yet not completely inaccessible, lest it mock our efforts. The mathematicians of past centuries ... knew the value of difficult problems. I remind you only of the 'problem of the [path] of quickest descent,' proposed by Johann Bernoulli. The calculus of variations owes its origin to this and to similar problems. ... it often happens also that the same special problem finds application in the most unlike branches of mathematical knowledge. So, for example, the problem of the shortest line plays a chief and historically important part in the foundations of geometry, in the theory of lines and surfaces, in mechanics, and in the calculus of variations.... And it seems to me that the numerous and surprising analogies and that apparently preestablished harmony which the mathematician so often perceives in the questions, methods and ideas of the various branches of his science, have their origin in [the] ever-recurring interplay between thought and experience....*"

"*It is an error to believe that rigor in the proof is the enemy of simplicity. The very effort for rigor forces us to discover simpler methods of proof.... But the most striking example of my statement is the calculus of variations. The treatment of the first and second variations of definite integrals required in part extremely complicated calculations, and the process applied by the old mathematicians had not the needful rigor. Weierstrass showed us the way to a new and sure foundation. By the examples of the single and double integral I will show briefly, at the close of my lecture, how this way leads at once to a surprising simplification....*"

"Mathematical science is in my opinion an indivisible whole, an organism whose vitality is conditioned upon the connection of its parts. The organic unity of mathematics is inherent in the nature of this science, for mathematics is the foundation of all exact knowledge of natural phenomena. That it may completely fulfill this high destiny, may the new century bring it gifted prophets and many zealous and enthusiastic disciples!"

DAVID HILBERT
At the Second International Congress of Mathematicians[1]

---

[1] These are fragments of the celebrated lecture in which Hilbert set forth 22 additional problems which have challenged mathematicians of all disciplines in this century. The translation of the complete text from which they were compiled will be found in Vol. 8 of the Bulletin of the A.M.S. (1902) pp. 437–445, 478, 479.

CHAPTER 7

Piecewise C^1 Extremal Functions

In many problems examined thus far we have required continuous differentiability of the function y (or Y) defining the classes for optimization. Already with the example of the minimal surface of revolution from §1.4(a) we have argued that for some configurations, the minimizing curve (if it exists) should exhibit "corners," and it is natural to wonder whether cornered curves $\hat{y}$ and $\hat{Y}$ such as those shown in Figure 7.1 might not give improved results for other problems. Such curves are represented readily by functions which are *piecewise continuously differentiable*, or *piecewise C^1*.

In the present chapter we shall extend our previous investigation to include this class of functions as possible extremals.[1] In §7.2, we show that it is only necessary to consider such extremals when C^1 extremals cannot be found, and in §7.3 and §7.5 prove that in general such extremals are stationary functions in intervals excluding their corner points, at each of which they must meet the transitional conditions of Weierstrass–Erdmann. The techniques are illustrated by application to a Sturm–Liouville problem where consideration of piecewise C^1 extremals provides valuable information. In §7.4, we use [strong] convexity (as in §3.2) to guarantee minimality for the local extremals of the associated integral functions and to study problems in which interval constraints preclude simple stationarity. (Although it is possible to introduce piecewise C^k functions for $k = 1, 2, \ldots$ even in a multidimensional setting—and obtain corresponding extensions to other results from Chapter 6, we shall not do so here. See [Mo].)

The above material is necessarily more technical but not appreciably more sophisticated than that which it generalizes. However, the results presented

---

[1] An older literature refers to these as "discontinuous" extremals. Extremals which exhibit actual discontinuities have been investigated by Krotov. See [Pe].

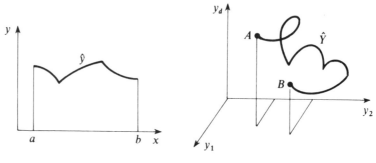

Figure 7.1

in the remainder of this chapter, namely, Hilbert's differentiability criterion (in §7.5) and the Weierstrass-Legendre conditions necessary for a minimum (in §7.6) are based on somewhat more difficult concepts. This concluding material should be regarded chiefly as a prelude to Chapter 9.

§7.1. Piecewise C^1 Functions

(7.1) Definition. A function $\hat{y} \in C[a, b]$ is *piecewise* C^1 (denoted $\hat{y} \in \hat{C}^1[a, b]$) provided that there is a finite (irreducible) partition $a = c_0 < c_1 < \cdots < c_{N+1} = b$ such that $\hat{y}$ may be regarded as a function in $C^1[c_k, c_{k+1}]$ for each $k = 0, 1, 2, \ldots, N$. When present, the *interior points* c_k for $k = 1, 2, \ldots, N$ are called *corner points* of $\hat{y}$.

When there are no corner points, $\hat{y} = y \in C^1[a, b]$. Moreover, if $y \in C^1[a, b]$, then $|y| \in \hat{C}^1[a, b]$ when y changes sign only finitely often.

Observe that $\hat{y}'$ is defined and continuous on $[a, b]$ except at a *corner point* c where it has *distinct* limiting values $\hat{y}'(c\pm)$. We shall use $\hat{y}'(c)$ to denote *both* values when the distinction is not essential. Discontinuities such as these of $\hat{y}'$ are said to be *simple*, and functions such as $\hat{y}'$ are said to be *piecewise continuous* on $[a, b]$, or to belong to $\hat{C}[a, b]$. Figure 7.2 illustrates the effect of the discontinuities of $\hat{y}'$ in producing "corners" on the graph of $\hat{y}$. Without these corner points, $\hat{y}$ might resemble the C^1 function, y, whose graph is presented for comparison.

A form of the fundamental theorem of calculus remains valid for $\hat{C}^1$ functions.

(7.2) Lemma. *If* $\hat{y} \in \hat{C}^1[a, b]$, *then* $\hat{y}(x) = \hat{y}(a) + \int_a^x \hat{y}'(t) \, dt$.

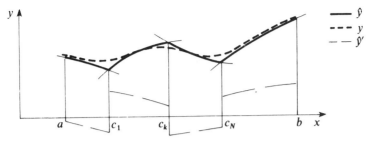

Figure 7.2

PROOF. When $x \in [c_k, c_{k+1}]$, then

$$\hat{y}(x) - \hat{y}(a) = \hat{y}(x) - \hat{y}(c_k) + \sum_{j=0}^{k-1} [\hat{y}(c_{j+1}) - \hat{y}(c_j)]$$

$$= \int_{c_k}^{x} \hat{y}'(t) \, dt + \sum_{j=1}^{k-1} \int_{c_j}^{c_{j+1}} \hat{y}'(t) \, dt = \int_{a}^{x} \hat{y}'(t) \, dt. \qquad \square$$

(Conversely, if $\hat{h}$ is *piecewise continuous* on $[a, b]$, then $\hat{y}(x) \stackrel{\text{def}}{=} \int_a^x \hat{h}(t) \, dt$ provides a $\hat{y} \in \hat{C}^1[a, b]$, with $\hat{y}' = \hat{h}$. See A.8 et seq.)

(7.3) **Proposition.** *Let* $\hat{y} \in \hat{C}^1[a, b]$:

(i) *If* $\int_a^b \hat{y}'(x)^2 \, dx = 0$, *then* $\hat{y}' = 0$ *on* $[a, b]$;

(ii) *If* $\hat{y}' = 0$ *where defined, then* $\hat{y} = $ const. *on* $[a, b]$.

PROOF. (i) Observe first that $(\hat{y}')^2$ is nonnegative and piecewise continuous so that the integral is defined and it may be represented as the finite sum $\sum_{k=0}^{N} \int_{c_k}^{c_{k+1}} \hat{y}'(x)^2 \, dx$. This sum vanishes *iff* each of its terms does, and then the *continuous* function $\hat{y}'(x)^2 = 0$ on (c_k, c_{k+1}) (from A.9), so that its continuous extension vanishes on $[c_k, c_{k+1}]$, for $k = 0, 1, 2, \ldots, N$.

(ii) This assertion is an immediate consequence of the preceding lemma.

$\qquad \square$

(a) Smoothing

As Figure 7.2 illustrates, each piecewise C^1 function is "almost" C^1; it is only necessary to *round out* the corners to produce the graph of a function y such as that shown. The next construction accomplishes this analytically and provides control over the approximation.

(7.4) Smoothing Lemma. *For each $\hat{y} \in \hat{C}^1[a, b]$ and $\delta > 0$, $\exists\ y \in C^1[a, b]$ such that $y \equiv \hat{y}$ except in a δ-neighborhood of each corner point of $\hat{y}$ where* $\max |y'(x)| \leq 4 \max |\hat{y}'(x)|$. *Thus* $\max |y(x) - \hat{y}(x)| \leq \hat{A}\delta$ *for a constant $\hat{A}$ determined by $\hat{y}$.*

PROOF. Since $\hat{y}$ has at most a finite number, N, of corner points, it suffices to explain its modification in the given δ-neighborhood of a typical corner point c. We suppose that δ is so small that this neighborhood excludes a δ-neighborhood of other corner points and the end points a, b. In this neighborhood we replace the discontinuous function $\hat{y}'$ by a continuous triangular function, such as that shown in Figure 7.3, which is determined by its "height" h at c and the values $\hat{y}'(c \pm \delta)$.

For any choice(s) of h at the corner point(s), the resulting function *denoted* y' is continuous by construction so that the function y defined by

$$y(x) \stackrel{\text{def}}{=} \hat{y}(a) + \int_a^x y'(t)\, dt \quad \text{is in } C^1[a, b]$$

and its derivative is the function y' just constructed. It remains to select the value(s) of h to effect the required estimates, and this choice is most readily understood when $\hat{y}$ has only the *single* corner point c. Then, clearly, when $x \leq c - \delta$: $y'(x) = \hat{y}'(x)$ so that $y(x) = \hat{y}(a) + \int_a^x \hat{y}'(t)\, dt = \hat{y}(x)$; to have $y(x) = \hat{y}(x)$, for $x \geq c + \delta$, it is only necessary to make

$$\int_{c-\delta}^{c+\delta} y'(t)\, dt = \hat{A}_\delta \stackrel{\text{def}}{=} \int_{c-\delta}^{c+\delta} \hat{y}'(t)\, dt. \quad \text{(Why?)}$$

Now, by elementary geometry, the *signed* area A_δ corresponding to the integral on the *left* is given by

$$A_\delta = h\delta + \frac{\delta}{2}[\hat{y}'(c - \delta) + \hat{y}'(c + \delta)],$$

and for the given δ, $A_\delta = \hat{A}_\delta$ provided that

$$h = \frac{\hat{A}_\delta}{\delta} - \frac{\hat{y}'(c - \delta) + \hat{y}'(c + \delta)}{2}.$$

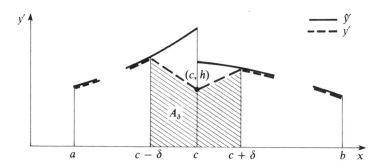

Figure 7.3

Thus with $M' = \max |\hat{y}'|$, it follows that

$$|\hat{A}_\delta| \le \int_{c-\delta}^{c+\delta} |\hat{y}'(t)| \, dt \le 2\delta M',$$

so that $|h| \le 3M'$ and $|y'(x)| \le M' + |h| \le 4M'$ on $[c - \delta, c + \delta]$, as is evident from the figure.

A similar choice for $h = h_k$ is possible at each successive corner point c_k, $k = 1, 2, \ldots, N$ so that by construction and obvious estimates:

$$|y(x) - \hat{y}(x)| \le \int_a^x |y'(t) - \hat{y}'(t)| \, dt \le \int_a^b |y'(t) - \hat{y}'(t)| \, dt$$

$$= \sum_{k=1}^N \int_{c_k-\delta}^{c_k+\delta} |y'(t) - \hat{y}'(t)| \, dt \le 2\delta(5M')N = \hat{A}\delta,$$

where $\hat{A} = 10M'N$ is determined by $\hat{y}$.   □

We require also $(\hat{C}^1[a, b])^d$, the d-dimensional vector valued analogue of $\hat{C}^1[a, b]$, consisting of those functions $\hat{Y} \in (C[a, b])^d$ with components $\hat{y}_j \in \hat{C}^1[a, b], j = 1, 2, \ldots, d$. The corner points of such $\hat{Y}$ are by definition those of any one of its components $\hat{y}_j$.[1] The above lemma can be applied to each component of a given $\hat{Y}$ and shows that $\hat{Y}$ can be approximated by a $Y \in (C^1[a, b])^d$ which agrees with it except in prescribed neighborhoods of its corner points.

In each case, these sets of piecewise C^1 functions form linear spaces of which the subsets of C^1 functions are subspaces. Indeed, it is obvious that the constant multiple of one of these functions is another of the same kind, and the sum of two such functions will exhibit the necessary piecewise continuous differentiability with respect to a suitable partition of the underlying interval $[a, b]$. Moreover, the *dot product* $\hat{Y} \cdot \hat{V}$ of two such functions will be in $\hat{C}^1[a, b]$. (Why?)

(b) Norms for $\hat{C}^1$

Since $\hat{C}^1[a, b] \subseteq C[a, b]$, it is evident that $|\hat{y}| \overset{\text{def}}{=} \max |\hat{y}(x)|$ is a norm for $\hat{C}^1[a, b]$ and $\|\hat{y}\| \overset{\text{def}}{=} \max(|\hat{y}(x)| + |\hat{y}'(x)|)$ can be shown to be another [Problem 7.1]. Although the first does not supply control over the differentiability properties of the functions $\hat{y}$, it is valuable when approximating $\hat{y}$ by y as in Lemma 7.4.

Another choice for a norm permitting two functions which agree except in small neighborhoods of their corner points to be "close" in norm, is the

---

[1] When $d > 1$, the curve represented parametrically by a $\hat{Y}$ with a *simultaneous* corner point in all of its components need *not* exhibit a corner. See Problem 7.13.

integral

$$\|\hat{y}\|_1 \stackrel{\text{def}}{=} \int_a^b (|\hat{y}(x)| + |\hat{y}'(x)|)\, dx.$$

Similarly, we shall supply $(\hat{C}^1[a, b])^d$ with the norms $|\hat{Y}| \stackrel{\text{def}}{=} \max |\hat{Y}(x)|$, $\|\hat{Y}\| \stackrel{\text{def}}{=} \max(|\hat{Y}(x)| + |\hat{Y}'(x)|)$, and $\|\hat{Y}\|_1 \stackrel{\text{def}}{=} \int_a^b (|\hat{Y}(x)| + |\hat{Y}'(x)|)\, dx$.

The maximum norms, $|\cdot|$ and $\|\cdot\|$, are called the *strong* and the *weak* norms, respectively, and the functions which are locally extremal with respect to the first [second] of these are said to be *strong* [*weak*] *extremal functions*.[1] Since there are many more functions in a strong norm neighborhood of a function than in a weak norm neighborhood [Why?], it is more difficult for a function to be a strong extremal than it is to be a weak extremal (locally). This classification of extremals was introduced by Kneser, a student of Weierstrass, in 1900; it is now a fundamental part of the theory ([K]).

The above norms are not independent, and they satisfy inequalities such as the following:

(75) (a) $A|\hat{y}| \le \|\hat{y}\|_1 \le (b - a)\|\hat{y}\|,$
    (b) $A|\hat{Y}| \le \|\hat{Y}\|_1 \le (b - a)\|\hat{Y}\|,$ $\Bigg\}$ *where* $A = \dfrac{b - a}{1 + b - a}$;
    (c) $\|\hat{Y}\| \le \displaystyle\sum_{j=1}^d \|\hat{y}_j\|$, *where* $\hat{Y} = (\hat{y}_1, \ldots, \hat{y}_d)$.

(Problems 7.1–7.2)

§7.2. Integral Functions on $\hat{C}^1$

When $f = f(x, Y, Z)$ is continuous on $[a, b] \times \mathbb{R}^{2d}$, and $\hat{Y} \in \hat{\mathcal{Y}} = (\hat{C}^1[a, b])^d$, then $f[\hat{Y}(x)] = f(x, \hat{Y}(x), \hat{Y}'(x))$ is *piecewise* continuous on $[a, b]$ and has in general (simple) discontinuities at the corner points of $\hat{Y}$. Thus

$$F(\hat{Y}) = \int_a^b f(x, \hat{Y}(x), \hat{Y}'(x))\, dx = \int_a^b f[\hat{Y}(x)]\, dx$$

is defined and finite, since a partition of $[a, b]$ reduces this integral to a finite sum of integrals considered previously. Now in general F is *not* continuous on $\hat{\mathcal{Y}}$ with respect to the strong norm $|\ |$ or the $\|\ \|_1$ norm. (F is continuous with respect to the *weak* norm $\|\ \|$. See Problem 7.3*.) However, $\mathcal{Y} = (C^1[a, b])^d \subseteq \hat{\mathcal{Y}}$, and the values of F on $\hat{\mathcal{Y}}$ can be approximated by those of F on $\mathcal{Y}$ as shown in the next

(7.6) Proposition. *If* $f \in C([a, b] \times \mathbb{R}^{2d})$, *then for each* $\hat{Y} \in \hat{\mathcal{Y}}$ *and* $\varepsilon > 0$, $\exists$ *a* $Y_\varepsilon \in \mathcal{Y}$, *with* $|\hat{Y} - Y_\varepsilon| < \varepsilon$, *for which* $|F(\hat{Y}) - F(Y_\varepsilon)| < \varepsilon$.

---

[1] Some authors *reverse* these designations for the norms, although there is uniform agreement about that for the extremals.

PROOF. We use the Smoothing Lemma 7.4, to replace each component $\hat{y}_j$ of $\hat{Y}$ by a C^1 function y_j which agrees with it except in nonoverlapping δ-neighborhoods of each corner point where

$$|y_j'(x)| \leq 4 \max |\hat{y}_j'(x)| \leq 4\|\hat{y}_j\|$$

and

$$|y_j(x) - \hat{y}_j(x)| \leq \hat{A}_j \delta,$$

so that $|y_j(x)| \leq \|\hat{y}_j\| + \hat{A}_j \delta$. Thus $\forall \ x\colon |y_j(x)|, |y_j'(x)|, |\hat{y}_j(x)|, |\hat{y}_j'(x)| \leq m_j = 4(\hat{A}_j \delta + \|\hat{y}_j\|)$. The resulting function $Y = (y_1, y_2, \ldots, y_d) \in \mathcal{Y}$ and it agrees with $\hat{Y}$ except in δ-neighborhoods of each of its corner points c_k, $k = 1, 2, \ldots, N$, where the previous estimates hold. Denoting $\max_{j=1,2,\ldots,d} m_j$ by m, we conclude from Proposition 5.3 that $|f|$ is bounded on the compact box $[a, b] \times [-m, m]^{2d}$, by $\bar{m}$, say, and $f[\hat{Y}(x)] = f[Y(x)]$ except when $|x - c_k| \leq \delta$ for some corner point c_k, $k = 1, 2, \ldots, N$. Hence

$$|F(\hat{Y}) - F(Y)| = \left| \sum_{k=1}^{N} \int_{c_k - \delta}^{c_k + \delta} (f[\hat{Y}(x)] - f[Y(x)]) \, dx \right|$$

$$\leq 2\delta N \max(|f[\hat{Y}(x)]| + |f[Y(x)]|) \leq 4N\bar{m}\delta.$$

Also $|\hat{Y} - Y| \leq \sum_{j=1}^{d} |\hat{y}_j - y_j| \leq \sum_{j=1}^{d} \hat{A}_j \delta = \hat{A}\delta$, say, by 7.5(c) and the above estimate. The proposition follows for $Y_\varepsilon = Y$ when δ is sufficiently small. □

It is straightforward to show that in the case of the geodesic problems for $\mathbb{R}^d$ and for the sphere, the arguments used in §1(a), (b) remain unchanged when the functions Y are replaced by their piecewise C^1 counterparts $\hat{Y}$. Thus we can assert that the geodesics in these cases remain the same, namely, the straight lines and the great circles, respectively.

There is in fact a general principle evidenced by these examples which may eliminate the search for piecewise C^1 extremals.

(7.7) **Theorem.** If $f \in C([a, b] \times \mathbb{R}^{2d})$ and Y_0 is a [local] extremal point for F on $\mathcal{D} = \{Y \in \mathcal{Y}\colon Y(a) = A, Y(b) = B\}$, then Y_0 is also a [local] extremal point for F on $\hat{\mathcal{D}} = \{\hat{Y} \in \mathcal{Y}\colon \hat{Y}(a) = A, \hat{Y}(b) = B\}$ [with respect to the same $|\cdot|$ or $\|\cdot\|$ norm].

PROOF. The first result is a direct consequence of the preceding proposition. Indeed, if $Y_0 \in \mathcal{D}$ supplies, say, a *minimum* for F on $\mathcal{D}$ in the strong neighborhood $S_\delta(Y_0) = \{Y \in \mathcal{Y}\colon |Y - Y_0| < \delta\}$, and $|\hat{Y} - Y_0| < \delta$, then with Y_ε as in the proposition, the triangle inequality shows that

$$F(\hat{Y}) \geq F(Y_\varepsilon) - |F(\hat{Y}) - F(Y_\varepsilon)|$$

$$\geq F(Y_\varepsilon) - \varepsilon \geq F(Y_0) - \varepsilon,$$

when ε is so small that $Y_\varepsilon \in S_\delta(Y_0)$ (and this can be achieved; see Problem 7.20(a)). Thus as $\varepsilon \searrow 0$, we conclude that $F(\hat{Y}) \geq F(Y_0)$.

The case where Y_0 is a *weak* local minimum point is left to Problem 7.20(c).

$\square$

(7.8) Remark. The previous characterizations of *local C^1* extremals given in Chapter 5 were with respect to an unspecified norm, but as there observed, *weak local* extremals Y_0 need not be *strong local* extremals; see Bolza's example in §7.6. However, in case Y_0 is a *global* extremal for F on $\mathcal{D}$, then the norm considerations are immaterial and we can assert that Y_0 is also a global extremal for F on $\hat{\mathcal{D}}$. In particular *all* of the minima obtained for the convex problems of Chapter 3 also minimize in the corresponding classes of piecewise C^1 functions.

(Problem 7.3)

§7.3. Extremals in $\hat{C}^1[a, b]$:
The Weierstrass–Erdmann Corner Conditions

Theorem 7.7 does not preclude a function F from being extremized by a $\hat{Y}_0$ which is *only piecewise C^1*. For example, the function (of §6.2, Example 3) $f(y, z) = y^2(1 - z)^2$ gives

$$F(\hat{y}) = \int_{-1}^{1} \hat{y}^2(x)(1 - \hat{y}'(x))^2 \, dx,$$

and F has a minimum value, 0, on

$$\mathcal{D} = \{\hat{y} \in \hat{C}^1[-1, 1]; \hat{y}(-1) = 0, \hat{y}(1) = 1\},$$

which is attained (*uniquely*) by the function

$$\hat{y}_0(x) = \begin{cases} 0, & -1 \leq x \leq 0, \\ x, & 0 \leq x \leq 1, \end{cases}$$

with a corner point at 0. [To establish the uniqueness, observe that if $F(\hat{y}) = 0$ for a $\hat{y} \in \hat{\mathcal{D}}$, then the associated nonnegative integrand must vanish on $[-1, 1]$ except possibly at the corner points of $\hat{y}$. Thus, at each such x, either $\hat{y}(x) = 0$ or $\hat{y}'(x) = 1$. Now $\hat{y}(1) = 1$, and in the largest subinterval $(a, 1]$ in which $\hat{y}(x) > 0$ we must have that $\hat{y}'(x) = 1$ and $\hat{y}(x) = x$. This interval must continue until $a = 0$; $\hat{y}(0) = 0$, and the only function in $\hat{C}^1[-1, 0]$ which vanishes at -1 and 0 and increases at each nonzero value is $\hat{y}(x) \equiv 0$.] On the other hand, by 7.6 we know that $F(\hat{y}_0)$ can be approximated as closely as we wish by values $F(y)$ for

$$y \in \mathcal{D} = \{y \in C^1[-1; 1]: y(-1) = 0, y(1) = 1\}.$$

Thus F *cannot* be minimized in $\mathcal{D}$, and this explains our previous lack of success in §6.2, Example 3.

When we seek conditions which are *necessary* in order that a function $\hat{y}_0 \in \hat{C}^1[a, b]$ should be a *local* extremal, we may assume at first, that it is a *weak* local extremal, i.e., that it is extremal with respect to some weak neighborhood of the form $S_r(\hat{y}_0) = \{\hat{y} \in \hat{C}^1[a, b]: \|\hat{y} - \hat{y}_0\| < r\}$. In fact, each local extremal with respect to the strong $|\cdot|$ (or the $\|\cdot\|_1$) norm is automatically a *weak* local extremal. (Why?)

Observe that $\hat{y} \in S_r(\hat{y}_0)$ iff $\hat{y} = \hat{y}_0 + \varepsilon\hat{v}$ for $\hat{v} \in \hat{C}^1[a, b]$, and a sufficiently small ε. In characterizing (weak) local extremals for the function

$$F(\hat{y}) = \int_a^b f(x, \hat{y}(x), \hat{y}'(x)) \, dx = \int_a^b f[\hat{y}(x)] \, dx$$

on

$$\hat{\mathcal{D}} = \{\hat{y} \in \hat{C}^1[a, b]: \hat{y}(a) = a_1, \hat{y}(b) = b_1\},$$

where $f = f(x, y, z)$ and its partials f_y, f_z are continuous on $[a, b] \times \mathbb{R}^2$, we duplicate the analysis of §6.1, and will only sketch the results.

For $\hat{y} \in \hat{\mathcal{D}}$ and $\hat{v} \in \hat{\mathcal{D}}_0 = \{\hat{v} \in \hat{C}^1[a, b]: \hat{v}(a) = \hat{v}(b) = 0\}$, we form

$$F(\hat{y} + \varepsilon\hat{v}) = \int_a^b f[\hat{y}(x) + \varepsilon\hat{v}(x)] \, dx;$$

then, taking into account the corner points of *both* $\hat{y}$ and $\hat{v}$, we represent it as a *finite* sum of integrals with *continuous* integrands, and differentiate each under the integral sign (A.13) to get upon reassembly that

$$\frac{\partial}{\partial \varepsilon} F(\hat{y} + \varepsilon\hat{v}) = \int_a^b \left[f_y[(\hat{y} + \varepsilon\hat{v})(x)]\hat{v}(x) + f_z[(\hat{y} + \varepsilon\hat{v})(x)]\hat{v}'(x)\right] dx.$$

Hence, in the limit as $\varepsilon \to 0$, we obtain

$$\delta F(\hat{y}; \hat{v}) = \int_a^b (\hat{f}_y(x)\hat{v}(x) + \hat{f}_{y'}(x)\hat{v}'(x)) \, dx, \tag{1}$$

where

$$\hat{f}_y(x) \stackrel{\text{def}}{=} f_y(x, \hat{y}(x), \hat{y}'(x)) = f_y[\hat{y}(x)],$$

and

$$\hat{f}_{y'}(x) \stackrel{\text{def}}{=} f_z(x, \hat{y}(x), \hat{y}'(x)) = f_z[\hat{y}(x)]$$

are again *piecewise* continuous on $[a, b]$ with at most (simple) discontinuities at the corner points of $\hat{y}$. Thus $\hat{g}(x) \stackrel{\text{def}}{=} \int_a^x \hat{f}_y(t) \, dt$, determines a function in $\hat{C}^1[a, b]$, and upon integrating by parts, we have as before that

$$\delta F(\hat{y}; \hat{v}) = \int_a^b [\hat{f}_{y'}(x) - \hat{g}(x)]\hat{v}'(x) \, dx.$$

In order that $\hat{y}$ be a local extremal function for F on $\hat{\mathcal{D}}$, it must make $\delta F(\hat{y}; \hat{v}) = 0$, $\forall \hat{v} \in \hat{\mathcal{D}}_0$, and, in particular, for that $\hat{v}$ defined by

$$\hat{v}(x) = \int_a^x (\hat{f}_{y'}(t) - \hat{g}(t) - c_0) \, dt,$$

which is in $\hat{\mathscr{D}}_0$ for an appropriately chosen constant c_0. As in the proof of Lemma 4.1 of du Bois-Reymond, we have that $\int_a^b \hat{v}'(x)^2\,dx = 0$, which gives $\hat{v}' = 0$ (Proposition 7.3(i)). Thus

$$\hat{f}_{y'}(x) = \int_a^x \hat{f}_y(t)\,dt + c_0, \tag{2}$$

is *continuous*, and in addition,

$$\frac{d}{dx}\hat{f}_{y'}(x) \text{ exists} = \hat{f}_y(x), \tag{2'}$$

except at each corner point c of $\hat{y}$ where from the aforementioned continuity:

$$\hat{f}_{y'}(c-) = \hat{f}_{y'}(c+). \tag{3}$$

(3) is the *first* of the *Weierstrass–Erdmann conditions* which must prevail at each corner point of a piecewise C^1 local extremal function. On each interval that excludes corner points, the local extremal function $\hat{y}$ must be C^1 and stationary for f in the sense of the previous chapter.

(7.9) Remark. Actually, the integral equation (2) could be used to characterize *stationarity* for F on $\hat{\mathscr{D}}$, and many of the subsequent properties obtained in this chapter for extremal functions $\hat{y}$ are also true for this larger class. See, in particular, Theorem 7.12 and its proof. For example, each solution $\hat{y}$ of (2) satisfies (3) at a corner point, and therefore (see Problem 7.25(a))

> At a corner point c of a solution $\hat{y}$ of (2), the double derivative $f_{zz}(c,\,\hat{y}(c),\,z)$, if *defined*, must vanish for some value of z.

[This is a simple consequence of the law of the mean (A.3) applied to the function $g(z) = f_z(c,\,\hat{y}(c),\,z)$, which by (3) has *equal* values at the *distinct* points $z = \hat{y}'(c\pm)$ and so must have a vanishing derivative $g'(z) = f_{zz}(c,\,\hat{y}(c),\,z)$ at an intermediate point.]

Similarly, when $f \in C^1([a, b] \times \mathbb{R}^2)$ we may duplicate from §6.3, the derivation of the *second* Euler–Lagrange equation in integral form, to conclude that a local extremal $\hat{y}$ for F on $\hat{\mathscr{D}}$ must satisfy

$$\hat{f}(x) - \hat{y}'(x)\hat{f}_{y'}(x) = \int_a^x \hat{f}_x(t)\,dt + \text{const.}, \tag{4}$$

where

$$\hat{f}(x) = f[\hat{y}(x)], \text{ and } \hat{f}_x(x) = f_x[\hat{y}(x)].$$

Thus $(d/dx)(\hat{f} - \hat{y}'\hat{f}_{y'})(x)$ exists $= \hat{f}_x(x)$, $\forall\, x \in (a, b)$ except at each corner point c of $\hat{y}$ at which holds the *second* Weierstrass–Erdmann condition:

$$(\hat{f} - \hat{y}'\hat{f}_{y'})(c-) = (\hat{f} - \hat{y}'\hat{f}_{y'})(c+). \tag{4'}$$

If we use (3), and let $w = \hat{y}'(c-)$, $z = \hat{y}'(c+)$, we can restate (4') as follows,

$$f(c,\,\hat{y}(c),\,w) - f(c,\,\hat{y}(c),\,z) - f_z(c,\,\hat{y}(c),\,z)(w - z) = 0; \tag{4''}$$

and since $w \neq z$, we see that *neither* $f(c, \hat{y}(c), \cdot$ (*nor* $-f(c, \hat{y}(c), \cdot)$ can be strictly convex. (Recall Definition 3.4). This fact can sometimes be used to locate or even preclude corner points.

For example, the function $f(\underline{x}, \underline{y}, z) = (x^2 + y^2)\sqrt{1 + z^2}$ is strictly convex (in z) except when $x^2 + y^2 = 0$. Therefore an associated local extremal $\hat{y}$ cannot have a corner point except possibly where $c = \hat{y}(c) = 0$. The function $f(y, z) = (1 + y^2)z^4$ cannot have local extremals with corner points, i.e., each local extremal is C^1, because $(1 + y^2)z^4$ is strictly convex in z. In this case, the test of Remark 7.9 fails since $f_{zz}(y, 0) = 0$. On the other hand, by the same test, when $f(y, z) = e^y\sqrt{1 + z^2}$ every solution of (2) is C^1, whether or not it gives a local extremal. (Recall Example 3 of §3.3)

With a generalization similar to that used to obtain Theorem 6.8, we get the following:

(7.10) **Theorem.** *For a domain D of* $\mathbb{R}^2$, *suppose that*

$$f = f(x, y, z) \in C^1([a, b] \times D),$$

and $\hat{y} \in \hat{\mathscr{Y}} = \hat{C}^1[a, b]$ *provides a (weak) local extremal value for*

$$F(\hat{y}) = \int_a^b f[\hat{y}(x)] \, dx$$

on

$$\hat{\mathscr{D}} = \{\hat{y} \in \hat{\mathscr{Y}}: \hat{y}(a) = a_1, \hat{y}(b) = b_1; (\hat{y}(x), \hat{y}'(x)) \in D\}.$$

Then except at its corner points, $\hat{y}$ *is* C^1 *and satisfies the first and second Euler–Lagrange equations (2′) and (4). At each corner point* c:

(i) $\hat{f}_{y'}(c-) = \hat{f}_{y'}(c+)$; *and*

(ii) $(\hat{f} - \hat{y}'\hat{f}_{y'})(c-) = (\hat{f} - \hat{y}'\hat{f}_{y'})(c+)$;

(iii) $\pm f(c, \hat{y}(c), z)$ *cannot be strictly convex in* z.

PROOF. See 7.13 in §7.5.   □

The Weierstrass–Erdmann conditions (i) and (ii) show that the discontinuities of $\hat{y}'$ which are permitted at corner points of a local extremal $\hat{y}$, are limited to those which preserve the *continuity of both* $\hat{f}_{y'}$ and $(\hat{f} - \hat{y}'\hat{f}_{y'})$. Observe that by (4), when $f_x \equiv 0$, then the latter term is *constant*.

Example 1. A piecewise C^1 extremal, $\hat{y}$, for the brachistochrone problem of §6.2(c), where

$$f(x, y, z) = \frac{\sqrt{1 + z^2}}{\sqrt{y}} \quad \text{and} \quad f_z(x, y, z) = \frac{z}{\sqrt{y}\sqrt{1 + z^2}}$$

should be a stationary function (and hence represent a cycloid) on each interval excluding corner points, at each of which we require the continuity of

$$\hat{f}_{y'} = \frac{\hat{y}'}{\sqrt{\hat{y}}\sqrt{1 + \hat{y}'^2}} \quad \text{and} \quad \hat{f} - \hat{y}'\hat{f}_{y'} = \frac{1}{\sqrt{\hat{y}}\sqrt{1 + \hat{y}'^2}},$$

while the latter function is in fact *constant* since $f_x \equiv 0$. Thus we require the continuity of $\hat{y}'$, so that even on mathematical grounds this problem can have only C^1 extremal functions. (An exception occurs at $x = 0$ where $\hat{y}(0) = 0$ and $\hat{y}'(0) = +\infty$. There, an extreme form of a corner is permitted at the cusp of the cylcoid.)

Example 2. For the function F which introduces this section, where

$$f(x, y, z) = y^2(1 - z)^2, \quad \text{and} \quad f_z(x, y, z) = 2y^2(z - 1).$$

a piecewise C^1 extremal function $\hat{y}$ must be stationary in intervals excluding corner points at which both

$$\hat{f}_{y'} = -2\hat{y}^2(1 - \hat{y}') \quad \text{and} \quad \hat{f} - \hat{y}'\hat{f}_{y'} = \hat{y}^2(1 - \hat{y}'^2)$$

are continuous, while the latter is *constant* over the interval. Since $\hat{y}$ is continuous, from the first condition it follows that $\hat{y}'$ is also continuous except *perhaps* at a point c where $\hat{y}(c) = 0$, and these are the only possible corner points. Thus unless $\hat{y}$ vanishes at some point in an interval $[a, b]$, it is not a local extremal; (moreover, as we have seen in §6.2, Example 3, there need not be any local extremals in $C^1[a, b]$ which satisfy certain boundary conditions).

However, if $\hat{y}$ vanishes at a single point in $[a, b]$, then from the second condition, $\hat{y}^2(1 - \hat{y}'^2) \equiv 0$ so that at each $x \in [a, b]$: either $\hat{y}(x) = 0$, or $\hat{y}'(x) = 1$, or $\hat{y}'(x) = -1$.

Unless $\hat{y} \equiv 0$, the subintervals in which $\hat{y}$ vanishes identically can only terminate at corner points beyond which $\hat{y}$ is linear with the slope 1 or the slope -1. Thus the most general piecewise C^1 (local) extremal function which vanishes at some point on an interval $[a, b]$ is one which vanishes on a *single* subinterval (which may reduce to a point), beyond which it rises or falls with unit slope. We have already encountered one function of this type on the interval $[-1, 1]$, viz., $\hat{y}_0(x) = \begin{cases} 0, & x \le 0 \\ x, & x \ge 0 \end{cases}$. Others are shown in Figure 7.4.

Observe that there cannot be more than one such function having given boundary values at a, b, but there are *none* with $|\hat{y}(a)|$ (or $|\hat{y}(b)|) > b - a$.

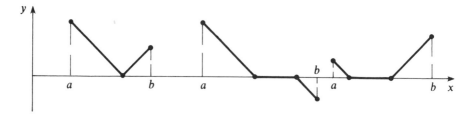

Figure 7.4

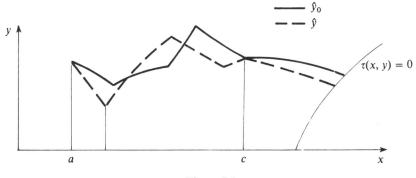

Figure 7.5

Natural boundary conditions corresponding to the various free end point constraints considered in §6.4 *remain the same*; viz., (12a), (12b), and (15), respectively, where, of course, $\hat{f}_{(\cdot)}$ replaces $f_{(\cdot)}$ in each instance.

To see this most easily when, say, freedom is permitted only at the right end point, as in Figure 7.5, suppose that $\hat{y}_0 \in \hat{C}^1[a, b]$ provides a local extremal for F, and let c be the right-most corner point of $\hat{y}_0$. Then restricting comparison to those competing $\hat{y}$ which also have their *last* corner point at c and satisfy $\hat{y}(c) = \hat{y}_0(c)$, it is seen that the corresponding directions $\hat{v}$ must utilize the end-point freedom *exactly* as in §6.4. Thus the resulting natural boundary conditions are the same.

(7.11) **Remark.** In order to apply the method of *Lagrangian multipliers* of §5.7 to piecewise C^1 extremals with *constraints*, it is, in general, necessary to use the weak norm $\| \ \|$ in order to have the requisite *weak continuity* of the variations $\delta F(\hat{y}; \hat{v}) = \int_a^b [\hat{f}_y(x)\hat{v}(x) + \hat{f}_{y'}(x)\hat{v}'(x)] \, dx$. Again, our initial observation that each local extremal of interest is necessarily a weak local extremal makes this permissible, and results in the now expected generalization of §6.5; viz., the existence of Lagrangian multipliers determining a modified function with respect to which an extremal function is stationary in intervals excluding corner points; at each corner point it satisfies the Weierstrass–Erdmann conditions for the modified function.

Application: A Sturm–Liouville Problem

Given functions ρ, q, $\tau \in C[a, b]$, with $\tau > 0$ on $[a, b]$, suppose that $\hat{y}_0 \in \hat{C}^1[a, b]$ minimizes

$$F(\hat{y}) \overset{\text{def}}{=} \int_a^b [\tau(x)\hat{y}'(x)^2 + q(x)\hat{y}(x)^2] \, dx \tag{5}$$

on the *subspace*

$$\hat{\mathcal{Y}}_0 = \{y \in \hat{C}^1[a, b]: \hat{y}(a) = \hat{y}'(b) = 0\},$$

under the *isoperimetric* constraint

$$G(\hat{y}) \stackrel{\text{def}}{=} \int_a^b \rho(x)\hat{y}^2(x)\, dx = 1.$$

Then, $\delta G(\hat{y}_0; \hat{y}_0) = 2\int_a^b \rho(x)\hat{y}_0^2(x)\, dx \neq 0$ and the first alternative of Theorem 5.15 is excluded. Thus with Remark 5.17, $\exists$ a constant (here denoted $-\lambda$) such that $\delta \tilde{F}(\hat{y}_0; \hat{v}) = 0$, $\forall \hat{v} \in \hat{\mathcal{Y}}_0$, where $\tilde{F} = F - \lambda G$.

Introducing

$$\tilde{f}(x, y, z) = \tau(x)z^2 + [q(x) - \lambda\rho(x)]y^2,$$

it follows from Theorem 7.10 that at a possible corner point c, $\tilde{f}_z[\hat{y}_0(x)] = 2\tau(x)\hat{y}_0'(x)$ is continuous. But since τ is continuous and *positive* at c, this implies that $\hat{y}_0'$ itself is continuous. Thus $\hat{y}_0 = y_0 \in C^1[a, b]$ is stationary for $\tilde{f}$ on (a, b), and so y_0 is a solution of the *homogeneous* linear differential equation

$$\frac{d}{dx}(\tau y')(x) = (q - \lambda\rho)(x)y(x), \qquad x \in (a, b), \tag{5$'$}$$

which satisfies homogeneous boundary conditions (such as) $y(a) = y'(b) = 0$.

This type of problem was studied extensively by Sturm and Liouville (c. 1835). It always has the solution $y = \mathcal{O}$. Each λ for which it has a *nontrivial* C^1 solution is called an *eigenvalue* (or characteristic value) of the problem, and each associated solution $y \neq \mathcal{O}$ is called an *eigenfunction*. Clearly, for each constant $c \neq 0$, cy is another eigenfunction for the same eigenvalue, and if $G(y) > 0$, then c may be chosen (within a sign) to make $G(cy) = 1$. From uniqueness of solution of (5$'$), y and y' cannot vanish simultaneously. See Problem 7.21.

In particular, if our supposed minimizing function $y_0 \neq \mathcal{O}$, then it is an eigenfunction for the above *Sturm–Liouville problem* and the Lagrangian multiplier λ is the associated eigenvalue. Now y_0 changes sign at most finitely often on $[a, b]$ (§A1) so that $|y_0| \in \hat{\mathcal{Y}}_0$. Since $|y_0|^2(x) = y_0^2(x)$, while $|y_0|'(x)^2 = (y_0')^2(x)$, it follows that $F(|y_0|) = F(y_0)$ and $G(|y_0|) = G(y_0)$. Thus $|y_0|$ is also a minimizing function, but then, $|y_0| \in C^1[a, b]$ by the argument used before, and hence $|y_0|$ is another eigenfunction for the *same* value λ.

From this fact we can conclude that a *minimizing* eigenfunction y_0 enjoys a special distinction: If $x \in (a, b)$ then $y_0(x) \neq 0$. [For if $y_0(x) = 0$, then also $y_0'(x) = 0$, since otherwise $|y_0|$ would have a corner point at x. But this is not possible.

Conversely, by using an ingenious computation from Picard (1896), we can prove that an eigenfunction y_1 *which is nonvanishing on (a, b)* does minimize F under the given conditions (Problem 7.19). The physical origin of the Sturm–Liouville problem and the significance of this fact will be discussed in §8.9.

(Problems 7.4–7.9)

§7.4. Minimization Through Convexity

When the function $f(\underline{x}, y, z)$ is convex on say $[a, b] \times \mathbb{R}^2$ as in §3.3 then in general, each local extremum becomes a *global minimum*, so that the distinction between weak and strong local extrema is superfluous. For it follows that

$$F(\hat{y}) = \int_a^b f[\hat{y}(x)]\, dx$$

is convex on

$$\hat{\mathscr{D}} = \{\hat{y} \in \hat{C}^1[a, b]: \hat{y}(a) = a_1;\, \hat{y}(b) = b_1\}$$

for given $a_1, b_1 \in \mathbb{R}$, since from 3.4 and (1),

$$F(\hat{y} + \hat{v}) - F(\hat{y}) - \delta F(\hat{y}; \hat{v})$$

$$= \int_a^b \left\{ f[\hat{y}(x) + \hat{v}(x)] - f[\hat{y}(x)] - (f_y[\hat{y}(x)]\hat{v}(x) + f_z[\hat{y}(x)]\hat{v}'(x)) \right\} dx$$

$$\ge 0, \qquad \forall\, \hat{y} \in \hat{\mathscr{D}} \quad \text{and} \quad \hat{v} \in \hat{\mathscr{D}}_0 = \{\hat{v} \in \hat{C}^1[a, b]: \hat{v}(a) = \hat{v}(b) = 0\}. \tag{6}$$

Moreover, when $f(\underline{x}, y, z)$ is *strongly* convex, then F is *strictly* convex on $\hat{\mathscr{D}}$. [Indeed, then *equality* in (6) is possible only if $\hat{v}(x)$ or $\hat{v}'(x) = 0$, $\forall\, x$ except at the corner points of $\hat{y}, \hat{v}$. This is seen by representing the integral in (6) as a finite sum of integrals with *continuous* nonnegative integrands to each of which may be applied the earlier argument from §3.2. Hence $\hat{v}^2(x)$ is continuous and piecewise constant on $[a, b]$. It follows that $\hat{v}^2(x) = \text{const.} = \hat{v}^2(a) = 0$, $\forall\, \hat{v} \in \hat{\mathscr{D}}_0$.]

By an analogous argument we may extend Theorem 3.5 as follows:

(7.12) **Theorem.** *If* $f(\underline{x}, y, z)$ *is* [*strongly*] *convex on* $[a, b] \times D$ *for a domain* $D \subseteq \mathbb{R}^2$ *then a*

$$\hat{y} \in \hat{\mathscr{D}} = \{\hat{y} \in \hat{C}^1[a, b]: \hat{y}(a) = a_1,\, \hat{y}(b) = b_1;\, (\hat{y}(x)\hat{y}'(x)) \in D\},$$

which is stationary for f *in intervals excluding corner points,*[1] *at each of which it satisfies condition 7.10(i), minimizes* F *on* $\hat{\mathscr{D}}$ [*uniquely*].

PROOF. Since the hypotheses on f imply that F is [strictly] convex on $\hat{\mathscr{D}}$, we need only verify that $\delta F(\hat{y}; \hat{v}) = 0$, $\forall\, \hat{v}$ for which $\hat{y} + \hat{v} \in \hat{\mathscr{D}}$. The hypotheses on $\hat{y}$ assure that $\hat{f}_{y'}$ is in $\hat{\mathscr{Y}} = \hat{C}^1[a, b]$ with the derivative $\hat{f}_y$. (Why?) Thus for

---

[1] Corner points *cannot* occur when $f(\underline{x}, y, z)$ is strongly convex by 7.10(iii).

each $\hat{v} \in \hat{\mathcal{Y}}$, the *product* $\hat{f}_{y'}\hat{v}$ is also in $\hat{\mathcal{Y}}$ and by (1), and 7.2,

$$\delta F(\hat{y}; \hat{v}) = \int_a^b [\hat{f}_y(x)\hat{v}(x) + \hat{f}_{y'}(x)\hat{v}'(x)] \, dx$$

$$= \int_a^b \frac{d}{dx}[\hat{f}_{y'}(x)\hat{v}(x)] \, dx$$

$$= \hat{f}_{y'}(x)\hat{v}(x)\Big|_a^b = 0,$$

since $\hat{v}(a) = \hat{v}(b) = 0$, when $\hat{y} + \hat{v} \in \hat{\mathcal{D}}_0$.   □

There is a corresponding version of Proposition 3.9. (Problem 7.12).

Internal Constraints

Convexity may play an important role in problems involving internal *point constraints* such as that of the following.

Example 1. To find the possible local extremal functions for:

$$F(\hat{y}) = \int_0^2 \hat{y}'(x)^2 \, dx$$

on

$$\hat{\mathcal{D}} = \{\hat{y} \in \mathcal{Y} = \hat{C}^1[0, 2]: \hat{y}(0) = \hat{y}(2) = 1; \hat{y}(1) = 0\},$$

which has the *internal* constraint $\hat{y}(1) = 0$, we recall from §3.3, Example 1, that $f(z) = z^2$ is strictly convex on $\mathbb{R}$.

Moreover, when $\hat{y} \in \hat{\mathcal{D}}$ then $\hat{y} + \hat{v} \in \hat{\mathcal{D}}$ iff $\hat{v} \in \hat{\mathcal{D}}_0 = \{\hat{v} \in \mathcal{Y}: \hat{v}(0) = \hat{v}(2) = 0; \hat{v}(1) = 0\}$ and then by the usual argument

$$F(\hat{y} + \hat{v}) - F(\hat{y}) \geq \delta F(\hat{y}; \hat{v}) = \int_0^2 2\hat{y}'(x)\hat{v}'(x) \, dx,$$

(with equality only if $\hat{v} = \mathcal{O}$). But clearly

$$\hat{y}'(x) = \begin{cases} c_1, & \text{on } [0, 1), \\ c_2, & \text{on } (1, 2], \end{cases}$$

gives

$$\delta F(\hat{y}; \hat{v}) = 2c_1 \hat{v}(x)\big|_0^1 + 2c_2 \hat{v}(x)\big|_1^2 = 0, \qquad \forall \, \hat{v} \in \hat{\mathcal{D}}_0.$$

Thus we conclude from the strict convexity of F on $\hat{\mathcal{D}}$, that the *piecewise linear* function $\hat{y}_0 \in \hat{\mathcal{D}}$ given by

$$\hat{y}_0(x) = \begin{cases} 1 - x, & 0 \leq x \leq 1, \\ x - 1, & 1 \leq x \leq 2, \end{cases}$$

is the only local extremal function for F on $\hat{\mathcal{D}}$, and it *minimizes* F on $\hat{\mathcal{D}}$ uniquely, by 7.12.

Observe that $\hat{y}_0$ is clearly in $\hat{C}^1$ but it does *not* satisfy the Weierstrass–Erdmann condition of equation (3) at its corner point $x = 1$. This is because the corner point in question is forced on the extremal function at the outset, while the analysis leading to equation (3) permitted the function to seek its own "natural" corner point(s).

The foregoing analysis extends in principle to any finite number of internal point constraints, and to other functions f (Problems 7.17, 7.18).

For some problems involving *inequality constraints* of the type considered in §2.3, the extremal function may be forced to satisfy a condition precluding its stationarity along an entire *subinterval*.

We suppose that $f(\underline{x}, y, z)$ is [strongly] convex as before, and that we wish to minimize

$$F(\hat{y}) = \int_a^b f[\hat{y}(x)]\, dx$$

on

$$\hat{\mathscr{D}} = \{\hat{y} \in \hat{C}^1[a, b]: \hat{y}(a) = a_1, \hat{y}(b) = b_1\},$$

subject to the *inequality* constraint

$$g[\hat{y}(x)] \leq 0, \qquad x \in (a, b),$$

where $g(\underline{x}, y)$ is also convex.

As suggested by Proposition 2.5 and Remark 3.17, we introduce an unspecified but *nonnegative* $\lambda \in C[a, b]$, and consider instead the problem of minimizing

$$\tilde{F}(\hat{y}) = \int_a^b (f[\hat{y}(x)] + \lambda(x)g[\hat{y}(x)])\, dx \quad \text{on } \hat{\mathscr{D}}.$$

Specifically, we seek $\hat{y}_0 \in \hat{\mathscr{D}}$ which accomplishes this for a $\lambda \geq 0$ with

$$\lambda(x)g[\hat{y}_0(x)] \equiv 0 \quad \text{on } (a, b).$$

Then it would follow from Proposition 2.5 and the [strong] convexity of

$$\tilde{f}(\underline{x}, y, z) = f(\underline{x}, y, z) + \lambda(\underline{x})g(\underline{x}, y)$$

that $\hat{y}_0$ minimizes F [uniquely] on $\hat{\mathscr{D}}$ under the given inequality constraint.

Thus we desire that in intervals excluding a corner point, $\hat{y}_0$ should satisfy the equation of stationarity for $\tilde{f}$; viz.,

$$\frac{d}{dx} f_z[\hat{y}(x)] - f_y[\hat{y}(x)] = \lambda(x)g_y[\hat{y}(x)], \tag{7}$$

where $\lambda \geq 0$ is required to make

$$\lambda(x)g[\hat{y}_0(x)] \equiv 0. \tag{7'}$$

In particular, in those intervals where $g[\hat{y}_0(x)] \neq 0$, λ must vanish and $\hat{y}_0$ is stationary for f. However, we now permit intervals with $g[\hat{y}_0(x)] \equiv 0$ and

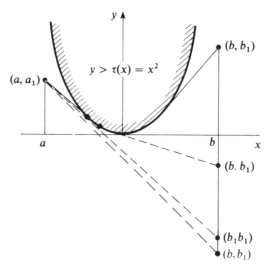

Figure 7.6

$\lambda \neq 0$, where $\hat{y}_0$ is *not* stationary for f. Moreover, since $\tilde{f}_z = f_z$, $\hat{y}_0$ has *only* those corner points permitted by f (Theorem 7.10).

To fix ideas let's consider the following obstacle problem:

Example 2. Minimize the strictly convex distance function

$$F(\hat{y}) = \int_a^b \sqrt{1 + \hat{y}'(x)^2}\, dx$$

on $\hat{\mathcal{D}}$ as above under the constraint $\hat{y}(x) \leq x^2$. Then, as is evident graphically from Figure 7.6, for some locations of the end points we must permit a portion of the minimizing curve to lie along the parabola defined by $g(x, y) = y - x^2 = 0$. Clearly $g(\underline{x}, y)$ is convex (as is $-g(\underline{x}, y)$), but the sign of λ is still important. Here $f(x, y, z) = \sqrt{1 + z^2}$ and $\tilde{f}(x, y, z) = \sqrt{1 + z^2} + \lambda(x)(y - x^2)$.

For this (nonstationary) part of the curve, $y(x) = x^2$, and since $g_y \equiv 1$ while $f_y \equiv 0$, we use (7), the resulting stationarity equation for $\tilde{f}$, to *define*

$$\lambda(x) = \frac{d}{dx}\frac{y'(x)}{\sqrt{1 + y'(x)^2}} = \frac{d}{dx}\left(\frac{2x}{\sqrt{1 + 4x^2}}\right) \geq 0.$$

For the remaining parts, which will be segments (Why?), we have $\lambda = 0$.[1] By taking these segments to be tangential to the parabola at their points of contact, the resulting function y_0 will in fact be C^1. The foregoing analysis

---

[1] The resulting λ is only piecewise continuous, but since $\tilde{f}_z = f_z$, the usual convexity arguments remain valid. See Problem 7.26.

guarantees that it is the *unique minimizing* function for the problem. (Alternatively, we may argue that since $\tilde{f}_{zz} = f_{zz} > 0$, $\hat{y}_0$ cannot have corner points by Theorem 7.10(iii).)

We have just shown how to prove that the natural conjecture for the curve of least length joining fixed points in the presence of a parabolic barrier is the correct one. The reader will find it instructive to analyze this geodesic problem with barriers of other shapes where more than one subarc may be required to lie along the barrier curve. It is more difficult to obtain necessary conditions for problems of this type, and in particular to investigate the behavior at the points of contact. See Problem 7.22 and the discussions in [Pe] and [Sm].

(Problems 7.23, 7.24)

§7.5. Piecewise C^1 Vector-Valued Extremals

When $f = f(x, Y, Z) \in C^1([a, b] \times \mathbb{R}^{2d})$ the function

$$F(\hat{Y}) = \int_a^b f[\hat{Y}(x)] \, dx = \int_a^b f(x, \hat{Y}(x), \hat{Y}'(x)) \, dx$$

is defined $\forall \hat{Y} \in \hat{\mathcal{Y}} = (\hat{C}^1[a, b])^d$, the linear space of piecewise C^1 vector-valued functions $\hat{Y} = (\hat{y}_1, \hat{y}_2, \dots, \hat{y}_d)$ with the weak norm $\|\hat{Y}\|$ of §7.1(b).

As in §6.7 we may analyze each component of a weak local extremal function, $\hat{Y}$, for F on

$$\hat{\mathcal{D}} = \{\hat{Y} \in \hat{\mathcal{Y}} : \hat{Y}(a) = A, \ \hat{Y}(b) = B\},$$

and conclude that at each corner point c of the component $\hat{y}_j$ we must have the (first) Weierstrass–Erdmann condition

$$\hat{f}_{y_j'}(c-) = \hat{f}_{y_j'}(c+),$$

where, of course,

$$\hat{f}_{y_j'}(x) = f_{z_j}[\hat{Y}(x)], \qquad j = 1, 2, \dots, d.$$

Thus at a corner point c, with the obvious notation:

$$\hat{f}_{Y'}(c-) = \hat{f}_{Y'}(c+). \tag{8}$$

In each subinterval excluding the corner points of a given component $\hat{y}_j$, $\hat{Y}$ satisfies the jth stationarity condition

$$\frac{d}{dx}\hat{f}_{y_j'}(x) = \hat{f}_{y_j}(x) = f_{y_j}[\hat{Y}(x)],$$

so that in the subintervals excluding its corner points, $\hat{Y}$ is C^1 and stationary and satisfies the Euler–Lagrange equation

$$\frac{d}{dx} \hat{f}_{Y'}(x) = \hat{f}_Y(x). \tag{8'}$$

Observe that (8) and (8') may be replaced by the equivalent vector valued integral equation:

$$\hat{f}_{Y'}(x) = \int_a^x \hat{f}_Y(t)\, dt + C_0, \tag{9}$$

for some constant vector C_0.

When $f \in C^1([a, b] \times \mathbb{R}^{2d})$ there is a corresponding version of the *second* Euler–Lagrange equation (§6.7) which now takes the form

$$\hat{f}(x) - \hat{Y}'(x) \cdot \hat{f}_{Y'}(x) = \int_a^x \hat{f}_x(t)\, dt + \text{const.}, \tag{10}$$

where

$$\hat{f}(x) = f(x, \hat{Y}(x), \hat{Y}'(x)) = f[\hat{Y}(x)]$$

and

$$\hat{f}_x(x) = f_x(x, \hat{Y}(x), \hat{Y}'(x)) = f_x[\hat{Y}(x)]$$

are well-defined when x is not a corner point of $\hat{Y}$. The continuity in (10) may be used to conclude that at a corner point c, a (local) extremal $\hat{Y}$ satisfies the *second* Weierstrass–Erdmann condition:

$$(\hat{f} - \hat{Y}' \cdot \hat{f}_{Y'})(c-) = (\hat{f} - \hat{Y}' \cdot \hat{f}_{Y'})(c+). \tag{11}$$

In each interval excluding corner points, the integrand in (10) is continuous so that $\hat{Y}$ satisfies the *second* equation:

$$\frac{d}{dx}(\hat{f} - \hat{Y}' \cdot \hat{f}_{Y'})(x) = \hat{f}_x(x). \tag{12}$$

Again with an obvious generalization there follows the

(7.13) **Theorem.** *For a domain D of $\mathbb{R}^{2d}$, let $f = f(x, Y, Z) \in C^1([a, b] \times D)$, and suppose that $\hat{Y}$ provides a local extremal value for*

$$F(\hat{Y}) = \int_a^b f[\hat{Y}(x)]\, dx$$

on

$$\hat{\mathscr{D}} = \{\hat{Y} \in (\hat{C}^1[a, b])^d \colon \hat{Y}(a) = A,\ \hat{Y}(b) = B;\ (\hat{Y}(x), \hat{Y}'(x)) \in D\}.$$

Then except at its corner points, $\hat{Y}$ is C^1 and satisfies the first and second Euler–Lagrange equations, (8') and (12). At each corner point, c, $\hat{Y}$ meets the Weierstrass–Erdmann conditions:

(i) $\hat{f}_{Y'}(c-) = \hat{f}_{Y'}(c+)$; and

(ii) $(\hat{f} - \hat{Y}' \cdot \hat{f}_{Y'})(c-) = (\hat{f} - \hat{Y}' \cdot \hat{f}_{Y'})(c+)$.

(iii) $\pm f(c, \hat{Y}(c), Z)$ cannot be strictly convex in Z.

Observe that when c is a corner point for only one component $\hat{y}_j$, then as before we may infer from (i) that *if defined*, each $f_{z_i z_j}(c, \hat{Y}(c), Z)$ must vanish, for some $Z \in \mathbb{R}^d$; (ii) reduces to requiring the continuity of only $(\hat{f} - \hat{y}_j' \cdot \hat{f}_{y_j'})$ at c; and (iii) becomes a statement of nonconvexity in z_j only. Finally, when $f_x \equiv 0$, then (10) shows that $(\hat{f} - \hat{Y}' \cdot \hat{f}_{Y'})$ is constant.

PROOF. When $Y_0 = \hat{Y}$, the radius function $r(x)$ used in proving Theorem 6.8 is *piecewise* continuous and so it again has a *positive* minimum value on $[a, b]$ guaranteeing a sufficient supply of $\mathcal{D}$-admissible directions at Y_0. For (iii) see Problem 7.25.   □

It is straightforward to extend to vector-valued piecewise C^1 extremals the method of Lagrangian multipliers (cf. Remark 7.11) and the use of convexity in minimization problems (§7.4). (Problem 7.10.)

Application: Minimal Surface of Revolution

In 1744, Euler published a solution to the following problem: Given positive numbers a_1 and b_1, find the planar curve joining points (a, a_1) and (b, b_1) which when revolved about the x axis will have the minimum surface area. (See §1.4(a).)

We suppose as we may that $a = 0$, $a_1 = 1$ and that $b_1 \geq 1$, so that the configuration is as shown in Figure 7.7(a).

Then, supposing that a meridianal curve admits parametric representation by a function $Y = (x, y) \in (C^1[0, 1])^2$, the resulting surface area is (from elementary calculus) given by

$$2\pi \int_0^1 y(t)\, ds(t) = 2\pi \int_0^1 y(t) |Y'(t)|\, dt.$$

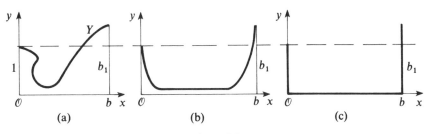

Figure 7.7

However, as we have seen with similar integrals, the minimizing curve (if it exists) may well have corners, and indeed, it is almost intuitively clear that when $b \gg b_1$, the minimum area arises from the limit of curves such as shown in Figure 7.7(b) and thus should be given by the degenerate three segment curve of Figure 7.7(c).

Hence, ignoring the constant 2π, we are led to formulate the problem as follows: Given positive numbers b, and $b_1 \geq 1$, minimize

$$F(\hat{Y}) = \int_0^1 \hat{y}(t)|\hat{Y}'(t)|\, dt$$

on

$$\hat{\mathcal{D}} \stackrel{\text{def}}{=} \{\hat{Y} = (\hat{x}, \hat{y}) \in (\hat{C}^1[0, 1])^2\colon \hat{Y}(0) = (0, 1),\ \hat{Y}(1) = (b, b_1);\ \text{with}$$

$$|\hat{Y}'| \neq 0 \quad \text{and} \quad \hat{y} \geq 0,\ \text{on}\ [0, 1]\}.$$

Here $f(t, Y, Z) = y(z_1^2 + z_2^2)^{1/2}$, so that if $\hat{Y} \in \hat{\mathcal{D}}$ minimizes F (locally) on $\hat{\mathcal{D}}$, we should expect that except at corner points it satisfies the Euler–Lagrange equation(s) (9); viz.,

$$\frac{d}{dx}\left[\frac{\hat{y}\hat{x}'}{|\hat{Y}'|}\right] = 0; \qquad \frac{d}{dt}\left[\frac{\hat{y}\hat{y}'}{|\hat{Y}'|}\right] = |\hat{Y}'|, \tag{13}$$

and at a corner point, c, each bracketed term in these equations is continuous. (However, see Problem 7.14(a).)

It follows that the first term

$$\frac{\hat{y}\hat{x}'}{|\hat{Y}'|} = \text{const.} = c_0 \quad \text{say } \textit{throughout}\ [0, 1].$$

Thus, if $\hat{y}$ or $\hat{x}'$ vanishes at any point in $[0, 1]$, then $c_0 = 0$ so that either $\hat{y}$ or $\hat{x}'$ must vanish at each noncorner point.

A little experimentation shows that the only curve of this type representable by functions in $\hat{\mathcal{D}}$ is that consisting of three segments, shown in Figure 7.7(c), for which the surface area is seen to degenerate to that of the two end disks of Figure 7.8(a). This solution is attributed to its discoverer

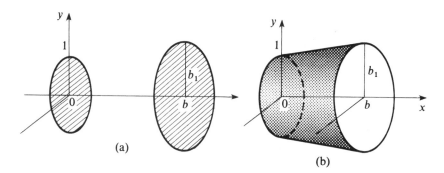

Figure 7.8

B. C. W. Goldschmidt (1831). When b is much larger than b_1 (≥ 1), we have already shown intuitively why it should supply the minimum sought.

However, when b is much smaller than 1 ($\leq b_1$), then the Goldschmidt solution should not provide the minimum area, since then the area of the two end disks will exceed that of the lateral area of the frustrum of the cone having them as bases (Figure 7.8(b).) Thus we must consider the remaining possibility: that $c_0 \neq 0$.

When $c_0 \neq 0$, then neither $\hat{y}$ nor $\hat{x}'$ can vanish anywhere, and by (13) the *unit* vector $\hat{Y}'/|\hat{Y}'|$ is continuous at each corner point. But this means that the *direction* of $\hat{Y}'$ is continuous, so that $\hat{Y}$ represents a curve without corners. We shall assume that $\hat{Y} = Y = (x, y) \in (C^1[0, 1])^2$. (However, see Problem 7.14).

Upon multiplying the resulting second equation (13)

$$(d/dt)[yy'/|Y'|] = |Y'|$$

by its bracketed term, it may be integrated to give $(yy'/|Y'|)^2 = y^2 + c_1$ for a constant c_1, which together with $(yx'/|Y'|)^2 = c_0^2$, shows that $c_1 = -c_0^2$ and

$$\left(\frac{y'}{x'}\right)^2 = \frac{y^2}{c_0^2} - 1. \tag{13'}$$

Since $x'(t)$ is continuous and nonvanishing, it must be positive in order to have $x(0) = 0$ and $x(1) = \int_0^1 x'(t)\, dt = b > 0$. It follows that $x(t)$ has an inverse in $C^1[0, b]$, and we can suppose the curve to be parametrized with respect to the variable $x = x(t)$.

For this parameter, $x' = 1$, and the resulting differential equation for $y = y(x)$ is with $c = c_0$,

$$y'(x)^2 = \left(\frac{y(x)}{c}\right)^2 - 1, \tag{14}$$

(which may be recognized as the *second* Euler–Lagrange equation of the function $f(x, y, z) = y\sqrt{1 + z^2}$ for the *nonparametric* minimal surface of revolution problem. See Problem 6.16.)

Since $y^2(x) \geq c^2$, it is convenient to make the hyperbolic substitution: $y(x) = c(\cosh \eta(x))$, so that $y'(x) = c \sinh \eta(x)\eta'(x)$, and the differential equation for η is

$$c^2(\sinh^2 \eta)(\eta')^2 = \cosh^2 \eta - 1 = \sinh^2 \eta,$$

or

$$c^2\eta'^2 = 1, \quad \text{for } \eta \neq 0.$$

Thus $\eta(x) = c^{-1}x + \mu$, so that the solution of (14) is

$$y(x) = c \cosh(c^{-1}x + \mu), \tag{14'}$$

for some constants $c \neq 0$ and μ to be determined. The boundary conditions for $y = y(x)$ are:

(i) $y(0) = 1$; and
(ii) $y(b) = b_1$.

To satisfy (i) we must have $c \cosh \mu = 1$, so that

$$y(x) = \frac{\cosh(x \cosh \mu + \mu)}{\cosh \mu}.$$

As defined, $y(x)$ is positive on $[0, b]$. We wish to choose μ, *if possible to* satisfy (ii), i.e., to have

$$y(b) = b_1 = \frac{\cosh(b \cosh \mu + \mu)}{\cosh \mu}. \tag{15}$$

However, since $\cosh t > |t|$, $\forall\, t \in \mathbb{R}$, in order that (15) be solvable, it is necessary to have

$$b_1 > \frac{(b \cosh \mu + \mu)}{\cosh \mu} \geq \frac{b \cosh \mu - |\mu|}{\cosh \mu} = b - \left(\frac{|\mu|}{\cosh \mu}\right), \quad \text{or} \quad b_1 > b - 1.$$

Hence when $b_1 \leq b - 1$, there are *no* C^1 local extremal functions, and the only possible minimizing functions for F on $\mathcal{D}$ are those which describe the Goldschmidt curve.

It is more delicate to find conditions on b, b_1 which permit solutions of (15) and hence local extremals other than the Goldschmidt curve. Sets of usable values surely exist. For example, given $b > 0$, the choice $\mu = 0$ will satisfy (15) when $b_1 \overset{\text{def}}{=} \cosh b$. For these values, the curve defined by

$$y(x) = \cosh x, \qquad 0 \leq x \leq b,$$

$$\left(\text{or parametrically by } \begin{cases} x(t) = bt, \\ y(t) = \cosh bt, \end{cases} 0 \leq t \leq 1\right) \text{ representing the catenary}$$

shown in Figure 9.10 will provide a possible local minimal surface area.

More general conditions are known, and some permit *two* distinct solutions of (15), and two possible minimizing catenaries such as those shown in Figure 9.10. (See Problem 7.15.) We reserve discussion of actual minimality of these curves until §9.6. (See [Bl].)

(Problems 7.13–7.15)

Hilbert's Differentiability Criterion*

We have already noted in 7.10(iii) that when present, second derivatives of f may provide useful information about the location of corner points of an extremal function. We shall now see that at a *noncorner* point a condition on the *matrix* f_{ZZ} of the second partial derivatives $f_{z_i z_j}$, $i, j = 1, 2, \ldots, d$, may guarantee higher differentiability of an extremal function. In fact we have the following result:

(7.14) Theorem (Hilbert). *If* $f_Z \in C^1$, *and* $\hat{Y} \in (C^1[a, b])^d$ *is a solution of the integral equation*

$$f_Z(x, \hat{Y}(x), \hat{Y}'(x)) = \int_a^x \hat{f}_Y(t)\, dt + C_0,$$

then $\hat{Y}$ is C^2 in a neighborhood of each noncorner point x_0 at which the matrix $f_{ZZ}[\hat{Y}(x_0)]$ is invertible.

PROOF. Through implicit function theory ([Ed]), the vector valued integral equation

$$f_Z(x, \hat{Y}(x), Z) = \int_a^x \hat{f}_Y(t) \, dt + C_0 \tag{16}$$

determines Z as a *unique* C^1 function of x with $Z(x_0) = \hat{Y}'(x_0)$, in a neighborhood of such x_0 provided that both sides of (16) are C^1 (in (x, Z)) in a neighborhood of $(x_0, \hat{Y}'(x_0))$ and $f_{ZZ}[\hat{Y}(x_0)]$ is invertible.

These latter conditions are assured by the hypotheses. [Indeed

$$\frac{\partial}{\partial x} f_Z(x, \hat{Y}(x), Z) = f_{Zx}(x, \hat{Y}(x), Z) + \sum_{j=1}^d f_{Zy_j}(x, \hat{Y}(x), Z)\hat{y}_j'(x)),$$

while

$$\frac{\partial}{\partial x} \int_a^x \hat{f}_Y(t) \, dt = \hat{f}_Y(x) = f_Y(x, \hat{Y}(x), \hat{Y}'(x)),$$

and both of these functions are continuous near the *noncorner* point x_0. Finally, $f_{z_i z_j}(x, \hat{Y}(x), Z)$ is continuous by hypothesis.]

Since $Z(x) = \hat{Y}'(x)$ will surely be one local solution to equation (16), uniqueness guarantees that it is the *only* solution, and since Z is C^1, it follows that Y is C^2 in this neighborhood.   □

Remark. The continuous differentiability of f_Z is required only in a neighborhood of $(x_0, \hat{Y}(x_0), \hat{Y}'(x_0))$.

Observe that when $d = 1$, the invertibility of the matrix f_{zz} is reduced to the nonvanishing of the single term $f_{zz}[\hat{y}(x_0)]$. For example, the only stationary functions for $f(x, y, z) = e^y \sqrt{1 + z^2}$ are necessarily C^2 since $f_{zz} > 0$ and by 7.9 they cannot have corner points. On the other hand, the example $f(x, y, z) = y^2(1 - z)^2$ of §7.3 shows that corner points can be present where $f_{zz}(x, y, z) = 2y^2 = 0$. A minimizing function which is *only* C^1 is given in Problem 7.16.

§7.6*. Conditions Necessary for a Local Minimum

Heretofore, the conditions obtained in this chapter which are necessary in order that a function minimize

$$F(\hat{Y}) = \int_a^b f[\hat{Y}(x)] \, dx = \int_a^b f(x, \hat{Y}(x), \hat{Y}'(x)) \, dx$$

locally on

$$\hat{\mathcal{D}} = \{\hat{Y} \in \hat{\mathcal{Y}} = (\hat{C}^1[a, b])^d : \hat{Y}(a) = A, \hat{Y}(b) = B\},$$

(in particular, (8), (9), and (10)), actually characterize stationarity and do not distinguish between maximal, minimal, or saddle point behavior—even locally.

The early efforts (notably by Legendre in 1786) to characterize local minimality utilized a second variation and resulted in conditions involving the second derivatives of f. However, in his lectures (c. 1879), Weierstrass showed that a condition involving the first derivatives was even more significant.

(a) The Weierstrass Condition

The Gâteaux variations of an integral function are obtained by comparing its values at a point $\hat{Y}$ with those at points $\hat{Y} + \varepsilon \hat{V}$ in a *weak* norm neighborhood. In contrast to these (weak) variations, we wish to consider a new type of (*strong*) variation by functions whose smallness does *not* imply that of their derivatives—at least not at a given point, $\xi \in (a, b)$.

Specifically, given $\hat{Y} \in \hat{\mathcal{Y}}$, take $(\xi, \hat{Y}(\xi)) = (0, \mathcal{O})$ for convenience. Then for *fixed* $W \in \mathbb{R}^d$, small $h > 0$ and $\varepsilon \in (0, 1)$, let $\tilde{\varepsilon} = \varepsilon/(1 - \varepsilon)$ and define the associated *strong variation* $\hat{W} \in \hat{\mathcal{Y}}$ as follows: when

$$
\begin{array}{llll}
0 \leq x \leq \varepsilon h\text{:} & \hat{W}(x) = xW & \text{so} & \hat{W}'(x) = W, \\
\varepsilon h \leq x \leq h\text{:} & \hat{W}(x) = \tilde{\varepsilon}(h - x)W & \text{so} & \hat{W}'(x) = -\tilde{\varepsilon}W, \\
x < 0 \text{ or } x > h\text{:} & \hat{W}(x) = 0 & \text{so} & \hat{W}'(x) = 0.
\end{array} \qquad (17)
$$

Graphs of $\hat{W}$ are indicated in Figure 7.9. Observe that as $h \searrow 0$, $|\hat{W}| \to 0$, but $\|\hat{W}\| \nrightarrow 0$ (unless $W = \mathcal{O}$). Therefore, if $\hat{Y}$ minimizes F on $\hat{\mathcal{D}}$ locally with

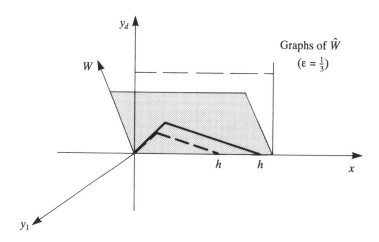

Figure 7.9

respect to the *strong* $\|\cdot\|$ norm, then for small h we have

$$0 \le F(\hat{Y} + \hat{W}) - F(\hat{Y}) = \int_0^h (f[\hat{Y}(x) + \hat{W}(x)] - f[\hat{Y}(x)]) \, dx$$

$$= h \int_0^1 (f[\hat{Y}(sh) + \hat{W}(sh)] - f[\hat{Y}(sh)]) \, ds,$$

where we have substituted $x = hs$ (so $dx = hds$) to obtain the last integral. Now, as $h \searrow 0$, we see from (17) that the first term of this integrand has different *constant* limiting values according to whether $s < \varepsilon$ or $s > \varepsilon$.

If we divide by h and take the limit we find that

$$0 \le \varepsilon[f(0, \mathcal{O}, E + W) - f(0, \mathcal{O}, E)] + (1 - \varepsilon)[f(0, \mathcal{O}, E - \tilde{\varepsilon}W) - f(0, \mathcal{O}, E)],$$

where $E \stackrel{\text{def}}{=} \hat{Y}'(0+)$.

If we now divide by ε and take the limit as $\tilde{\varepsilon} \searrow 0$, we can use the chain rule to see that

$$0 \le [f(0, \mathcal{O}, E + W) - f(0, \mathcal{O}, E)] - f_Z(0, \mathcal{O}, E) \cdot W.$$

Finally, if we replace W by $W - E$ and recall our simplifications we see that $\mathscr{E}(0, \hat{Y}(0), \hat{Y}'(0+), W) \ge 0$, where $\mathscr{E}$ is the *excess function* of Weierstrass defined by

$$\mathscr{E}(x, Y, Z, W) = f(x, Y, W) - f(x, Y, Z) - f_Z(x, Y, Z) \cdot (W - Z), \quad (18)$$

for $x \in [a, b]$ and $Y, Z, W \in \mathbb{R}^d$. Clearly, a similar construction is possible to the left of $\xi = 0$, and for each $(\xi, \hat{Y}(\xi))$ along the extremal curve. Hence we have proven the

(7.15) **Theorem** (Weierstrass). *Let $f = f(x, Y, Z)$ and its derivatives f_Y and f_Z be continuous on $[a, b] \times \mathbb{R}^{2d}$. Suppose that $\hat{Y}$ minimizes*

$$F(\hat{Y}) = \int_a^b f(x, \hat{Y}(x), \hat{Y}'(x)) \, dx$$

locally on

$$\mathcal{D} = \{\hat{Y} \in (\hat{C}^1[a, b])^d \colon \hat{Y}(a) = A, \hat{Y}(b) = B\}$$

for given $A, B \in \mathbb{R}^d$, with respect to the strong $\|\cdot\|$ norm. Then $\hat{Y}$ satisfies the Weierstrass condition:

$$\mathscr{E}(x, \hat{Y}(x), \hat{Y}'(x), W) \ge 0, \quad \forall x \in [a, b], \quad W \in \mathbb{R}^d, \quad (19)$$

(where at a corner point, either one-sided derivative of $\hat{Y}$ is permitted). □

(7.16) **Remarks.** Weierstrass' condition (19) also holds when in the hypotheses, $\mathbb{R}^{2d}$ is replaced by a subdomain D, but now only for those W for which it is defined. Further restrictions may be required. See Problem 7.27.

If we recall the definition 3.4 of partial convexity as extended in Problem 3.33, we see that (19) is satisfied automatically when the function $f(x, Y, Z)$ is convex on say $[a, b] \times \mathbb{R}^{2d}$. Moreover, if this function is strictly convex, a locally minimizing $\hat{Y}$ cannot have corner points. We will return to such functions in Chapter 9.

(7.17) Corollary. *If f is as above and Y minimizes F on*

$$\mathcal{D} = \{Y \in (C^1[a, b])^d \colon Y(a) = A, \; Y(b) = B\}$$

locally with respect to the strong $|\cdot|$ norm, then Y satisfies the Weierstrass condition:

$$\mathscr{E}(x, Y(x), Y'(x), W) \geq 0, \qquad \forall \, x \in [a, b], \quad W \in \mathbb{R}^d. \tag{20}$$

PROOF. By Theorem 7.7 we know that Y also minimizes F on $\hat{\mathscr{D}}$ locally with respect to the $|\cdot|$ norm so that the above theorem is applicable.   □

(7.18) Remark. The proofs of the *local* results given above fail if, in the hypotheses the strong $|\cdot|$ norm is replaced by the weak $\|\cdot\|$ norm. [The Weierstrass construction of $\hat{Y} + \hat{W}$ is *not* possible within the given $\|\cdot\|$ neighborhood of $\hat{Y}$.] However, if $\hat{Y}$ actually minimizes F on $\hat{\mathscr{D}}$, or Y actually minimizes F on $\mathscr{D}$, then these norm distinctions are irrelevant and the appropriate Weierstrass conditions are satisfied.

(b) The Legendre Condition

For *fixed* x, Y, Z, consider the excess function (18)

$$e(W) \stackrel{\text{def}}{=} f(x, Y, W) - f(x, Y, Z) - f_Z(x, Y, Z) \cdot (W - Z), \tag{21}$$

as a function of $W \in \mathbb{R}^d$ alone.

   Both $e(W)$ and its gradient $e_W(W) = f_Z(x, Y, W) - f_Z(x, Y, Z)$ vanish when $W = Z$. Moreover, *when defined*, the second partials are given by

$$e_{w_i w_j}(Z) = f_{z_i z_j}(x, Y, Z), \qquad i, j = 1, 2, \dots, d,$$

at this *stationary point* $W = Z$ of e where $e(Z) = 0$.

   From (0.12), it follows that for small $V = W - Z \neq 0$, $e(W)$ is positive or negative with the *quadratic form*

$$q(V) \stackrel{\text{def}}{=} \sum_{i,j=1}^{d} f_{z_i z_j}(x, Y, Z) v_i v_j,$$

provided that each $f_{z_i z_j}$ is continuous (in Z). This leads to the

(7.19) Proposition (Legendre). *If f, f_Y, and f_Z are continuous on $[a, b] \times \mathbb{R}^{2d}$, and $\hat{Y}$ minimizes F on $\hat{\mathscr{D}}$ locally with respect to the strong $|\cdot|$ norm, then $\hat{Y}$ satisfies the Legendre condition:*

$$q(x, V) \stackrel{\text{def}}{=} \sum_{i,j=1}^{d} f_{z_i z_j}[\hat{Y}(x)] v_i v_j \geq 0, \qquad \forall \, V \in \mathbb{R}^d, \tag{22}$$

at each x at which the coefficient functions $f_{z_i z_j}(x, \hat{Y}(x), Z)$ are defined and continuous (in Z) at $Z = \hat{Y}'(x)$, $i, j = 1, 2, \ldots, d$.

PROOF. Were $q(x, V) < 0$ for *some* V, then the Weierstrass condition (19) would be violated at x for *some* values of W, contradicting Theorem 7.15. $\square$

For example, when $f(x, Y, Z) = x|Z|^2$ so that

$$f_{z_i z_j}(x, Y, Z) = \begin{cases} 2x, & i = j, \\ 0 & i \neq j, \end{cases} \quad i, j = 1, 2, \ldots, d,$$

and $q(x, V) = 2x|V|^2$, Legendre's condition is satisfied trivially on the interval $[0, 2]$, but cannot be satisfied on any interval containing an $x < 0$. Thus, there can be no (local) minimizing functions on such an interval.

Remarks. In contrast to the Weierstrass condition, (22) can also be shown to hold when $\hat{Y}$ is only a weak local minimum for F on $\hat{\mathscr{D}}$; i.e., when $\hat{Y}$ only minimizes F locally with respect to $\| \cdot \|$. See [S].

The Weierstrass and Legendre conditions are *not* equivalent. However, if at some $x \in (a, b)$ we have the *strict* Legendre condition: $q(x, V) > 0$ when $V \neq 0$, then from the above remarks, it follows that $\mathscr{E}(x, \hat{Y}(x), \hat{Y}'(x), W) > 0$ when $0 < |W - \hat{Y}'(x)|$ is sufficiently small. Observe also that the *strict* Legendre condition at x implies (as in 0.13) that the matrix $f_{ZZ}[\hat{Y}(x)]$ is invertible, and hence that $\hat{Y}$ is C^2 in a neighborhood of each *non*corner point x by Theorem 7.14. Conversely, (22) and this invertibility implies the strict Legendre condition [S].

There is also an analogous version of Corollary 7.17.

The following example of Bolza (1902) makes clear the relative strengths of these conditions:

Bolza's Problem

Consider

$$f(z) = z^2(z + 1)^2, \quad z \in \mathbb{R}^1.$$

Here

$$f_z(z) = 4z^3 + 6z^2 + 2z \quad \text{and} \quad f_{zz}(z) = 2(6z^2 + 6z + 1). \tag{23}$$

Clearly the linear function $y_0(x) = mx + c$ is stationary for f since $y_0'(x) = m = \text{const.}$ (§6.2(a)), and each

$$\hat{\mathscr{D}} = \{\hat{y} \in C^1[a, b]: \hat{y}(a) = a_1, \hat{y}(b) = b_1\}$$

contains precisely one such y_0, that with $m = (b_1 - a_1)/(b - a)$.

Moreover, by (23) the parabolic function $f_{zz}(z) = 0$ precisely when $z = m_\pm = -\frac{1}{2} \pm 12^{-1/2}$; also $-1 < m_\pm < 0$. When $m \leq m_-$ (or $m \geq m_+$) we see that y_0 does satisfy the *Legendre condition* $f_{zz}(y_0'(x)) = f_{zz}(m) \geq 0$. Actually, we can use the strict convexity of f on $(-\infty, m_-]$ or $[m_+, \infty)$ (Proposition 3.10) to assert that in these ranges, by Theorem 7.12, y_0 provides the *unique* minimum for

$$F(\hat{y}) = \int_a^b f(\hat{y}'(x))\, dx$$

on

$$\hat{\mathcal{D}}_\pm = \{\hat{y} \in \hat{\mathcal{D}} : \hat{y}'(x) \leq m_-(\hat{y}'(x) \geq m_+)\}.$$

(See also Corollary 3.8.) When $m < m_-$ (or $m > m_+$) then y_0 is a *weak local minimum* point for F on $\hat{\mathcal{D}}$, since all those $\hat{y} \in \hat{\mathcal{D}}$ which belong to the weak neighborhood where $\max |\hat{y}'(x) - y_0'(x)| = \max |\hat{y}'(x) - m| < m - m_-$ (or $m_+ - m$) will be in $\hat{\mathcal{D}}_-$ (or $\hat{\mathcal{D}}_+$). (Why?)

However, if $-1 < m < 0$, then y_0 *cannot* provide a *strong* local minimum for F on $\hat{\mathcal{D}}$, since obviously $F(y_0) = \int_a^b m^2(m+1)^2\, dx > 0$, while each strong norm neighborhood of y_0 contains a $\hat{y}_0 \in \hat{\mathcal{D}}$ for which $\hat{y}_0'(x) = 0$ or -1, so that $F(\hat{y}_0) = 0$. See Figure 7.10. In this range, the Weierstrass condition is violated for y_0. Indeed, from (19), (23) and subsequent simplification,

$$\begin{aligned}
\mathcal{E}(x, y_0(x), y_0'(x), w) &= f(w) - f(y_0'(x)) - f_z(y_0'(x))(w - y_0'(x)) \\
&= f(w) - f(m) - f_z(m)(w - m) \\
&= w^2(w+1)^2 - m^2(m+1)^2 \\
&\quad - 2m(2m^2 + 3m + 1)(w - m) \\
&= (w - m)^2[(w + m + 1)^2 + 2m(m+1)],
\end{aligned}$$

and if $-1 < m < 0$, so that $m(m+1) < 0$, the bracketed expression is *negative* when, say, $w = -(m+1)$.

Thus, in particular, when $-1 < m < m_-$ or $m_+ < m < 0$, y_0 provides a weak local minimum which is *not* a strong local minimum.

Actually, when $-1 < m < 0$, the bracketed expression is positive whenever $(w + m + 1)^2$ is sufficiently large, and hence y_0 also *cannot* give a *strong* local *maximum* for F, since the Weierstrass condition for $-f$ is violated.

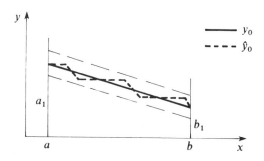

Figure 7.10

But again, when $m_- < m < m_+$, then $-f_{zz}(m) > 0$ and we can use the convexity of $-f$ exactly as before to conclude that y_0 does provide a unique *weak local maximum* for F on $\hat{\mathscr{D}}$.

PROBLEMS

7.1.   (a) Show that
$$\|\hat{y}\| = \max_{x \in [a,b]} (|\hat{y}(x)| + |\hat{y}'(x)|)$$
and
$$\|\hat{y}\|_1 = \int_a^b (|\hat{y}(x)| + |\hat{y}'(x)|)\, dx$$

define norms on $\hat{C}^1[a, b]$.
   (b) Can you provide other norms for $\hat{C}^1[a, b]$?

7.2.   Verify the inequalities (7.5). Hint: Use 7.2 for $a = x_0$ where $|\hat{y}(x_0)| \leq |\hat{y}(x)|$.

7.3*.   Assume that $f \in C([a, b] \times \mathbb{R}^2)$. Show that $F: \hat{C}^1([a, b]) \to \mathbb{R}$ defined by
$$F(\hat{y}) = \int_a^b f(x, \hat{y}(x), \hat{y}'(x))\, dx$$

is continuous with respect to the weak norm $\|\cdot\|$. (See Example 4 in §5.3.)

7.4.   Show that the following functions admit only C^1 extremals for an associated integral function:
   (a) $f(x, y, z) = \sqrt{1 + z^2}$.
   (b) $f(x, y, z) = e^x(1 + y^2 + z^2)$.
   (c)* $f(x, y, z) = \sqrt{y^2 + z^2}$.
   (d) $f(x, y, z) = 4z^2 + 2yz - y^2 + x$.

7.5.   Find all possible corner points for extremals corresponding to
   (a) $f(x, y, z) = x\sqrt{1 + z^2}$.
   (b) $f(x, y, z) = (z - 1)^2(z + 1)^2$.
   (c)* $f(x, y, z) = (x^2 + y^2)e^z$.
   (d) $f(x, y, z) = y^2 - (\cos x)z^3$,
   using (i) $f_{zz} \neq 0$; (ii) the Weierstrass–Erdmann conditions.

7.6.   (a) Find the possible (local) extremal functions for
$$F(\hat{y}) = \int_{-1}^1 [\hat{y}'(x)^2 - 1]^2\, dx$$
on
$$\hat{\mathscr{D}} = \{\hat{y} \in \hat{C}^1[-1, 1]: \hat{y}(-1) = \hat{y}(1) = 0\},$$

which have *exactly one* corner point.
   (b) What is the value of F for each of these extremals?
   (c) Are there any extremals for F belonging to $\hat{\mathscr{D}}$ which have more than one corner point?

7.7*.   (a) Find the possible (local) extremal functions for
$$F(\hat{y}) = \int_0^3 [\hat{y}'(x)^4 - 8\hat{y}'(x)^2]\, dx$$

on
$$\hat{\mathscr{D}}_1 = \{\hat{y} \in \hat{C}^1[0, 3]: \hat{y}(0) = 0, \hat{y}(3) = 2\},$$
which have *exactly one* corner point.

(b) What is the value of F for each of these extremals?

(c) Are there any extremals with corner points for F on
$$\hat{\mathscr{D}}_2 = \{\hat{y} \in \hat{C}^1[0, 3]: \hat{y}(0) = 0, \hat{y}(3) = 8\}?$$

7.8. Discuss the possible local extremal functions for
$$F(\hat{y}) = \int_a^b [2x\hat{y}(x) + x^2\hat{y}'(x)]\, dx$$
on
$$\hat{\mathscr{D}} = \{\hat{y} \in \hat{C}^1[a, b]: \hat{y}(a) = a_1, \hat{y}(b) = b_1\}.$$

7.9. Consider the function
$$F(\hat{y}) = \int_0^2 ([\hat{y}'(x)]^2 + \hat{y}(x)^2][\hat{y}(x) - x]^2 - \tfrac{4}{3}\hat{y}(x)^3 + 2x\hat{y}(x)^2)\, dx.$$

(a) Find the corresponding first Euler–Lagrange equation.

(b) Show that $y = x$ and $y = \alpha e^x$ are solutions of the differential equation found in part (a), where α is an arbitrary constant.

(c) Find all possible corner points for extremals of F.

(d) Find a possible extremal for F on
$$\hat{\mathscr{D}} = \{\hat{y} \in \hat{C}^1[0, 2]: \hat{y}(0) = 0, \hat{y}(2) = e\}.$$

7.10. Formulate and prove an analogue of Theorem 7.12 for piecewise C^1 vector valued extremals. (See Problem 3.33 for the relevant definition of convexity.)

7.11. Show that if $\alpha, \beta, \gamma \in C[a, b]$, then when $\alpha < 0$:
$$F(\hat{y}) = \int_a^b (\alpha(\hat{y}')^2 + \beta\hat{y}\hat{y}' + \gamma\hat{y}^2)(x)\, dx$$
cannot have a strong local minimum on
$$\hat{\mathscr{D}} = \{\hat{y} \in \hat{C}^1[a, b]: \hat{y}(a) = a_1, \hat{y}(b) = b_1\}.$$

7.12. State and prove an analogue of Proposition 3.9 for piecewise C^1 functions, using the methods employed in proving Theorem 7.12.

7.13*. When $\hat{Y} \in (\hat{C}^1[a, b])^d$ is used to represent a curve parametrically in $\mathbb{R}^d$ for $d > 1$, it can happen that this curve does *not* exhibit a corner at a point $\hat{Y}(c)$ when c is a corner point of *each* component $\hat{y}_j, j = 1, 2, \dots, d$.

(a) To understand this, let $d = 2$, and show that the curve represented by
$$\hat{Y}_\alpha = (\hat{x}_\alpha, \hat{y}_\alpha) \in \hat{\mathscr{Y}} = (\hat{C}^1[0, 2])^2,$$
where for $1 \le \alpha < +\infty$:
$$\hat{x}_\alpha(t) = \hat{y}_\alpha(t) = \begin{cases} t^\alpha, & 0 \le t \le 1, \\ t, & 1 \le t \le 2, \end{cases}$$
has no corners, but when $\alpha \ne 1$, $\hat{x}_\alpha$ and $\hat{y}_\alpha$ have the corner point $c = 1$ at

which the ratio $(\hat{y}'_\alpha/\hat{x}'_\alpha)$ *is* continuous. The *curve* is also represented by $\hat{Y}_1 = Y_1 \in (C^1[0, 2])^2$.

(b)* Argue conversely that near a simultaneous corner point $c = 1$ of a general $\hat{Y} \in \mathcal{Y}$ where $\hat{y}'/\hat{x}'$ is continuous, we should be able to use functions similar to those in part (a) to reparametrize the curve by C^1 functions. Hint: Argue geometrically, using (a) as an example.

(c)* Generalize (a) and (b) for $d > 2$.

7.14.   Minimal Surface of Revolution Problem. (See §7.5.)

(a) What is the second Weierstrass–Erdmann condition for $f(t, Y, Z) = y\sqrt{z_1^2 + z_2^2}$? Is it useful?

(b) Make a careful analysis of the second partial derivatives of f to obtain information about possible corner points. Can there be a $\hat{Y}$ with a corner point in only *one* component? Hint: Theorem 7.13 et seq.

(c) When $c_0 \neq 0$, duplicate the derivation of (13′) for a *general* $\hat{Y}$ satisfying equations (13) to conclude that $(\hat{y}'/\hat{x}')^2 = (\hat{y}/c_0)^2 - 1$ on (0.1).

(d)* Since the ratio in (c) is continuous, argue as in Problem 7.13, that the curve represented by $\hat{Y}$ could also be represented by $Y \in (C^1[0, 1])^2$ for which (13′) is satisfied.

(e) What does this problem accomplish?

7.15.   Symmetric Minimal Surface of Revolution (§7.5).

If $b_1 = 1$, we should consider a symmetric form of equation (14′) when $b < 2$. (Why?) To facilitate analysis, translate the interval $[0, b]$ to $[-b/2, b/2]$.

(a) Show that in order that y defined by (14′) satisfy the symmetric boundary conditions $y(\pm b/2) = 1$, we must take $\mu = 0$ and find $\beta = 2c/b > 0$ to make $\varphi(\beta) \stackrel{\text{def}}{=} \beta \cosh(1/\beta) = 2/b (> 1)$.

(b) Prove that $\beta > 0 \Rightarrow \varphi''(\beta) > 0$, and that the (unique) minimum point $\beta_0 (\approx 0.83)$ gives $\varphi(\beta_0) = \varphi_0 (\approx 1.51) \leq \varphi(\beta)$.

(c) Conclude that for $1 < 2/b < \varphi_0$, there can be no *smooth* minimal surface; while for $\varphi_0 < 2/b$ there are *two* possible surfaces. What happens when $\varphi_0 = 2/b$?

(d) In view of Problem 3.30(c), (d), how can there be two solutions to this problem?

(e)* Can you make a similar analysis when $b_1 > 1$?

7.16.   For the function $f(y, z) = z^4 + 3y^2$:

(a) Show that the *second* equation for $y \in C^1[-2, 2]$ is $(y'^2 - y)(y'^2 + y) = c_0 = \text{const.}$

(b) Take $c_0 = 0$ in (a), and find a function which satisfies this equation in $\mathcal{D} = \{y \in C^1[-2, 2]: y(\pm 2) = \pm 1\}$. Hint: Patch solutions which make a factor in (a) vanish.

(c) Use the convexity of f, to conclude that the function y_0 found in (b) must minimize $F(y) = \int_{-2}^{2} f[y(x)]\, dx$ on $\mathcal{D}$ uniquely (Chapter 3.)

(d) Show that y_0 is *not* C^2; does this contradict Theorem 7.14? Explain.

(e)* Can you find solutions to the equation in (a) for *any* $c_0 \neq 0$?

7.17.   (a) Show that a piecewise linear function $\hat{y}_0$ uniquely minimizes

$$F(\hat{y}) = \int_a^b \hat{y}'(x)^2\, dx$$

on

$$\hat{\mathscr{D}} = \{\hat{y} \in \hat{C}^1[a, b] \colon \hat{y}(a) = a_1, \hat{y}(b) = b_1; \hat{y}(c_j) = \hat{c}_j, j = 1, 2, \ldots, N\}$$

for given $a_1, b_1, \hat{c}_j$ and c_j with $a < c_1 < c_2 < \cdots < c_N < b$. Must $\hat{y}_0$ have a corner point at each c_j?

(b) When $f(\underline{x}, y, z)$ is strongly convex on $[a, b] \times \mathbb{R}^2$, and $\hat{\mathscr{D}}$ is as in (a) prove that each $\hat{y}_0 \in \hat{\mathscr{D}}$ which is stationary for f in the intervals excluding the c_j, must minimize $F(\hat{y}) = \int_a^b f[\hat{y}(x)] \, dx$ on $\hat{\mathscr{D}}$ uniquely.

(c) Use (b) to conclude that a polygonal curve will supply the geodesic between fixed points which can be joined *in order* by a function in $\hat{\mathscr{D}}$.

7.18.   Point Constraints. If $f \in C^1([a, b] \times \mathbb{R}^2)$:

(a)   Prove that each direction $\hat{v}$ in

$$\hat{\mathscr{D}}_0 = \{\hat{v} \in \hat{C}^1[a, b] \colon \hat{v}(a) = \hat{v}(b) = \hat{v}(c_j) = 0, j = 1, 2, \ldots, N\}$$

is $\hat{\mathscr{D}}$-admissible for $F(\hat{y}) = \int_a^b f[\hat{y}(x)] \, dx$ where $\hat{\mathscr{D}}$ is as in Problem 7.17(a).

(b)* Conclude that if $\hat{y}_0$ is a local extremal point for F on $\hat{\mathscr{D}}$, then $\hat{y}_0$ must be stationary for f on each interval which excludes the c_j. Will the Weierstrass–Erdmann conditions be satisfied at any of these c_j?

7.19.   Sturm–Liouville problem. (See §7.3.)

(a) If λ is an eigenvalue for (5′) and y_0 is an associated eigenfunction, show that $F(y_0) = \lambda G(y_0)$ where F and G are defined in (5) et seq. Hint: Integrate the term involving τ by parts.

(b) Conclude that if $\rho > 0$ on (a, b) while $q \geq 0$, then each eigenvalue $\lambda > 0$.

(c) Suppose that y_1 is an eigenfunction for (5′) *which is nonvanishing on* (a, b). When $y \in \mathscr{Y}_0 = \hat{\mathscr{Y}}_0 \cap C^1[a, b]$, consider $\tau y_1'$ as a single term, and verify the following:

$$F(y) - \lambda G(y) = \int_a^b \left[\tau(y')^2 + (\tau y_1')' \frac{y^2}{y_1} \right](x) \, dx$$

$$= \int_a^b \left[\tau \left(y' - y \frac{y_1'}{y_1} \right)^2 (x) + r'(x) \right] dx \geq r(x) \Big|_a^b,$$

where

$$r(x) \overset{\text{def}}{=} \left(\frac{\tau y_1' y^2}{y_1} \right)(x), \qquad x \in (a, b).$$

(d) Since $y \in \mathscr{Y}_0$, we know that $y(a) = y_1(a) = 0$. Use L'Hôpital's rule on the factor y/y_1, to conclude that $r(x) \to 0$ as $x \searrow a$.

(e) Since $y_1 \in \mathscr{Y}_0 \Rightarrow y_1'(b) = 0$, with $\tau(b) \neq 0$, think of a *theoretical* basis to claim that $y_1(b) \neq 0$ so that $\tau(x)$ also vanishes as $x \nearrow b$. (If $\tau(b) = 0$, but $\tau'(b)$ exists, show how to reach the same conclusion. Hint: See Problem 7.21.)

(f) Use (a), (c), (d), and (e) to conclude that when $G(y) = G(y_1) = 1$ say, then $F(y) \geq F(y_1), \forall y \in \mathscr{D}_0$.

(g) Use theorems to extend the conclusion of (f) to $\hat{y} \in \hat{\mathscr{Y}}_0$ with $G(\hat{y}) = G(y_1)$.

(h) Give other sets of boundary conditions defining $\hat{\mathscr{Y}}_0$ for which the conclusions of this problem will hold.

7.20.* (Theorem 7.7)

(a) For that part of the proof of Theorem 7.7 considered in the text, explain why it is possible to choose ε so small that $Y_\varepsilon \in S_\delta(Y_0)$. Hint: the triangle inequality.

(b) Suppose $\|\hat{Y} - Y_0\| < \delta$ for a fixed $\delta > 0$. Set $\hat{V} = \hat{Y} - Y_0$ and use Lemma 7.4 to approximate $\hat{V}$ by a V_ε with $|V_\varepsilon'| \leq 4|\hat{V}'|$. Explain why for sufficiently small ε, $\|Y_\varepsilon - Y_0\| < 5\delta$, where $Y_\varepsilon \equiv Y_0 + V_\varepsilon$.

(c) Now, suppose that Y_0 minimizes F on $\mathscr{D}$ in the weak neighborhood $S_{5\delta}(Y_0)$. Argue that Y_0 minimizes F on $\hat{\mathscr{D}}$ in the *smaller* weak neighborhood $\hat{S}_\delta(Y_0) = \{\hat{Y} \in \hat{\mathscr{Y}}: \|\hat{Y} - Y_0\| < \delta\}$. Hint: Observe that Y_ε of (b) approximates $\hat{Y}$ in the sense of Proposition 7.6.

(d*) Explain why this approximation does *not* contradict the fact that a function $\hat{Y}$ with a corner point has a weak neighborhood which *excludes* all C^1 functions.

7.21. Suppose that y is a C^1 solution of $(5')$ on (a, b) with $y(a) = y'(a) = 0$. For $x > a$, set $\alpha(x) = (q - \lambda p)(x)$ and

$$\sigma(x) = ((\tau y')^2 + y^2)(x).$$

(a) Show that $\sigma' = 2[\alpha(\tau y')y + (1/\tau)(\tau y')y]$

(b) Use Proposition 5.3 to verify that when $x \in [a, b]$: $\sigma'(x) \leq M[(\tau y')^2 + y^2] = M\sigma(x)$, for some positive constant M.

(c) Conclude that $(e^{-Mx}\sigma(x))' \leq 0$ so that $0 \leq \sigma(x) \leq e^{M(x-a)}\sigma(a) = 0$; i.e., that $y = 0$.

7.22*. To obtain *necessary* conditions associated with minimizing

$$F(\hat{y}) = \int_a^b f[\hat{y}(x)]\,dx = \int_a^b f(x, \hat{y}(x), \hat{y}(x))\,dx$$

locally on

$$\hat{\mathscr{D}} = \{\hat{y} \in \hat{C}^1[a, b]: \hat{y}(a) = a_1, \hat{y}(b) = b_1\}$$

subject to an inequality constraint $\hat{y}(x) \leq \tau(x)$, $\forall\, x \in (a, b)$, where $\tau \in C^1[a, b]$ is given, introduce the "slack" variable $\hat{\eta}$ by $\hat{y}(x) = \tau(x) - \hat{\eta}^2(x)$.

(a) Argue that it suffices to consider

$$F^*(\hat{\eta}) = \int_a^b f^*[\hat{\eta}(x)]\,dx$$

(locally) on an appropriate $\hat{\mathscr{D}}^*$ where

$$f^*(x, \eta, \zeta) \equiv f(x, \tau(x) - \eta^2, \tau'(x) - 2\eta\zeta).$$

(b) Show that a minimizing $\hat{\eta}_0$ should satisfy the equation

$$\hat{\eta}_0(x)\left(\left(\frac{d}{dx}\right)f_z[\hat{y}_0(x)] - f_y[\hat{y}_0(x)]\right) = 0.$$

except at possible corner points where $f_z[\hat{y}_0(x)]$ is required to be continuous. (Here $\hat{y}_0(x) \equiv \tau(x) - \hat{\eta}_0^2(x)$.)

(c) Conclude that $\hat{y}_0$ will be stationary for f on subintervals where $\hat{y}_0(x) < \tau(x)$, but intervals of nonstationarity where $\hat{y}_0(x) \equiv \tau(x)$ (or $\hat{\eta}_0(x) \equiv 0$) may occur.

(d) Can you draw any conclusions about the behavior of $\hat{y}_0$ at points of transition between the intervals described in (c)? Could these be corner points of $\hat{y}_0$?

7.23.   An Optimal Control Problem.
To solve the control problem presented in Example 3 of §3.5 under the additional Lagrangian inequality $0 \le u \le \beta$ as discussed in Remark 3.20:

(a)   Explain why it would be appropriate to consider the convexity of the integrand

$$f(t, u) = u^2 + \lambda(t)(u^2 - \beta^2) + \lambda_1(T - t)u,$$

where λ_1 is constant.

(b)   Argue that a $u_0 \in \mathscr{D} = C[0, T]$ which solves the equation
   (i) $u(1 + \lambda) = \lambda_0(T - t)$ with $\lambda_0 = -\lambda_1/2, \lambda \ge 0$, will minimize

$$F(u) = \int_0^T u^2(t)\, dt$$

uniquely on $\mathscr{D}$ subject to the isoperimetric constraint
   (ii) $G(u) = \int_0^T (T - t)u(t) = k = h + gT^2/2$, and the Lagrangian inequality $u^2 \le \beta^2$, provided that
   (iii) $\lambda(t)(u_0^2(t) - \beta^2) \equiv 0$.

(c)   Examine (i) and (iii) carefully to conclude that a solution

$$u_0(t) = \begin{cases} \beta, & t \le \tau \\ \lambda_0(T - t), & t \ge \tau \end{cases}$$

is possible if $\lambda_0 = \beta/(T - \tau)$, and τ can be found so that u_0 satisfies (ii). (Remember that we want $\lambda \ge 0$ and $u_0 \ge 0$.) How is λ defined?

(d)*   For $g = 2, \beta = 3$, find τ and compute $F(u_0) = F(u_0, T)$ say.

(e)*   Minimize $F(u_0, T)$ over the allowable range of T to obtain an optimal flight time T_0.

7.24*.   Repeat the analysis of the previous problem when the isoperimetric condition of part (b) is replaced by

$$G(u) \overset{\text{def}}{=} \int_0^T (1 - e^{-\alpha(T-t)})(u(t) - g)\, dt = \alpha h,$$

for constant $\alpha > 0$, which incorporates the aerodynamic damping discussed in Problem 3.38. (In the final computations, set $\alpha = 1$.)

7.25.   (a) Use the mean value theorem on $f_z(c, \hat{y}(c), z)$ to establish the assertion in Remark 7.9.

(b) Show that 7.13(iii) holds. Hint: See the derivation of (4″) and Problem 3.33.

7.26*.   (a) In Example 2 of §7.4, show why the convexity arguments carry through even though λ is only piecewise continuous.

(b) Formulate and prove a more general result of this type involving a strongly convex $f(x, y, z)$, a convex $g(x, y)$, and a piecewise continuous $\lambda \ge 0$.

7.27**. (Newton's minimum drag problem.)

In §3.4(c), we used convexity to determine minimum drag profiles for a projectile moving through a medium under Newton's resistance law within the restricted class where $y' \leq -1/\sqrt{3}$. (See Figure 3.4.) Physically it is clear that we only need to consider profiles with $y' \leq 0$ but Newton argued that for a given h, a cornered profile $\hat{y}$ incorporating a segment where $\hat{y}' = 0$ is more efficient than one in which $0 < -\hat{y}' < 1$. To test his insight, suppose $\hat{y}_0 \in \hat{C}^1[0, 1]$ minimizes

$$F(y) = \int_0^1 \frac{x}{1 + \hat{y}'(x)^2} \, dx$$

on

$$\mathscr{D}^* = \{\hat{y} \in \hat{C}^1[0, 1]: \hat{y}(0) = h, \; \hat{y}(1) = 0, \; \hat{y}' \leq 0\}.$$

We *cannot* claim that $\hat{y}_0$ satisfies (2) because at points where $\hat{y}'_0$ vanishes we are not free to vary $\hat{y}_0$ in the y-direction and remain in $\hat{\mathscr{D}}^*$. However, we can vary the *curve* represented by $\hat{y}_0$ in the x-direction, and conclude that $\hat{y}_0$ satisfies (4),[1] and hence (4') at a corner point. Hereafter denote $\hat{y}_0$ by y for simplicity.

(a) Show that (4') requires continuity of $(1 + 3y'^2)/(1 + y'^2)$ at a corner point where $x = c > 0$, and conclude that the only such points are ones where y' has one-sided values of $m = 0$ and $n = -1$.

(b) Away from corner-points y is C^1 and where $y' \neq 0$, argue that y is C^2 by replacing f by $f - zf_z$ in the proof of Theorem 7.14. Differentiate (4) and conclude that in such intervals our y satisfies equation (14) of §3.4, so that by the earlier analysis, $-y'$ is strictly increasing when $-y' \geq 1/\sqrt{3}$.

(c) Use the previous information to establish that y can have at most one corner point, and that it occurs at $x = a$ where $y'(a-) = 0$, $y'(a+) = -1$, resulting in a single-segment profile predicted by Newton!

(d) Along the curved part of y where $y' \neq 0$ the Legendre test (22) is applicable. Conclude that necessarily $-y' \geq 1/\sqrt{3}$.

(e)* Along the same arc as in (d), the Weierstrass condition (19) holds—but only when restricted to $y' \leq 0$ and $w \leq 0$. (Why?) Conclude that we must have $y'^2 - 1 + 2wy' \geq 0$, so that $y'^2 \geq 1$; i.e., Newton was correct! and in Problem 9.27, we will see that his suggested profile does minimize.

---

[1] This is best accomplished by representing the curve parametrically as in [P], but recall the footnote following Theorem 6.12.

CHAPTER 8

Variational Principles in Mechanics

The recognition that minimizing an integral function through variational methods (as in the last chapters) leads to the second-order differential equations of Euler–Lagrange for the minimizing function made it natural for mathematicians of the eighteenth century to ask for an integral quantity whose minimization would result in Newton's equations of motion. With such a quantity, a new principle through which the universe acts would be obtained. The belief that "something" should be minimized was in fact a long-standing conviction of natural philosophers who felt that God had constructed the universe to operate in the most efficient manner—but how that efficiency was to be assessed was subject to interpretation. However, Fermat (1657) had already invoked such a principle successfully in declaring that light travels through a medium along the path of least time of transit. Indeed, it was by recognizing that the brachistochrone should give the least time of transit for light in an appropriate medium that Johann Bernoulli "proved" that it should be a cycloid in 1697. (See Problem 1.1.) And it was Johann Bernoulli who in 1717 suggested that static equilibrium might be characterized through requiring that the work done by the external forces during a small displacement from equilibrium should vanish. This "principle of virtual work" marked a departure from other minimizing principles in that it incorporated *stationarity*—even local stationarity—(tacitly) in its formulation. Efforts were made by Leibniz, by Euler, and most notably, by Lagrange to define a *principle of least action* (kinetic energy), but it was not until the last century that a truly satisfactory principle emerged, namely, Hamilton's principle of stationary action (c. 1835) which was foreshadowed by Poisson (1809) and polished by Jacobi (1848) and his successors into an enduring landmark of human intellect, one, moreover, which has survived transition to both relativity and quantum mechanics. (See [L], [Fu] and Problems 8.11 8.12.)

This expository chapter is concerned chiefly with an introduction to the theory of Hamilton's principle as obtained from variational considerations. The action integral for a single particle is generated in §8.1, and it is extended to Hamilton's principle for a dynamical system in §8.2 through the introduction of generalized coordinates consistent with the constraints on the system. In §8.3, the total energy function is obtained as a consequence of the second Euler–Lagrange equation.

Unfortunately, applications of Hamilton's principle are usually quite complicated requiring the solution of a highly nonlinear system of differential equations of the second order. In §8.3 we illustrate by example the feasibility of linearization of these equations, making possible their solution in restricted cases. In the remaining sections we present the Hamilton–Jacobi theory for obtaining at least partial solutions (integrals of motion) of the nonlinear system. This is accomplished by replacing the original equations by an equivalent first-order system—the canonical equations (§8.4), which for some coordinates may admit immediate integration (§8.5). The search for new coordinates for which this *must* occur culminates in §8.7 with the Hamilton–Jacobi equation—a single partial differential equation of the first order whose complete solution *if available* would provide such coordinates. In §8.8 we apply Hamilton–Jacobi theory to the general problem of finding stationary functions for a given f, and use it to uncover convexity of certain f that enables us to resolve the brachistochrone problem among others. Finally, in §8.9, we consider the extension of Hamilton's principle to simple continuous media.

§8.1. The Action Integral

For the motion of a single particle of mass m which at time t has the position vector $Y(t) \in \mathbb{R}^3$, we have Newton's equations of motion in the form $(d/dt)(m\dot{Y}(t)) = F(t)$. Here, F represents the force impressed on the particle to produce the motion, and the dot denotes differentiation with respect to time in Newton's notation. We should like these equations to be the Lagrange equations for a function $L = L(t, Y, \dot{Y})$; viz.,

$$\frac{d}{dt}(L_{\dot{Y}}(t)) = L_Y(t), \tag{1}$$

as in §6.7. Thus if $Y = (y_1, y_2, y_3)$, we require that

$$m\dot{y}_i = \frac{\partial L}{\partial \dot{y}_i} \quad \text{for } i = 1, 2, 3,$$

which would occur if

$$L(t, Y, \dot{Y}) = \tfrac{1}{2}m|\dot{Y}|^2 - U(t, Y)$$

for some function U.

But we also require that with $F = (f_1, f_2, f_3)$:

$$f_i = \frac{\partial L}{\partial y_i} = -\left(\frac{\partial U}{\partial y_i}\right), \qquad i = 1, 2, 3,$$

which means that the vector force F should be derivable from the "scalar potential" $-U$.

Introducing $T = \frac{1}{2}m|\dot{Y}|^2$ called (after Leibniz) the *vis viva*[1] or *kinetic energy* of the particle, we see that when $F = -U_Y$ for $U = U(t, Y)$, then indeed Newton's equations are precisely those which characterize the stationary functions for $L = T - U$. Thus in *each* time interval $[a, b]$, they would be satisfied by a function Y which minimizes the *action integral*

$$A(Y) = \int_a^b L(t, Y(t), \dot{Y}(t)) \, dt$$

on

$$\mathscr{D} = \{Y \in (C^1[a, b])^3 : Y(a), \, Y(b) \text{ fixed}\}.$$

However, they would also be satisfied by a function which maximizes (locally) or more generally, *makes stationary*, A on $\mathscr{D}$, in the sense that

$$\delta A(Y; V) = 0, \qquad \forall \, V \in \mathscr{D}_0 = \{V \in (C^1[a, b])^3 : V(a) = V(b) = \mathcal{O}\}. \quad (2)$$

Indeed, as we know, the Euler–Lagrange equations (1) are both necessary and sufficient that (2) holds. (See 6.9 et seq.)

The function $L = L(t, Y, \dot{Y}) = T - U$ is called the *Lagrangian* for the motion, and the first term of the action integral $A(Y) = \int_a^b (T - U) \, dt$ reflects the kinetic energy during the motion, while the second may be regarded as arising from the potential energy injected into the system as a result of the work done by the external forces during the motion. However interpreted, the resulting equations of motion are those of Euler–Lagrange which must agree with those of Newton if the principle is to be valid.

§8.2. Hamilton's Principle: Generalized Coordinates

Dynamical systems may be thought of as consisting of a large finite number (N) of particles of masses m_j which occupy positions $Y_j(t) \in \mathbb{R}^3$, moving with velocities $\dot{Y}_j(t)$ at time $t, j = 1, 2, \ldots, N$. The associated *kinetic energy* of the system is

$$T = \frac{1}{2} \sum_{j=1}^N m_j |\dot{Y}_j|^2, \quad (3)$$

and in 1835, Hamilton postulated that if the system occupies certain positions at times a and b, then between those times, it should move along those

---

[1] Leibniz applied this term to $2T$.

admissible trajectories which make "stationary" the action integral

$$A(Y) = \int_a^b L(t, Y(t), \dot{Y}(t)) \, dt,$$

where $Y = (Y_1, Y_2, \ldots, Y_N) \in (C^1[a, b])^{3N}$ and $L = T - U$.

Here, U represents the *potential energy* given to the system (at time t) through the work done by the *external* forces acting on the system, and the function L is called the *Lagrangian* of the system.

The *admissible trajectories* are those which are consistent with the constraints on the system and which have the prescribed end point values at times a, b, but stationarity for constrained systems is difficult to define.

Unfortunately, the motions of most dynamical systems must satisfy certain geometrical (or other) constraints. For example, in order that the system move as a rigid body, the position functions $Y_j(t)$, must preserve at all times the individual distances between the particles. Even when such constraints can be treated through Lagrangian multipliers as in §6.7, their number makes this approach impractical.

There is the following alternative approach through the use of *generalized coordinates* introduced by Lagrange (c. 1782):

We suppose that the state of the *constrained* system at time t admits description by n *independent* kinematic variables $q_1, q_2, \ldots, q_n$ which constitute the position vector $Q \in \mathbb{R}^n$. [For example, each position of the rigid planar pendulum of Figure 8.1 is determined by specifying the angle θ which it makes with a fixed axis, rather than the pair of cartesian coordinates (x, y) of its tip together with the constraining relation $x^2 + y^2 = l^2$. Of course, there are the familiar equations $x = l \cos \theta$, $y = l \sin \theta$, which permit retrieval of the original cartesian description.]

Similarly, in the general case, the position vector Y should be determinable from Q by means of a transformation $Y = G(Q)$, say. Thus during a motion described by $Q(t)$, $\dot{Y} = G_Q \dot{Q}$, where G_Q represents the Jacobian *matrix* of the transformation. The kinetic energy of the system as given by (3) now assumes the general quadratic form

$$T = \frac{1}{2} \sum_{i,j=1}^n a_{ij}(Q) \dot{q}_i \dot{q}_j \tag{4}$$

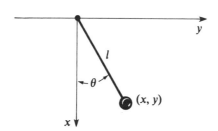

Figure 8.1

for certain functions $a_{ij}(Q) = a_{ji}(Q)$; $i, j = 1, 2, \ldots, n$, which are determined from G_Q and the masses m_j.

[For the pendulum,

$$\dot{x} = -l(\sin \theta)\dot{\theta},$$

$$\dot{y} = l(\cos \theta)\dot{\theta},$$

and

$$T = \tfrac{1}{2}m(\dot{x}^2 + \dot{y}^2) = \frac{m}{2}l^2(\sin^2 \theta + \cos^2 \theta)\dot{\theta}^2 = \frac{m}{2}l^2\dot{\theta}^2.]$$

The Lagrangian of the motion may also be regarded as a *new* function $L = L(t, Q, \dot{Q})$ from $\mathbb{R}^{2n+1}$, and now *Hamilton's principle* takes the following more attractive form:

Between fixed times a, b, the system should move along those trajectories represented by generalized coordinates $Q \in (C^1[a, b])^n$ with prescribed values at a, b, which make stationary the action integral $A(Q) = \int_a^b L(t, Q(t), \dot{Q}(t))\, dt$.

Thus we should have

$$\delta A(Q; V) = 0, \forall\, V \in \mathcal{D}_0 = \{V \in (C^1[a, b])^n: V(a) = V(b) = \mathcal{O}\}.$$

From the general theory of §6.7, on (a, b) the admissible Q should satisfy the Euler–Lagrange equations

$$\frac{d}{dt}L_{\dot{Q}}(t) = L_Q(t) \quad \text{or} \quad \frac{d}{dt}\left(\frac{\partial L}{\partial \dot{q}_j}\right) = \frac{\partial L}{\partial q_j}, \qquad j = 1, 2, \ldots, n, \qquad (5)$$

and these may be regarded as the equations of motion of the system.

Hamilton's principle, expressed in generalized coordinates, provides a straightforward method of determining equations of motion for a general dynamical system. Moreover, it is macroscopic in nature and does not involve the specific action of the position of each individual particle (which for a true rigid pendulum would be astronomical in number!). A difficulty in its application is in the determination of the proper function U to represent the potential energy of the system (which is definable only within an additive reference constant). In some cases, this is decided by choosing that function necessary to recover already agreed upon equations of motion in (5). This was in effect our approach to the problem for the motion of a single particle in the previous section. And as in that case where we required that the force $F = -U_Y$, there need *not* be a choice which renders the principle valid. In particular, the presence of irreversible phenomena such as friction usually precludes the use of Hamilton's principle as stated.

(8.0) **Remark.** In using Hamilton's principle in generalized coordinates we assume implicitly that the *number n* of *independent* generalized coordinates required to describe the configuration of the constrained system is fixed by the system, and would be independent of the particular set of generalized coordinates employed. For then, two such sets of coordinates Q, $\tilde{Q}$, say, should be relatable through an invertible transformation of the form (38) of §6.8; we can conclude from Theorem 6.12, that those C^2 functions $Q(t)$ which make stationary Hamilton's action integral for the Lagrangian $L(t, Q, \dot{Q})$, will transform into functions $\tilde{Q}(t)$, which will make stationary the corresponding action integral when the Lagrangian is expressed in terms of t, $\tilde{Q}$, and $\dot{\tilde{Q}}$. Thus, *in this sense*, Hamilton's principle of stationary action may be said to be an *invariant of the system*. (When a C^1 function $Q(t)$ actually minimizes (or maximizes) its action integral (locally) then simpler considerations show that $\tilde{Q}(t)$ must do so as well for its action integral.)

Bernoulli's Principle of Static Equilibrium

Hamilton's principle represents an attempt to characterize the manner in which the transfer between kinetic and potential energies takes place during an actual motion. It would be satisfying to assert that actual motions minimize this transfer in some sense (and in §9.8(a), we shall see that during small time intervals, the action integral *is* minimized in the classical situation), but even for a single mass particle, the actual motion along stationary trajectories may not always minimize the action integral (Problem 9.1). (See, however, the discussion of Jacobi's principle of least action in §8.5.)

If motion of the system does not occur, the system is in *static equilibrium*. Then its kinetic energy $T \equiv 0$, and if its potential energy, U, depends only on position, Hamilton's principle reduces to *Bernoulli's principle*, which states that this equilibrium state is one which makes stationary the potential energy, U, of the system. Here $U = U(Q)$ may be regarded as the work done by the external forces in bringing the system from a reference state to the position Q. That this equilibrium state may not minimize U is evidenced by the example of a marble balanced on top of a sphere. However, as with the case of this marble, actual physical disturbances will transfer it from its state of (unstable) equilibrium, to one of *stable* equilibrium in which it rests, say, on a table supporting the sphere. And in this state its potential energy is minimized, at least, locally, relative to small displacements. (With larger displacements it could fall to the floor and further reduce its potential energy, or alternatively, change its reference state.)

Thus we may expect that the *stable equilibrium* states, those capable of being sustained indefinitely during small disturbances, should provide a *local minimum* for the potential energy function U. It was appropriate to invoke this principal of *minimum potential energy* in §3.5, when we were seeking the stable equilibrium shape of a cable hanging under its own weight. Moreover, as we shall see in §8.9, some physical systems can be in equilibrium only when they minimize their potential energy (relative to its reference value); i.e., they cannot exhibit unstable equilibrium states.

§8.3. The Total Energy

From Hamilton's principle we suppose that the functions $Q \in (C^1[a, b])^n$ which describe the motions of a dynamical system are the stationary functions for a suitable Lagrangian $L = L(t, Q, \dot{Q})$. Thus when Q is C^2 (and this assumption is usually made in dynamics) they should satisfy also the *second* Euler–Lagrange equation of §6.7, in the form of equation (35):

$$L_{\dot{Q}}(t) \cdot \dot{Q}(t) - L(t) = -\int_a^t L_t(\tau) \, d\tau + c;$$

or introducing

$$E = E(t, Q, \dot{Q}) \overset{\text{def}}{=} L_{\dot{Q}}(t, Q, \dot{Q}) \cdot \dot{Q} - L(t, Q, \dot{Q}),$$

then   (6)

$$E(t) = -\int_a^t L_t(\tau, Q(\tau), \dot{Q}(\tau)) \, d\tau + c.$$

Now, when $L = T(Q, \dot{Q}) - U(t, Q)$, and T is given by (4), then

$$\frac{\partial L}{\partial \dot{q}_j} = \frac{\partial T}{\partial \dot{q}_j} = \sum_{i=1}^n a_{ij}(Q)\dot{q}_i,$$

so that

$$E = \sum_{i,j=1}^n a_{ij}(Q)\dot{q}_i\dot{q}_j - (T - U)$$

$$= 2T - (T - U) = T + U,$$

and E is the sum of the kinetic and potential energies of the system.

In any case, we call E as defined by (6) the *total energy* of the system. When the Lagrangian, L, does not depend explicitly on time so that $L_t \equiv 0$, then the second part of (6) shows that motion can occur only along those trajectories represented by Q for which

$$E(t) = E(t, Q(t), \dot{Q}(t)) = \text{const.},$$

i.e., for which the *total energy is conserved*. In general, we have

$$E(t) - E(a) = -\int_a^t L_t(\tau, Q(\tau), \dot{Q}(\tau)) \, d\tau,$$ (7)

or

$$\dot{E}(t) = -L_t(t, Q(t), \dot{Q}(t)),$$

which may be regarded as a kind of conservation law for the system. (7) is also an *integral of motion*. When $Q = q_1$ alone, then it is equivalent to the single Euler–Lagrange equation of motion (see §6.3). However, in general, there will be other integrals of motion.

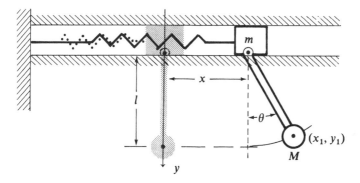

Figure 8.2

Application: Spring–Mass–Pendulum System

In the system shown in Figure 8.2 which consists of a rigid pendulum of length l supported from a frictionless spring–mass system, the motion is quite difficult to describe in terms of standard rectangular or polar coordinates. However, the configuration is completely specified by the pair of *generalized coordinates* (x, θ) where x denotes the position of the mass m and θ denotes the deflection angle of the pendulum, both measured from the equilibrium position illustrated.

We assume that the mass m is constrained to move horizontally by frictionless guides with its motion opposed by a spring force $k(x)$. Moreover, we shall suppose that the mass of the pendulum arm is negligible in comparison to that, M, at its tip. Then, the potential energy function $U(x, \theta)$ is the work required to bring the system from its reference state of equilibrium $(x = \theta = 0)$ to the configuration (x, θ), i.e.,

$$U(x, \theta) = \int_0^x k(s)\, ds + Mgl(1 - \cos \theta), \tag{8}$$

where the first term represents the work done in stretching the spring (as given by elementary calculus) and the second term is the work done in raising the pendulum mass M against gravity. (g is the familiar gravitational constant.)

To obtain the kinetic energy function T it is best to express it first in terms of the rectangular coordinates (x_1, y_1) of the mass M as shown in Figure 8.2, and then to use the geometrical relations

$$
\begin{aligned}
x_1 &= x + l \sin \theta, \\
y_1 &= l \cos \theta,
\end{aligned}
\quad \text{so that} \quad
\begin{aligned}
\dot{x}_1 &= \dot{x} + l(\cos \theta)\dot{\theta}, \\
\dot{y}_1 &= -l(\sin \theta)\dot{\theta},
\end{aligned}
\tag{9}
$$

to convert to the generalized coordinates. Thus from (3),

$$T = \tfrac{1}{2}m(\dot{x}^2) + \tfrac{1}{2}M(\dot{x}_1^2 + \dot{y}_1^2),$$

or, after substitution of (9),

$$T = \tfrac{1}{2}(m + M)\dot{x}^2 + Ml(\cos\theta)\dot{x}\dot{\theta} + \tfrac{1}{2}Ml^2\dot{\theta}^2. \tag{10}$$

Observe that T is in the form (4). From (8) and (10) we see that the Lagrangian for this system is given by

$$L = T - U = \tfrac{1}{2}(m + M)\dot{x}^2 + Ml(\cos\theta)\dot{x}\dot{\theta} + \tfrac{1}{2}Ml^2\dot{\theta}^2$$

$$- \int_0^x k(s)\,ds - Mgl(1 - \cos\theta); \tag{11}$$

it is seen that L does not depend on time explicitly so that $L_t \equiv 0$, and from (6), the total energy $E = T + U$ is *constant* for this system. We now assume that k is continuous.

From Hamilton's principle, there are two equations of motion (5) for this system, one for each of the generalized coordinates x, θ.
    That for x is

$$\frac{d}{dt}(L_{\dot{x}}) = L_x$$

or from (11) and A.8:

$$\frac{d}{dt}[(m + M)\dot{x} + Ml(\cos\theta)\dot{\theta}] = -k(x); \tag{12}$$

while that for θ is

$$\frac{d}{dt}(L_{\dot{\theta}}) = L_\theta,$$

or

$$\frac{d}{dt}Ml[(\cos\theta)\dot{x} + l\dot{\theta}] = -Ml\sin\theta(g + \dot{x}\dot{\theta}). \tag{13}$$

Dividing both equations by Ml and introducing $\mu = (m + M)/(Ml)$, $\kappa(x) = k(x)/Ml$, there results the coupled second-order system:

$$(\mu\dot{x} + (\cos\theta)\dot{\theta})^{\displaystyle\cdot} = -\kappa(x),$$

$$(\cos\theta)\ddot{x} + l\ddot{\theta} = -g\sin\theta, \tag{14}$$

for which we know one integral of motion, namely, that $E = T + U$ is constant. The particular constant depends upon the initial configuration. For example, if the mass m is given an initial velocity $\dot{x}_0$ when the system is in equilibrium, then

$$E = \tfrac{1}{2}(m + M)\dot{x}_0^2.$$

However, there are three remaining integrals of motion of the nonlinear system (14) with the associated initial conditions, and without further simplifications, these cannot be obtained explicitly. Because linear systems can in general be solved, the further assumptions usually employed are those which "linearize" (14). If we suppose $|\theta| \ll 1$ so that $\sin \theta \approx \theta$, $\cos \theta \approx 1$, then (14) becomes

$$\mu \ddot{x} + \ddot{\theta} = -\kappa(x), \qquad \ddot{x} + l\ddot{\theta} = -g\theta. \tag{15}$$

If, in addition, we suppose $\kappa(x) \approx \kappa x$ which will describe a so-called linear spring with the spring *constant* κ, then (14) is approximated by the fully linearized system

$$\mu \ddot{x} + \ddot{\theta} = -\kappa x, \qquad \ddot{x} + l\ddot{\theta} = -g\theta. \tag{16}$$

These linearizing assumptions must be retained and examined in conjunction with any solution of (16). Observe, though, that the assumptions are different in character. The first is a definite restriction on the allowable amplitude of the pendulum swings, while the second is satisfied in some springs for quite large extensions x. If these same assumptions are used in approximating the Lagrangian (11) (with $1 - \cos \theta \approx \theta^2/2$), then the linearized equations (16) follow directly from Hamilton's principle. (See Problem 8.1.) The further solution and discussion of these equations is left to Problem 8.2.

(Problems 8.1–8.3, 8.12)

§8.4. The Canonical Equations

The equations of motion for a dynamical system in the Euler–Lagrange form (5)

$$\frac{d}{dt}\left(\frac{\partial L}{\partial \dot{q}_j}\right) = \frac{\partial L}{\partial q_j}, \qquad j = 1, 2, \ldots, n, \tag{17}$$

constitute a system of n *second*-order differential equations which are usually nonlinear and difficult, if not impossible, to integrate directly. Hamilton and later mathematicians sought transformations which would simplify this task.

First, we note that in terms of the *conjugate momenta* $p_j = \partial L/\partial \dot{q}_j$, $j = 1, 2, \ldots, n$, the equations (17) take the form

$$p_j = \frac{\partial L}{\partial \dot{q}_j}, \qquad \dot{p}_j = \frac{\partial L}{\partial q_j}, \qquad j = 1, 2, \ldots, n, \tag{18}$$

which is now a *first-order* system in the $2n$ variables $Q = (q_1, q_2, \ldots, q_n)$, $P = (p_1, p_2, \ldots, p_n)$.

To transform this system into a form more amenable to analysis, Hamilton made the *Legendre transformation* in which it is supposed that the n equations

$$p_j = \frac{\partial L}{\partial \dot{q}_j}(t, Q, \dot{Q}), \qquad j = 1, 2, \ldots, n, \tag{19}$$

can be solved for the $\dot{q}_j$ as C^1 functions of the *remaining* variables t, Q, P, say, $\dot{Q} = G(t, Q, P)$.

[If we regard equations (19) as constituting a transformation from $\mathbb{R}^{2n+1}$ to $\mathbb{R}^n$, then G is locally determined implicitly as a C^1 function of its variables provided that the $n \times n$ Jacobian determinant $|\partial^2 L/\partial \dot{q}_i \partial \dot{q}_j| \neq 0$. (See, for example, [Ed].) When $L = T(Q, \dot{Q}) - U(t, Q)$ with T given by (4), then $\partial^2 L/\partial \dot{q}_i \partial \dot{q}_j = a_{ij}(Q)$, $i, j = 1, 2, \ldots, n$; and the nonvanishing of the determinant $|a_{ij}(Q)|$ is assured by the positive definiteness of the quadratic form

$$\sum_{i,j=1}^{n} a_{ij}(Q)\dot{q}_i\dot{q}_j = 2T = \sum_{i=1}^{N} m_i |\dot{Y}_i|^2, \tag{19'}$$

when expressed in the original Cartesian coordinates of equation (3); see §0.13. In this simple but important case, (19) is just the *linear* system

$$p_j = \sum_{i=1}^{n} a_{ij}(Q)\dot{q}_i, \qquad j = 1, 2, \ldots, n,$$

which has an explicit solution of the *same* form.]

    With

$$\dot{Q} = G(t, Q, P), \quad \text{or} \quad \dot{q}_j = g_j(t, Q, P), \qquad j = 1, 2, \ldots, n,$$

we have

$$L_{q_j}(t, Q, \dot{Q}) = L_{q_j}(t, Q, G(t, Q, P)) = f_j(t, Q, P),$$

say, for $j = 1, 2, \ldots, n$, and now the system (18) becomes

$$\begin{aligned} \dot{q}_j(t) &= g_j(t, Q(t), P(t)), \\ \dot{p}_j(t) &= f_j(t, Q(t), P(t)), \end{aligned} \qquad j = 1, 2, \ldots, n, \tag{20}$$

which is in the so-called *normal* form. Moreover, when expressed in these variables, the total energy function of (6) is given by

$$\begin{aligned} E(t, Q, \dot{Q}) &= P \cdot \dot{Q} - L(t, Q, G(t, Q, P)) \\ &= P \cdot G(t, Q, P) - L(t, Q, G(t, Q, P)) = H(t, Q, P), \end{aligned} \tag{21}$$

say, if we use (21) to introduce the new function H called the *Hamiltonian* of the system. Observe that

$$H(t, Q, P) = \sum_{j=1}^{n} p_j g_j(t, Q, P) - L(t, Q, G(t, Q, P)) \tag{21'}$$

so that by the chain rule

$$\frac{\partial H}{\partial p_i} = g_i + \sum_{j=1}^{n} \left(p_j - \frac{\partial L}{\partial \dot{q}_j} \right) \frac{\partial g_j}{\partial p_i}$$

and   (22)

$$\frac{\partial H}{\partial q_i} = \sum_{j=1}^{n} \left(p_j - \frac{\partial L}{\partial \dot{q}_j} \right) \frac{\partial g_j}{\partial q_i} - \frac{\partial L}{\partial q_i}, \quad i = 1, 2, \ldots, n.$$

Now, since

$$p_j \equiv \frac{\partial L}{\partial \dot{q}_j}(t, Q, \dot{Q}) \equiv \frac{\partial L}{\partial \dot{q}_j}(t, Q, G(t, Q, P)),$$

each summation in equations (22) vanishes; also

$$\frac{\partial L}{\partial q_j} = \frac{\partial L}{\partial q_j}(t, Q, G(t, Q, P)) = f_j(t, Q, P) \quad \text{for } j = 1, 2, \ldots, n, \text{ as above;}$$

thus we obtain the following equations:

$$\frac{\partial H}{\partial p_i} = g_i \quad \text{and} \quad \frac{\partial H}{\partial q_i} = -f_i, \quad i = 1, 2, \ldots, n,$$

which complete the Legendre transformation. These equations do *not* express laws of dynamics, but with their help, the equations of motion (20) assume the form

$$\dot{q}_j(t) = \frac{\partial H}{\partial p_j}(t, Q(t), P(t)),$$
$$j = 1, 2, \ldots, n. \quad (23)$$
$$\dot{p}_j(t) = -\frac{\partial H}{\partial q_j}(t, Q(t), P(t)),$$

Thus, along a *stationary trajectory* where $H(t, Q(t), P(t)) = H(t)$, say,

$$\frac{dH}{dt}(t) = \frac{\partial H}{\partial t} + \sum_{j=1}^{n} \left(\frac{\partial H}{\partial q_j} \dot{q}_j + \frac{\partial H}{\partial p_j} \dot{p}_j \right) = \frac{\partial H}{\partial t}(t, Q(t), P(t)), \quad (24)$$

in view of (23). This is the transformed version of the conservation law (7) and it may be considered as a $(2n + 1)$st equation of motion.

Equations (23) and (24) may be expressed in the condensed form

$$\dot{Q} = H_P, \quad \dot{P} = -H_Q, \quad \dot{H} = H_t.$$

or   (25)

$$\dot{q}_j = H_{p_j}, \quad \dot{p}_j = -H_{q_j}, \quad \dot{H} = H_t, \quad j = 1, 2, \ldots, n,$$

and they are known as the *canonical equations* of motion for the system. They are attributed usually to Hamilton (1835), although they first appear in work of Lagrange (1809) and were also used by Cauchy (1831). They show that instead of thinking of a dynamical system as being described by its Lagrangian as a function of positions and velocities, it is preferable (analytically) to regard the system as determined by its *Hamiltonian* expressed in terms of *positions* and *conjugate momenta*.[1]

Now, formally, (21) may be restated as $L = P \cdot \dot{Q} - H$, and it is straightforward to verify that equations (25) are those of Euler–Lagrange for the *generalized action integral*

$$A(Q, P) = \int_a^b [P(t) \cdot \dot{Q}(t) - H(t, Q(t), P(t))] \, dt. \tag{26}$$

Moreover, functions satisfying (25) make this integral stationary on sets such as $\{(Q, P) \in (C^1[a, b])^{2n}: Q(a) = A, Q(b) = B\}$; in particular, this stationarity does *not* involve the specification of P at a or b. (See Problem 8.4.) Observe that the new integrand is "naturally" of the form $2T(P, \dot{Q}) - H(t, Q, P)$, is independent of $\dot{P}$, and depends explicitly (and simply) on $\dot{Q}$.

Now, when $T = \frac{1}{2}\sum_{i,j=1}^n a_{ij}(Q)\dot{q}_i\dot{q}_j$ and $U = U(t, Q)$ as in §8.3, then

$$E = T + U \quad \text{and} \quad p_j = L_{\dot{q}_j} = T_{\dot{q}_j} = \sum_{i=1}^n a_{ij}(Q)\dot{q}_i, \qquad j = 1, 2, \ldots, n,$$

so that

$$P \cdot \dot{Q} = \sum_{j=1}^n p_j\dot{q}_j = \sum_{i,j=1}^n a_{ij}(Q)\dot{q}_i\dot{q}_j = 2T,$$

and from (21), the Hamiltonian is,

$$H(t, Q, P) = E = T + U = \tfrac{1}{2}P \cdot \dot{Q} + U(t, Q)$$
$$= \tfrac{1}{2}P \cdot G(t, Q, P) + U(t, Q). \tag{27}$$

For example, in the spring–mass–pendulum system of the previous section where $Q = (q_1, q_2) = (x, \theta) \in \mathbb{R}^2$, the conjugate momenta are from (8) and (10)

$$p_1 = L_{\dot{x}} = T_{\dot{x}} = (m + M)\dot{x} + Ml(\cos \theta)\dot{\theta},$$
$$p_2 = L_{\dot{\theta}} = T_{\dot{\theta}} = Ml(\cos \theta)\dot{x} + Ml^2\dot{\theta}. \tag{28}$$

We may solve this simple linear system to get

$$\dot{q}_1 = \dot{x} = [Ml^2 p_1 - Ml(\cos \theta)p_2]/\Delta,$$
$$\dot{q}_2 = \dot{\theta} = [-Ml(\cos \theta)p_1 + (m + M)p_2]/\Delta, \tag{28'}$$

where $\Delta = Ml^2[m + M \sin^2 \theta] > 0$. It follows from substituting (28') into

---

[1] The Hamiltonian can be defined purely geometrically when the Lagrangian $L(t, Q, Z)$ is strictly convex. See Problem 8.22.

(27) that the Hamiltonian for this system is

$$H = H(Q, P) = \tfrac{1}{2}(p_1\dot{q}_1 + p_2\dot{q}_2) + U(t, Q)$$
$$= [Ml^2 p_1^2 - 2Ml(\cos\theta)p_1 p_2 + (m + M)p_2^2]/2\Delta$$
$$+ \int_0^x k(s)\,ds + Mgl(1 - \cos\theta). \tag{29}$$

The canonical equations (25) can now be obtained. The first of these, namely $\dot{q}_1 = \dot{x} = H_{p_1}$ and $\dot{q}_2 = \dot{\theta} = H_{p_2}$, simply recover equations (28'). However, $\dot{p}_1 = -H_{q_1} = -H_x = -k(x)$ and $\dot{p}_2 = -H_{q_2} = -H_\theta$ are *new* equations, the latter being rather complicated by the dependence of Δ on θ. These four first-order equations replace the two second-order equations (14) obtained earlier, and they constitute the canonical equations of motion for the system. (The final equation $\dot{H} = H_t \equiv 0$ shows that $H = E = $ const. along the stationary trajectories.)

The local solution of this first-order nonlinear system with given initial conditions may be obtained by numerical approximation. See [I–K]. This is one of the reasons that the canonical equations are superior to the Euler–Lagrange equations (14).

The canonical equations also offer significant advantages to those concerned with constructing mathematical models for a dynamical system. It should be observed that only usage in the last three centuries has made terms such as "Kinetic energy of rotation" seem familiar. Hamilton's equations suggest that the "true" dynamical variables are the momenta p_j, and for a system, it is only necessary to know—*or postulate*—a suitable Hamiltonian expressed in these variables. Experience may suggest some of the terms for a Hamiltonian, H, and the effects of modifying an assumed H on the resulting equations of motion (25) can be studied easily. Moreover, in general, Hamilton's equations cannot have more than one solution $(Q(t), P(t))$ with $(Q(a), P(a))$ prescribed, so that with any given Hamiltonian, the solution curves in (Q, P) space *cannot* intersect. (See A.17.) This makes them much more stable with respect to numerical integration schemes than those of Lagrange (14).

(Problems 8.4–8.5)

§8.5. Integrals of Motion in Special Cases

The canonical equations (25); viz.,

$$\dot{q}_j = H_{p_j}, \qquad \dot{p}_j = -H_{q_j}, \qquad \dot{H} = H_t, \qquad j = 1, 2, \ldots, n,$$

provide immediate integrals of motion when the Hamiltonian is independent of some of the $2n + 1$ variables $t, q_1, q_2, \ldots, q_n, p_1, p_2, \ldots, p_n$. For example, if $H_{p_i} \equiv 0$ for some $i = 1, 2, \ldots, n$, then $q_i = $ const. is one integral of motion, and the motion is confined to such hyperplanes. Similarly, if $H_{q_i} \equiv 0$ for some $i = 1, 2, \ldots, n$ then $p_i = $ const. $= c_i$, say, is an integral of motion. Moreover, in this case the ith variable q_i may be effectively deleted or "ignored" by considering the stationary functions for the modified action integral

$$\tilde{A}(\tilde{Q}, \tilde{P}) = \int_a^b [\tilde{P}(t) \cdot \dot{\tilde{Q}}(t) - H(t, \tilde{Q}(t), \tilde{P}(t))] \, dt, \tag{30}$$

where

$$\tilde{Q} = (q_1, q_2 \cdot q_{i-1}, q_{i+1}, \ldots, q_n),$$

and

$$\tilde{P} = (p_1, p_2, p_{i-1}, c_i, p_{i+1}, \ldots, p_n),$$

each vary only in $\mathbb{R}^{n-1}$. (See Problem 8.9.)

Jacobi's Principle of Least Action

Of special interest is the case when $H_t \equiv 0$. Then $H = H(Q, P)$, and from the last of (25), $H(Q(t), P(t)) = $ const. is an integral of motion of the system. Thus the motion takes place along curves confined to hypersurfaces in the Q, P space which conserve the total energy E as in Figure 8.3. In this case, t is simply a parameter for the motion of the system and time becomes ignorable on replacing it with $t = t(\tau)$ where τ is a new variable awaiting specification.

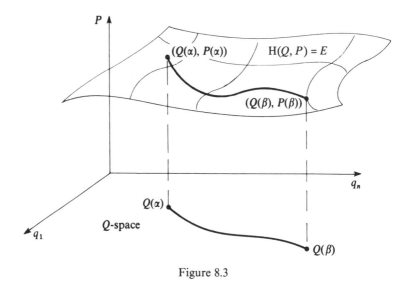

Figure 8.3

Then formally, $\dot{Q}(t)\,dt$ becomes $Q'(\tau)\,d\tau$, and we consider instead of (26), the problem of finding extremal functions Q_0, P_0 for *Jacobi's action integral*

$$\tilde{A}(Q, P) = \int_\alpha^\beta P(\tau) \cdot Q'(\tau)\,d\tau \tag{31}$$

on

$$\mathcal{D} = \{(Q, P) \in (C^1[\alpha, \beta])^{2n} : Q(\alpha) = A, Q(\beta) = B\},$$

subject to the *Lagrangian constraint* $H(Q(\tau), P(\tau)) = E$. Here, α and β are *fixed*, and τ is used to parametrize *each* curve on the hypersurface $H(Q, P) = E$ so that this is possible. (Earlier investigators had formulated this principle improperly *without* reparametrizing, a fact first criticized by Jacobi.) It can be shown (in principle) that such functions Q_0, P_0 do satisfy equations (25) for a properly chosen τ (Problem 8.10).

Jacobi's principle can be given an even more geometrical flavor in the classical case where $U = U(Q)$ and as in (19′)

$$2T = \sum_{i,j=1}^n a_{ij}(Q)\dot{q}_i\dot{q}_j = \left(\sum_{i,j=1}^n a_{ij}(Q)q_i'q_j'\right)\Big/(t')^2.$$

Then, as in (27), $T + U = E = H = $ const., so that $E - U(Q) = T \geq 0$, while also

$$2T = P \cdot \dot{Q} = (P \cdot Q')/t'$$

Thus $P \cdot Q' = 2Tt' = 2\sqrt{E - U(Q)}\sqrt{T(t')^2}$, and Jacobi's integral reduces to

$$\tilde{A}(Q) = \int_\alpha^\beta 2Tt'\,d\tau = \int_\alpha^\beta ds(\tau), \tag{32}$$

say, where s is the "arc length" of element whose square is

$$ds^2 = \sum_{i,j=1}^n A_{ij}(Q)\,dq_i dq_j, \tag{32'}$$

with

$$A_{ij}(Q) = 2(E - U(Q))a_{ij}(Q) = A_{ji}(Q), \qquad i, j = 1, 2, \ldots, n.$$

Because the associated quadratic form is positive definite, (32) defines a Riemannian metric *in Q space* whose coefficients A_{ij} are mass dependent. For this case, the curves which satisfy *Jacobi's principle of least action* by minimizing (32) are those on the constant energy surface whose *projections* in the physical Q space will be the *geodesics* with respect to the metric. (See Figure 8.3.) Such metrics will not be spatially homogeneous in general, and Q space may be thought of as being "curved" appropriately by the *mass dependent metric* to reflect this fact. However, these motions in Q space are physical and involve changes which can be observed.

Symmetry and Invariance

As we have seen, coordinates q_i for which $H_{q_i} = 0$, are ignorable, and yield $p_i = $ const. as associated integrals of motion. Since there are many possible coordinate systems, it is natural to ask whether we can detect the presence of such ignorable variables and their integrals of motion in some other coordinate system. For example, when expressed in cartesian coordinates x, y, can we already "see" that H_θ would be zero if we transform to polar coordinates (r, θ)? Perhaps. We would surely know this if $H(x, y) = k(x^2 + y^2)$, i.e., if H is *symmetric* with respect to the origin.

More generally, if our entire dynamical system has an axis of symmetry, then we should expect a corresponding ignorable coordinate and an associated constant momentum—in this case angular momentum about the axis of symmetry. However, to establish the presence of the axis of symmetry, it would suffice to show that the system is descriptively *invariant* with respect to the family of rotations about that axis. This led E. Noether (1918) to an examination of invariance as a source of integrals of motion; for her principal result, see [G–F].

§8.6. Parametric Equations of Motion

Thus far in our development, time has occupied a distinguished position. Suppose we simply regard it as one more generalized coordinate q_0, say, and enlarge our position vector to $Q = (q_0, q_1, \ldots, q_n) \in \mathbb{R}^{n+1}$.

Then, if we consider motions as given in *parametric form* $t = q_0(\tau), q_1(\tau)$, $\ldots, q_n(\tau), \alpha \leq \tau \leq \beta$, Hamilton's action integral (26) is transformed into the following:

$$A(Q, P) = \int_\alpha^\beta \left[\sum_{j=1}^n p_j(\tau)q_j'(\tau) - H(Q(\tau); p_1(\tau), \ldots, p_n(\tau))q_0'(\tau) \right] d\tau$$

or

$$A(Q, P) = \int_\alpha^\beta \left(\sum_{j=0}^n p_j(\tau)q_j'(\tau) \right) d\tau = \int_\alpha^\beta P(\tau) \cdot Q'(\tau) \, d\tau, \tag{33}$$

where we have introduced the *new* conjugate momentum

$$p_0 = -H(Q, p_1, p_2, \ldots, p_n), \tag{34}$$

and enlarged the momentum vector to

$$P = (p_0, p_1, \ldots, p_n) \in \mathbb{R}^{n+1}.$$

Now, suppose that this new action integral is to be made locally *extremal* among all trajectories in $\mathbb{R}^{2n+2}$ which join fixed points in $\mathbb{R}^{n+1}$ subject to the Lagrangian constraint of the simple form analyzed in Theorem 6.10:, viz.,

$$K(Q, P) \equiv p_0 + H(Q, p_1, p_2, \ldots, p_n) = 0. \tag{35}$$

For a general K of this form we may *in principle* choose τ so that the resulting equations of motion become

$$q'_j = \frac{\partial K}{\partial p_j}, \qquad p'_j = -\frac{\partial K}{\partial q_j}, \qquad j = 0, 1, 2, \ldots, n. \tag{36}$$

(See Problem 8.10.)

Observe that the integrand in (33) does not depend explicitly on τ and its form recalls Jacobi's action integral (31) from the preceding section. However, were we to attempt an analogous geodesic interpretation of least-action principle, we would be led to examine a form

$$ds^2 = \sum_{i,j=0}^{n} A_{ij}(Q)\, dq_i\, dq_j,$$

which is *not* positive definite and hence could only be associated with a nonstandard metric. Such metrics were studied by Ricci (1892) and they form a basis for the general relativity theory of Einstein (1916). The resulting geodesics are not purely spatial in character as were Jacobi's, but occur in the space-time world suggested by our new coordinates.

§8.7*. The Hamilton–Jacobi Equation

There are, in practice, many sets of generalized coordinates which may be used to describe the motion of a given dynamical system; (but see Remark 8.0 of §8.2). It is natural to ask for one, $(\bar{Q}, \bar{P})$, say, in terms of which the equations of motion (36) admit simple integration. Following Jacobi, we seek that for which the associated constraining function (35) simplifies to

$$\bar{K}(\bar{Q}, \bar{P}) \equiv \bar{q}_0, \tag{37}$$

since then the equations of motion (36) reduce to $\bar{q}_0 = 0$, $\bar{p}_0(\tau) = -\tau + c_0$ with *all* other $\bar{p}_j$, $\bar{q}_j$, constant.

To obtain this coordinate system from an original one (Q, P), Hamilton had proceeded indirectly by recognizing that the trajectories which could make stationary the integral (33) with $Q(\alpha)$, $Q(\beta)$ prescribed, subject to the constraint $K(Q, P) = \text{const.}$, will also make stationary a modified integral

$$\int_\alpha^\beta \left[(P \cdot Q')(\tau) - \frac{d}{d\tau} S(Q(\tau), \bar{Q}(\tau)) \right] d\tau, \tag{38}$$

where $S = S(Q, \bar{Q})$ is a *single* real valued C^1 function of the variables $(Q, \bar{Q}) \in \mathbb{R}^{2n+2}$; [the second term is actually a constant (after integration) so that S does *not* participate in the variations]. In particular, if we could find S with $P = S_Q$ and *define* $\bar{P} = -S_{\bar{Q}}$, then, by the chain rule:

$$\frac{d}{d\tau} S(Q(\tau), \bar{Q}(\tau)) = S_Q \cdot Q' + S_{\bar{Q}} \cdot \bar{Q}' = P \cdot Q' - \bar{P} \cdot \bar{Q}',$$

and the *new* integrand of (38) is just $\overline{P} \cdot \overline{Q}'$. Thus our search for Jacobi's coordinates is reduced to finding an $S = S(Q, \overline{Q})$, for which

$$\overline{q}_0 = \overline{K}(\overline{Q}, \overline{P}) = K(Q, P), \quad \text{when} \quad P = S_Q = S_Q(Q, \overline{Q}).$$

Moreover, since motion in the new system occurs with $\overline{q}_0 \equiv 0$, it suffices to find a *complete* solution $S = S(Q)$ to the first-order *partial* differential equation

$$K(Q, S_Q) \equiv 0.$$

Relative to the *original* generalized coordinates where

$$q_0 = t, \quad \text{and} \quad K(Q, P) = p_0 + H(t, q_1, \ldots, q_n, p_1, \ldots, p_n)$$

(as in (35)), this equation for $S = S(t, q_1, \ldots, q_n)$ is

$$\frac{\partial S}{\partial t} + H\left(t, q_1, \ldots, q_n, \frac{\partial S}{\partial q_1}, \frac{\partial S}{\partial q_2}, \frac{\partial S}{\partial q_n}\right) \equiv 0, \tag{39}$$

which is called the *Hamilton–Jacobi equation*.

For the spring–mass–pendulum problem of §8.4 where the Hamiltonian function is given by (29), the Hamilton–Jacobi equation is

$$S_t + \frac{1}{2\Delta}[Ml^2 S_x^2 - 2Ml(\cos\theta)S_x S_\theta + (m + M)S_\theta^2]$$

$$+ \int_0^x k(s)\, ds + Mgl(1 - \cos\theta) \equiv 0. \tag{40}$$

As this example illustrates, the Hamilton–Jacobi equation (39) is not simple, nor are its solutions readily available. However, if a complete solution to this *single* equation is known, then *in principle* it provides a solution to the entire system of equations of motion. By definition, a *complete* solution to this equation is a function $S = S(t, Q; c_1, c_2, \ldots, c_n) + c_0$ which satisfies the equation for each choice of the *essential* constants $c_1, c_2, \ldots, c_n$. Replacing c_j by $\overline{q}_j, j = 1, 2, \ldots, n$, it follows that $S = S(t, Q; \overline{Q})$ satisfies (39) for each $\overline{Q} \in \mathbb{R}^n$.

Next, the n equations $P = S_Q = S_Q(t, Q, \overline{Q})$ *in principle* determine $\overline{Q}$ as functions of t, P, Q. Then

$$\overline{P} \overset{\text{def}}{=} -S_{\overline{Q}} = -S_{\overline{Q}}(t, Q, \overline{Q}) \tag{41}$$

completes the transformation from the original (t, Q, P) coordinates to the $(t, \overline{Q}, \overline{P})$ coordinates of Jacobi in which the equations of motion are simply $\overline{Q}(t) = C = \text{const.} \in \mathbb{R}^n$ and $\overline{P}(t) = B = \text{const.} \in \mathbb{R}^n$, so that (41) becomes

$$B = -S_{\overline{Q}}(t, Q, \overline{Q})\bigg|_{\overline{Q}=C} = -S_C(t, Q, C). \tag{42}$$

Finally, these last equations can also *in principle* be solved for Q as a function of t and the vector constants B, C; say,

$$Q = Q(t) = G(t, B, C),\tag{43}$$

and these are the equations of motion in the *original* space–time coordinates. The vector constants B and C are to be chosen (if possible) to satisfy given initial conditions on the system.

Observe that when $H_t \equiv 0$, then it suffices to find a complete solution $s(Q; C)$ to the *reduced equation*

$$H(Q, s_Q) = E = \text{const.},\tag{44}$$

since then $S = -Et + s$ will provide a complete solution to (39).

We illustrate the method by a simple example which permits a complete solution, namely that for a Lagrangian of a single particle given by

$$L(t, q, \dot{q}) = \dot{q}^2.$$

Then $p \stackrel{\text{def}}{=} L_{\dot{q}} = 2\dot{q}$, and the resulting Hamiltonian is defined by

$$H(t, q, p) = p\dot{q} - \dot{q}^2, \quad \text{with } \dot{q} = \frac{p}{2},$$

so that

$$H = H(p) = \frac{p^2}{4}.$$

Here $H_t \equiv 0$, and the *reduced equation* (44) for $s = s(q)$ is now

$$H(s_q) = E \quad \text{or} \quad s_q^2 = 4E,$$

which has the complete solution $s = 2c_1 q + c_0$, where $c_1^2 = E$. Then $S = -c_1^2 t + 2c_1 q + c_0$ will be a complete solution to the corresponding Hamilton–Jacobi equation (39),

$$S_t + \frac{S_q^2}{4} = 0.$$

For this example, $n = 1$ and equation (42) is simply

$$b_1 = -\frac{\partial S}{\partial c_1}(t, q; c_1) = 2c_1 t - 2q,$$

which when solved for q becomes

$$q = q(t) = c_1 t - \frac{b_1}{2}.\tag{45}$$

We contend that this is a solution to the *original* Lagrangian equation for L, namely, $(d/dt)(2\dot{q}) = 0$ (Proposition 6.1). Now, (45) does indeed solve this equation, and, conversely, each solution of this equation is of the form (45) for appropriate choice of the constants b_1, c_1.

The method of separation used to obtain the complete solution in this simple case admits extension for more complicated Hamiltonians. [We]

(Problems 8.5–8.8)

§8.8. Saddle Functions and Convexity; Complementary Inequalities

In the last sections of this chapter we have shown how the problem of finding stationary functions for a Lagrangian $L = L(t, Q, \dot{Q})$ may be systematically replaced by the simpler problem of finding the stationary functions for a Hamiltonian $H = H(t, Q, P)$, which in turn can be accomplished by solving a related first-order partial differential equation

$$S_t + H(t, Q, S_Q) \equiv 0.$$

Although motivated by physical considerations and couched in the terminology of analytical mechanics—velocity, momenta, energy, etc.—this transformation is purely mathematical, and it is equally applicable to the problem of finding stationary functions for *any* $f = f(x, Y, Y')$. (In the next chapter we shall see the connection between the corresponding Hamilton–Jacobi equation and the existence of a field used in establishing the minimality of a given stationary function for f.)

For example, the *integrand* $f(x, y, y') = y'^2$ corresponds to the *Lagrangian* $L(t, q, \dot{q}) = \dot{q}^2$ which we have examined at the conclusion of the preceding section. As a result, we know that the stationary functions for this f must be of the form corresponding to (45), namely,

$$y(x) = c_1 x - \frac{b_1}{2}.$$

When the integrand function $f(\underline{x}, Y, Z)$ is convex, then we know from Chapter 3 (Problem 3.33) that each stationary function for f (as "Lagrangian"), in fact *minimizes* an associated (action) integral. To see how this convexity relates to the behavior of the "Hamiltonian" of f, assume that the equations $P = f_Z(x, Y, Z)$ determine the components of Z as C^1 functions of x, Y, and P. Then with this Z, as in (21), the associated Hamiltonian is defined by

$$H(x, Y, P) = P \cdot Z - f(x, Y, Z) \tag{46}$$

and exactly as before we conclude that at corresponding points:

$$H_Y = -f_Y, \qquad H_P = Z. \tag{46'}$$

Consequently, at a fixed x, let (Y, Z, P) and (Y_0, Z_0, P_0) be triples in $\mathbb{R}^{3d}$ related through the above transformations. Then it is straightforward to verify the following identity:

$$f(x, Y, Z) - f(x, Y_0, Z_0) - f_Y(x, Y_0, Z_0) \cdot (Y - Y_0) - f_Z(x, Y_0, Z_0) \cdot (Z - Z_0)$$
$$= [H(x, Y, P_0) - H(x, Y, P) - H_P(x, Y, P) \cdot (P_0 - P)]$$
$$- [H(x, Y, P_0) - H(x, Y_0, P_0) - H_Y(x, Y_0, P_0) \cdot (Y - Y_0)], \qquad (47)$$

which shows that in general (see Problem 8.24)

> $f(\underline{x}, Y, Z)$ is [strictly] *convex precisely when both* $H(\underline{x}, \underline{Y}, P)$
> *and* $- H(\underline{x}, Y, \underline{P})$ *are* [strictly] *convex on appropriate sets.*[1]

When a function $H = H(x, Y, P)$ has these convexity properties on a set $S \subseteq \mathbb{R}^{2d+1}$, we will say that $H(\underline{x}, Y, P)$ is a [strict] *saddle function* on S [Sew]. For example,

$$H(\underline{x}, Y, P) = \underline{x}^3(|P|^2 - |Y|^2)$$

is a strict saddle function on $[1, 2] \times \mathbb{R}^{2d}$. Saddle functions have their own significance as we see in the next results.

(8.1) Theorem (Arthurs). *Let* $H = H(x, Y, P) \in C^1(\mathbb{R}^{2d+1})$ *and suppose that* $(Y_0, P_0) \in \mathcal{Y} = (C^1[a, b])^{2d}$ *is stationary for the corresponding action integral*

$$A(Y, P) = \int_a^b [P(x) \cdot Y'(x) - H(x, Y(x), P(x))] \, dx \qquad (48)$$

on

$$\mathcal{D} = \{(Y, P) \in \mathcal{Y} : Y(a) = Y_0(a), \ Y(b) = Y_0(b)\}.$$

If $H(\underline{x}, Y, P)$ *is a* [strict] *saddle function on* $[a, b] \times \mathbb{R}^{2d}$, *then* (Y_0, P_0) *minimizes* A [uniquely] *on*

$$\mathcal{D}_{\min} = \{(Y, P) \in \mathcal{D} : Y' = H_P\}$$

and maximizes A [uniquely] *on*

$$\mathcal{D}_{\max} = \{(Y, P) \in \mathcal{D} : P' = - H_Y\}.$$

PROOF. We shall establish the first implication and leave that for the second to Problem 8.13. From its definition, we may write (with obvious abbreviations)

$$A(Y, P) - A(Y_0, P_0) = \int_a^b [H(x, Y, P_0) - H(x, Y, P) - Y' \cdot (P_0 - P)] \, dx$$

$$- \int_a^b [H(x, Y, P_0) - H(x, Y_0, P_0) - P_0 \cdot (Y' - Y_0')] \, dx. \qquad (48')$$

Now, $Y \in \mathcal{D}_{\min} \Rightarrow Y' = H_P(x, Y, P)$ and with the convexity of $H(\underline{x}, \underline{Y}, P)$, it follows that the *first* integral is nonnegative. Consequently, for such Y, we

---

[1] It does not seem possible to use (47) to characterize *strong* convexity of f. See Problem 8.28(b).

have after an integration by parts, that

$$A(Y, P) - A(Y_0, P_0) \geq - \int_a^b [H(x, Y, P_0) - H(x, Y_0, P_0) + P_0' \cdot (Y - Y_0)] \, dx,$$

$$(48'')$$

since the boundary term $P_0 \cdot (Y - Y_0)|_a^b = 0$. But for the *stationary function*, $P_0' = -H_Y(x, Y_0, P_0)$ (Why?), and now the convexity of $-H(\underline{x}, Y, \underline{P})$ will ensure that the *last* integral is nonpositive. When *strict* convexity obtains in both cases, then from the usual arguments of Chapter 3, we conclude that $P = P_0$ and $Y = Y_0$. □

(8.2) Remarks. Theorem 8.1 does *not* require that H be a Hamiltonian associated with a Lagrangian f. In applications, we could take *any* function $P \in (C^1[a, b])^d$, and seek a solution

$$Y \in \mathscr{D}_0 = \{Y \in (C^1[a, b])^d : Y(a) = Y_0(a), Y(b) = Y_0(b)\}$$

of the first-order system $Y'(x) = H_P(x, Y(x), P(x))$, to obtain a $(Y, P) \in \mathscr{D}_{\min}$, and hence an *upper bound* to $A(Y_0, P_0)$. Similarly, choosing a $Y \in \mathscr{D}_0$, we could use any solution P of the system $P'(x) = -H_Y(x, Y(x), P(x))$, to provide a *lower bound* for $A(Y_0, P_0)$. For more information on this method of *complementary inequalities*, see [Ar].

Now, in general, it is difficult to find $(Y, P) \in \mathscr{D}_{\min}$ unless H *is* the Hamiltonian for a Lagrangian function $f = f(x, Y, Z)$. Then each $Y \in \mathscr{D}_0$ provides a $P(x) \stackrel{\text{def}}{=} f_{Y'}(x)$ for which $Y'(x) = H_P(x, Y(x), P(x))$ since this equation is a consequence of the Legendre transformation; (recall (46')). This pair $(Y, P) \in \mathscr{D}_{\min}$, and from Theorem 8.1 we would conclude that

$$F(Y) \stackrel{\text{def}}{=} \int_a^b f[Y(x)] \, dx = A(Y, P) \geq A(Y_0, P_0) = F(Y_0),$$

as we would expect from (47) and its convexity implications. However, this same Y also provides (in principle) a solution $P_1 \neq P$ of the first-order system $P'(x) = -H_Y(x, Y(x), P(x))$ for which $(Y, P_1) \in \mathscr{D}_{\max}$. Thus $A(Y, P_1)$ is a *lower bound* for $F(Y_0)$, and this could be of use in approximating an $F(Y_0)$ that cannot be determined exactly.

It is easier to assess saddle function behavior of H on a product domain. For example, in the one-dimensional case,

$$H(y, p) = -\frac{1}{y}\sqrt{1 - p^2}$$

is a strict saddle function on the product domain $\{y > 0\} \times \{p^2 < 1\}$ since $H_{yy} < 0 < H_{pp}$ there; (recall Example 5 in §3.3). To find an associated "Lagrangian" $f = f(y, z)$, we use (46') to define

$$z = H_p = \frac{1}{y}\frac{p}{\sqrt{1 - p^2}} \quad \text{(which has the sign of } p\text{),}$$

and solve this equation algebraically to obtain

$$p = \frac{z}{\sqrt{y^{-2} + z^2}}.$$

Then with this p and z, from (46) we see that

$$f(y, z) \overset{\text{def}}{=} pz - H(y, p) = \frac{1}{y} \frac{p^2}{\sqrt{1 - p^2}} + \frac{1}{y}\sqrt{1 - p^2} = \frac{1}{y\sqrt{1 - p^2}} = \frac{z}{p},$$

so that

$$f(y, z) = \sqrt{y^{-2} + z^2} \quad \text{is strictly convex on } \{y > 0\} \times \mathbb{R}.$$

We can produce other functions in the same manner whose convexity is not easy to establish by, say, the second derivative test of Problem 3.5. See Problem 8.25. Now, let's utilize this ability and give new solutions to old problems.

Example 1. The Cycloid Is the Brachistochrone.
    To solve the brachistochrone problem of §1.2(a), we must find the $y_0 \in C[0, x_1]$ that minimizes the time-of-travel integral

$$T(y) = \frac{1}{\sqrt{2g}} \int_0^{x_1} \sqrt{\frac{1 + y'^2}{y}} \, dx$$

on

$$\mathscr{D}^* = \{y \in C[0, x_1]: y(0) = 0, y(x_1) = y_1; T(y) < +\infty\}.$$

The integrand function $\sqrt{(1 + z^2)/y}$ is not convex, but if we let $y(x) = \tilde{y}^2(x)/2$, so that $y' = \tilde{y}\tilde{y}'$ the integral becomes

$$\tilde{T}(\tilde{y}) = \frac{1}{\sqrt{g}} \int_0^{x_1} \sqrt{\tilde{y}^{-2} + \tilde{y}'^2} \, dx,$$

and we now recognize the *new* integrand function

$$f(\tilde{y}, z) = \sqrt{\tilde{y}^{-2} + z^2}$$

as being strictly convex on $\{\tilde{y} > 0\} \times \mathbb{R}$!
    Thus if $y_0 = y_0(x)$ describes the cycloid joining the origin to the point (x_1, y_1) (see Example 4 of §6.2) then $\tilde{y}_0(x) = \sqrt{2y_0(x)}$ describes a curve that is *stationary* for f on $(0, x_1)$. (This can be verified either directly or by appeal to the invariance of stationarity in §6.8.) Finally $|f_z| = |z|/\sqrt{\tilde{y}^{-2} + z^2} \leq 1$, and we can use Theorem 3.5 as extended in Problem 3.21(b) to conclude that if $y \in \mathscr{D}^*$, then $T(y) = \tilde{T}(\tilde{y}) \geq \tilde{T}(\tilde{y}_0) = T(y_0)$; hence $\tilde{y}_0$ (so y_0) is the unique minimizing function. We see that the cycloid *is* the brachistochrone because a suitable time-of-travel integral *is* strictly convex. Apparently, enough speed is attained in nearly vertical fall from $\mathcal{O}$ to compensate for lower speeds in the flatter parts of the cycloid even if we must travel upward to reach our destination!

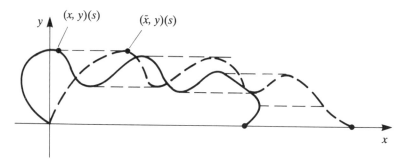

Figure 8.4

As this example shows, sometimes a simple transformation can produce an integral having a convex integrand function from one that does not. Moreover, it may be easier to establish convexity of an integrand f by showing that the associated Hamiltonian is a saddle function.

Example 2*. Dido's Problem.

One of the most important assertions of classical origin is the isoperimetric inequality of plane geometry (§1.3). It is closely related to the problem of Queen Dido, concerning how to arrange a curve of length l in the upper half-plane $\{y > 0\}$ with ends on the x-axis that together with this axis encloses the maximal area. (See Problem 3.32.)

We can suppose a typical curve to be parametrized *with respect to its arc-length s* by functions $x, y \in C^1[0, l]$ with $x'^2 + y'^2 = 1$, $x(0) = 0$, $y(0) = y(l) = 0$ and $y(s) > 0$,[1] $s \in (0, l)$. (See Figure 8.4.) By using Green's theorem [Ed], it is not difficult to see that this curve encloses an area of size

$$A(x, y) = \left| \int_0^l y(s)x'(s)\, ds \right| \leq \int_0^l y(s)|x'(s)|\, ds.$$

Moreover, the last integral is just $A(\tilde{x}, y)$ when we define

$$\tilde{x}(s) = \int_0^s |x'(t)|\, dt \quad \text{so that} \quad \tilde{x}'(s) = |x'(s)| = \sqrt{1 - y'(s)^2};$$

i.e., we can replace the given curve by a related curve that encloses an area at least as large for which $\tilde{x}' \geq 0$. Thus we only need to solve the *nonisoperimetric* problem of *minimizing*

$$F(y) = -\int_0^l y\sqrt{1 - y'^2}\, ds$$

on

$$\mathcal{D} = \{y \in C^1[0, l]: y(0) = y(l) = 0, y \geq 0\}.$$

[1] This restriction will be relaxed later.

To learn whether $f(y, z) = -y\sqrt{1 - z^2}$ is convex we can use the second derivative test of Problem 3.5 which in this case is very simple because $f_{yy} = 0$. Since $f_{yz} \neq 0$, $f_{yy}f_{zz} - (f_{yz})^2 < 0$, and we see that this f is not convex, (nor is $-f$). However, $f(y, z)$ is strictly convex (Example 5 of §3.3), and our previous success leads us to look for a simple transformation. But which one? The fact that $y\sqrt{1 - y'^2} = \sqrt{y^2 - (yy')^2}$, suggests trying $y^2(x) = 2\tilde{y}(x)$ so that $yy' = \tilde{y}'$ (which is just the reverse of the transformation used in Example 1!) This leads us to consider minimizing

$$\tilde{F}(\tilde{y}) = -\int_0^l \sqrt{2\tilde{y} - \tilde{y}'^2}\, ds \tag{49}$$

on

$$\tilde{\mathscr{D}} = \{\tilde{y} \in C^1[0, l]: \tilde{y}(0) = \tilde{y}(l) = 0,\ \tilde{F}(\tilde{y}) < +\infty\}$$

and so asking whether $\tilde{f}(\tilde{y}, z) = -\sqrt{2\tilde{y} - z^2}$ is convex.

To find out, let's introduce the associated momentum

$$p = \tilde{f}_z(\tilde{y}, z) = \frac{z}{\sqrt{2\tilde{y} - z^2}}, \quad \text{solve for} \quad z = \frac{\sqrt{2\tilde{y}}\,p}{\sqrt{1 + p^2}},$$

and substitute in (46) to obtain the Hamiltonian

$$H(\tilde{y}, p) = pz + \sqrt{2\tilde{y} - z^2} = pz + \frac{z}{p} = \left(p + \frac{1}{p}\right)\frac{\sqrt{2\tilde{y}}\,p}{\sqrt{1 + p^2}} = \sqrt{2\tilde{y}}\sqrt{1 + p^2}$$

Now, when $\tilde{y} > 0$, $H(\tilde{y}, p)$ is strictly convex as is $-H(\tilde{y}, p)$, so that H is a strict saddle function, and therefore by (47), $\tilde{f}$ is strictly convex on $D = \{2\tilde{y} < z^2\}$!

Dido believed that the semicircle of length l with ends on the x-axis gives the minimizing curve, and it is easy to show that relative to *arc-length*, it is parametrized by

$$y_0(s) = \frac{l}{\pi}\sin\left(\frac{\pi s}{l}\right), \quad 0 \leq s \leq l.$$

This function is stationary for $f(y, z) = -y\sqrt{1 - z^2}$ since

$$(f - y_0'f_{y_0'})(s) = -\frac{y_0'}{\sqrt{1 - y_0'^2}}(s) = -1,$$

and it follows that

$$\tilde{y}_0(s) = \frac{y_0^2(s)}{2} = \frac{l^2}{2\pi^2}\sin^2\left(\frac{\pi s}{l}\right)$$

is stationary for $\tilde{f}(y, z) = -\sqrt{2y - z^2}$ on $(0, l)$. Here

$$\tilde{f}_z[\tilde{y}_0(s)] = \frac{\tilde{y}_0'}{\sqrt{2\tilde{y}_0 - \tilde{y}_0'^2}}(s) = \frac{\cos(\pi s/l)}{\sin(\pi s/l)}$$

is unbounded near $s = 0$ and $s = l$. However, $s\tilde{f}_z[y_0(s)]$ and $(l - s)\tilde{f}_z[y_0(s)]$ are bounded near $s = 0$ and $s = l$ respectively while to be in $\tilde{\mathscr{D}}$, $\tilde{y}$ must have

$\tilde{y}'(0) = \tilde{y}'(l) = 0$ (since $\tilde{y}'^2 < 2\tilde{y}$). Thus we can use Theorem 3.5 as extended in Problem 3.21(d) to conclude that the semicircle is the unique minimizing curve for F, and so we verify Dido's conjecture that it alone encloses the maximal area against the x-axis.

(8.3) **Remarks.** 1. By Theorem 7.7, the semicircle also gives the maximal area among polygonal or other piecewise C^1 curves $(\hat{x}, \hat{y})$ with $\hat{y}(s) > 0$ on $(0, l)$, and this conclusion remains valid when we permit curves $(\hat{x}, \hat{y})$ with $\hat{y}(s) = 0$ at some points in $(0, l)$. Indeed at such points $\tilde{y}'(s) = \tilde{y}(s) = 0$, where $\tilde{y} = \hat{y}^2/2$, and when $\tilde{y}_0 = \tilde{y}_0(s)$ is as before,

$$\tilde{f}(\tilde{y}, \tilde{y}') - \tilde{f}(\tilde{y}_0, \tilde{y}'_0) - \tilde{f}_y(\tilde{y}_0, \tilde{y}'_0)(\tilde{y} - \tilde{y}_0) - \tilde{f}_z(\tilde{y}_0, \tilde{y}'_0)(\tilde{y}' - \tilde{y}'_0)$$

$$= \sqrt{2\tilde{y}_0 - \tilde{y}'^2_0} - \frac{\tilde{y}_0}{\sqrt{2\tilde{y}_0 - \tilde{y}'^2_0}} + \frac{\tilde{y}'^2_0}{\sqrt{2\tilde{y}_0 - \tilde{y}'^2_0}} = \frac{\tilde{y}_0}{\sqrt{2\tilde{y}_0 - \tilde{y}'^2_0}} > 0.$$

Thus the fundamental inequality holds even at these points.

2. It is possible to use these results to establish the isoperimetric inequality (Problem 8.26). (A different derivation of this inequality will be given in §9.5, Example 3.) However, we can also use this approach to solve other significant problems, including that of Zenodoros in Problem 1.9. See Problems 8.27–8.30.

§8.9. Continuous Media

Thus far we have analyzed dynamical systems consisting of a given *finite* number of "point" masses, and applied Hamilton's principle to obtain the equations of motion. Obviously the finiteness becomes questionable when we attempt to describe the motions of an elastic medium which is undergoing dynamic deformation as occurs, for example, when a drum head is struck. Fortunately, in many cases of importance, Hamilton's principle still applies to an appropriate Lagrangian function, which is now represented by spatial integrals. We shall illustrate how this is accomplished with two standard elastic media, the taut string and the membrane. (The elastic bar and plate are considered in Problems 8.20 and 8.21.) The approach remains valid for more general continuua including elastic solids, fluids, and gases, as well as for electromagnetic phenomena. [See C–H, We, Bi, W, O–R.] Moreover, in the absence of motion, we obtain Bernoulli's principle which characterizes static equilibrium for these systems.

(a) Taut String

In order to analyze the transverse *planar* motions of an elastic string (such as a guitar string) which is stretched horizontally between fixed supports sepa-

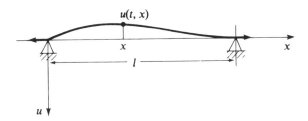

Figure 8.5

rated a distance l as shown in Figure 8.5, we suppose that the "vertical" position of the string at any time t is described by $u = u(t, x), 0 \leq x \leq l$. Then the resulting velocity history $u_t(t, x)$ gives rise to a kinetic energy at time t given by

$$T = \frac{1}{2} \int_0^l \rho u_t^2 \, dx, \tag{50}$$

where $\rho = \rho(t, x)$ is the (local) linear mass density of the string which is assumed continuous but may be nonuniform. (T may be considered as the limit of the ordinary kinetic energy of a large *finite* number of masses $m_i \approx \rho(t, x_i) \Delta x_i$ located at x_i with $0 = x_0 < x_1 < \cdots < x_n = l$, and moving with vertical velocity $u_t(t, x_i)$, when the $\Delta x_i = x_i - x_{i-1}$ tend toward zero.)

As usual, it is less apparent how to select the potential energy function U to complete the Lagrangian, and we must now make additional assumptions about the motion. First, we shall suppose that the string is so thin that the only work required to deform it from the straight line ($u = 0$), to a position described by $u(t, x)$ is that necessary to *enlongate* it against a local tensile force $\tau = \tau(t, x)$. In particular, we consider the work in bending the string to be negligible. (Alternatively, we could suppose that the curvature of the deflected string is so small that W_B as determined in §6.6, is small in comparison with W_C.) In any case, we shall assume that the potential energy stored in the string as a result of its deformation is the work required to stretch it and hence by the same analysis as in §6.6, take

$$U = -W_C = \int_0^l \tau(\sqrt{1 + u_x^2} - 1) \, dx, \tag{50'}$$

at time t. We are also assuming that there is no external loading on the string and thereby disregard the effects of gravity. (However, see Problem 8.17(c), (d), and [Se] for less restrictive assumptions.)

From (49) and (50) we obtain the Lagrangian

$$L = T - U = \int_0^l [\tfrac{1}{2}\rho u_t^2 - \tau(\sqrt{1 + u_x^2} - 1)] \, dx, \tag{51}$$

and Hamilton's principle now requires that between times 0 and b, say, at which the positions are known, the motion will make stationary the action integral

$$A(u) = \int_0^b L \, dt = \int_0^b dt \int_0^l [\tfrac{1}{2}\rho u_t^2 - \tau(\sqrt{1 + u_x^2} - 1)] \, dx \qquad (52)$$

on

$$\mathscr{D} = \{u \in C^2([0, b] \times [0, l]): u(t, 0) = u(t, l) = 0; u(0, x), u(b, x) \text{ prescribed}\}.$$

Upon introducing

$$f(x, u, \nabla u) = \tfrac{1}{2}\rho u_t^2 - \tau(\sqrt{1 + u_x^2} - 1), \qquad (53)$$

we see that this integrand function depends on $X = (t, x) \in \mathbb{R}^2$ through ρ and τ, and on $\nabla u = (u_t, u_x)$, but does not depend explicitly on u. (See §6.9.) Thus by Theorem 6.13, we should seek a C^2 solution u to the equation

$$\nabla \cdot f_{\nabla u} = 0,$$

or upon substitution of (53), the equation

$$(\rho u_t)_t - (\tau u_x(1 + u_x^2)^{-1/2})_x = 0, \qquad (54)$$

which meets the geometrical conditions

$$u(t, 0) = u(t, l) = 0, \qquad 0 \le t \le b. \qquad (55)$$

In addition, we suppose that the *initial position* is prescribed by

$$u(0, x) = u_0(x), \qquad 0 \le x \le l, \qquad (56)$$

so that from (55), $u_0(0) = u_0(l) = 0$. However, instead of attempting to specify the position at a later time b, we seek a solution valid for *all* later times, with the prescribed *initial velocity*

$$u_t(0, x) = v_0(x), \qquad 0 \le x \le l. \qquad (57)$$

We observe that (54) is nonlinear, and for example, will *not* admit a pure time-oscillatory solution of the form $u(t, x) = u(x) \cos \omega t$. Numerical methods must be used to obtain approximate solutions to this equation. However, as in §6.6, we can easily linearize it by supposing that the slopes $|u_x| \ll 1$.

Then (54) reduces to

$$(\rho u_t)_t = (\tau u_x)_x, \qquad (58)$$

which for constant ρ, τ becomes the *one-dimensional wave equation*

$$u_{tt} = \sigma^2 u_{xx}, \quad \text{with } \sigma^2 = \tau/\rho. \qquad (58')$$

Because of its importance to the description of simple acoustical phenomena, (58′) was the first partial differential equation to receive serious attention. It admits the (unique) solution with given initial data first obtained by

D'Alembert (1746) in the form

$$u(t, x) = \frac{u_0(x + \sigma t) + u_0(x - \sigma t)}{2} + \frac{1}{2\sigma} \int_{x-\sigma t}^{x+\sigma t} v_0(s) \, ds. \tag{59}$$

(Problem 8.14.)

Here we must suppose that the initial functions u_0 in C^2 and v_0 in C^1 are given on all of $\mathbb{R}$. (However, to satisfy the end point conditions (55), the initial data must be extended *periodically*.) In particular, when the initial velocity $v_0 \equiv 0$, this solution at time t may be interpreted as the average of the translate of the initial shape u_0 which is travelling to the right at velocity σ, with that of the same shape travelling to the left at velocity σ. Moreover, for constant ρ and τ, (58′) does admit time-oscillatory solutions in the form

$$u(t, x) = u_0(x) \cos \omega t, \tag{60}$$

provided that u_0 in C^2 is a solution of the linear *ordinary* differential equation

$$u_0'' + (\omega/\sigma)^2 u_0 = 0, \tag{61}$$

as is easily verified by direct substitution.

Now, for constant $\mu = \omega/\sigma$, the general solution to (61) is well known to be given by

$$u_0(x) = A \cos \mu x + B \sin \mu x.$$

In addition, u_0 must satisfy the end conditions $u_0(0) = 0$, $u_0(l) = 0$. The first of these conditions requires that $A = 0$, so that $u_0(x) = B \sin \mu x$. However, the *second* can now only be satisfied (for $u_0 \neq \mathcal{O}$), when $\sin \mu l = 0$; i.e., for $\mu = \mu_n = n\pi/l$, $n = 1, 2, \ldots$.

It follows that simple oscillatory solutions of the form (60), can occur only for specific "*natural*" *frequencies*

$$\omega = \omega_n = \sqrt{\tau/\rho} \left(\frac{n\pi}{l} \right) = n\omega_1, \qquad n = 1, 2, \ldots.$$

Observe that these natural frequencies increase with the tension, τ, and decrease as the density, ρ, or the length, l, are increased, facts confirmed by experience with, say, a guitar string. Moreover, from experimental evidence, Mersenne, in his *Harmonie Universelle* of 1636 predicted the exact square root dependence on τ/ρ, while other properties of ω_1 were known to the Pythagorean school.

The *mode shape* associated with oscillations at frequency ω_n is

$$u_n(x) = \sin \mu_n x = \sin \left(\frac{n\pi x}{l} \right),$$

demonstrating that the sinusoidal shape accompanying higher frequencies has more nodal or fixed points. The resulting "pure" motion is described by

$$u_n(t, x) = u_n(x) \cos \omega_n t = \sin(n\pi x/l) \cos \omega_n t. \tag{62}$$

These natural vibrations are to be understood in the sense that a motion which is begun with initial deflection of pure sinusoidal shape $u_n(x)$ and initial velocity 0, should exercise small amplitude oscillations in this shape indefinitely at frequency ω_n, as a result of precise interchange between kinetic and potential energies. (Of course, a real motion would die out in time as a consequence of the fact that the string is not perfectly elastic.)

Now, what happens, if the string is not released in one of these pure shapes, but is instead plucked; that is, released in a triangular (or some other) shape? In 1753, D. Bernoulli argued that since overtones could be detected audibly, it was reasonable to suppose that the resulting motion might be represented by superposition of the natural modes in the form of an infinite series. However, Bernoulli's arguments were criticized by Euler, D'Alembert, Lagrange, and Laplace so severely, that when Fourier reawakened interest in such series representation as possible solutions to problems in heat conduction (c. 1807), his work was also regarded with suspicion. The resulting questions of Fourier series representation exerted a profound influence on the subsequent development of mathematical analysis, and interest continues to this day. We state *without proof*, the following result (which is *not* the best possible):

(8.4) **Proposition.** *If* $u_0 \in C^4[0, l]$, *with* $u_0(0) = u_0(l) = u_0''(0) = u_0''(l) = 0$, *then the coefficients* $b_n \overset{\text{def}}{=} (2/l) \int_0^l u_0(x) \sin(n\pi x/l)\, dx$, *satisfy for some constant M, an estimate of the form* $|b_n| \le M/n^4$, $n = 1, 2, \dots$. *Hence, the series*

$$u(t, x) \overset{\text{def}}{=} \sum_{n=1}^{\infty} b_n u_n(t, x)$$

converges to the unique solution of (58') *which satisfies*

$$u(0, x) = u_0(x), \qquad u_t(0, x) = 0, \qquad x \in [a, b].$$

(See Problem 8.15.)   □

(It is a consequence of D'Alembert's result (59) that these initial conditions (as extended periodically) do determine the solution uniquely. An alternate proof of this fact is given in the next proposition.)

The Nonuniform String

We shall discuss briefly the case of a nonuniform string in which ρ and τ are permitted to vary with x but *not* with t. Then (58) becomes

$$\rho u_{tt} = (\tau u_x)_x,$$

which would also follow as the equation of motion for the modified Lagrangian (51) in which U is replaced by $\tilde{U} = \frac{1}{2} \int_0^l \tau u_x^2\, dx$.

(8.5) **Proposition.** *When ρ and τ are positive with $\rho_t = \tau_t = 0$, there is at most one C^2 solution u to the wave equation $\rho u_{tt} = (\tau u_x)_x$ on $[0, \infty) \times [0, l]$ with given initial data (56) and (57), which satisfies the boundary conditions (55): $u(t, 0) = u(t, l) \equiv 0$.*

PROOF. If u_1 and u_2 are two solutions to this problem, then $u \overset{\text{def}}{=} u_1 - u_2$ provides a solution to the *same* equation with *zero* initial data. Introducing the associated *total energy* at time t

$$E(t) \overset{\text{def}}{=} T + \tilde{U} = \tfrac{1}{2} \int_0^l (\rho u_t^2 + \tau u_x^2) \, dx,$$

it follows from A.13 that since $\rho_t = \tau_t = 0$:

$$E'(t) = \int_0^l (\rho u_t u_{tt} + \tau u_x u_{xt}) \, dx$$

$$= \int_0^l u_t(\rho u_{tt} - (\tau u_x)_x) \, dx + (\tau u_x)u_t \bigg|_0^l ,$$

after an obvious integration by parts using the fact that $u_{xt} = u_{tx}$.

The integral vanishes (Why?) and from (55) follows $u_t(t, 0) = u_t(t, l) \equiv 0$. Thus $E'(t) = 0$, so that $E(t) = E(0) = 0$, since $u_t(0, x)$ is known to vanish while $u(0, x) \equiv 0$ implies the vanishing of $u_x(0, x)$. But then by the standard argument we conclude that the nonnegative continuous integrand of $E(t)$ must vanish identically on $[0, l]$. From the positivity of ρ and τ, we infer that at time $t > 0$: $u_t(t, x) = u_x(t, x) \equiv 0$, so that $u(t, x) = \text{const.} = u(t, 0) = 0$. Thus $u_1 \equiv u_2$ as we wished. ☐

This argument admits modification. (See Problem 8.17.)

If we again consider oscillatory solutions of the form (60), we see that the mode shape u_0 is now required to satisfy the equation

$$(\tau y')' = -\omega^2 \rho y \tag{63}$$

and the homogeneous boundary conditions

$$y(0) = y(l) = 0.$$

This is a problem of the *Sturm–Liouville* type encountered in §7.3 for a positive *eigenvalue* $\lambda = \omega^2$. In Problem 7.19, it is shown that a solution u_1 which is *positive* on $(0, l)$ should *minimize*

$$F(y) = \int_0^l \tau y'^2 \, dx$$

on $\mathcal{Y}_0 = \{y \in C^1[0, l]: y(0) = y(l) = 0\}$ under the (achievable) constraint

$$G(y) = \int_0^l \rho y^2 \, dx = 1. \tag{64}$$

It follows that the corresponding eigenvalue λ_1 determines the *fundamental* or least frequency $\omega_1 = \sqrt{\lambda_1}$, and from Problem 7.19, that

$$\lambda_1 = R(u_1) \overset{\text{def}}{=} \frac{F(u_1)}{G(u_1)}$$

minimizes the *Rayleigh* ratio

$$R = F/G \quad \text{on } \mathscr{D} = \{y \in C^1[0, l]: y(0) = y(l) = 0; y \neq \mathcal{O}\}.$$

[Indeed, when $y \neq \mathcal{O}$, then $R(cy) = R(y)$, while $G(cy) = c^2 G(y) = 1$, for some choice of the constant $c \neq 0$. Thus on $\mathscr{D}$, R assumes the same values as does F under the additional constraint (64).]

This reformulation of the problem initiated by Lord Rayleigh (c. 1873) together with a corresponding characterization of the (necessarily) higher eigenvalues λ_n provides the basis for most results in this subject. [See §A.6, C–H, W–S.] We observe here that each y in $\mathscr{D}$ provides an *upper bound* $\sqrt{R(y)}$ to the unknown fundamental frequency ω_1. Moreover, direct comparison can be made between the frequencies ω_n and those, $\tilde{\omega}_n$, of an appropriately chosen uniform string (See A.23). Finally, we may replace the fixed end conditions by other homogeneous boundary conditions as in Problem 7.19(h)). Again superposition of the pure motions is possible and there emerges a theory of generalized Fourier series. Indeed, the Sturm–Liouville problem considered here and its generalizations have played a vital role in developing mathematics and its applications. See Kline, [Tr], [Wei], and [C–H].

(b) Stretched Membrane

A two-dimensional analogue of the stretched string is provided by a thin elastic membrane or skin which is stretched tightly over a (horizontal) frame in the shape of the boundary of a domain D of $\mathbb{R}^2$ as illustrated in Figure 8.6. We may think of the membrane as the head of a drum for which both the

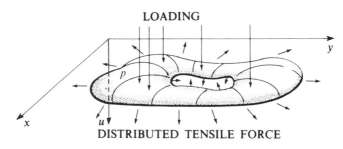

Figure 8.6

local density and tensile stress may vary with time, as well as position, as would occur when a kettle drum is tightened just after it is struck.

Let $X = (x, y)$. Then we shall assign the membrane a local areal density $\rho = \rho(t, X)$ so that if $u = u(t, X)$ denotes the *vertical* displacement at time t, then the associated kinetic energy is given by

$$T = \frac{1}{2} \int_D \rho u_t^2 \, dX, \tag{65}$$

where $dX = dx \, dy$ denotes the element of area.

We require that at the boundary ∂D

$$u|_{\partial D} = 0, \qquad \forall \, t \geq 0.$$

Again we shall neglect the strain energy of bending in comparison with that of areal stretching against the distributed local tension $\tau = \tau(t, X)$. However, we shall also suppose that there is a *downward* distributed loading giving rise to the pressure $p = p(t, X)$ on the membrane, possibly, its own weight.

Then the resulting potential energy at time t is given by

$$U = \int_D \tau(\sqrt{1 + u_x^2 + u_y^2} - 1) \, dX - \int_D pu \, dX. \tag{66}$$

Here, the first term represents the work done in stretching the membrane as in §3.4(e), while the second term is the work done in moving the distributed load under deflection. In both cases we take as reference state that described by $u \equiv 0$.

The resulting Lagrangian is, of course,

$$L = T - U,$$

and Hamilton's principle requires that between times $t = 0$ and $t = b$ at which the positions of the membrane are prescribed, it should execute a motion which makes stationary the action integral

$$A(u) = \int_0^b L \, dt = \int_0^b dt \int_D (\tfrac{1}{2}\rho u_t^2 - \tau(\sqrt{1 + u_x^2 + u_y^2} - 1) + pu) \, dX \tag{67}$$

on

$$\mathscr{D} = \{u \in C^2([0, b] \times D): u|_{\partial D} = 0, \, u(0, X), \, u(b, X) \text{ prescribed}\},$$

Thus, denoting the integrand by

$$f = f((t, X), u, \nabla u),$$

we have from Theorem 6.13, that u must satisfy the differential equation $\nabla \cdot f_{\nabla u} = f_u$; upon substituting, and subsequently utilizing the linearizing assumption that $u_x^2 + u_y^2 \ll 1$, it follows that u should satisfy (approximately) the partial differential equation:

$$(\rho u_t)_t - (\tau u_x)_x - (\tau u_y)_y = p, \qquad t \geq 0, \, x \in D. \tag{68}$$

(Problem 8.18).

In addition, we require that $u|_{\partial D} \equiv 0$, while $u(0, X) = u_0(X)$ is prescribed, as is $u_t(0, X) = v_0(X)$, in lieu of $u(b, X)$. (68) is linear, but it is nonhomogeneous because of the presence of the external forcing term $p = p(t, X)$. It can have at most one solution for the given data when ρ and τ are positive and time independent (Problem 8.18).

When $p \equiv 0$, ρ and τ are *constant*, and $v_0 \equiv 0$, we may attempt an oscillatory solution of the form $u(t, X) = u_0(X) \cos \omega t$, and see that in D, u_0 must satisfy the equation

$$\Delta u_0 + \lambda u_0 = 0, \tag{69}$$

with

$$\lambda = (\rho/\tau)\omega^2, \quad \text{and} \quad \Delta u \overset{\text{def}}{=} u_{xx} + u_{yy}.$$

Of course, we require that $u_0|_{\partial D} \equiv 0$. (Problem 8.18(e)).

This is known as the *membrane eigenvalue problem* for the domain D. Again, only certain values of the *eigenvalues* $\lambda = \lambda_n \geq 0$, $n = 1, 2, \ldots$, will permit a solution, and these will determine the natural frequencies, $\omega_n = (\tau \lambda_n / \rho)^{1/2}$, of free oscillations. The associated mode shapes $u_n(X)$ determine pure motions $u_n(t, X) = u_n(X) \cos \omega_n t$, corresponding to those of a drum head which has been struck or excited in an appropriate manner. Again we see that these natural frequencies increase with the tension τ and decrease with the density ρ of the material as experience dictates. Moreover, the frequencies together with the mode shapes depend on the shape of the bounding curve(s). The interesting long-open problem of whether we can "hear" the shape of a drum; i.e., of whether we can recover the shape of the bounding curve from knowledge of these natural frequencies, has been resolved quite recently. We can't![1]

The pure oscillations $u_n(t, x)$ can again be combined to obtain (formal) series solutions with undetermined coefficients which may in principle be chosen to make the resulting solution describe a given initial displacement, $u_0 = u_0(X)$, provided that u_0 satisfies certain regularity conditions. Theoretically, in this manner we can solve the *homogeneous* two-dimensional *wave* equation

$$\rho u_{tt} = \tau(u_{xx} + u_{yy}), \tag{70}$$

for constant ρ, τ, with given initial displacement u_0, and initial velocity v_0, to obtain the resulting *unforced* motions ($p \equiv 0$) of the membrane.

There are associated series methods which may be employed to find *forced* motions, making further use of the linearity of (68), but the actual computations, even for the homogeneous case, are quite difficult. They have been

---

[1] See "One Cannot Hear the Shape of a Drum" by C. Gordon, D. Webb, and S. Wolpert, *Bulletin of the A.M.S.*, Volume 27, No. 4, July, 1992 pp. 134–137.

carried out only for a few simple domains, D, such as the disk, the rectangle, and the annulus. See [Tr], [Wei], and [C–H].

Static Equilibrium of (Nonplanar) Membrane

In the absence of motion, $u_t \equiv 0$ and hence from (65), $T = 0$. Moreover, when $p = p(X)$, U as defined by (66) depends only on the position $u = u(X)$. Thus the action integral (67) becomes

$$A(u) = -\int_0^b U \, dt = -bU(u), \quad \text{say,}$$

which should now be made stationary on

$$\mathcal{D} = \{u \in C^2(D): u|_{\partial D} = \gamma\},$$

where $\gamma \in C(\partial D)$ is prescribed.

Inspection of the integrand function of U in (66) in the form

$$f(\underline{X}, u, \nabla u) = \tau(\underline{X})(\sqrt{1 + u_x^2 + u_y^2} - 1) - p(\underline{X})u \tag{71}$$

shows that it is strongly convex. (This is true for the first term by the argument of §3.4(e), while the second term is linear in u.) Hence, the only u which can make A or U stationary on $\mathcal{D}$ is that which *minimizes* uniquely the potential energy U. (Problem 8.19.) Under a prescribed static loading p, the membrane assumes a *unique* shape u_0 of *stable* equilibrium.

For a homogeneous membrane, $\tau = $ constant, and u_0 will satisfy (approximately) the *linearized* equation (68) with $u_{tt} \equiv 0$; namely,

$$\Delta u = -p/\tau,$$

with

$$u|_{\partial D} \equiv \gamma.$$

This is *Poisson's* equation with a *Dirichlet* boundary condition. When the external loading $p \equiv 0$, it is *Laplace's equation* which is clearly satisfied by $u \equiv 0$. Hence, *when* $\gamma \equiv 0$, this is the unique solution reflecting the fact that we have ignored the actual weight of the membrane.

The above analysis applies also to the *nonplanar* membrane in which $u|_{\partial D} = \gamma \neq 0$. Then, in the absence of loading, we are simply solving the *Dirichlet problem* for D. With a smooth boundary function γ, it always has a unique solution (as does the corresponding Poisson problem [C–H]).

However, should we seek a solution to the actual *nonlinear* equation governing the equilibrium position of the nonplanar membrane, then in the absence of loading, we have in effect the minimal area problem which as we have discussed in §3.4(e), could have at most one solution, but need *not* have any solution.

(Problems 8.14–8.21)

PROBLEMS

8.1. (a) Show that if the approximations

$$Ml(\cos \theta)\dot{x}\dot{\theta} \approx Ml\dot{x}\dot{\theta},$$

$$Mgl(1 - \cos \theta) \approx Mgl\frac{\theta^2}{2},$$

(for small θ) are made in the Lagrangian (11), and we suppose that $\kappa(x) = \kappa x$, then the corresponding equations of motion are given by (16).

(b) Explain why it is *not* inconsistent to use both of the approximations $\cos \theta \approx 1$ and $\cos \theta \approx 1 - \theta^2/2$ in the same equation in part (a).

8.2. Let $Q = \begin{pmatrix} x \\ \theta \end{pmatrix}$.

(a) Show that equations (16) can be written in the form

$$\ddot{Q} = \mathbb{A}Q, \tag{72}$$

where the coefficient matrix $\mathbb{A}$ is given by

$$\mathbb{A} = \frac{M}{m}\begin{bmatrix} -\kappa l & g \\ \kappa & -\mu g \end{bmatrix}.$$

(b) Define $Q_1(t) = V \cos \omega t$ and $Q_2(t) = V \sin \omega t$, where $V \neq 0$ is a constant vector and ω is a (real) constant. Show that Q_1 and Q_2 are solutions of (72) iff $\mathbb{A}V = -\omega^2 V$; i.e., V is an *eigenvector* of $\mathbb{A}$ with corresponding *eigenvalue* $-\omega^2$.

(c) Show that for any positive choices of m, M, l, κ, the eigenvalues λ of $\mathbb{A}$ are always negative. Are the eigenvalues necessarily distinct? (An eigenvalue λ is a root of the equation $\det[\mathbb{A} - \lambda I] = 0$.)

(d) Conclude that if $\mathbb{A}$ has real eigenvalues $-\omega_1^2 < -\omega_2^2 < 0$ with corresponding eigenvectors V_1 and V_2, then the solution of the initial value problem $\ddot{Q} = \mathbb{A}Q$ with $Q(0) = Q_0$ and $\dot{Q}(0) = \tilde{Q}_0$ can always be expressed in the form

$$Q(t) = (c_1 \cos \omega_1 t + \tilde{c}_1 \sin \omega_1 t)V_1 + (c_2 \cos \omega_2 t + \tilde{c}_2 \sin \omega_2 t)V_2$$

where c_1, $\tilde{c}_1$, c_2, $\tilde{c}_2$ are constants. (Hint: Eigenvectors associated with distinct eigenvalues are always linearly independent.)

8.3. A double pendulum consists of two light inextensible rods of length l and two bobs of mass m_1 and m, respectively, which are constrained to move in a vertical plane as shown in Figure 8.7. Assume that the pivots are frictionless. Use the generalized coordinates θ_1 and θ.

(a) Express T, U, and L in terms of θ_1, θ, $\dot{\theta}_1$, $\dot{\theta}$.

(b) Determine the differential equations of motion.

(c)* Find solutions for the linearized equations by the method of the previous problem.

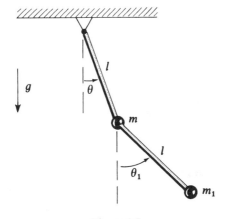

Figure 8.7

8.4. (a) Verify that the canonical equations (25) are indeed the first Euler–Lagrange equations for the generalized action integral (26).

(b) Show that the variation of A at (Q, P) in the direction (V, W) is given by

$$\delta A((Q, P); (V, W)) = \int_a^b [(\dot{Q} - H_P)\cdot W - (\dot{P} + H_Q)\cdot V](t)\, dt + (P\cdot V)(t)\Big|_a^b.$$

(c) Conclude that functions (Q, P) satisfying (25) make A stationary on a set where only $Q(a)$ and $Q(b)$ are prescribed. Is the converse true?

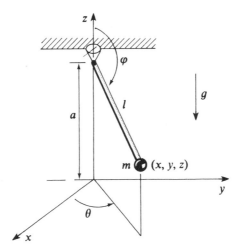

Figure 8.8

8.5. (Spherical Pendulum.) A bob of mass m is attached to a light inextensible rod of length l which is free to swing in any direction about a fixed pivot as shown in Figure 8.8.

(a) Use the relations

$$x = l \sin \varphi \cos \theta,$$
$$y = l \sin \varphi \sin \theta,$$
$$z = a + l \cos \varphi,$$

to express the Lagrangian in terms of the generalized coordinates $Q \equiv (\varphi, \theta)$.

(b) Find the associated differential equations of motion and deduce that $ml^2(\sin^2 \varphi)\dot{\theta} = \text{const.}$ (which corresponds to conservation of angular momentum about the z-axis).

(c) Determine the conjugate momenta $P \equiv (p_\varphi, p_\theta)$.

(d) Express $\dot{Q}$ in terms of t, Q, P, and find the Hamiltonian $H(t, \varphi, \theta, p_\varphi, p_\theta)$.

(e) What are the canonical equations of motion? (Compare with part (b).)

(f) What is the Hamilton–Jacobi equation (39) for this system?

8.6. Consider a dynamical system with one degree of freedom. If the Hamiltonian does not depend explicitly on time, the Hamilton–Jacobi equation takes the form

$$\frac{\partial S}{\partial t} + H\left(q, \frac{\partial S}{\partial q}\right) = 0. \tag{73}$$

The following argument can sometimes be used to find a complete solution of (73).

(a) Show that $S = \varphi(t) + s(q)$ is a solution of (73) iff

$$\varphi(t) = c_1 t + c_0$$

and

$$H(q, s'(q)) = -c_1, \tag{74}$$

where c_0 and c_1 are constants. Observe that (74) is a first-order ordinary differential equation for s.

(b) Conclude that if $s = s(q; c_1)$ satisfies (74) for each choice of the constant c_1, then

$$S = c_1 t + s(q; c_1) + c_0$$

is a complete solution of (48).

(c) Find a complete solution of

$$\frac{\partial S}{\partial t} + e^{-q}\left(\frac{\partial S}{\partial q}\right)^3 = 0.$$

8.7. Assuming that the sun is fixed at the origin, the *planar* motion of a single planet of mass m about the sun may be specified by giving the polar coordinates (r, θ) of its position at each time t. The potential energy function which recovers the *inverse square law* is $U = -mk/r$ for an appropriate constant k.

(a) Show that the associated kinetic energy function for $Q \equiv (r, \theta)$ is

$$T = (m/2)(\dot{r}^2 + r^2\dot{\theta}^2).$$

(b) Obtain the Hamiltonian H in terms of the conjugate momentum $P \equiv (p_r, p_\theta)$, and write the reduced Hamilton–Jacobi equation $H(Q, s_Q) = E = \text{const.}$

(c)* Attempt a solution of this equation by separation in the form

$$s = s(r, \theta) = R(r) + \theta(\theta), \text{ say.}$$

(d)* Compare the equations for the separated variables with those of Lagrange for this problem.

8.8.   The *non*planar motion of the planet of mass m about the sun (assumed fixed at an origin), may be described in terms of spherical coordinates (r, φ, θ) (as in Figure 1.1), with the *same* energy function $U = -mk/r$ as in the previous problem.

(a)   Prove that the kinetic energy function is $T = (m/2)(\dot{r}^2 + r^2\dot{\varphi}^2 + r^2(\sin^2 \varphi)\dot{\theta}^2)$.

(b)   Derive the Hamiltonian

$$H = \frac{1}{2m}\left(p_r^2 + \frac{p_\varphi^2}{r^2} + \frac{p_\theta^2}{r^2 \sin^2 \varphi}\right) - \frac{mk}{r}$$

in terms of $P \equiv (p_r, p_\varphi, p_\theta)$. Hint: Use (27).

(c)* Show that the reduced Hamilton–Jacobi equation $H(Q, s_Q) = E = \text{const.}$ may be separated *successively* as follows:

$$\frac{\partial s}{\partial \theta} = \alpha; \quad \left(\frac{\partial s}{\partial \varphi}\right)^2 + \frac{\alpha^2}{\sin^2 \varphi} = \beta^2; \quad \left(\frac{\partial s}{\partial r}\right)^2 + \frac{\beta^2}{r^2} - \frac{2m^2k}{r} = 2mE$$

for constants α, β.

8.9.   Verify that the functions which are stationary for the modified action integrand of equation (30) satisfy the remaining equations of motion (25), for $j \neq i$.

8.10. (a) Apply the result of Theorem 6.10 to obtain equations for the stationary functions minimizing the action integral of equation (33) subject to the general constraint $K(Q, P) = 0$, in the form

$$\frac{d}{d\tau}Q(\tau) = \lambda(\tau)K_P(\tau); \quad \frac{d}{d\tau}P(\tau) = -\lambda(\tau)K_Q(\tau)$$

for an *appropriate function* $\lambda = \lambda(\tau)$.

(b) Prove that in terms of the *new* parameter $\tau \stackrel{\text{def}}{=} \int \lambda(\tau)\, d\tau$, the equations in (a) will take the form (36).

(c) Show that for the special choice of constraining function

$$K(Q, P) \equiv p_0 + H(Q, p_1, p_2, \ldots, p_n),$$

equations (36) are the canonical equations of Hamilton corresponding to taking $\tau = t$.

(d) Explain how this same analysis would apply to Jacobi's action integral of equation (31) subject to the constraint $H(Q, P) = \text{const.}$

8.11. For a single particle of mass m moving freely under the action of a force of potential $U = U(X)$, we may use as generalized coordinates Q, the cartesian coordinates $X \in \mathbb{R}^3$ of position.

(a) Show that the associated momenta $P = m\dot{X}$.

(b) Conclude that the Hamiltonian is

$$H = H(X, P) = |P|^2/2m + U(X).$$

(c) Make the Schrödinger substitution $s = k \log \psi$ in the corresponding *reduced* equation (44) and prove that ψ satisfies the equation

$$f(X, \psi, \nabla\psi) \equiv (k^2/2m)(\nabla\psi^2) + U(X)\psi^2 = E\psi^2,$$

for a constant E.

(d) Consider for this f,

$$F(u) = \int_{\mathbb{R}^3} f(X, u, \nabla u)\, dX,$$

and ignoring questions of convergence, show *formally* that a function ψ which makes F extremal on $\mathcal{D} = \{u \in C^2(\mathbb{R}^3): F(u) < \infty\}$ subject to the constraint

$$G(u) \equiv \int_{\mathbb{R}^3} u^2\, dX = 1,$$

should satisfy the three-dimensional equation similar to (69),

$$(k^2/2m)\nabla^2\psi + (E - U)\psi = 0,$$

where now E is to be regarded as a Lagrangian multiplier.

(This indicates Schrödinger's method (1926) of associating a wave character to a mass particle. The permissible eigenvalues of E which permit a solution ψ of the equation meeting certain homogeneous boundary conditions define the energy levels of the particle. [We].)

8.12. In the theory of *special relativity* in which we ignore the effects of gravitation and postulate the constancy of c, the speed of light in a vacuum, we may modify the Lagrangian so that the *form* of Hamilton's principle remains valid.

For the case of a single particle of constant (rest) mass m_0, moving freely in an electromagnetic field of *vector* potential $\mathfrak{U}(t, X)$ and scalar potential $U = q\varphi = q\varphi(t, X)$, at a speed $v = |\dot{X}|$, we may use as the Lagrangian

$$\tilde{L} = -m_0 c^2(1 - v^2/c^2)^{1/2} - U + q\dot{X} \cdot \mathfrak{U}$$

(where q is a constant) and take the generalized coordinates $Q = X \in \mathbb{R}^3$.

(a) Show that when $\mathfrak{U} = \mathcal{O}$, the *particle* has momentum

$$\tilde{P} = \tilde{L}_{\dot{X}} = m_0 \beta \dot{X}, \quad \text{where } \beta = (1 - v^2/c^2)^{-1/2},$$

and hence even when $U = 0$, it has an *apparent* mass $m = m_0\beta$ as a result of its motion. When is $m = m_0$?; $m \approx m_0$?; $m \gg m_0$?.

(b) Prove that the associated *total* energy

$$\tilde{E} \stackrel{\text{def}}{=} \tilde{L}_{\dot{X}} \cdot \dot{X} - \tilde{L} = mc^2 + U = m_0 c^2 + U + \frac{m_0 v^2}{1 + \sqrt{1 - v^2/c^2}}$$

so that when $v^2 \ll c^2$,

$$\tilde{E} \approx m_0 c^2 + \frac{m_0 v^2}{2} + U.$$

Conclude that even when at rest with $U = 0$, the particle should contain the enormous store of energy $E_0 = m_0 c^2$, which is potentially available for release.

(c)* Show that the corresponding equations of motion may be written

$$\frac{d}{dt}(m_0 \beta \dot{X}) = -q[\mathfrak{U}_t + \nabla\varphi - (V \times \mathfrak{B})],$$

where $V = \dot{X}$, $\mathfrak{B} = \text{curl } \mathfrak{U}$, and the last term is a *vector* product in $\mathbb{R}^3$.

8.13. Complementary Inequalities (§8.8). To establish the inequality for $\mathcal{D}_{max}$ given in Theorem 8.1:
   (a) Consider $A(Y_0, P_0) - A(Y, P)$ by simply interchanging the appropriate subscripts in (48′).
   (b) Conclude that since $Y_0' = H_P(x, Y_0, P_0)$, the first integral in (48′) *after interchange* is nonnegative.
   (c) Integrate the last term by parts as in the proof of 8.1 to prove that when $Y \in \mathcal{D}_{max}$, then also the second integral (with its minus sign) is nonnegative.
   (d) Formulate a corresponding theorem when the convexity properties for H and $-H$ are reversed.
   (e) Apply Theorem 8.1 to the function

$$H(x, Y, P) = |P|^2 - |Y|^2 + cY.$$

8.14. D'Alembert's Solution.
   (a) Verify that when ρ and τ are constant, then $u(t, x)$ as defined by (59) is a solution of the wave equation (58′), satisfying (56) and (57).
   (b) Take $v_0 \equiv 0$, and decide how the displacement $u_0(x)$ defined initially on $[0, l]$ must be extended to $\mathbb{R}$ if (59) is to give $u(t, 0) = u(t, l) = 0$, for $t \geq 0$. Give an example of a function $u_0 \neq \mathcal{O}$ so extended.

8.15. Fourier Series Solution. (Take $l = 1$, for simplicity.)
   (a) Verify that the coefficients b_n as defined in Proposition 8.4, do satisfy an estimate of the form $|b_n| \leq M/n^4$, $n = 1, 2, \ldots$. Hint: Integrate by parts *four* times.
   (b) Use Proposition 8.4 and the hint in (a) to find a formal series solution to (59) when $u_0(x) = x^3(1 - x)^3$.
   (c) Show that the series $u(t, x) = \sum_{n=1}^{\infty} b_n u_n(t, x)$ converges *uniformly* on $[0, 1]$, as does each of the differentiated series $\sum_{n=1}^{\infty} b_n \ddot{u}_n(t, x)$, $\sum_{n=1}^{\infty} b_n u_n''(t, x)$ where the dots and primes denote differentiations with respect to t and x, respectively, and $u_n(t, x)$ is defined by (62) with $l = 1$.
   (d) The convergence established in (c) proves that u, as defined, is C^2. Verify that formally at least it must satisfy (58′), with constant ρ, τ.
   (e) Observe that $u_t(0, x) \equiv 0$ and $u(0, x) = \sum_{n=1}^{\infty} b_n \sin n\pi x$. Conclude that $b_n = 2\int_0^1 u(0, x) \sin(n\pi x)\, dx$, $n = 1, 2, \ldots$. Hint: Integrate term-by-term using integral tables if necessary to eliminate all but one of the terms.
   (f) We would like to prove that $g(x) \equiv u(0, x) - u_0(x) = 0$. (Why?) Try to invent an argument which could establish this using the fact that from (e)

$$\int_0^l g(x) \sin n\pi x\, dx = 0, \qquad n = 1, 2, \ldots.$$

8.16. The vertical column of water rotating at *constant* angular velocity ω (from §2.3) may be analyzed via Hamilton's principle.

(a) Using the coordinates of Figure 2.1, show that at time t, the kinetic energy (of rotation) is $T = \frac{1}{2}\rho \int_0^1 2\pi x (\int_0^y (\omega x)^2 \, dz) \, dx = \pi \rho \omega^2 \int_0^1 x^3 y(t, x) \, dx$, where $y = y(t, x)$ describes the profile of the upper free surface.

(b) Demonstrate that the potential energy (due to gravity) associated with the deflected upper surface is $U = \rho g \int_0^1 2\pi x (\int_0^y z \, dz) \, dx = \pi \rho g \int_0^1 x y^2(t, x) \, dx$.

(c) Assuming that the positions of the rotating system are specified at times 0 and b, apply Hamilton's principle to the action integral

$$A = A(y) = \int_0^b L \, dt = \int_0^b (T - U) \, dt,$$

subject to the *volume constraint*

$$G(y) = \int_0^b dt \int_0^1 2\pi x y(t, x) \, dx = \text{const.}$$

(d) Conclude that the only profile of stationarity; viz.,

$$y = y(x) = (-\lambda + \omega^2 x^2)/2g$$

for an appropriate constant λ, *maximizes* the action integral. Hint: Use convexity.

(e)* Contrast the conclusion of (d) with the principle of minimum potential energy used to obtain the same profile in §2.3.

(f) Repeat the entire analysis when ω is permitted to vary with t.

8.17. Uniqueness of Solutions to the Wave Equation.

(a) Show that the conclusion of Proposition 8.5 is unaffected if (55) is replaced by the conditions $u(t, 0) = c_1$, $u(t, l) = c_2$, for constants c_1, c_2.

(b)* Find other end point conditions which preserve the uniqueness argument and describe the physical situation represented by your choices.

(c) The presence of an additional (downward) distributed loading of intensity $p(t, x)$ adds a term $-\int_0^l pu \, dx$ to U. Conclude that the resulting equation of motion is $(\rho u_t)_t - (\tau u_x)_x = p$, when $u_x^2 \ll 1$.

(d) Argue that the conclusion of Proposition 8.5 will be unaffected by the addition of a prescribed loading term p as in (c). Hint: What equation will the *difference* $u = u_1 - u_2$ satisfy?

8.18*. The Membrane Problem.

(a) When $u_x^2 + u_y^2 \ll 1$, why is $\tilde{U} = \frac{1}{2}\int [\tau(u_x^2 + u_y^2) - pu] \, dX$ a good approximation for U of (66)?

(b) Verify that with $L = T - \tilde{U}$, the equation of motion is (68).

(c) Using the total energy function at time t given by

$$E(t) = T + \tilde{U},$$

establish a uniqueness result for solutions to (68) modelled after Proposition 8.5. Hint: Use Green's theorem to show that for the *difference* u of solutions,

$$E'(t) = \int_{\partial D} \tau u_t \partial_N u \, ds,$$

where N is the outward pointing unit normal to the boundary curve from D.

(d) Find other boundary conditions which would also guarantee uniqueness of solution, and describe the membranes so supported.

(e) Verify that when ρ and τ are constant while $p = 0$, and u_0 is a solution of (69), then $u(t, X) \equiv u_0(X) \cos \omega t$ will be a solution of (68) satisfying the condition $u_t(0, X) = v_0(X) = 0$.

8.19. (a) Verify the strong convexity of $f(\underline{X}, u, \nabla u)$ as defined in (71). (Problem 3.26).

(b) Obtain the associated (nonlinear) equation characterizing a function which is stationary for this f (Theorem 6.13).

(c) Conclude that each solution $u_0 \in \mathscr{D}$ of the equation found in (b), must in fact *minimize* U on $\mathscr{D}$ uniquely.

(d)* Can convexity be used to give a uniqueness argument for solutions to the *time dependent* wave equation in any of the forms considered in §8.9?

8.20*. Transverse Motion of an Elastic Bar. For a horizontal thick elastic bar of constant rectangular cross-section and length l in which the energy of stretching may be neglected in comparison with that of bending, let $u(t, x)$ denote the "vertical" position of the center line at time t.

(a) Argue that for a suitable density ρ there should be an associated kinetic energy $T = \frac{1}{2}\int_0^l \rho u_t^2 \, dx$.

(b) If the bar is subjected to a distributed downward loading of intensity $p(t, x)$, argue that the resulting potential energy is approximately

$$\tilde{U} = \int_0^l (\tfrac{1}{2}\mu u_{xx}^2 - pu) \, dx,$$

for a suitable material stiffness function $\mu = \mu(t, x)$. Hint: See the discussion of W_B in §6.6.

(c)* For the action integral $A(u) = \int_0^b (T - \tilde{U}) \, dt$, use the definition $\delta A(u; v) = \lim_{\varepsilon \to 0} (d/d\varepsilon) A(u + \varepsilon v)$ to prove that

$$\delta A(u; v) = \int_0^b dt \int_0^l (\rho u_t v_t - \mu u_{xx} v_{xx} + puv) \, dx.$$

(d) Suppose that $u \in C^4$, and integrate the expression in (c) by parts as required to show that

$$\delta A(u; v) = \int_0^b dt \int_0^l [-(\rho u_t)_t - (\mu u_{xx})_{xx} + pu]v \, dx$$

$$+ \int_0^l \rho u_t v \Big|_0^b dx - \int_0^b (\mu u_{xx})v_x \Big|_0^l dt + \int_0^b (\mu u_{xx})_x v \Big|_0^l dt.$$

(e) Conclude that stationarity of A for prescribed u at times 0 and b, and, say, the *cantilever* support conditions $u(t, 0) = u_x(t, 0) = 0$; $u(t, l) = u_x(t, l) = 0$, requires that

$$(\rho u_t)_t + (\mu u_{xx})_{xx} = -p. \tag{75}$$

(f)* When $\rho_t = \mu_t = 0$, use the total energy function $E(t) = T + \tilde{U}$ to give a uniqueness result modelled after Proposition 8.5 as extended in Problem 8.17.

(g)   Study the expression in (d) to obtain alternate sets of boundary conditions which would lead to the same differential equation as in (e).

(h)   When $\rho = \mu = $ const. and $p = 0$, determine an ordinary differential equation for $u_0 = u_0(x)$, if $u(t, x) \equiv u_0(x) \cos \omega t$ is to be a solution of (75).

(i)   Can you find or guess a nontrivial solution u_0 of the equation in (h) which will permit u to meet the cantilever conditions in (e) at least for certain values of ω? Hint: See §6.6.

8.21*.   Transverse Motion of a Uniform Plate. If the membrane discussed in §8.9(b) is replaced by a plate of uniform thickness and material, we may neglect the energy of stretching in comparison with that of bending which is now given (approximately) by

$$U = \frac{\mu}{2} \int_D [(u_{xx}^2 + u_{yy}^2) - 2(1 - \tau)(u_{xx}u_{yy} - u_{xy})^2] \, dX, \qquad (76)$$

where, of course, $u(t, X)$ denotes the vertical position of a center section at time t, and μ and $\tau < 1$ are positive material *constants*.

(a)   Argue that for an appropriate constant density ρ, the kinetic energy of motion at time t should be approximately $T = \frac{1}{2} \int_D \rho u_t^2 \, dX$.

(b)   Set $A(u) = \int_0^b (T - U) \, dt$, and, neglecting external loading, reason that for *some* boundary conditions, stationarity of A at $u \in C^4$ requires that u should satisfy the equation

$$\rho u_{tt} + \mu \Delta^2 u = 0, \qquad (77)$$

where

$$\Delta^2 u = \Delta(\Delta u) = \Delta(u_{xx} + u_{yy}).$$

(c)   Which equation is $u_0 = u_0(X)$ required to satisfy in order that $u(t, X) = u_0(X) \cos \omega t$ be a solution of (77)?

(d)   For static equilibrium of the loaded plate with pressure $p = p(X)$, when all functions are time independent, use convexity of the integrand of

$$\tilde{U} \overset{\text{def}}{=} U - \int_D pu \, dX$$

to conclude that even for a nonplanar plate, only *stable* equilibrium is possible, and it is uniquely characterized by a u_0 which satisfies the equation:

$$\mu \Delta^2 u = p.$$

Hint: The term $u_{xx}u_{yy} - u_{xy}^2 = \mathrm{div}(u_x u_{yy}, -u_x u_{xy})$.

8.22.   (Geometric Hamiltonian.)
If $L(\underline{t}, \underline{Q}, Z)$ is strictly convex, argue that for each *fixed* t, Q, P: $l(Z) \overset{\text{def}}{=} P \cdot Z - L(t, Q, \overline{Z})$ will be maximized by that Z_0 for which $P = L_{\dot{Q}}(t, Q, Z_0)$. Hence $H(t, Q, P) \overset{\text{def}}{=} \max_Z [P \cdot Z - L(t, Q, Z)]$ defines this Hamiltonian without "solving" equations (19) for $\dot{Q}$.

8.23.   Find the Hamiltonian H associated with the "Lagrangian" f and decide whether your H is a saddle function; when

(a) $f(x, y, z) = z^2/4y$.

(b) $f(x, y, z) = \sqrt{y^{-2} + z^2} \ (y > 0)$.

(c) $f(x, y, z) = y^2 + z^2$.

(d) $f(x, y, z) = y\sqrt{1 + z^2}$.

(e) $f(x, y, z) = -\sqrt{y}\sqrt{1 - z^2}\,(y > 0, z^2 < 1)$.

(f) $f(x, y, z) = \sqrt{1 + y^2 + z^2}$.

8.24. (a) In (47) if $H(\underline{x}, Y, P)$ is a [strict] saddle function show that $f(\underline{x}, Y, Z)$ is [strictly] convex. Hint: If the left side of (47) is zero, then both bracketed expressions on the right must vanish.

(b) Establish the converse result to that in part (a).

(c)* Could these assertions hold with [strictly] replaced by [strongly]?

(d) Verify (47).

8.25. Show that $H(\underline{x}, y, p)$ is a strict saddle function and find a related convex "Lagrangian" f when:

(a) $H(x, y, p) = \sqrt{2y}\sqrt{1 + p^2}$, $\qquad y > 0$.

(b) $H(x, y, p) = \sqrt{y + p^2}$, $\qquad y + p^2 > 0$.

(c) $H(x, y, p) = -\sqrt{1 + y^2}\sqrt{1 - p^2}$, $\quad p^2 < 1$.

8.26. (The isoperimetric inequality, §1.3.)

Let $\mathscr{C}$ be a closed simple (nonintersecting) curve of length $2l$ in $\mathbb{R}^2$, parametrized with respect to arc-length s by functions $x, y \in C^1[0, 2l]$. Assume that $(x, y)(0) = 0 \in \mathscr{C}$, and that $P = (x, y)(l)$ lies on the positive x-axis as shown in Figure 8.9. By Green's theorem, we can suppose that $\mathscr{C}$ encloses a region of area $A(x, y) = \left| \displaystyle\int_0^{2l} y(s)x'(s)\, ds \right|$.

(a) Show that as in the derivation of (49)

$$A_1(x, y) \overset{\text{def}}{=} \left| \int_0^l yx'\, ds \right| \le \int_0^l |y||x'|\, ds = \int_0^l \sqrt{y^2 - (yy')^2}\, ds$$

$$= -\tilde{F}(\tilde{y}) \le l^2/2\pi,$$

with equality *iff* on $[0, l]$, (x, y) parametrizes a semicircle. Hint: $\tilde{y} = y^2/2 \in C^1(0, l)$. See Remarks 8.3.

(b) Define A_2 similarly and conclude that $A \le A_1 + A_2 \le l^2/\pi$ with equality *iff* (x, y) parametrizes a circle.

(c) Can you extend the last result to piecewise C^1 curves $\tilde{\mathscr{C}}$?

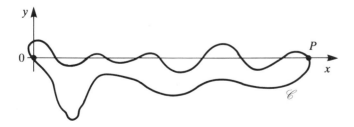

Figure 8.9

8.27. A Zenodoros' problem (see Problem 1.9(b))

(a) Verify that $f(y, z) = -y\sqrt{1 - y^2 z^2}$ is not convex.

(b) Make the substitution $\tilde{y} = y^2/2$, so $\tilde{y}' = yy'$ and show that this Zenodoros problem reduces to *minimizing* the integral

$$\tilde{F}(\tilde{y}) = -\int_0^T \sqrt{\tilde{y}(1 - \tilde{y}'^2)}\, dt$$

on

$$\mathcal{D}^* = \{\tilde{y} \in C[0, T]: \tilde{y}(0) = \tilde{y}(T) = 0, |\tilde{F}(\tilde{y})| < +\infty\}$$

(c) Show that $\tilde{f}(y, z) = -\sqrt{y(1 - z^2)}$ is strictly convex on $\{y > 0\} \times \{z^2 < 1\}$ by considering its Hamiltonian.

(d) Establish that $\tilde{y}_0(t) = t - (t^2/T)$ is the unique minimizing function. Hint: Remarks 8.3.

(e) Return to the original variables in Problem 1.9 and conclude that the curve which can produce the maximum value is given by a semicircle. Hint: $x_0'^2 + y_0'^2 = 1/y_0^2(t)$ (why?), so that with part (d), $x_0'(t) = \sqrt{2/T}$ is constant. Eliminate t.

8.28. (a) Use the substitution $\tilde{y}^2 = 2y$ to transform the (nonconvex) problem of minimizing

$$\tilde{F}(\tilde{y}) = \int_0^b (\tilde{y}'^2 - \tilde{y}^2)(x)\, dx \quad \text{on} \quad \tilde{\mathcal{D}} = \left\{ \begin{matrix} \tilde{y} \in C^1[0, b]: \tilde{y}(0) = 0 \\ \tilde{y}(b) = \sin b \end{matrix} \right\}$$

to the problem of minimizing

$$F(y) = \int_0^b \left(\frac{y'^2}{2y} - 2y \right)(x)\, dx$$

on

$$\mathcal{D} = \left\{ y \in C^1(0, b) \cap C[0, b]: y(x) \geq 0, F(y) < +\infty, \right.$$

$$\left. y(0) = y'(0) = 0; y(b) = \frac{\sin^2 b}{2} \right\}.$$

(b) Show that $f(y, z) = z^2/2y - 2y$ is (only) convex on $\{y > 0\}$. Hint: Look at its Hamiltonian. (for which $H(\underline{x}, y, p)$ is strictly convex).

(c) When $b < \pi$ conclude that $y(x) = (\sin^2 x)/2$ minimizes F on $\mathcal{D}$ uniquely.

(d) When $b = \pi$, conclude that $y(x) = (\sin^2 x)/2$ minimizes F on $\mathcal{D}$ (but not uniquely), and obtain the Wirtinger inequality

$$\int_0^\pi \tilde{y}^2\, dx \leq \int_0^\pi (\tilde{y}')^2\, dx \quad \text{when } \tilde{y} \in C^1[0, \pi] \text{ with } \tilde{y}(0) = \tilde{y}(\pi) = 0.$$

Hint: We can assume that $y(x) > 0$, when $0 < x < \pi$. (Why?)

(e)* Show that $\tilde{y}(x) = (\sin^2 x)/2$ does not minimize F on $\mathcal{D}$ when $b > \pi$, even though it appears to satisfy the appropriate conditions.

8.29. (a) For the seismic wave problem 1.8(b), make a transformation (such as that in Problem 8.28) for which $y'(x)/y(x) = \tilde{y}'(x)$, and show that the new integrand function $\tilde{f}(y, z) = \sqrt{e^{-2y} + z^2}$ is strictly convex on $\mathbb{R}^2$.

(b) Conclude that a circular arc provides the (only) path of travel for such a seismic wave.

8.30. (a) If $g = g(y)$ is [strictly] convex and positive on an interval I, then show that $f(y, z) = \sqrt{g^2(y) + z^2}$ is [strictly] convex on $I \times \mathbb{R}$.

(b) Under what conditions on $g = g(x, y)$ will $f(\underline{x}, y, z) = -\sqrt{g^2(\underline{x}, y) - z^2}$ be [strictly] convex on a domain of definition? Give some nontrivial examples of such g.

CHAPTER 9*

Sufficient Conditions for a Minimum

As we have noted repeatedly, the equations of Euler–Lagrange are necessary but not sufficient to characterize a minimum value for the integral function

$$F(Y) = \int_a^b f(x, Y(x), Y'(x))\, dx = \int_a^b f[Y(x)]\, dx$$

on a set such as

$$\mathscr{D} = \{Y \in C^1([a, b])^d \colon Y(a) = A,\, Y(b) = B\},$$

since they are only conditions for the stationarity of F. However, in the presence of [strong] convexity of $f(\underline{x}, Y, Z)$ these conditions do characterize [unique] minimization. [Cf. §3.2, Problem 3.33 et seq.] Not all such functions are convex, but we have also seen in §7.6 that a minimizing function Y_0 must necessarily satisfy the Weierstrass condition $\mathscr{E}(x, Y_0(x), Y_0'(x), W) \geq 0$, $\forall\, W \in \mathbb{R}^d$, $x \in [a, b]$, where

$$\mathscr{E}(x, Y, Z, W) \stackrel{\text{def}}{=} f(x, Y, W) - f(x, Y, Z) - f_Z(x, Y, Z) \cdot (W - Z), \tag{1}$$

and this is recognized as a convexity statement for $f(\underline{x}, \underline{Y}, Z)$ along a trajectory in $\mathbb{R}^{2d+1}$ defined by Y_0.

This chapter is devoted to showing that conversely, when $f(\underline{x}, \underline{Y}, Z)$ is [strictly] convex (§9.2) in the presence of an appropriate *field*, then each stationary Y_0 in $\mathscr{D}$ does minimize F on $\mathscr{D}$ [uniquely] (§9.3, §9.4) and this will afford a solution for the brachistochrone problem. The method extends in principle to problems with variable end point conditions (§9.4) and to those on which constraints are imposed (§9.5).

However, the field in question requires an entire family of stationary functions with special properties which may, or may not, exist. A central field will suffice (§9.6) and this provides further insight into the problem of finding the minimal surface of revolution. In §9.7 we provide conditions which assure that a given stationary trajectory Γ_0 may be considered as that of a central field. We encounter a new criterion, that of Jacobi, but in §9.9 we demonstrate that it, too, is almost essential for local minimization. This embedding of Γ_0 together with the appropriate convexity of f supplies sufficient conditions for the local minimization of integral functions F, both in the weak and in the strong senses (§9.8).

To motivate the inquiry, let's look at the original method used to attack such problems.

§9.1. The Weierstrass Method

In his lectures of 1879, Weierstrass presented the following approach to prove that a given stationary function $Y_0 \in (C^1[a, b])^d$ minimizes

$$F(Y) = \int_a^b f[Y(t)] \, dt = \int_a^b f(t, Y(t), Y'(t)) \, dt$$

on

$$\mathscr{D} = \{Y \in (C^1[a, b])^d : Y(a) = Y_0(a); Y(b) = Y_0(b)\},$$

(where for simplicity we suppose that $f \in C^1([a, b] \times \mathbb{R}^{2d})$).

Let Y in $\mathscr{D}$ be a competing function and assume that each $x \in (a, b]$ determines a *unique* function $\Psi(\cdot\,; x)$, stationary for f on (a, x) as in §6.7, whose graph joins $(a, Y_0(a))$ to $(x, Y(x))$ as shown in Figure 9.1.

Then, in particular, $\Psi(t; b) = Y_0(t)$, (Why?), and we consider the integral function

$$\sigma(x) \overset{\text{def}}{=} -\int_a^x f[\Psi(t; x)] \, dt - \int_x^b f[Y(t)] \, dt \qquad (a \le x \le b), \qquad (2)$$

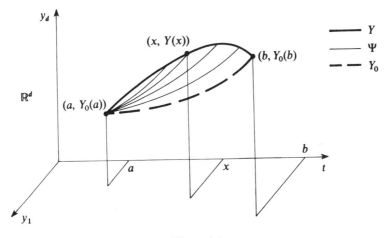

Figure 9.1

which interpolates between

$$\sigma(a) = -\int_a^b f[Y(t)] \, dt = -F(Y)$$

and

$$\sigma(b) = -\int_a^b f[\Psi(t; b)] \, dt = -\int_a^b f[Y_0(t)] \, dt = -F(Y_0),$$

so that $F(Y) - F(Y_0) = \sigma(b) - \sigma(a)$. Were $\sigma'(x) \geq 0$, it would follow by the mean value theorem that $F(Y) - F(Y_0) \geq 0$. Moreover, if also σ' is continuous on $[a, b]$, then equality holds *iff* $\sigma' \equiv 0$. (§A.1, §A.2.)

For example, when $d = 1$, consider the problem of minimizing the (non-convex) function

$$F(y) = \int_0^b [y'(t)^2 - y(t)^2] \, dt$$

on

$$\mathcal{D} = \{y \in C^1[0, b]: y(0) = y(b) = 0\} \quad \text{for } b < \pi.$$

Here, the stationary functions ψ satisfy the Euler–Lagrange equation

$$\frac{d}{dt} 2\psi'(t) = -2\psi(t) \quad \text{or} \quad \psi''(t) + \psi(t) = 0,$$

with the well-known general solution

$$\psi(t) = c_0 \cos t + c_1 \sin t.$$

Since $b < \pi$, the only solution in $\mathcal{D}$ is $y_0 = \mathcal{O}$, and for a given y in $\mathcal{D}$, it is seen by inspection that for each $x \in (0, b)$:

$$\psi(t; x) \overset{\text{def}}{=} y(x) \frac{\sin t}{\sin x}$$

is the unique function which is stationary for f and satisfies $\psi(0; x) = 0$ with $\psi(x; x) = y(x)$. (See Figure 9.2.)

Thus, for this example, equation (2) becomes

$$\sigma(x) = -\int_0^x [\psi'(t; x)^2 - \psi(t; x)^2] \, dt - \int_x^b [y'(t)^2 - y(t)^2] \, dt$$

$$= -\frac{y^2(x)}{\sin^2 x} \int_0^x (\cos^2 t - \sin^2 t) \, dt - \int_x^b [y'(t)^2 - y(t)^2] \, dt,$$

so that after integration,

$$\sigma(x) = -y^2(x) \cotan x - \int_x^b [y'(t)^2 - y(t)^2] \, dt. \tag{3}$$

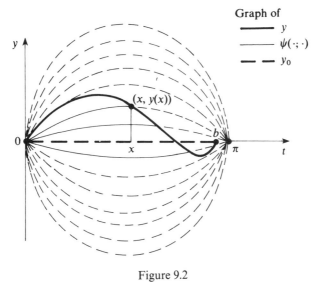

Figure 9.2

Then by the fundamental theorem of calculus, for $x \in (0, b)$:

$$\sigma'(x) = -2y(x)y'(x) \cotan x + y^2(x) \cosec^2 x - y^2(x) + y'(x)^2$$
$$= (y(x) \cotan x - y'(x))^2 \geq 0,$$

with equality *iff* $y'(x) - y(x) \cotan x = 0$.

Now, an apparent difficulty occurs when $x = 0$ as a result of the multiplicity of stationary functions ψ which pass through $(0, 0)$; namely, that $\sigma(0)$ is *not* defined by (3). However, by L'Hôpital's rule,

$$\lim_{x \searrow 0} \frac{y^2(x)}{\sin x} = \lim_{x \searrow 0} \frac{2y(x)y'(x)}{\cos x} = 0,$$

for y in $\mathscr{D}$, so that $\sigma(0+) = -F(y)$. Thus $F(y) - F(y_0) = \sigma(b) - \sigma(0+) \geq 0$; or $F(y) \geq F(y_0) = 0$, with equality *iff* $y'(x) = y(x) \cotan x$, $\forall\, x \in (0, b)$. But for y in $\mathscr{D}$, this implies that $y'(b) = y(b) = 0$ (since $b < \pi$), and hence that $y(x) \equiv 0 = y_0(x)$.

Thus we have proven that for $b < \pi$, $y_0 = \mathcal{O}$ is the unique minimizing function for F on $\mathscr{D}$. When $b = \pi$, the method is still applicable, (and provides a proof for Dido's conjecture of Problem 1.5), but the minimizing function y_0 is no longer unique. However, when $b > \pi$, y_0 fails to minimize. (See Problem 9.1.)

Having illustrated the effectiveness of the Weierstrass approach in a simple case, we return to the general problem and equation (2). As defined, $\Psi(t; x)$ depends on two variables, and as above, we use the prime to denote t differentiation. Supposing that Ψ is C^2, it follows that

$$(\Psi')_x = \Psi_{tx} = \Psi_{xt} = (\Psi_x)'. \tag{4}$$

Then, from Leibniz's rule (A.14) applied to (2), we obtain

$$\sigma'(x) = f[Y(x)] - f[\Psi(x; x)] - \int_a^x \frac{\partial}{\partial x} f[\Psi(t; x)] \, dt, \tag{5}$$

and the integrand is from the chain rule, (4), and stationarity, given by

$$\frac{\partial}{\partial x} f(t, \Psi(t; x), \Psi'(t; x)) = f_Y[\Psi(t; x)]\Psi_x(t; x) + f_Z[\Psi(t; x)](\Psi')_x(t; x)$$

$$= \frac{\partial}{\partial t}\{f_Z[\Psi(t; x)] \cdot \Psi_x(t; x)\}.$$

But $\Psi(x; x) \equiv Y(x)$ by construction, so that $\Psi_x(x; x) = Y'(x) - \Psi'(x; x)$ (Why?); while $\Psi_x(a; x) = 0$, since $\Psi(a; x) = Y_0(a)$, is constant. Hence, after integration and substitution, (5) becomes

$$\sigma'(x) = f(x, Y(x), Y'(x)) - f(x, Y(x), \Psi'(x; x))$$

$$- f_Z(x, Y(x), \Psi'(x; x)) \cdot (Y'(x) - \Psi'(x; x)),$$

or upon utilizing (1),

$$\sigma'(x) = \mathcal{E}(x, Y(x), \Psi'(x; x), Y'(x)). \tag{6}$$

(For $d = 1$, the reader should verify each step of this derivation by purely formal calculations.) Thus finally, we obtain *Weierstrass' formula*:

$$F(Y) - F(Y_0) = \int_a^b \sigma'(x) \, dx = \int_a^b \mathcal{E}(x, Y(x), \Psi'(x; x), Y'(x)) \, dx, \tag{7}$$

which proves that $\mathcal{E} \geq 0$ will imply that $F(Y) \geq F(Y_0)$, provided that an appropriate family of stationary functions $\Psi(\cdot; \cdot)$ having *all* of the assumed properties is available. Unfortunately, it is quite difficult to prescribe conditions which ensure the existence of such families (one for each competing $Y \in \mathcal{D}$), and instead in §9.3 et seq. we shall concentrate on a less direct approach of Hilbert, which yields Weierstrass' result even for piecewise C^1 functions $\hat{Y}$.

(Problems 9.1–9.2, 9.5–9.6)

§9.2. [Strict] Convexity of $f(x, Y, Z)$

The definition (1) of the Weierstrass excess function for a given $f = f(x, Y, Z)$; viz.,

$$\mathcal{E}(x, Y, Z, W) = f(x, Y, W) - f(x, Y, Z) - f_Z(x, Y, Z) \cdot (W - Z),$$

and our wish to consider $\mathcal{E} \geq 0$, suggests in comparison with 3.4, the following:

(9.1) **Definition.** $f(\underline{x}, \underline{Y}, Z)$ is said to be [strictly] *convex* on a set $S \subseteq \mathbb{R}^{2d+1}$, when f and f_Z are defined and continuous in S and satisfy the inequality $\mathscr{E}(x, Y, Z, W) \geq 0$, or equivalently,

$$f(x, Y, W) - f(x, Y, Z) \geq f_Z(x, Y, Z) \cdot (W - Z), \tag{8}$$

when $(x, Y, Z) \in S$ and $(x, Y, W) \in S$, [with equality at (x, Y) iff $W = Z$].

Usually, the set S will be of the form $S = D \times \mathbb{R}^d$ for a *domain* D of $\mathbb{R}^{d+1}$.

Since the [strict] convexity of $f(\underline{x}, \underline{Y}, Z)$ is identical with the [strong] convexity of $f(\underline{x}, \underline{Y}, Z)$ as defined in Chapter 3 (Problem 3.33), the usual linear combinations of such functions remain [strictly] convex. See Proposition 3.2 and Facts 3.11.

Example 1. $f(\underline{x}, Y, Z) = -|Y|^2 + |Z|^2$ is not convex, but $f(\underline{x}, \underline{Y}, Z)$ is strictly convex in $\mathbb{R}^{2d+1}$ since $f_Z(x, Y, Z) = 2Z$ so that

$$f(x, Y, W) - f(x, Y, Z) = |W|^2 - |Z|^2 = (W + Z) \cdot (W - Z)$$
$$= |W - Z|^2 + 2Z \cdot (W - Z)$$
$$\geq f_Z(x, Y, Z) \cdot (W - Z),$$

with equality at (x, Y) iff $|W - Z|^2 = 0$ or $W = Z$.

Example 2. For $d = 1$, the brachistochrone function of §1.2(a)

$$f(\underline{x}, y, z) = \sqrt{\frac{1 + z^2}{y}}$$

is not convex, but $f(\underline{x}, \underline{y}, z)$ is strictly convex for the half space

$$\{(x, y, z) \in \mathbb{R}^3 : y > 0\}$$

since $f_{zz}(x, y, z) > 0$. (See Proposition 3.10 and the next example.)

Example 3. For $d = 2$, when $Y = (x, y)$, the function $f(\underline{t}, \underline{Y}, Z) = \sqrt{y}|Z|^2$ is [strictly] convex on the half-space

$$\{(t, x, y, Z) \in \mathbb{R}^5 : y \geq 0\}, \; [\{(t, x, y, Z) \in \mathbb{R}^5 : y > 0\}].$$

For, by the computation of Example 1,

$$f(t, Y, W) - f(t, Y, Z) = \sqrt{y}(|W - Z|^2 + 2Z \cdot (W - Z))$$
$$\geq f_Z(t, Y, Z) \cdot (W - Z),$$

[with equality for $y > 0$ iff $W = Z$].

Example 4. For $d = 1$, the function $f(\underline{x}, \underline{y}, z) = -\sqrt{1 - z^2}/y$ is strictly convex on the set $S = \{(x, y, z) \in \mathbb{R}^3 : y > 0, |z| < 1\}$. (See Example 5 of §3.3, and Example 2 above.)

From §0.13, there is the following generalization of Proposition 3.10:

(9.2) Proposition. *If $f = f(x, Y, Z)$ together with its partials f_{z_i} and $f_{z_i z_j}$, $i, j = 1, 2, \ldots, d$, is continuous in a Z-convex set $S \subseteq \mathbb{R}^{2d+1}$ (one which contains the segment joining each pair of its points (x, Y, Z_0), (x, Y, Z_1)) and the matrix f_{ZZ} is positive semidefinite [positive definite] in S, then $f(\underline{x}, \underline{Y}, Z)$ is [strictly] convex in S.*

PROOF. For (x, Y, Z_0) and (x, Y, Z_1) in S and $t \in [0, 1]$, the point

$$Z_t \overset{\text{def}}{=} (1 - t)Z_0 + tZ_1$$

lies on a segment contained in S by hypothesis. Integrating by parts, we get

$$f(x, Y, Z_1) - f(x, Y, Z_0) = \int_0^1 \frac{d}{dt} f(x, Y, Z_t)\, dt$$

$$= (Z_1 - Z_0) \cdot \int_0^1 f_Z(x, Y, Z_t) d(t - 1)$$

$$= (Z_1 - Z_0) \cdot (t - 1) f_Z(x, Y, Z_t) \Big|_{t=0}^{t=1}$$

$$+ \int_0^1 (1 - t) \left(\sum_{i,j=1}^{d} f_{z_i z_j}(x, Y, Z_t) v_i v_j \right) dt,$$

where $V = Z_1 - Z_0$. The last term is nonnegative when $Z_1 \neq Z_0$ as a consequence of the assumed semidefiniteness of f_{ZZ} [and with positive definiteness of f_{ZZ}, it vanishes iff $V = \mathcal{O}$]. Hence

$$f(x, Y, Z_1) - f(x, Y, Z_0) \geq f_Z(x, Y, Z_0) \cdot (Z_1 - Z_0)$$

[with equality iff $Z_1 = Z_0$]. $\qquad\square$

A (spherical) neighborhood of a point (x_0, Y_0, Z_0) is Z-convex. If D is a domain of $\mathbb{R}^{d+1}$, then $S = D \times \mathbb{R}^d$ is Z-convex.

§9.3. Fields

The Weierstrass construction in §9.1, when possible, results in a *family* of *stationary trajectories* (the graphs of the functions $\Psi(\cdot; x)$) which is consistent in that one and only one member of the family passes through a given point $(x, Y(x))$. Suppose more generally, that for a given f we have a single family of stationary functions whose trajectories cover a domain D of $\mathbb{R}^{d+1}$ *consistently* in that through each point $(x, Y) \in D$ passes one and only one trajectory of the family, say that represented by $\Psi(\cdot; (x, Y)) \in (C^1[a, b])^d$. Then the *direction* of the *tangent line* to the *trajectory at* (x, Y) given by

$$\Phi(x, Y) \overset{\text{def}}{=} \Psi'(x; (x, Y))$$

determines a vector valued function in D (and one, moreover, whose values are required for Weierstrass' formula (7) along each competing trajectory).

For a domain D of $\mathbb{R}^{d+1}$, we shall call any C^1 function $\Phi\colon D \to \mathbb{R}^d$ a *field in* D. When $\Phi = \Phi(x, Y)$ the differential equation

$$Y'(x) = \Phi(x, Y(x)) = \Phi[Y(x)]$$

is called the *field equation*.[1] It is a consequence of the theory of differential equations, that through each point $(x_1, Y_1) \in D$, passes exactly one *field trajectory*, the graph of the *unique* solution Y_0 of the field equation with $Y_0(x_1) = Y_1$. (See A.19.)

For a given f, the field will be called *stationary* when each solution of the field equation is a stationary function, so that each field trajectory is a stationary trajectory. Conversely, when the field Φ is determined from the tangents to a *consistent* family of stationary trajectories Ψ as above, then it is stationary. (Why?)

While reformulating the Weierstrass theory in 1900, Hilbert observed that the principal requirement of a usable field was that it be *exact* in the sense of 9.3 which follows. (His motivation for this definition is indicated in Problem 9.13, and the corresponding multidimensional generalization—also given by Hilbert—is taken up in Problems 9.14 and 9.15.)

(9.3) **Definition.** A C^1 function Φ from a domain D of $\mathbb{R}^{d+1}$ to $\mathbb{R}^d$ is said to be an *exact field* for $f = f(x, Y, Z)$ in D, if there it makes the differential $h\,dx + P \cdot dY$ exact when

$$P(x, Y) \equiv f_Z(x, Y, \Phi(x, Y)),$$

and   (9)

$$h(x, Y) \equiv f(x, Y, \Phi(x, Y)) - P(x, Y) \cdot \Phi(x, Y);$$

i.e., there exists a real valued function $S \in C^1(D)$ with partials $S_x = h$ and $S_Y = P$, so that $dS = h\,dx + P \cdot dY$. (S is determined within an additive constant.)

The significance of an exact field will be demonstrated in the next section. First, however, observe that when P and h are C^1, and $h\,dx + P \cdot dY = dS$ is exact, then equality of the mixed partials of S implies that for $P = (p_1, p_2, \ldots, p_d)$

$$P_x = h_Y \quad \text{and} \quad (p_i)_{y_j} = (p_j)_{y_i}, \qquad i, j = 1, 2, \ldots, d. \tag{10}$$

(When $d = 1$, the second condition is always fulfilled.)

Conversely, it is known that when the exactness conditions (10) are satisfied in a *simply connected* domain D of $\mathbb{R}^{d+1}$; (i.e., a domain without

---

[1] The literature in this subject does not provide a uniform definition for fields. That given here seems most convenient for our purposes.

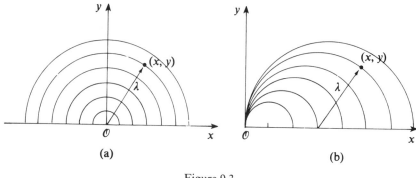

Figure 9.3

"holes") then there exists a function S as required to make the differential $h\,dx + P\cdot dY$ exact. [Ed]. (See also Problem 9.13.)

Example 1. The time of travel function for a seismic wave (Problem 1.8); viz., $f(y, z) = \sqrt{1 + z^2}/y$, is C^2 for $y > 0$, and has as its stationary trajectories the semicircles with centers on the x axis (Problem 9.3). Geometrically, as shown in Figure 9.3(a), it is evident that those with *fixed center* (say the origin) indexed by the radius λ, form a consistent stationary family, in the upper half plane $D = \{(x, y) \in \mathbb{R}^2, y > 0\}$. Analytically, these are defined by

$$\psi(t, \lambda) = \sqrt{\lambda^2 - t^2}, \qquad |t| < \lambda;$$

and that passing through the point (x, y) is obtained for $\lambda = \sqrt{x^2 + y^2}$. Thus the associated (stationary) field in D is

$$\varphi(x, y) = \psi'(x, \sqrt{x^2 + y^2}) = -\frac{x}{y}.$$

From (9) we form

$$p(x, y) = f_z(x, y, \varphi(x, y)) = \frac{-x/y}{y\sqrt{1 + x^2/y^2}} = \frac{-x}{y\sqrt{x^2 + y^2}},$$

and

$$h(x, y) = f(x, y, \varphi(x, y)) - f_z(x, y, \varphi(x, y))\varphi(x, y)$$

$$= \left[\frac{\sqrt{1 + z^2}}{y} - \frac{z^2}{y\sqrt{1 + z^2}}\right]_{z=\varphi(x, y)} = \frac{1}{y\sqrt{1 + x^2/y^2}} = \frac{1}{\sqrt{x^2 + y^2}}.$$

Here, it may be verified directly that $p = S_y$ and $h = S_x$ when $S(x, y) = \log(\sqrt{x^2 + y^2} + x) - \log y$; (or use (10)). Thus $\varphi(x, y) = -x/y$ defines an exact field for f in D. (Another consistent stationary family is given by those semicircles with fixed (left) end point, as in Figure 9.3(b). The resulting field is also exact, but it is defined only in a quarter plane (Problem 9.3).)

In this example, stationary fields provided the exact fields. We examine the intimate relations between these fields in the next two results.

(9.4) **Proposition.** *When f_z is C^1, then each exact field Φ for f is a stationary field; i.e., each C^1 solution Y_0 of the field equation $Y'(x) = \Phi(x, Y(x))$ is a stationary function for f.*

PROOF*. We have by (9) and the hypothesis that when $Y_0'(x) = \Phi[Y_0(x)]$, then

$$f_z[Y_0(x)] = f_z(x, Y_0(x), \Phi[Y_0(x)]) = P(x, Y_0(x)) \tag{11}$$

is C^1. Also both P and h are C^1 and again from (9)

$$h_Y = f_Y + \Phi_Y f_z - \Phi_Y P - P_Y \Phi$$

$$= f_Y - P_Y \Phi,$$

when the partials of f are evaluated at $(x, Y, \Phi(x, Y))$. (In the *matrices* P_Y and Φ_Y, the *rows* are indexed by Y; moreover, by the second set of equations (10), the Jacobian matrix P_Y is symmetric and so is equal to its *transpose* $\bar{P}_Y$.) Thus

$$f_Y[Y_0(x)] = h_Y(x, Y_0(x)) + P_Y(x, Y_0(x))\Phi(x, Y_0(x)), \tag{12}$$

so that finally by substitution of (11) and (12), and the chain rule:

$$\frac{d}{dx} f_z[Y_0(x)] - f_Y[Y_0(x)] = \frac{d}{dx} P(x, Y_0(x)) - f_Y[Y_0(x)]$$

$$= P_x(x, Y_0(x)) + \bar{P}_Y(x, Y_0(x)) Y_0'(x)$$

$$- h_Y(x, Y_0(x)) - P_Y(x, Y_0(x))\Phi(x, Y_0(x)).$$

The terms involving P_Y cancel (Why?), and we obtain the equation

$$\frac{d}{dx} f_z[Y_0(x)] - f_Y[Y_0(x)] = (P_x - h_Y)(x, Y_0(x)) \tag{13}$$

$$= 0, \quad \text{by (10)}.$$

Thus Y_0 satisfies the Euler–Lagrange equation of §6.7 and so is stationary.

□

Hereafter we shall suppose f_z is C^1 in all cases of interest.

The converse of Proposition 9.4 is not true in general. (See [C].) However, if φ is a stationary field in a domain D of $\mathbb{R}^2$, then, by definition, each field trajectory is a stationary trajectory; since the "matrix" p_y is trivially symmetric we may again conclude that equation (13) holds and the first exactness condition $p_x - h_y = 0$ of (10), is met at each point $(x, y_0) \in D$. When $d = 1$ and D is *simply* connected, it is the *only* requirement and we have established the following:

292   9*. Sufficient Conditions for a Minimum

(9.5) **Corollary.** If f_z is C^1 and φ is a stationary field for $f = f(x, y, z)$ in a simply connected domain D of $\mathbb{R}^2$, then φ is an exact field for f. □

(9.6) **Remark.** In higher dimensions, more than simple stationarity is required to produce an exact field, and this leads to the study of Mayer fields [S]. However, in §9.6 we shall prove that an appropriately constructed *central* field determined by a consistent family of stationary trajectories emanating from a common point is always exact.

Example 2. From the discussion for the brachistochrone in §6.2(c) we know that the function

$$f(x, y, z) = \sqrt{\frac{1 + z^2}{y}}$$

provides precisely one stationary curve joining the origin $(0, 0)$ to a given "lower" point (x, y); namely, the cycloid which is represented parametrically by the equations

$$t = \lambda(\tau - \sin \tau)$$
$$\psi = \lambda(1 - \cos \tau), \qquad (0 \le \tau \le \theta < 2\pi), \qquad (14)$$

where $\lambda > 0$ and θ are determined uniquely by the boundary conditions

$$x = \lambda(\theta - \sin \theta),$$
$$y = \lambda(1 - \cos \theta). \qquad (15)$$

(The previous notation has been replaced by one more amenable to our present requirements.) The resulting family of cycloids, denoted $\psi(t, \lambda)$, is shown in Figure 9.4. The associated field is defined by the direction of $\psi(t, \lambda)$ at $t = x$; i.e.,

$$\varphi(x, y) = \psi'(t, \lambda)\Big|_{t=x} = \frac{\psi_\tau}{t_\tau}\Big|_{\tau=\theta} = \frac{\lambda \sin \tau}{\lambda(1 - \cos \tau)}\Big|_{\tau=\theta},$$

or $\varphi(x, y) = \cotan \theta/2$, where, of course, θ is determined implicitly by equations (15) as a C^1 function of (x, y). As defined, φ is a stationary field for f in

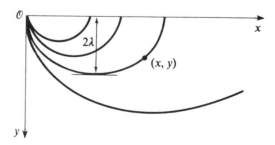

Figure 9.4

the quarter plane $D = \{(x, y) \in \mathbb{R}^2 : x > 0, y > 0\}$ which *is* simply connected. Thus φ is exact, by Corollary 9.5. (Another exact field is discussed in Problem 9.9.)

Exact Fields and the Hamilton–Jacobi Equation*

The notation of P and h used to represent the exact differential $dS = h\, dx + P \cdot dY$ in Definition 9.3 suggests that used for the conjugate momenta and the Hamiltonian in Chapter 8. This is not accidental, for if we regard $f(x, Y, Z)$ as the Lagrangian function for a dynamical system at time "x", and, suppose as in §8.8, that the equations $P = f_Z(x, Y, Z)$ determine $Z = G(x, Y, P)$ for an appropriate C^1 function G, then the associated Hamiltonian is defined by

$$- H(x, Y, P) = f(x, Y, G(x, Y, P)) - P \cdot G(x, Y, P).$$

In particular, if for some field Φ, we set $\tilde{P}(x, Y) = f_Z(x, Y, \Phi(x, Y))$ (as in (9)), then $Z = \Phi(x, Y) \equiv G(x, Y, \tilde{P}(x, Y))$, so that

$$- H(x, Y, \tilde{P}(x, Y)) = f(x, Y, \Phi(x, Y)) - \tilde{P}(x, Y) \cdot \Phi(x, Y)$$
$$= h(x, Y), \quad \text{as defined by (9).} \tag{16}$$

Moreover, when Φ makes the differential $h\, dx + \tilde{P} \cdot dY = dS$, say, exact in a domain D of $\mathbb{R}^{d+1}$, then, of course,

$$\tilde{P} = S_Y \quad (\text{i.e., } \tilde{p}_j = S_{y_j}, \, j = 1, 2, \ldots, d),$$

and $h = S_x$, so that upon substituting in (16) we get that

$$S_x + H(x, Y, S_Y) \equiv 0 \quad \text{in } D. \tag{17}$$

This is recognized as the partial differential equation of Hamilton–Jacobi (cf. §8.7), and we see that *each exact field Φ for f determines a solution S of this equation.*

Conversely, if S is any C^1 solution of (17) in a domain D of $\mathbb{R}^{d+1}$, then we may *define* $\tilde{P}(x, Y) = S_Y$, and

$$h(x, Y) = - H(x, Y, \tilde{P}(x, Y)) = S_x,$$

and conclude that $\Phi(x, Y) \overset{\text{def}}{=} G(x, Y, S_Y(x, Y))$ determines an exact field for f.

In making these assertions we are assuming that the equations $P = f_Z(x, Y, Z)$ determine $Z = G(x, Y, P)$. From implicit function theory we know this to be true in a neighborhood of each point $(x, Y, Z) \in \mathbb{R}^{2d+1}$ in which f is C^2 and the Jacobian matrix f_{ZZ} is invertible. Thus *locally* at least, we may expect a direct relation between the solvability of the Hamilton–Jacobi equation (17) and the existence of an exact field. (See also Problem 8.22.)

§9.4. Hilbert's Invariant Integral

If $\Phi(x, Y)$ is an exact field for a given f in a domain D of $\mathbb{R}^{d+1}$ as defined in 9.3, then for *piecewise* C^1 curves $\hat{\Gamma}$ in D (those which admit parametrization by piecewise C^1 functions), the value of the *line integral*

$$I(\hat{\Gamma}) \overset{\text{def}}{=} \int_{\hat{\Gamma}} (h \, dx + P \cdot dY) = \int_{\hat{\Gamma}} dS, \tag{18}$$

say, depends only on the end points of $\hat{\Gamma}$; i.e., its value is *invariant* with the particular curve in D which joins given end points. (This familiar consequence of exactness is independent of particular choices for h, P [Ed].) I is called *Hilbert's invariant integral* for Φ in D.

However, with our choices (9), we see that when $\hat{\Gamma}$ is represented as the graph of a function $\hat{Y} \in (\hat{C}^1[a, b])^d$, then formally, on $\hat{\Gamma}$, $dY = \hat{Y}'(x) \, dx$ and (18) becomes

$$I(\hat{\Gamma}) = \int_a^b [f(x, \hat{Y}(x), \Phi(x, \hat{Y}(x))) + P(x, \hat{Y}(x)) \cdot (\hat{Y}'(x) - \Phi(x, \hat{Y}(x)))] \, dx$$
$$\left(= \int_a^b f[\hat{Y}(x)] \, dx - \int_a^b \mathscr{E}(x, \hat{Y}(x), \Phi(x, \hat{Y}(x)), \hat{Y}'(x)) \, dx, \text{ by (1)} \right). \tag{19}$$

Moreover, when another such curve Γ_0, say, is *a trajectory of the field* Φ, represented by $Y_0 \in (C^1[a, b])^d$, then by definition $Y_0'(x) = \Phi(x, Y_0(x))$ so that (19) reduces to

$$I(\Gamma_0) = \int_a^b f(x, Y_0(x), Y_0'(x)) \, dx = \int_a^b f[Y_0(x)] \, dx = F(Y_0), \tag{20}$$

where $F(\hat{Y}) \overset{\text{def}}{=} \int_a^b f[\hat{Y}(x)] \, dx$.

It follows that when both $\hat{\Gamma}$ and Γ_0 have the *same end points* and lie in D as shown in Figure 9.5, then by (20):

$$F(\hat{Y}) - F(Y_0) = \int_a^b f[\hat{Y}(x)] \, dx - I(\Gamma_0)$$
$$= \int_a^b f[\hat{Y}(x)] \, dx - I(\hat{\Gamma}), \quad \text{by the invariance of } I;$$

hence, from (19), we obtain *Hilbert's formula*

$$F(\hat{Y}) - F(Y_0) = \int_a^b \mathscr{E}(x, \hat{Y}(x), \Phi[\hat{Y}(x)], \hat{Y}'(x)) \, dx. \tag{21}$$

This formula of Hilbert should be compared with that of Weierstrass (7). Observe that it applies to piecewise C^1 functions as promised. However, as before, we see that $\mathscr{E} \geq 0 \Rightarrow F(\hat{Y}) \geq F(Y_0)$.

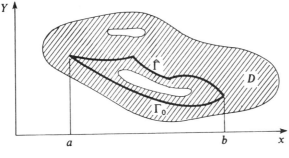

Figure 9.5

As we have stated, in §9.6 we will prove that every central field is exact.[1] To avoid duplication we anticipate this result in formulating the next

(9.7) **Theorem** (Hilbert). *Let $f(\underline{x}, \underline{Y}, Z)$ be* [strictly] *convex on $D \times \mathbb{R}^d$ where D of $\mathbb{R}^{d+1}$ is the domain of Φ, an exact field (a central field) for f, which contains the stationary field trajectory Γ_0 represented by $Y_0 \in (C^1[a, b])^d$. Then Y_0 minimizes*

$$F(\hat{Y}) = \int_a^b f[\hat{Y}(x)]\, dx$$

[uniquely] *on*

$$\hat{\mathcal{D}} = \{\hat{Y} \in \hat{C}^1[a, b]^d : \hat{Y}(a) = Y_0(a);\ \hat{Y}(b) = Y_0(b);\ (x, \hat{Y}(x)) \in D\}.$$

PROOF. Since from convexity, $\mathscr{E} \ge 0$, it follows from (21) that $F(\hat{Y}) \ge F(Y_0)$ with equality *iff*

$$\mathscr{E}(x, \hat{Y}(x), \Phi[\hat{Y}(x)], \hat{Y}'(x)) = 0 \quad \text{(where defined)}.$$

[With strict convexity this implies that where defined, $\hat{Y}'(x) = \Phi(x, \hat{Y}(x))$ so that $\hat{Y} = Y \in (C^1[a, b])^d$ represents a field trajectory. However, the field trajectory through the point $(a, Y_0(a))$ is unique, and thus $\hat{Y} \equiv Y_0$.]   □

(9.8) **Remarks.** Since Y_0 minimizes F on $\mathcal{D}$, it is necessarily stationary (§6.7); however, not every stationary function will represent a trajectory of the given field. Indeed, it is only by construction that a given stationary trajectory Γ_0 can be established as a field trajectory for an exact field, and it is this fact which makes the Weierstrass–Hilbert theory so difficult to apply. Observe also that the size of the minimization class $\hat{\mathcal{D}}$ is determined by the extent of the field domain D, but within $\hat{\mathcal{D}}$, Hilbert's theorem furnishes a *strong* minimum in the sense of Chapter 7. In particular, it is geometrical in character. On the other hand, in order to admit arbitrary $\hat{Y}' \in \mathbb{R}^d$, $f(x, Y, Z)$ must be defined on the set $S = D \times \mathbb{R}^d$, and this would exclude some f with restricted

---

[1] See Remark 9.6.

convexity such as $-\sqrt{1-z^2}$ (for $d=1$). A corresponding theorem obtains for such f, but it is more awkward to state (Problem 9.7).

A further characteristic difficulty is encountered in the next example.

Application: The Brachistochrone*

For the brachistochrone function

$$f(x, y, z) = \sqrt{\frac{1+z^2}{y}}, \quad \text{with} \quad f_z(x, y, z) = \frac{z}{\sqrt{y}\sqrt{1+z^2}},$$

we know from Example 2 of §9.3 that we have an exact field φ in the quadrant $D = \{(x, y) \in \mathbb{R}^2, x > 0, y > 0\}$ for which the field trajectories are the cycloids with cusp at the origin. Moreover, from §3.3, Example 3, it follows that $f(\underline{x}, \underline{y}, z)$ is strictly convex in $D \times \mathbb{R}$. Hence, from Hilbert's theorem, we can conclude that the cycloid, Γ_0, joining the origin to a given point $(b, b_1) \in D$ is represented by $y_0 \in C^1(0, b]$ which minimizes *uniquely* for each $a > 0$, the time-of-descent integral given within a constant factor by

$$T_a(\hat{y}) = \int_a^b \sqrt{\frac{1+\hat{y}'(x)^2}{\hat{y}(x)}} \, dx \tag{22}$$

on

$$\hat{\mathscr{D}}_a = \{\hat{y} \in \hat{C}^1[a, b] \colon \hat{y}(a) = y_0(a), \hat{y}(b) = b_1 ; \hat{y}(x) > 0\}.$$

See Figure 9.6(a). When $a = 0$ in (22), denote T_a by T.

Now, this is not quite what we wish; namely, to establish the (unique) minimality when $a = 0$. The resulting problems are two-fold: first, the origin is not in the domain of the field; and second, the cycloid (and perhaps other curves of interest) have an infinite slope at the origin.

We can circumvent these difficulties by the following arguments:

With Γ_0 as the above cycloid, let $\hat{\Gamma}$ be a piecewise C^1 curve joining the origin to (b, b_1) contained except for its end points in D, and represented by

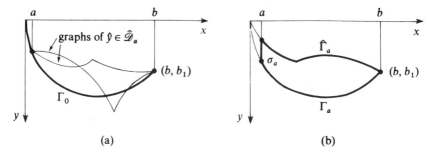

(a)   (b)

Figure 9.6

$\hat{y} \in C^1(0, b]$ for which $T(\hat{y}) < +\infty$. Then for each $a \in (0, b)$, let σ_a be the segment (in D) joining the (possibly distinct) points $(a, y_0(a))$ and $(a, \hat{y}(a))$ as shown in Figure 9.6(b).

Denote by Γ_a and $\hat{\Gamma}_a$ the parts of the curves Γ_0 and $\hat{\Gamma}$, respectively, corresponding to $x \geq a$, and observe that with proper orientation of σ_a, $\hat{\Gamma}_a + \sigma_a$ constitutes a *piecewise* C^1 curve with the *same end points* as Γ_a. By the invariance of I, Hilbert's integral (18) for φ, we obtain

$$I(\Gamma_a) = I(\hat{\Gamma}_a + \sigma_a) = I(\hat{\Gamma}_a) + I(\sigma_a)$$

Next, with T_a as defined by (22) above, we may reproduce the analysis leading to Hilbert's theorem, utilizing in particular (19) and (20), to conclude that

$$T_a(\hat{y}) - T_a(y_0) = \int_a^b \mathcal{E}(x, \hat{y}(x), \varphi[\hat{y}(x)], \hat{y}'(x)) \, dx - I(\sigma_a) \qquad (23)$$

(Problem 9.10)

Since $f \geq 0$, and $\mathcal{E} \geq 0$ in $D \times \mathbb{R}^2$, we have

$$T(\hat{y}) = \int_0^b \sqrt{\frac{1 + \hat{y}'(x)^2}{\hat{y}(x)}} \, dx \geq \int_a^b \sqrt{\frac{1 + \hat{y}'(x)^2}{\hat{y}(x)}} \, dx = T_a(\hat{y}) \geq T_a(y_0) - I(\sigma_a).$$

$$(23')$$

Now as $a \searrow 0$, $T_a(y_0) \nearrow T(y_0)$, while on σ_a: $dx = 0$, and $p(a, y) = f_z(a, y, \varphi(a, y))$, so that from (18),

$$|I(\sigma_a)| = \left| \int_{a_0}^{\hat{a}} \frac{\varphi(a, y)}{\sqrt{1 + \varphi^2(a, y)}} \frac{dy}{\sqrt{y}} \right| \leq \int_{a_0}^{\hat{a}} \frac{dy}{\sqrt{y}},$$

(supposing that $a_0 \equiv y_0(a) \leq \hat{y}(a) = \hat{a}$)

$$\leq 2|\sqrt{\hat{y}(a)} - \sqrt{y_0(a)}| \to 0,$$

(and this holds also when $\hat{y}(a) \leq y_0(a)$).

Hence, in the limit as $a \searrow 0$, (23') gives

$$T(\hat{y}) \geq T(y_0),$$

which establishes the minimality of the cyloid. The uniqueness requires further analysis; [see Problem 6.15].

(Problems 9.7–9.10, 9.13–9.15)

Variable End-Point Problems

When an exact field Φ for $f = f(x, Y, Z)$ is available in a domain D of $\mathbb{R}^{d+1}$, then Hilbert's approach admits modification to provide theoretical access to

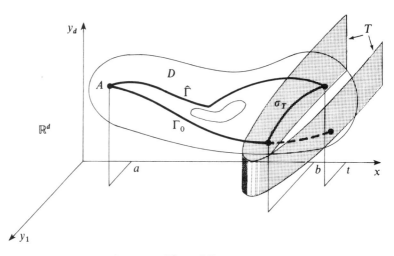

Figure 9.7

the problem of minimizing an integral function

$$F(\hat{Y}, t) = \int_a^t f[\hat{Y}(x)] \, dx$$

on

$$\hat{\mathcal{D}}_\tau = \{\hat{Y} \in (\hat{C}^1[a, t])^d \colon \hat{Y}(a) = A, (t, \hat{Y}(t)) \in T; (x, \hat{Y}(x)) \in D\},$$

where the right end point is confined to the *transversal* T defined as the *zero level set* of a C^1 function τ (whose gradient, $\nabla\tau$, is nonvanishing on T) provided that T is C^1 *arcwise connected* in D.

For then, when Γ_0 is a trajectory of the field represented by $Y_0 \in \hat{\mathcal{D}}_\tau$, and $\hat{\Gamma}$ is the curve represented by a competing function $\hat{Y} \in \hat{\mathcal{D}}_\tau$, under this assumption, we may join their *right* end points by a C^1 arc σ_T in $D \cap T$ as illustrated in Figure 9.7.

With proper orientation we may consider Γ_0 and $\hat{\Gamma} + \sigma_T$ as piecewise C^1 curves in D having the same end points. Since Φ is exact we have as before from (20)

$$F(Y_0, b) = \int_a^b f[Y_0(x)] \, dx = I(\Gamma_0)$$

$$= I(\hat{\Gamma} + \sigma_T) = I(\hat{\Gamma}) + I(\sigma_T),$$

(where I is Hilbert's integral for Φ), so that for the appropriate t:

$$F(\hat{Y}, t) - F(Y_0, b) = \int_a^t \mathcal{E}(x, \hat{Y}(x), \Phi[\hat{Y}(x)], \hat{Y}'(x)) \, dx - I(\sigma_T), \qquad (24)$$

which may be considered as Hilbert's formula in the presence of the transversal defined by τ.

To recover the minimizing inequality with the convexity of $f(\underline{x}, Y, Z)$ in $D \times \mathbb{R}^d$, we evidently should want that $I(\sigma_T) = 0$. Upon parametrizing σ_T by $(\bar{x}(s), \bar{Y}(s))$, $0 \le s \le 1$, we have from (18)

$$I(\sigma_T) = \int_0^1 [h(\bar{x}, \bar{Y})\bar{x}'(s) + P(\bar{x}, \bar{Y}) \cdot \bar{Y}'(s)]\, ds, \tag{24'}$$

and since $(\bar{x}', \bar{Y}')$ is a tangent vector to a curve in T (and hence to T itself), which will vary with σ_T, we must demand the vanishing of the *integrand* in (24'); i.e., that the field functions h and P defined by (9) provide a *vector* (h, P) which is *normal* to T at each point. But, by the well-known argument (reproduced at the end of §5.6), the gradient $\nabla \tau = (\tau_x, \tau_Y)$ is normal to T at each point. Hence we should require that (h, P) is proportional to $\nabla \tau$ at each point of T, or equivalently, we require that the *field* Φ meet the *transversal condition*

$$(h\tau_Y - \tau_x P)|_T \equiv 0. \tag{25}$$

When $d = 1$, recalling that $y_0'(x) = \varphi(x, y_0(x))$, we can verify that (25) is precisely the transversal condition (equation (15) of §6.4) which is *necessary* for a local extremal of this problem. (25) may be regarded as its vector valued generalization.

(9.9) **Theorem.** *Let $f(\underline{x}, Y, Z)$ be [strictly] convex on $D \times \mathbb{R}^d$ where $D \subseteq \mathbb{R}^{d+1}$ is the domain of Φ, an exact field (a central field) for f, which meets the transversal condition* (25) *on the set T where:*

(i) *$T = \{(x, Y) \in D: \tau(x, Y) = 0\}$, for a given C^1 function τ with $\nabla \tau|_T \neq 0$; and*
(ii) *T is C^1 arcwise connected.*

If Γ_0 is a field trajectory represented by

$$Y_0 \in \hat{\mathcal{D}}_\tau = \{\hat{Y} \in (\hat{C}^1[a, t])^d: \hat{Y}(a) = A, (t, \hat{Y}(t)) \in T; (x, \hat{Y}(x)) \in D\},$$

then Y_0 minimizes $F(\hat{Y}, t) = \int_a^t f[\hat{Y}(x)]\, dx$ on $\hat{\mathcal{D}}_\tau$ [uniquely within the specification of its interval of definition].

PROOF. Since the transversal condition (25) forces the vanishing of $I(\sigma_T)$ for all C^1 arcs $\sigma_T \subseteq T$, Hilbert's formula (24) is applicable. The convexity of $f(\underline{x}, Y, Z)$ makes $\mathscr{E} \ge 0$, and gives the minimizing inequality

$$F(\hat{Y}, t) \ge F(Y_0, b),$$

(where we suppose $(b, Y_0(b) \in T)$, with equality only if

$$\mathscr{E}(x, \hat{Y}(x), \Phi[\hat{Y}(x)], \hat{Y}'(x)) \equiv 0.$$

[With *strict* convexity this requires that $\hat{Y}'(x) = \Phi(x, \hat{Y}(x))$. Hence $\hat{\Gamma} = \Gamma_1$ is a trajectory of the field, and since both Γ_1 and Γ_0 pass through (a, A), we know that their representing functions Y_1 and Y_0 will agree *on each common interval* $[a, x]$ *of definition*. As Figure 9.7 indicates, it is possible that the transversal T cuts a given field trajectory more than once, and then we must accept the consequence that $F(Y_0, b) = F(Y_1, t_1)$ for $b \neq t_1$, even in the presence of strict convexity (Problem 9.17).]   □

The preceding arguments admit various simplifications with the form of τ, which will be taken up in Problems 9.16 and 9.17. However, it must be admitted that field construction is difficult at best without the additional complications accompanying the imposition of boundary conditions such as (25). Hence our applications will be confined to those covered by Hilbert's theorem 9.7. For a more complete discussion, see [S].

§9.5. Minimization with Constraints

We may also extend the Weierstrass–Hilbert theory (in principle) to embrace minimization problems involving constraints of either the isoperimetric or the Lagrangian form. We shall obtain a result covering both cases utilizing Lagrangian multiplier function(s) (which in this chapter will be denoted by μ to avoid confusion). Our result is an amalgam of Hilbert's theorem 9.7 and the approach taken in §2.3.

We assume that functions $f = f(x, Y, Z)$ and $g_j = g_j(x, Y, Z)$, $j = 1, 2, \ldots,$ N, are suitably defined, and set $\tilde{f} = f + \sum_{j=1}^{N} \mu_j g_j$ where $\mu_j = \mu_j(x, Y)$ is continuous for $j = 1, 2, \ldots, N$.

(9.10) **Theorem.** *Suppose that* $\tilde{f}(\underline{x}, \underline{Y}, Z)$ *is* [strictly] *convex on* $D \times \mathbb{R}^d$ *where* D *of* $\mathbb{R}^{d+1}$ *is the domain of* Φ, *an exact field (a central field) for* $\tilde{f}$ *which contains a trajectory of the field represented by* $Y_0 \in (C^1[a, b])^d$. *Then* Y_0 *minimizes*

$$F(\hat{Y}) = \int_a^b f[\hat{Y}(x)] \, dx$$

[uniquely] *on*

$$\hat{\mathscr{D}} = \{\hat{Y} \in (\hat{C}^1[a, b])^d \colon \hat{Y}(a) = Y_0(a), \hat{Y}(b) = Y_0(b), (x, \hat{Y}(x)) \in D\},$$

under the constraining relations:

$$\int_a^b \mu_j[\hat{Y}(x)]g_j[\hat{Y}(x)] \, dx = \int_a^b \mu_j[Y_0(x)]g_j[Y_0(x)] \, dx, \qquad j = 1, 2, \ldots, N.$$

$$\tag{26}$$

PROOF. Set $\tilde{F}(\hat{Y}) = \int_a^b \tilde{f}[\hat{Y}(x)] \, dx$. By Hilbert's theorem, $\tilde{Y}_0$ minimizes

$$\tilde{F}(\hat{Y}) = F(\hat{Y}) + \sum_{j=1}^{N} \int_a^b \mu_j[\hat{Y}(x)]g_j[\hat{Y}(x)] \, dx$$

[uniquely] on $\hat{\mathscr{D}}$, and the conclusion is obvious.   □

Remarks. In view of Proposition 3.2, the [strict] convexity for $\tilde{f}$ will follow from the convexity of $f(\underline{x}, \underline{Y}, Z)$ and $\mu_j(\underline{x}, \underline{Y})g_j(\underline{x}, \underline{Y}, Z)$, $j = 1, 2, \ldots, N$; [with the strict convexity of one of these terms]. $\tilde{Y}_0$ also minimizes on the larger class of functions $\hat{Y}$ in $\hat{\mathscr{D}}$ which satisfy (26) when "$\leq$" replaces the equality sign as in Proposition 2.5.

In application, there are three separate cases of interest:

(9.10a) *If the μ_j are constant, $j = 1, 2, \ldots, N$, then we obtain* [unique] *minimization for F on $\hat{\mathscr{D}}$ under the isoperimetric constraints*

$$G_j(\hat{Y}) \overset{\text{def}}{=} \int_a^b g_j[\hat{Y}(x)]\, dx = G_j(Y_0), \qquad j = 1, 2, \ldots, N. \tag{27}$$

Since there are as many multipliers μ_j, as constraining relations (27), there is some hope that we may find the μ_j which retain the convexity while permitting the $G_j(Y_0)$ to be specified.

(9.10b) *When the $\mu_j = \mu_j(x)$, $j = 1, 2, \ldots, N$, then we obtain* [unique] *minimization on $\hat{\mathscr{D}}$ under the Lagrangian form of constraining equations*

$$g_j[\hat{Y}(x)] = g_j[Y_0(x)], \quad x \in (a, b), \qquad j = 1, 2, \ldots, N. \tag{28}$$

Now, however, there is little reason to believe that the μ_j can be found which permit preassigned functions $g_j[Y_0(x)]$ (say $g_j[Y_0(x)] \equiv 0$), $j = 1, 2, \ldots, N$.

Indeed, each Lagrangian constraint of the form $g_j[Y(x)] = 0$, in general restricts the dimensionality of the Euclidean space available for the solution trajectories, so that unless $N < d$ (which precludes the case $d = 1$) we should not suppose that even these constraining relations can be satisfied. However,

(9.10c) *When the $\mu_j = \mu_j(x, Y)$ and $g_j[Y_0(x)] \equiv 0$, $j = 1, 2, \ldots, N \leq d$, then we obtain* [unique] *minimization for Y_0 under the preassigned Lagrangian constraints*

$$g_j[\hat{Y}(x)] = g_j[Y_0(x)] \equiv 0, \qquad j = 1, 2, \ldots, N, \qquad \forall\, x \in [a, b]. \tag{29}$$

Observe that since Y_0 represents a trajectory of the field Φ, (29) yields $g_j(x, Y_0(x), \Phi(x, Y_0(x))) \equiv 0$ and this will hold *a fortiori* if we require of the field that $g_j(x, Y, \Phi(x, Y)) \equiv 0$, in D, for $j = 1, 2, \ldots, N$. As we shall see in Example 3, this additional field requirement may actually assist in the determination of the multiplier functions μ_j.

A problem with multiple constraints, each of one of these types, is covered by Theorem 9.10 provided that each constraint has associated with it a μ_j of the *corresponding form permitted* above.

Example 1. To minimize

$$F(\hat{y}) = \int_0^1 [\hat{y}'(x)^2 + \hat{y}(x)\hat{y}'(x)^4]\, dx$$

on

$$\hat{\mathcal{D}} = \{\hat{y} \in \hat{C}^1[0, 1]: \hat{y}(0) = 1, \hat{y}(1) = 0, \hat{y} \geq 0\},$$

when subject to the isoperimetric constraining relation

$$G(\hat{y}) \overset{\text{def}}{=} \int_0^1 x\hat{y}'^4(x) \, dx = \tfrac{1}{2},$$

we take $f(x, y, z) = z^2 + yz^4$, $g(x, y, z) = xz^4$, and $\tilde{f}(x, y, z) = z^2 + yz^4 + \mu xz^4$ for an unknown *constant* μ so that

$$\tilde{f}_{zz}(x, y, z) = 2 + 12yz^2 + 12\mu xz^2 > 0, \quad \text{when } x, y, \mu \geq 0.$$

It follows from Proposition 9.2 that for $\mu \geq 0$, $f(\underline{x}, \underline{y}, z)$ is strictly convex in the quarter space $D \times \mathbb{R}$ where

$$D = \{(x, y) \in \mathbb{R}^2: x \geq 0, y \geq 0\}.$$

The stationary functions for $\tilde{f}$ are those which satisfy the Euler–Lagrange equation

$$\frac{d}{dx}(2y' + 4yy'^3 + 4\mu xy'^3) = y'^4,$$

and those which are C^2 satisfy

$$(2 + 12yy'^2 + 12\mu xy'^2)y'' + 3y'^4 + 4\mu y'^3 = 0. \tag{30}$$

By inspection, this equation is satisfied by the family of functions $y(x) = $ const., which do *not* belong to $\hat{\mathcal{D}}$; and also by the family $y(x) = mx + x_0$ (for which $y'(x) = m$) provided that $3m^4 + 4\mu m^3 = 0$, which is possible for the nontrivial choice $m = -\tfrac{4}{3}\mu$. The stationary family of lines with slope $m = -1$ is consistent in the plane, and so in D, which is simply connected; and $\mu \overset{\text{def}}{=} -\tfrac{3}{4}m = \tfrac{3}{4} > 0$. Thus by Corollary 9.5, it provides an exact field for $\tilde{f} = f + \tfrac{3}{4}g$ in D. Moreover, $y_0(x) = 1 - x$ defines a function in $\hat{\mathcal{D}}$, whose graph has slope -1.

We conclude from Theorem 9.10a that y_0 minimizes F on $\hat{\mathcal{D}}$ uniquely, when further constrained to the $G(y_0)$-level set of G. But $G(y_0) = \tfrac{1}{2}$.

Example 2. We may also use the field of Example 1 to conclude that $y_0(x) = 1 - x$ minimizes F [uniquely] on $\hat{\mathcal{D}}$ under the *Lagrangian* constraint:

$$g_0[\hat{y}(x)] \equiv \hat{y}'(x)^4 = y_0'(x)^4 \equiv 1,$$

or equivalently,

$$g_1[\hat{y}(x)] \equiv \hat{y}'(x)^4 - 1 \equiv 0,$$

if we take $\mu(x) = (\tfrac{3}{4})x$ in 9.10b.

We have observed frequently that a solution to one optimization problem may solve others. A general result of this type is the following:

(9.11) **Proposition.** *Isoperimetric problems may be reformulated as Lagrangian problems in higher-dimensional space.*

PROOF. To transform the problem of minimizing

$$F(\hat{Y}) = \int_a^b f[\hat{Y}(x)]\, dx$$

on

$$\hat{\mathscr{D}} = \{\hat{Y} \in (\hat{C}^1[a, b])^d \colon \hat{Y}(a) = A,\ \hat{Y}(b) = B;\ (x, \hat{Y}(x)) \in D\},$$

under the *isoperimetric constraints*

$$G_j(\hat{Y}) \stackrel{\text{def}}{=} \int_a^b g_j[\hat{Y}(x)]\, dx = l_j, \qquad j = 1, 2, \ldots, N, \tag{31}$$

we proceed as follows:

To each $\hat{Y} \in \hat{\mathscr{D}}$ we associate N new constraint variables $\hat{\kappa}_j$ defined by

$$\hat{\kappa}_j(x) = \int_a^x g_j[\hat{Y}(t)]\, dt, \qquad j = 1, 2, \ldots, N;$$

observe that $\hat{\kappa}_j \in \hat{C}^1[a, b]$, and $\hat{\kappa}_j'(x) = g_j[\hat{Y}(x)]$. Set

$$\hat{K}(x) = (\hat{\kappa}_1(x), \hat{\kappa}_2(x), \ldots, \hat{\kappa}_N(x)) \quad \text{and} \quad L = (l_1, l_2, \ldots, l_N).$$

Then $\hat{Y}^* \stackrel{\text{def}}{=} (\hat{Y}, \hat{K}) \in (\hat{C}^1[a, b])^{d+N}$ and the original *isoperimetric problem is equivalent to the Lagrangian problem* of minimizing

$$F^*(\hat{Y}^*) = \int_a^b f[\hat{Y}(x)]\, dx = F(\hat{Y})$$

on

$$\hat{\mathscr{D}}^* = \{\hat{Y}^* \in (\hat{C}^1[a, b])^{d+N} \colon \hat{Y}(a) = A,\ \hat{Y}(b) = B;\ (x, \hat{Y}) \in D;$$
$$\hat{K}(a) = \mathcal{O},\ \hat{K}(b) = L\},$$

under the *Lagrangian constraints*

$$g_j^*[\hat{Y}^*(x)] \stackrel{\text{def}}{=} g_j[\hat{Y}(x)] - \hat{\kappa}_j'(x) \equiv 0 \quad \text{on } (a, b), \qquad j = 1, 2, \ldots, N. \tag{32}$$

For if $Y_0^* = (Y_0, K_0)$ minimizes F^* on $\hat{\mathscr{D}}^*$ [uniquely] under these Lagrangian constraints, then $Y_0 \in \hat{\mathscr{D}}$, while for each $\hat{Y} \in \hat{\mathscr{D}}$ we have $\hat{Y}^* \in \hat{\mathscr{D}}^*$ and

$$G_j(\hat{Y}) = \int_a^b g_j[\hat{Y}(x)]\, dx = \int_a^b \hat{\kappa}_j'(x)\, dx$$
$$= \hat{\kappa}_j(b) - \hat{\kappa}_j(a) = l_j. \tag{33}$$

Thus Y_0 minimizes F on $\hat{\mathscr{D}}$ [uniquely] under the isoperimetric constraints (31).

Conversely, if Y_0 in $\hat{\mathscr{D}}$ solves the latter problem, then its counterpart $Y_0^* \in \hat{\mathscr{D}}^*$ satisfies the Lagrangian constraints, (32), while for each $\hat{Y}^* = (\hat{Y}, \hat{K}) \in \hat{\mathscr{D}}^*$ which satisfies (32), $\hat{Y} \in \hat{\mathscr{D}}$ and it satisfies the isoperimetric constraints (31) as above. Thus Y_0^* solves the Lagrangian problem for F^* [uniquely]. □

Theorem 9.10(c) permits the construction of a more usable field for the isoperimetric problem as reformulated than does Theorem 9.10(a). However, it is more difficult to construct higher-dimensional exact fields, and moreover, as we have observed in §3.5, it is especially difficult to use the freedom in selecting μ effectively. Despite these complications, this approach will yield a solution to the classical isoperimetric problem.

The Wirtinger Inequality

Example 3*. In Problem 1.6 it was shown that the classical isoperimetric inequality would follow from the *Wirtinger inequality*

$$F(y) = \int_0^{2\pi} [y'(x)^2 - y(x)^2] \, dx \geq 0$$

on

$$\mathscr{D} = \{y \in C^1[0, 2\pi]: y(0) = y(2\pi) = 0\},$$

subject to the isoperimetric constraint

$$G(y) \equiv \int_0^{2\pi} y(x) \, dx = 0. \tag{34}$$

Since $y = \mathcal{O}$ is admissible, this is a minimization problem.

To transform this to a Lagrangian problem as in 9.11, we introduce for $\hat{y} \in \hat{\mathscr{D}} = \{\hat{y} \in \hat{C}^1[0, 2\pi]: \hat{y}(0) = \hat{y}(2\pi) = 0\}$, the constraint variable, (here) denoted

$$\hat{y}_1(x) \equiv \int_0^x \hat{y}(t) \, dt,$$

and consider instead the problem of minimizing

$$\int_0^{2\pi} [\hat{y}'(x)^2 - \hat{y}(x)^2] \, dx$$

on

$$\hat{\mathscr{D}}^* = \{\hat{Y} = (\hat{y}, \hat{y}_1) \in (\hat{C}^1[0, 2\pi])^2: \hat{Y}(0) = \hat{Y}(2\pi) = \mathcal{O}\},$$

subject to the *Lagrangian* constraint

$$g[\hat{Y}(x)] \equiv \hat{y}(x) - \hat{y}'_1(x) = 0. \tag{35}$$

(Asterisks have been suppressed where possible.)

To do this, we shall find a central field[1] for the modified function

$$\tilde{f}(x, Y, Z) = z^2 - y^2 + 2\mu(x, Y)(y - z_1), \tag{36}$$

[1] See Remark 9.6.

where μ is as yet unspecified, and the factor of 2 is introduced for convenience. The stationary functions for $\tilde{f}$ satisfy the usual equations, (§6.7) which in this case are

$$\frac{d}{dx} 2y' = -2y + 2\mu + 2\mu_y(y - y_1'),$$

and

$$-\frac{d}{dx} 2\mu = 2\mu_{y_1}(y - y_1').$$

In addition, we require that the constraining equation, $y - y_1' \equiv 0$, be satisfied along each stationary trajectory, so that the *terms in parentheses vanish*. From the second of the resulting equations we conclude that μ is *constant* along each stationary trajectory. Then, the first becomes $y'' + y = \mu$, which has the well-known general solution

$$y(x) = \gamma \cos x + \lambda \sin x + \mu, \tag{37}$$

for trajectory constants μ, λ, γ.

We shall construct a stationary field whose trajectories emanate from the origin. To have $y(0) = 0$ in (37), we must take $\gamma = -\mu$, so that with $\Lambda \equiv (\mu, \lambda)$:

$$\psi(x; \Lambda) \stackrel{\text{def}}{=} y(x) = \mu(1 - \cos x) + \lambda \sin x,$$

$$\psi_1(x; \Lambda) \stackrel{\text{def}}{=} y_1(x) = \int_0^x y(t)\, dt = \mu(x - \sin x) + \lambda(1 - \cos x). \tag{38}$$

Equations (38) determine the *stationary family*, $\Psi(x; \Lambda)$, *indexed by the parameter* Λ. By construction, $\Psi(0; \Lambda) = \mathcal{O}$. Moreover, for each $x \in (0, 2\pi)$: the Jacobian matrix,

$$\Psi_\Lambda(x, \Lambda) \stackrel{\text{def}}{=} \begin{bmatrix} \psi_\mu & \psi_\lambda \\ \psi_{1_\mu} & \psi_{1_\lambda} \end{bmatrix} (x; \Lambda) = \begin{bmatrix} 1 - \cos x & \sin x \\ x - \sin x & 1 - \cos x \end{bmatrix},$$

so that its determinant,

$$\Delta(x) \stackrel{\text{def}}{=} \det \Psi_\Lambda(x; \Lambda) = (1 - \cos x)^2 - \sin x(x - \sin x)$$

$$= \begin{cases} 2(1 - \cos x) - x \sin x > 0, & (\pi \leq x < 2\pi) \\ 2 \sin x(\tan x/2 - x/2) > 0, & (0 < x < \pi). \end{cases} \tag{39}$$

Hence, $\Delta(x) > 0$ on $(0, 2\pi)$, and it follows that through each point $(x, Y) \in D = (0, 2\pi) \times \mathbb{R}^2$ passes a unique trajectory of the family—namely, that with parameter Λ given as the *unique* solution of the *linear* system (38). An easy inversion gives $\Lambda = \Lambda(x, Y)$, where

$$\mu(x, Y) = [(1 - \cos x)y - (\sin x)y_1]/\Delta(x),$$

$$\lambda(x, Y) = [-(x - \sin x)y + (1 - \cos x)y_1]/\Delta(x). \tag{40}$$

The stationary field $\Phi = (\varphi, \varphi_1)$ so determined is given from (38) and

(40) by

$$\varphi(x, Y) = \psi'(x; \Lambda(x, Y)) = \mu(x, Y) \sin x + \lambda(x, Y) \cos x$$

$$= [(\sin x - x \cos x)y + (\cos x - 1)y_1]/\Delta(x), \tag{41a}$$

while in view of (35),

$$\varphi_1(x, Y) = \psi_1'(x; \Lambda(x, Y)) = y. \tag{41b}$$

In the next section we shall verify that this is a central field for $\tilde{f}$ in D, and from (36) we observe that $\tilde{f}(\underline{x}, \underline{Y}, Z)$ is convex (but not strictly convex) in D, since this is true of the term z^2 while the remaining term is linear in z_1. Hence Hilbert's theorem (9.7) is applicable to $\tilde{f}$ in $D \times \mathbb{R}^2$.

It follows that if Γ_{ab} is a trajectory of the field represented by a $Y_0 \in (C^1[a, b])^2$, with $0 < a < b < 2\pi$, and $\hat{\Gamma}_{ab} \subseteq D$ has the *same* end points as Γ_{ab} and is represented by $\hat{Y} \in (\hat{C}^1[a, b])^2$, then

$$\int_a^b \tilde{f}[\hat{Y}(x)] \, dx \geq \int_a^b \tilde{f}[Y_0(x)] \, dx = \int_a^b f[y_0(x)] \, dx \quad \text{(Why?)}.$$

As usual, this is not the desired inequality which must be obtained by a limiting argument made rather delicate by the fact that $\Delta(0) = \Delta(2\pi) = 0$.

We first note that $Y_0 = \mathcal{O} \in \hat{\mathcal{D}}^*$, and it has a graph Γ_0 on the x axis which (except for its end points) *is* a trajectory of the field in D, since $Y_0 = \Psi(x; \mathcal{O})$, as given by (38).

Next, if $\hat{y} \in \hat{\mathcal{D}}$ satisfies (34), define $\hat{y}_1(x) = \int_0^x \hat{y}(t) \, dt$, so that $\hat{Y} = (\hat{y}, \hat{y}_1) \in \hat{\mathcal{D}}^*$, and the constraining relation $\hat{y}_1' = \hat{y}$ hold. We also note for future use that since $\hat{y}(0) = 0$, then

$$|\hat{y}(x)| = \left| \int_0^x \hat{y}'(t) \, dt \right| \leq \hat{M}x,$$

and hence

$$|\hat{y}_1(x)| \leq \hat{M} \int_0^x t \, dt \leq \hat{M}x^2.$$

Here, $\hat{M} = \max|\hat{y}'|$ is fixed by $\hat{y}$.

Let $\hat{\Gamma}$ be the graph of $\hat{Y}$, and for $0 < a < b < 2\pi$, let $\hat{\Gamma}_{ab}$ be that part of $\hat{\Gamma}$ for $x \in [a, b]$. Finally, let σ_a and σ_b be the segments joining the end points of $\hat{\Gamma}_{ab}$ to $(a, \mathcal{O})$ and $(b, \mathcal{O})$, respectively, as shown in Figure 9.8.

With proper orientation of these segments we may use Hilbert's integral I for the field Φ of $\tilde{f}$ (as in the analysis of the brachistochrone in §9.4) to conclude that

$$\int_a^b \tilde{f}[\hat{Y}(x)] \, dx \geq \int_a^b \tilde{f}[Y_0(x)] \, dx + I(\sigma_a) + I(\sigma_b),$$

or

$$\int_a^b [\hat{y}'(x)^2 - \hat{y}(x)^2] \, dx \geq I(\sigma_a) + I(\sigma_b).$$

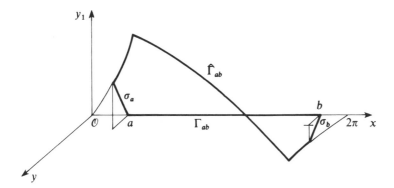

Figure 9.8

The desired inequality will follow if we prove that $I(\sigma_x) \to 0$ when $x = a \searrow 0$, and when $x = b \nearrow 2\pi$. We shall concentrate on the first and more difficult of these assertions. Since $\tilde{f}_Z(x, Y, Z) = 2(z, -\mu)$, by (36), then from (9), the field function $\tilde{P}(x, Y) = \tilde{f}_Z(x, Y, \Phi(x, Y)) = 2(\varphi, -\mu)(x, Y)$, where μ and φ are given by (40) and (41a).

Since $dx = 0$ on σ_x, we have by (18):

$$I(\sigma_x) = \int_{\sigma_x} \tilde{P} \cdot dY = 2 \int_{\sigma_x} (\varphi \, dy - \mu \, dy_1)$$

$$= \pm\{[(\sin x - x \cos x)y^2 + 2(\cos x - 1)yy_1 + (\sin x)y_1^2]/\Delta(x)\},$$

$$\text{when } Y = \hat{Y}(x). \quad (42)$$

This last follows because $2(\varphi \, dy - \mu \, dy_1)$ is (for fixed x) an *exact* differential of the function in braces, as may be verified by differentiation and comparison.

Using the estimates for $\hat{Y}(x)$ obtained previously, we have for small x, that

$$|I(\sigma_x)| \le \hat{M}^2 x^2 [(\sin x - x \cos x) + 2(1 - \cos x)x + x^2 \sin x]/\Delta(x).$$

By using, say, L'Hôpital's rule, it can be shown that $\lim_{x \searrow 0} \Delta(x)/x^4$ is non-zero, while $\lim_{x \searrow 0}$ ([term in brackets]$/x^2$) = 0. It follows that $|I(\sigma_x)| \to 0$ as $x \searrow 0$.

To establish the corresponding results as $x \nearrow 2\pi$, we note that since $\hat{Y}(2\pi) = \mathcal{O}$ we have

$$|\hat{Y}(x)| \le \left| \int_x^{2\pi} \hat{Y}'(t) \, dt \right| \le \hat{M}(2\pi - x), \quad \text{while } \lim_{x \nearrow 2\pi} \frac{\Delta(x)}{2\pi - x} \text{ is nonzero.}$$

Thus in particular, we have established the Wirtinger inequality: $F(y) \ge 0$, $\forall \, y \in \mathcal{D}$ for which $\int_0^{2\pi} y(x) \, dx = 0$. Moreover, if equality holds and $F(y_0) = 0$, then by the methods of §6.7, we know that y_0 is stationary for $\tilde{f}$ with $\mu =$ const., and hence must be of the form y of (38) with $\int_0^{2\pi} y_0(x)\, dx = 2\pi\mu = 0$. Thus $y_0(x) = \lambda \sin x$. Conversely, any function of this form gives $F(y_0) = \lambda \int_0^{2\pi}(\cos^2 x - \sin^2 x)\, dx = 0$. (A direct ad hoc proof of the Wirtinger inequality is given in [H–L–P], while a field for the isoperimetric problem itself is presented in Problem 9.20.)

(Problems 9.18–9.20)

§9.6*. Central Fields

With the brachistochrone problem as analyzed in §9.3, we saw that we could find a unique stationary trajectory (cycloid) joining the point $\alpha = (0, 0)$ to a given point in $D = \{(x, y): x > 0, y > 0\}$ by making a suitable choice of the parameter $\lambda = \lambda(x, y)$, say. The functions $\psi(\cdot\,; \lambda)$ representing these trajectories determine in D, the stationary field

$$\varphi(x, y) = \psi'(x; \lambda(x, y)), \tag{43}$$

which, by Corollary 9.5, is exact. As we have observed, in higher dimensions, more than stationarity is required for exactness of a field, and we shall now obtain this for a field determined by a suitable generalization of the family ψ.

(9.12) **Definition.** For a given $f = f(x, Y, Z)$, a family of stationary functions $\Psi(\cdot\,; \Lambda)$ indexed by a parameter $\Lambda \in \mathbb{R}^d$, is a *central family* with *center* $\alpha = (a, A) \in \mathbb{R}^{d+1}$, if $\Psi(a; \Lambda) = A$, $\forall \Lambda$, and, in addition, both $\Psi = \Psi(x; \Lambda)$ and Ψ_x are C^1.

(9.12′) **Definition.** For a given $f = f(x, Y, Z)$, a *central field* in a domain D is a stationary field whose trajectories are defined by a central family $(\Psi; \Lambda)$ for which Ψ_Λ is invertible "in D" (i.e., at each (x_1, Λ_1) for which $(x_1, Y_1) \in D$ when $Y_1 = \Psi(x_1; \Lambda_1)$).
$\Lambda = \Lambda(x, Y)$ is then called *a parameter for the field*.

Having Λ in $\mathbb{R}^d$ makes possible the following result:

(9.13) **Lemma.** Let $\Psi(\cdot\,; \Lambda)$ be a central family for f. If the matrix Ψ_Λ with elements $\partial\psi_i/\partial\lambda_j$, $i, j = 1, 2, \ldots, d$ is invertible at (x_1, Λ_1), then in a neighborhood D_1 of $(x_1, Y_1 \equiv \Psi(x_1, \Lambda_1))$, for values of Λ near Λ_1, $\Lambda = \Lambda(x, Y)$ is defined and C^1, and

$$\Phi(x, Y) = \Psi'(x; \Lambda(x; Y)) \overset{\text{def}}{=} \Psi_t(t; \Lambda(x, Y))|_{t=x} \tag{44}$$

is a central field in D_1.

PROOF. By implicit function theory, the *equation* $\Psi(x; \Lambda) = Y$ determines $\Lambda = \Lambda(x, Y)$ as a (unique) C^1 function in a neighborhood D_1 of (x_1, Y_1) such that $\Lambda_1 = \Lambda(x_1, Y_1)$ and

$$\Psi(x, \Lambda(x, Y)) \equiv Y. \tag{45}$$

It follows that through each $(x, Y) \in D_1$, passes one and only one trajectory of the possibly smaller family, namely, that with parameter $\Lambda = \Lambda(x, Y)$. Thus Φ is defined in D_1 by (44) and it is C^1 by the chain rule and our differentiability requirements on Ψ. It is stationary by construction, and Ψ_Λ is invertible in a possibly smaller neighborhood D_1 by continuity. □

This lemma shows that *however indexed, a central family will determine a central field in a neighborhood of each point associated with an invertible matrix* Ψ_Λ. Outside this neighborhood, the associated trajectories may intersect in a quite complicated manner, and in particular, they *must* intersect at the center α. Moreover, other trajectories of the original family might intersect these even in this neighborhood. The situation is illustrated in Figure 9.9.

Example 1. In the last section, we saw that with $\Lambda = (\mu, \lambda) \in \mathbb{R}^2$, the functions of equations (38); viz.,

$$\Psi(x; \Lambda) = \begin{cases} \mu(1 - \cos x) + \lambda \sin x, \\ \mu(x - \sin x) + \lambda(1 - \cos x), \end{cases} \quad x \in [0, 2\pi],$$

are stationary for a certain $\tilde{f}$, and clearly $\Psi(0; \Lambda) = \mathcal{O} \in \mathbb{R}^2$, while both Ψ and Ψ_x are C^1 everywhere. It follows that $\Psi(\cdot; \Lambda)$ is a central family for this $\tilde{f}$ with center $\alpha = \mathcal{O}$.

Moreover, by the computation of (39), we know that for $x \in (0, 2\pi)$ the Jacobian

$$\det \Psi_\Lambda(x; \Lambda) = \frac{\partial(\psi, \psi_1)}{\partial(\mu, \lambda)}(x; \Lambda) \neq 0,$$

and hence Ψ_Λ is invertible.

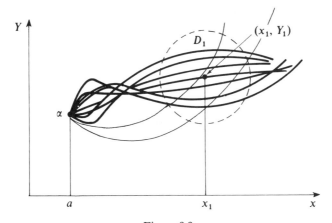

Figure 9.9

We already know that each point $(x, Y) \in D = (0, 2\pi) \times \mathbb{R}^2$, determines a unique Λ, so that the field $\Phi \equiv (\varphi, \varphi_1)$ is well defined by (41). It follows that Φ is a central field for $\tilde{f}$ in the domain D which is that required for the application.

Lemma 9.13 shows that the field parameter *function* Λ is in $C^1(D)$, and this can now be used to show that Φ is exact. More generally, we have the following.

(9.14) **Theorem.** *A central field is exact.*

PROOF. For a given $f = f(x, Y, Z)$, let Φ be a central field for f in a domain D of $\mathbb{R}^{d+1}$ with center $\alpha = (a, A)$, where we may suppose $a < x$, $\forall (x, Y) \in D$. By definition, Φ is determined by a central family $\Psi(\cdot; \Lambda)$ (with center α) where by Lemma 9.13, the field parameter $\Lambda \in C^1(D)$. Thus to each point $(x, Y) \in D$ is associated the stationary function $\Psi(\cdot; \Lambda(x, Y)) \in (C^1[a, x])^d$; then

$$S(x, Y) \stackrel{\text{def}}{=} \int_a^x f[\Psi(t; \Lambda(x, Y))] \, dt \tag{46}$$

is defined in D, and depends on (x, Y) only through $\Lambda(x, Y)$ and the upper limit of the integral. We shall prove that Φ is exact by demonstrating that in D, S has partials of the correct form (9). It is here that we shall use the differentiability of Ψ and Λ guaranteed by 9.12 and 9.13.

Recall that $f[Y(t)] \equiv f(t, Y(t), Y'(t))$ and introduce the abbreviation $\tilde{t} = (t, \Lambda(x, Y))$. Then by A.13 and the chain rule, using the prime to denote "t" differentiation, we get

$$S_Y(x, Y) = \int_a^x \{f_Y[\Psi(\tilde{t})]\Psi_\Lambda(\tilde{t}) + f_Z[\Psi(\tilde{t})](\Psi')_\Lambda(\tilde{t})\}\Lambda_Y(x, Y) \, dt, \tag{47}$$

where Ψ_Λ and Λ_Y are the matrices of partial derivatives with, respectively, columns and rows indexed by Λ, and $(\Psi')_\Lambda$ is the matrix of t derivatives corresponding to Ψ_Λ. Here S_Y, f_Y, and f_Z are *row* matrices.

Now, by assumption $(\Psi')_\Lambda = (\Psi_t)_\Lambda = (\Psi_\Lambda)_t = (\Psi_\Lambda)'$, and $\Psi(\tilde{t}) = \Psi(t; \Lambda(x, y))$ is stationary on $(a, x]$. Hence, $f_Y[\Psi(\tilde{t})] = (d/dt)f_Z[\Psi(\tilde{t})]$, and so (47) becomes

$$S_Y(x, Y) = (f_Z[\Psi(\tilde{t})]\Psi_\Lambda(\tilde{t}))\Big|_{t=a}^{t=x} \Lambda_Y(x, Y).$$

But

$$\Psi_\Lambda(\tilde{a}) = \mathbb{O}, \quad \text{since} \quad \Psi(a; \Lambda) = A, \forall \Lambda;$$

and if I_d denotes the $d \times d$ identity matrix,

$$\Psi_\Lambda(\tilde{x})\Lambda_Y(x, Y) = I_d, \text{ since } \Psi(\tilde{x}) = \Psi(x; \Lambda(x, Y)) = Y.$$

Finally, since

$$\Phi(x, Y) \stackrel{\text{def}}{=} \Psi'(x; \Lambda(x, Y)) = \Psi'(\tilde{x});$$

$$S_Y(x, Y) = f_Z[\Psi(\tilde{x})] = f_Z(x, Y, \Phi(x, Y)) \tag{48}$$

$$= P(x, Y), \quad \text{as in (9)}.$$

Similarly, by A.14, from (46):

$$S_x(x, Y) = f[\Psi(\tilde{x})] + \int_a^x \{f_Y[\Psi(\tilde{t})]\Psi_\Lambda(\tilde{t}) + f_Z[\Psi(\tilde{t})](\Psi')_\Lambda(\tilde{t})\}\Lambda_x(x, Y)\, dt$$

$$= f[\Psi(\tilde{x})] + f_Z[\Psi(\tilde{t})]\Psi_\Lambda(\tilde{t})\Big|_{t=a}^{t=x}\Lambda_x(x, Y).$$

Here Λ_x is a column matrix. Again $\Psi_\Lambda(\tilde{a}) = \mathbb{O}$, but now differentiating $\Psi(x; \Lambda(x, Y)) \equiv Y$ with respect to x shows that

$$\Psi_\Lambda(\tilde{x})\Lambda_x(x, Y) = -\Psi'(x; \Lambda(x, Y)) = -\Phi(x, Y).$$

Hence

$$S_x(x, Y) = f(x, Y, \Phi(x, Y)) - f_Z(x, Y, \Phi(x, Y)) \cdot \Phi(x, Y)$$

$$= h(x, Y), \quad \text{as in (9)}. \tag{49}$$

Equations (48) and (49) prove that Φ is exact by Definition 9.3. $\qquad\square$

The reader should verify each step of this derivation in the simple case $d = 1$, at least by formal operations.

Theorem 9.14 justifies the incorporation of the term "central field" in the statements of Theorems 9.7, 9.9, and 9.10, which should now be reexamined. Observe that both fields considered in Example 1 of §9.3 are exact, but only one, that obtained from the semicircles with fixed (left) end point α is central (Problem 9.3); moreover, it is the one with smaller domain. (Attempts to enlarge this domain by including say the semicircles with the *same* fixed right end point α result in a pair of *disjoint* quarter planes.)

In order to apply Hilbert's theorem (9.7) for central fields, we must be able to find or construct them. For a given f, and center $\alpha = (a, A) \in \mathbb{R}^{d+1}$, the collection of "all" stationary functions $\Psi \in (C^1[a, b])^d$ with $\Psi(a) = A$ (and *possibly variable b*) constitutes a potential central family. It may always be parametrized by $\Lambda = \Psi'(a)$, or, some other choice may be more natural. However, whatever choice is made, the resulting family, $\Psi = \Psi(x; \Lambda)$, with its derivatives, $\Psi_x = \Psi_x(x; \Lambda)$, must be C^1 in all variables. In explicit cases this technical requirement is usually met automatically.

Supposing that $\Psi(\cdot; \Lambda)$ is a *central family*, then each domain D covered consistently by its trajectories constitutes the domain of a *stationary field* Φ. However, in order to claim that it is a *central field*, we must establish that $\Lambda \in C^1(D)$, or by Lemma 9.13 that the matrix Ψ_Λ is invertible "in D."

Alternatively, we may simply examine the matrix Ψ_Λ for (x, Y) domains of invertibility. Each point (x_1, Y_1) of invertibility provides through Lemma 9.13, a *local* central field, but each domain of invertibility provides only a *possible* central field. Coverage of the domain is assured, but consistency must still be established (and need not be present).

An additional complication to applications is the fact that the center α must be located on the extension of a trajectory of interest but be outside the domain of a usable central field containing this trajectory. In our analysis of the brachistochrone in §9.4 we have presented one method of confronting this problem. However, the usual approach is to extend the given trajectory slightly, make its new end point $\tilde{\alpha}$, say, the center, and show that a resulting central family can still produce a central field in a domain which contains the original trajectory. We shall adopt this approach in the following application and in the embedding construction of the next section.

Smooth Minimal Surface of Revolution

Example 2. We know from our discussion in §7.5 that when the positive configuration parameters b, b_1 permit, there may be a smooth curve Γ_0 parametrized by $y_0 \in C^1[0, b]$ which minimizes the area of the surface obtained by revolving it around the x axis. y_0 is stationary for f on $[0, b]$, where $f(x, y, z) = y\sqrt{1 + z^2}$ is strictly convex on $[0, b] \times (0, \infty) \times \mathbb{R}$ from Example 3 of §3.3.

By equation (14') of §7.5, each positive function y, stationary for this f, must be of the form

$$y(x) = c \cosh(\bar{c}x + \mu), \tag{50}$$

for some choice of the constants $c > 0$, $\bar{c} \equiv c^{-1}$, and μ, and hence must graph as a catenary. We seek a central family $\psi(\cdot; \lambda)$ of such functions with center $\alpha = (a, a_1)$, say, with $a < 0$, whose graphs are consistent on $[0, b]$. On geometrical grounds, it is evident that this is possible only in domains D of $\mathbb{R}^2$, such as that shown in Figure 9.10. D is unbounded above, but it is surely bounded below by that catenary through α which "first" touches the x axis.

A central family through α in the form (50) is given by

$$\psi(x; \lambda) = c \cosh(\bar{c}(x - a) + \lambda), \quad \text{for } a_1 = c \cosh \lambda, \tag{51}$$

since ψ as defined, is C^2. Observe that $\psi'(a; \lambda) = \sinh \lambda$.

Using the first of the addition laws

$$\cosh(t + \tau) = \cosh t \cosh \tau + \sinh t \sinh \tau,$$
$$\sinh(t + \tau) = \sinh t \cosh \tau + \cosh t \sinh \tau, \tag{52}$$

equation (51) may be rewritten as follows:

$$\psi(x; \lambda) = a_1(\cosh u + \sinh u \tanh \lambda), \tag{53}$$

where $u = \bar{a}_1(x - a) \cosh \lambda$, and $\bar{a}_1 \equiv (a_1)^{-1}$.

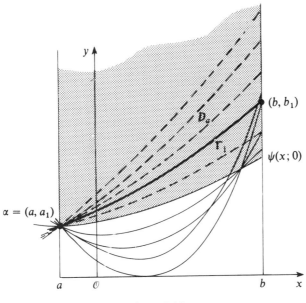

Figure 9.10

For fixed $x > a$, u, and hence, $\psi(x; \lambda)$, increases *strictly* to $+\infty$ as λ traverses the *positive* real axis. It follows that for each such x, $\psi(x; \lambda)$ assumes each value $\geq \psi(x; 0) = a_1 \cosh \bar{a}_1(x - a)$, precisely *once* for some choice of $\lambda \geq 0$. Thus for $\lambda \geq 0$, $\psi(\cdot; \lambda)$ provides *a central family* of catenaries which remain consistent in the domain

$$D_a = \{(x, y) \in \mathbb{R}^2 : x > a, y > a_1 \cosh \bar{a}_1(x - a)\}.$$

To show that $\psi(\cdot; \lambda)$ determines a *central field* in D_a, it only remains to establish that the *scalar* ψ_λ is invertible in D_a, i.e., that it is nonvanishing. But, from (53) and the chain rule, we have

$$\psi_\lambda(x; \lambda) = a_1\{u \tanh \lambda(\sinh u + \cosh u \tanh \lambda) + \sinh u(\cosh \lambda)^{-2}\};$$

after some manipulation involving (52), we get

$$\psi_\lambda(x; \lambda) = a_1(\cosh \lambda)^{-2}[u \sinh \lambda \sinh(u + \lambda) + \sinh u], \tag{54}$$

which for $\lambda \geq 0$, *is positive* with u; so, $\forall x > a$.

Now fix $\lambda_1 \geq 0$ and set $y_1 = \psi(\cdot; \lambda_1)$. Using Hilbert's theorem (9.7), we conclude with the

(9.15) **Proposition.** *The catenary Γ_1 represented by y_1 provides (uniquely) the minimum surface of revolution among those admissible piecewise smooth curves joining its end points in D_a.* □

When $\lambda_1 > 0$, y_1 provides a *strong* local minimum for the surface area function, in the sense of Chapter 7. (Why?)

It is considerably more delicate to introduce trajectories for $\lambda < 0$ in order to enlarge D_a since they may intersect each other, as well as those already present. However, although, as in Problem 7.15, some configurations may permit another catenary $\tilde{\Gamma}_1$ to join these same end points, it is clear from the above that it cannot provide a lesser surface area if it is contained in D_a, or in any legitimate enlargement of D_a. For a complete discussion, see [B1]. (A similar problem is analyzed in Example 1 of §9.9.)

Note that we still have not obtained a result for the given catenary Γ_0 unless $y_0 = \psi(\cdot; \lambda_0)$ for some λ_0. To ensure this, we must enlarge the interval of definition of y_0, and choose $a_1 = y_0(a)$ for some $a < 0$. However, to construct the family with this center as above requires that $y_0'(a) = \sinh \lambda_0 \geq 0$. Geometrically, it is seen that this is assured if and only if $y_0'(0) > 0$. In the next section we shall construct such families for a general f.

(Problems 9.11–9.12)

§9.7. Construction of Central Fields with Given Trajectory: The Jacobi Condition

As we have seen in this chapter, in order to obtain a complete solution for a minimization problem involving a convex integrand $f(\underline{x}, \underline{Y}, \underline{Z})$ with given end point conditions we should first obtain a stationary trajectory Γ_0 which meets these conditions and then show that Γ_0 is a trajectory of a central field Φ for f (or of some other exact field for f). In the previous examples this was accomplished usually by appeal to geometric arguments which enabled us to "see" the central family producing the field. However, it is clearly of importance to characterize if possible, those f and Γ_0 which afford such fields, at least in a neighborhood of Γ_0. (The minimal surface example of §9.6 shows that even on geometrical grounds this may be the best achievable.) In the present section we shall obtain sufficient conditions to guarantee this local embedding of Γ_0 under further assumptions on f.

First, we utilize results from the theory of systems of ordinary differential equations to construct a central family $\Psi(\cdot; \Lambda)$ of stationary functions containing a representative Y_0 for Γ_0. This, in principle, is always possible. Second, we examine the matrix Ψ_Λ and show that its invertibility along Γ_0 guarantees the field construction. This invertibility is not always present; however, as we shall prove in §9.9, it is almost essential if Y_0 is to provide even a weak local minimum.

We operate with a *fixed* $Y_0 \in (C^1[a, b])^d$ which is *stationary* for a given

$f = f(x, Y, Z)$ assumed to be C^3 at least in a neighborhood of

$$\mathscr{C}_0 = \{(x, Y_0(x), Y_0'(x)), x \in [a, b]\} \subseteq \mathbb{R}^{2d+1};$$

We further suppose that f_{ZZ}, the matrix of second partial z_j derivatives, is invertible along $\mathscr{C}_0$, and so by continuity, in this neighborhood. Then by Theorem 7.14, Y_0 is C^2, and so it satisfies the Euler–Lagrange equations of §6.7 for f in the condensed form

$$Y'(x) = Z(x),$$

$$\frac{d}{dx} f_Z(x) = f_{xZ}(x) + f_{ZY}(x)Z(x) + f_{ZZ}(x)Z'(x) = f_Y(x), \tag{55}$$

where f_{ZY} is the matrix with elements $f_{z_i y_j}$, $i, j = 1, 2, \ldots, d$ having its *rows* indexed by i, while the remaining matrices are columns.

Using the hypothesized invertibility of f_{ZZ}, we conclude that $(Y_0, Z_0 = Y_0')$ satisfies the system

$$Y'(x) = Z(x),$$

$$Z'(x) = G(x, Y(x), Z(x)), \tag{56}$$

where

$$G(x, Y, Z) = f_{ZZ}^{-1}[f_Y - f_{xZ} - f_{ZY}Z], \tag{57}$$

with the partial derivatives of f evaluated at (x, Y, Z).

Now, (56) is a first-order system for the vector *function* $T \equiv (Y, Z) \in (C^1[a, b])^{2d}$. Moreover having f in C^3 ensures that G is C^1 in the neighborhood of invertibility of f_{ZZ}. Upon setting $A = Y_0(a)$ and $\Lambda_0 = Y_0'(a) = Z_0(a)$, we can use the theory of such systems as presented in §A.5, to assert the following:

(9.16) **Proposition.** *Under the above conditions, Y_0 is contained in a central family $\Psi(\cdot; \Lambda)$ for f, defined on $[a, b]$. Moreover, $\forall \Lambda: \Psi(a; \Lambda) = A = Y_0(a)$ and $\Psi'(a; \Lambda) = \Lambda$.*

PROOF. The theory provides a $\delta > 0$ so small that for each $\Lambda \in \mathbb{R}^d$ with $|\Lambda - \Lambda_0| < \delta$, there exists a unique solution (Y, Z) of the system (56) on $[a, b]$ which meets the initial conditions $Y(a) = A$, $Z(a) = \Lambda$; set this $Y = \Psi(\cdot; \Lambda)$. The differentiability of Ψ and Ψ_x required by Definition 9.12, also follow from the general theory, since $\Psi_x(\cdot; \Lambda) = Y' = Z$.   □

Of course, the trajectories of this family need not cover a domain consistently. They obviously intersect at $\alpha = (a, A)$, and we are indexing the family by their slopes at α. The situation is illustrated in Figure 9.11 where it is seen that the trajectories appear to remain distinct, at least for *small* $x - a$.

In view of Lemma 9.13, the determinant of the matrix Ψ_Λ plays a crucial role, and it suffices to consider it when $\Lambda = \Lambda_0$. Upon differentiating (56) with respect to Λ and using the chain rule, we obtain for the matrices, Ψ_Λ and Z_Λ,

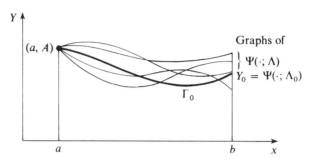

Figure 9.11

evaluated at Λ_0, the system

$$\Psi'_\Lambda(x) = Z_\Lambda(x),$$
$$Z'_\Lambda(x) = G_Y(x)\Psi_\Lambda(x) + G_Z(x)Z_\Lambda(x),$$
(58)

where the partials of G are evaluated *along* $\mathscr{C}_0$ (Why?) and so are known *a priori*. In addition, we have that

$$\Psi_\Lambda(a) = \mathcal{O}, \quad \text{since } \Psi(a; \Lambda) = A, \quad \forall \Lambda;$$

and
(59)

$$Z_\Lambda(a) = I_d, \quad \text{since } Z(a; \Lambda) = \Psi'(a; \Lambda) = \Lambda, \quad \forall \Lambda.$$

Observe that (59) may be written as the *linear* system

$$T'(x) = \begin{bmatrix} \mathcal{O} & I_d \\ G_Y(x) & G_Z(x) \end{bmatrix} T(x) = \mathbb{G}(x)T(x),$$
(60)

say, where $T \equiv (\Psi_\Lambda, Z_\Lambda) \in \mathbb{R}^N$, with $N = 2d^2$. I_d is the identity matrix of order d.

(9.17) **Proposition.** *If for $\varepsilon > 0$ the coefficient functions in the matrix $\mathbb{G}$ are continuous on an interval $I_\varepsilon = [a - \varepsilon, b + \varepsilon]$, then on this interval, each solution T of the system (60) admits an estimate*

$$|T(x)| \le e^{M|x-a|}|T(a)|, \quad \text{for a constant } M = M(\varepsilon) > 0.$$
(61)

PROOF. Let $N = 2d^2$. Each coefficient function g_{ij} in the matrix $\mathbb{G}$ is bounded on I_ε by M/N^2, say. (Proposition 5.3). The dot product of (60) by $T(x) = (t_1(x), t_2(x), \dots, t_N(x))$ is

$$\frac{1}{2}\frac{d}{dx}|T(x)|^2 = T(x) \cdot T'(x) = \sum_{i,j=1}^N g_{ij}(x)t_i(x)t_j(x)$$

$$\le M|T(x)|^2,$$

since

$$|t_j(x)| \le |T(x)|, \quad j = 1, 2, \dots, N.$$

Thus for $x > a$:

$$\frac{d}{dx}(e^{-2Mx}|T(x)|^2) \leq 0,$$

so that

$$|T(x)|^2 \leq e^{2M(x-a)}|T(a)|^2,$$

or

$$|T(x)| \leq e^{M(x-a)}|T(a)|.$$

A similar argument gives the desired inequality (61), when $x < a$. $\square$

Recall that our coefficient functions f, and hence, G, together with its partials, G_Y and G_Z, are defined in a neighborhood of $\mathscr{C}_0$. Thus we may suppose that Y_0, and, by Proposition 9.16, that $\Psi(\cdot; \Lambda)$, together with Ψ_Λ and Z_Λ, are defined on $[a - \varepsilon, b + \varepsilon]$ for *some* $\varepsilon > 0$. Then in view of (59), Proposition 9.17 proves that each solution to (58) on this interval of length $l = b - a + 2\varepsilon$ has components bounded by e^{Ml}, for $M = M(\varepsilon)$. (Why?) Furthermore, each component $b_{ij}(x)$ of the matrix

$$B(x) \stackrel{\text{def}}{=} G_Y(x)\Psi_\Lambda(x) + G_Z(x)Z_\Lambda(x) \tag{62}$$

is also bounded by *some* $M = M(\varepsilon)$. Integrating (58) and incorporating the initial conditions (59) gives

$$\Psi_\Lambda(x) = \Psi_\Lambda(a) + \int_a^x Z_\Lambda(t)\, dt,$$

or

$$\Psi_\Lambda(x) = (x - a)I_d + \int_a^x dt \int_a^t B(\tau)\, d\tau, \tag{63}$$

where B is given by (62).

(9.18) **Lemma.** $\Delta(x, a) = \det[\Psi_\Lambda(x)] \neq 0$, *for* $0 < |x - a| < 2/Md$, *where* $M = M(\varepsilon)$.

PROOF. If $\Delta(x, a) \stackrel{\text{def}}{=} 0$ for some $x \neq a$, then $\Psi_\Lambda(x)$ is *not* invertible and there exists a $U = (u_1, u_2, \ldots, u_d) \in \mathbb{R}^d$ with $\|U\| = \max_{j=1,\ldots,a}|u_j| = 1$, for which $\Psi_\Lambda(x)U = \mathcal{O}$. From (63), we have

$$(x - a)U = -\int_a^x dt \int_a^t B(\tau)U\, d\tau,$$

so that for $x > a$:

$$|x - a| = |x - a|\,\|U\| \leq \int_a^x dx \int_a^t \|B(\tau)U\|\, d\tau \leq Md\frac{(x-a)^2}{2};$$

i.e., $(x - a) \geq 2/Md$, and again the proof for $x < a$ is similar. $\square$

(9.19) **Remark.** Since the constants $M = M(\varepsilon)$ used in the preceding arguments apply to each solution of (58) on the interval $[a - \varepsilon, b + \varepsilon]$, it follows that the solution Y_0 can be extended to a *larger* interval $[\tilde{a}, b]$, where $\tilde{a} < a$, such that a corresponding central family with center $\tilde{\alpha} = (\tilde{a}, Y_0(\tilde{a}))$ will have $\Delta(a, \tilde{a}) \neq 0$. Hence, by Lemma 9.18, that portion of Γ_0 near $\alpha = (a, Y_0(a))$ *is a trajectory of a central field for f.*

To discuss a total embedding of Γ_0, we shall require one additional fact about Δ.

(9.20) **Proposition.** $\Delta(x, \tilde{a})$ *is (jointly) continuous on* $[a - \varepsilon, b + \varepsilon]^2$.

PROOF. For each fixed $\tilde{a}$, $\Delta(x, \tilde{a})$ is, by definition, the determinant of the solution $\tilde{\Psi}_\Lambda$ of the linear system (58) with the boundary conditions $\Psi_\Lambda(\tilde{a}) = \mathcal{O}$, $\Psi'_\Lambda(\tilde{a}) = I_d$. Clearly, each component of $\tilde{\Psi}_\Lambda$ and, hence, its determinant, (which is just a sum of products of these components) is continuous in x. Moreover, by analogy with (63):

$$\tilde{\Psi}_\Lambda(x) - \tilde{\Psi}_\Lambda(x_0) = (x - x_0)I_d + \int_{x_0}^x dt \int_{\tilde{a}}^t \tilde{B}(\tau)\, d\tau,$$

so that

$$\|\tilde{\Psi}_\Lambda(x) - \tilde{\Psi}_\Lambda(x_0)\| \leq |x - x_0| + M|x - x_0|(b - a + 2\varepsilon)$$

$$\leq M_1|x - x_0|, \quad \text{for } M_1 = M_1(\varepsilon).$$

(In estimating the inner integral, we observe that $|\tilde{a} - t| \leq b - a + 2\varepsilon$.)

To obtain the continuity in $\tilde{a}$ for fixed $x = x_0$, it is simpler to return to the general linear system. (See Problem 9.21.) However, once this is established, the joint continuity of each component of Ψ_Λ (and hence of Δ) follows from the triangle inequality:

$$\|\Psi_\Lambda(x, \tilde{a}) - \Psi_\Lambda(x_0, \tilde{a}_0)\|$$

$$\leq \|\Psi_\Lambda(x, \tilde{a}) - \Psi_\Lambda(x_0, \tilde{a})\| + \|\Psi_\Lambda(x_0, \tilde{a}) - \Psi_\Lambda(x_0, \tilde{a}_0)\|$$

$$\leq M_1|x - x_0| + \|\Psi_\Lambda(x_0, \tilde{a}) - \Psi_\Lambda(x_0, \tilde{a}_0)\|,$$

since the right side may be made as small as we wish, when

$$|x - x_0| + |\tilde{a} - \tilde{a}_0| \quad \text{is sufficiently small.} \qquad \square$$

(9.21) **Proposition.** *If* $\Delta(x, a) \neq 0$ *for* $x \in (a, b]$, *then* Γ_0 *is a trajectory of a central field for f.*

PROOF. It follows from the hypothesis that $\Delta(x, \tilde{a}) \neq 0$ on $[a, b]$ for some $\tilde{a} < a$. Suppose to the contrary, that for a sequence $\tilde{a}_n \nearrow a$, as $n \to \infty$, we had $\Delta(x_n, \tilde{a}_n) = 0$ for some $x_n \in [a, b]$, $n = 1, 2, \ldots$. Then by Lemma 9.18, $|x_n - \tilde{a}_n| \geq 2/Md$, so that $|x_n - a| \geq 1/Md$, if n is large enough, i.e., $x_n \in I_0 \overset{\text{def}}{=} [a + (Md)^{-1}, b]$. But this interval I_0 is compact and as established in §A.0, there is a *subsequence* $x_{n_k} \to x_0 \in I_0$, as $k \to \infty$. Finally, by the joint continuity of Δ, we have $\Delta(x_0, a) = \lim_{k \to \infty} \Delta(x_{n_k}, \tilde{a}_{n_k}) = 0$, which contradicts the hypothesis, since $x_0 > a$.

Then by Lemma 9.13, the trajectories of a central family $\Psi(\cdot\,; \Lambda)$ with Λ near Λ_0 and center at $\tilde{\alpha} = (\tilde{a}, Y_0(\tilde{a}))$, cover *consistently* a neighborhood of each point on the trajectory Γ_0 of $Y_0 = \Psi(\cdot\,; \Lambda_0)$. Ψ_Λ is invertible in each of these neighborhoods and their union constitutes a domain D of a central field; obviously $\Gamma_0 \in D$. $\qquad\qquad\qquad\qquad\qquad\qquad\qquad\qquad\qquad\qquad\square$

(9.22) **Definition.** When $\Delta(x, a) \neq 0$, $\forall\, x \in (a, b]$, then Γ_0 is said to satisfy the *Jacobi condition* for f.

By Lemma 9.18, it is always satisfied if $b - a$ is sufficiently small. Unfortunately, this condition is *not* always satisfied on larger intervals even when f is simple.

For example, $f(y, z) = z^2 - y^2$ has stationary functions $\psi(x; \lambda) = \lambda \sin x$ on $[0, 3\pi/2]$, satisfying $\psi(0; \lambda) = 0$ and $\psi'(0; \lambda) = \lambda$. But $\Delta(x, 0) = \psi_\lambda(x; \lambda_0) = \sin x$ and $\Delta(\pi, 0) = 0$.

On the other hand, $\Delta(x, 0) \neq 0$ on $(0, b]$ for each $b < \pi$. Hence, for each λ_0, the associated trajectory Γ_0 *is* a trajectory of a central field, but the extent of the field will depend on the particular λ_0.

In §9.9, we will see that the nonvanishing of $\Delta(x, a)$ on (a, b) is in fact *necessary*, if Y_0 is to provide a weak local minimum.

§9.8. Sufficient Conditions for a Local Minimum

We now have the ingredients required to provide sufficient conditions for local minimization. As in §9.7, we assume that Y_0 in $(C^1[a, b])^d$ is stationary for a given $f = f(x, Y, Z)$ which is C^3 in a neighborhood of

$$\mathscr{C}_0 = \{(x, Y_0(x), Y_0'(x)) \in \mathbb{R}^{2d+1} : x \in [a, b]\}.$$

We know from Proposition 7.19 and subsequent remarks, that if Y_0 provides a local minimum, then necessarily

$$\sum_{i,j=1}^{d} f_{z_i z_j}(x, Y_0(x), Y_0'(x))v_i v_j \geq 0, \qquad \forall\, V \in \mathbb{R}^d, \quad x \in [a, b].$$

Conversely, if the matrix f_{ZZ} of these second partials is positive definite (i.e., equality holds iff $V = \mathcal{O}$) at a *single point* on $\mathscr{C}_0$, then, by continuity, it remains so in a spherical neighborhood of this point. Thus as in §0.13, it is invertible, and by Proposition 9.2, $f(\underline{x}, \underline{Y}, Z)$ is *strictly* convex in this neighborhood. Although this will suffice to establish that in the neighborhood Y_0 is a *weak local minimum*, i.e., a local minimum with respect to the weak norm $\|Y\| = \max(|Y(x)| + |Y'(x)|)$, we must permit $\hat{Y}'$ with arbitrary values in $\mathbb{R}^d$ in order that Y_0 could minimize with respect to the strong norm, $|\hat{Y}| = \max|\hat{Y}(x)|$.

It is convenient to treat first the results which can be obtained from conditions specified at a point. The analogous results for the full trajectory require the Jacobi condition.

(a) Pointwise Results

(9.23) **Theorem.** *If Y_0 is stationary for a C^3 function $f = f(x, Y, Z)$ and f_{ZZ} is positive definite at $\gamma = (a, Y_0(a), Y_0'(a))$, then for sufficiently small $b - a$, Y_0 provides a unique weak local minimum for*

$$F(\hat{Y}) = \int_a^b f(x, \hat{Y}(x), \hat{Y}'(x)) \, dx$$

on

$$\hat{\mathcal{D}} = \{\hat{Y} \in (\hat{C}^1[a, b])^d : \hat{Y}(a) = Y_0(a), \hat{Y}(b) = Y_0(b)\}.$$

If, in addition, $f(\underline{x}, \underline{Y}, Z)$ is [strictly] convex on $D \times \mathbb{R}^d$, where D is a neighborhood of $\alpha = (a, Y_0(a))$, then Y_0 provides a [unique] strong local minimum.

PROOF. As noted above, $f(\underline{x}, \underline{Y}, Z)$ is strictly convex in a neighborhood of γ (which we may suppose to be Z-convex). We may also assume that f_{ZZ} is invertible in this neighborhood and, hence, along $\mathscr{C}_0$, for $b - a$ sufficiently small, so that by 9.18 and 9.21, $\alpha = (a, Y_0(a))$ is contained in the domain D_0 of a useable central field. Hilbert's comparison, $F(\hat{Y}) > F(Y_0)$, is possible for all $\hat{Y}$ in $\hat{\mathcal{D}}$, $(\hat{Y} \neq Y_0)$, with graphs $\hat{\Gamma}$ in D_0, for which $(x, \hat{Y}(x), \hat{Y}'(x))$ is in the given neighborhood of γ, $\forall x \in [a, b]$. (At the corner points, we require this for *both* one-sided derivatives.) But this defines a weak neighborhood of Y_0 in $\hat{\mathcal{D}}$.

If, in addition, $f(\underline{x}, \underline{Y}, Z)$ is [strictly] convex in $D_0 \times \mathbb{R}^d$, then we may remove these restrictions on $\hat{Y}'$, and apply Hilbert's theorem to all $\hat{Y}$ in $\hat{\mathcal{D}}$ which are in a *strong* neighborhood of Y_0. $\qquad\square$

Observe that the positive definiteness of f_{ZZ} at a point ensures that on a *small* interval determined by the point, Y_0 is the unique weak local minimizing function. This result was known to Legendre (1786) who believed erroneously that it would extend to the full trajectory. By contrast, the similar claim for Y_0 as a *strong* local minimum is not guaranteed unless $f(\underline{x}, \underline{Y}, Z)$ is strictly convex on some $D \times \mathbb{R}^d$, and this *cannot* be inferred from a pointwise condition. However, Proposition 9.2 should be recalled. (Problem 9.23.)

Application: Hamilton's Principle

When the Lagrangian expressed in generalized coordinates is given by

$$L(t, Q, \dot{Q}) = \frac{1}{2} \sum_{i,j=1}^n a_{ij}(t, Q)\dot{q}_i\dot{q}_j - U(t, Q) \quad \text{where } a_{ij} = a_{ji}, i, j = 1, 2, \ldots, n.$$

then $[L_{ZZ}] = [L_{\dot{Q}\dot{Q}}] = [a_{ij}]$ is always positive definite. (See §8.4.) Thus $L(t, Q, \dot{Q})$ is strictly convex on $D \times \mathbb{R}^n$ for some domain D of $\mathbb{R}^{n+1}$, by 9.2.

Hence, supposing that L is C^3, it follows that each point on a stationary trajectory is contained in the domain of a useable central field. By Theorem

9.23, each stationary trajectory provides a strong local minimum for the action integral

$$A(\hat{Q}) \equiv \int_a^b L(t, \hat{Q}(t), \dot{\hat{Q}}(t))\, dt.$$

when b − a is sufficiently small, among nearby trajectories with the same end points at *a, b*.

Thus Hamilton's principle of stationary action becomes in this case a principle of (strong) *local* minimal action, even though the total action might be *maximized* by this same stationary function.

(b) Trajectory Results

An extension of Theorem 9.23 which provides corresponding information about the entire trajectory Γ_0, for a given stationary function $Y_0 \in (C^1[a, b])^d$, is more difficult to obtain, and must be nonexistent in some cases. Even those stationary functions which minimize in neighborhoods of each point may not do so on an interval, and so we cannot hope to make their trajectories those of central fields. We must invoke the Jacobi condition of the last section (which depends only on information along $\mathscr{C}_0$). Remarkably, this is the only additional requirement.

(9.24) **Theorem.** *If* $Y_0 \in (C^1[a, b])^d$ *is stationary for a* C^3 *function* $f = f(x, Y, Z)$ *and*

(i) f_{ZZ} *is positive definite along* $\mathscr{C}_0$;
(ii) $\Delta(x, a) \neq 0$, $x \in (a, b]$;

then Y_0 *provides the unique weak local minimum for*

$$F(\hat{Y}) = \int_a^b f(x, \hat{Y}(x), \hat{Y}'(x))\, dx$$

on

$$\hat{\mathscr{D}} = \{\hat{Y} \in (\hat{C}^1[a, b])^d \colon \hat{Y}(a) = Y_0(a),\ \hat{Y}(b) = Y_0(b)\}.$$

If, in addition, $f(\underline{x}, \underline{Y}, Z)$ *is [strictly] convex in* $D \times \mathbb{R}^d$, *where D is a domain containing* Γ_0 *(the graph of* Y_0), *then* Y_0 *provides a [unique] strong local minimum.*

PROOF. (i) implies by Proposition 9.2 that $f(\underline{x}, \underline{Y}, Z)$ is strictly convex *in a spherical* neighborhood of each point on $\mathscr{C}_0$. The union, $\mathscr{U}_0$, of these spheres will be a Z-convex domain containing $\mathscr{C}_0$ in which $f(\underline{x}, \underline{Y}, Z)$ is strictly convex by Proposition 9.2. Moreover, the projection of $\mathscr{U}_0$ on the (x, Y) space is a domain containing Γ_0 as illustrated in Figure 9.12.

Then, (i) and (ii) together assure, by Proposition 9.21, that Γ_0 is a trajectory of a central field (in perhaps a *smaller* domain D_0). Again we may

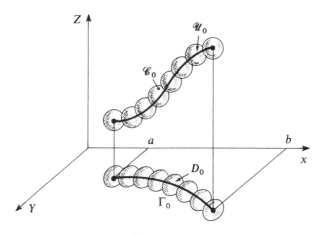

Figure 9.12

apply Hilbert's comparison $F(\hat{Y}) > F(Y_0)$ to all $\hat{Y} \in \hat{\mathcal{D}} \sim \{Y_0\}$ which are in a weak neighborhood of Y_0. If, in addition, $f(\underline{x}, \underline{Y}, Z)$ is [strictly] convex in $D_0 \times \mathbb{R}^d$, then Hilbert's theorem (9.7) applies to all $\hat{Y} \in \hat{\mathcal{D}}$ which are in a strong neighborhood of Y_0. That such a neighborhood exists requires an appeal to compactness.   □

We defer application of this result until the end of the next section.

§9.9*. Necessity of the Jacobi Condition

In the previous section, the Jacobi condition, $\Delta(x, a) \neq 0$ on $(a, b]$, was used to embed a given stationary trajectory Γ_0 in the domain of a central field on which Hilbert's theorem could act to show that Y_0 provides a local minimum. If we give up the nonvanishing at $x = b$, the reduced condition is necessary for Y_0 to be a weak local minimum (or a weak local maximum).

(9.25) Theorem (Jacobi). *Let Y_0 be a weak local extremum for*

$$F(Y) = \int_a^b f[Y(x)]\, dx$$

on

$$\mathcal{D} = \{Y \in (C^1[a, b])^d : Y(a) = A, Y(b) = B\}$$

and suppose that f is C^3 in a neighborhood of $\mathcal{C}_0$, while f_{ZZ} is invertible along $\mathcal{C}_0$. Then $\Delta(x, a) \neq 0$ on (a, b).

PROOF. (We may assume that Y_0 is a weak local minimum.) Suppose to the contrary, that $\Delta(a^*, a) = 0$, for *some* point $a^* \in (a, b)$. Then the matrix $\Psi_\Lambda(a^*; \Lambda_0)$ is *not* invertible, and it follows that $\Psi_\Lambda(a^*; \Lambda_0)U = \mathcal{O}$ for *some* (column) vector $U \in \mathbb{R}^d$ of unit length.

Set
$$V(x) = \Psi_\Lambda(x; \Lambda_0)U, \quad \text{so that } V(a) = V(a^*) = \mathcal{O},$$

and observe that since Ψ_x is C^1, then
$$V'(x) = \Psi_\Lambda'(x; \Lambda_0)U = \Psi_\Lambda'(x; \Lambda_0)U.$$

Now, it is convenient to return to the original Euler–Lagrange equations for the stationary family $\Psi(x; \Lambda)$ as in §6.7; viz.;

$$\frac{d}{dx} f_Z[\Psi(x; \Lambda)] = f_Y[\Psi(x; \Lambda)]. \tag{64}$$

Upon differentiating (64) with respect to Λ and subsequently evaluating at Λ_0, we obtain

$$\frac{d}{dx} [f_{ZY}(x)\Psi_\Lambda(x) + f_{ZZ}(x)\Psi_\Lambda'(x)] = f_{YY}(x)\Psi_\Lambda(x) + f_{YZ}(x)\Psi_\Lambda'(x), \tag{64'}$$

where the double partials of f are matrices with *columns* indexed by the *second* subscript, and rows by the first, evaluated on $\mathscr{C}_0$. Post-multiplying this equation by the *constant* column vector U and incorporating the properties of V as above, we obtain

$$\frac{d}{dx} [f_{ZY}(x)V(x) + f_{ZZ}(x)V'(x)] = f_{YY}(x)V(x) + f_{YZ}(x)V'(x). \tag{65}$$

It may be verified that (65) is the first Euler–Lagrange equation for the quadratic function

$$q(x, V, W) \overset{\text{def}}{=} \overline{V}f_{YY}(x)V + 2\overline{V}f_{YZ}(x)W + \overline{W}f_{ZZ}(x)W, \tag{66}$$

where V, W are (column) vectors in $\mathbb{R}^d$ and $\overline{V}$, $\overline{W}$ are the corresponding row vectors. (See Problem 9.22.) However, for our purposes, it suffices to observe that since $\overline{V}f_{YZ}(x)W = \overline{W}f_{ZY}(x)V$,

$$\int_a^{a^*} q(x, V(x), V'(x))\, dx = \int_a^{a^*} \{\overline{V}(x)[f_{YY}(x)V(x) + f_{YZ}(x)V'(x)]$$
$$+ \overline{V}'(x)[f_{ZY}(x)V(x) + f_{ZZ}(x)V'(x)]\}\, dx = 0, \tag{66'}$$

if we utilize (65) together with the fact that $\overline{V}(a) = \overline{V}(a^*) = \mathcal{O}$.

Next, for $\hat{V} \in \hat{\mathscr{D}}_0 = \{\hat{V} \in (\hat{C}^1[a, b])^d \colon \hat{V}(a) = \hat{V}(b) = \mathcal{O}\}$, let
$$h(\varepsilon) = F(Y_0 + \varepsilon\hat{V}), \quad \text{for small } \varepsilon.$$

Then by Taylor's formula with remainder, for fixed $\varepsilon > 0$,

$$F(Y_0 + \varepsilon\hat{V}) - F(Y_0) = h(\varepsilon) - h(0) = h'(0)\varepsilon + h''(\tilde{\varepsilon})\frac{\varepsilon^2}{2},$$

where $0 < \tilde{\varepsilon} < \varepsilon$. But $h'(0) = \delta F(Y_0; \hat{V}) = 0$, since as hypothesized, Y_0 minimizes F locally on $\mathscr{D}$ (weakly) and hence by Theorem 7.7, on

$$\hat{\mathscr{D}} = \{\hat{Y} \in (\hat{C}^1[a, b])^d: \hat{Y}(a) = A, \hat{Y}(b) = B\}.$$

Now, $h''(0) = Q(\hat{V}) \stackrel{\text{def}}{=} \int_a^b q(x, \hat{V}(x), \hat{V}'(x))\, dx$ (Problem 9.22). Were $h''(0) < 0$, then by continuity of h'', we would have $h''(\tilde{\varepsilon}) < 0$ for sufficiently small $\varepsilon > 0$, implying that $F(Y_0 + \varepsilon\hat{V}) < F(Y_0)$ contradicting the assumed minimality. It follows that $h''(0) \geq 0$, or equivalently, that $Q(\hat{V}) \geq 0$, $\forall\, \hat{V} \in \hat{\mathscr{D}}_0$. However,

$$\hat{V}_0(x) \stackrel{\text{def}}{=} \begin{cases} V(x), & x \leq a^*, \\ \mathscr{O}, & x \geq a^*, \end{cases} \quad \text{is in } \hat{\mathscr{D}}_0, \text{ since } V(a^*) = \mathscr{O}, \tag{67}$$

and $Q(\hat{V}_0) = \int_a^{a^*} q(x, V(x), V'(x))\, dx = 0$, from (66'). Hence $\hat{V}_0$ minimizes Q on $\hat{\mathscr{D}}_0$ and so, at the (possible) corner point a^*, it must satisfy the Weierstrass–Erdmann condition (7.13), which from (65) and (67) is

$$\lim_{x \nearrow a^*} [f_{ZY}(x)V(x) + f_{ZZ}(x)V'(x)] = \lim_{x \searrow a^*} [f_{ZY}(x)\mathscr{O} + f_{ZZ}(x)\mathscr{O}] = \mathscr{O},$$

or, since $V(a^*) = \mathscr{O}$, we conclude that $f_{ZZ}(a^*)V'(a^*) = \mathscr{O}$.

But $f_{ZZ}(a^*)$ is invertible by hypothesis so that $V'(a^*) = V(a^*) = \mathscr{O}$. Since V is the unique solution of the homogeneous second-order *linear* equation (65) on $[a, a^*]$, satisfying these conditions, it follows that when $x \in [a, a^*]$, $\Psi_\Lambda(x)U = V(x) = \mathscr{O}$. Thus, for x near a, $\Psi_\Lambda(x)$ is *not* invertible and so $\Delta(x, a) = 0$, contradicting Lemma 9.18. (See A.17.) $\square$

(9.26) **Remark.** Observe that the Weierstrass–Erdmann argument *cannot* be invoked if $a^* = b$. The case considered in §9.1 of $f(y, z) = z^2 - y^2$ where $\Delta(x, 0) = \sin x$ while $F(y) = \int_0^\pi [y'(x)^2 - y(x)^2]\, dx \geq 0$ shows that the vanishing of $\Delta(x, a)$ at $b(=\pi)$ must be permitted.

(9.27) **Remark.** (65) is called *Jacobi's equation*. Its linearity (with f_{ZZ} invertible on $\mathscr{C}_0$) assures that for each $j = 1, 2, \ldots, d$, it has a *unique* solution V_j on $[a, b]$, with $V_j(a) = \mathscr{O}$ and $V_j'(a) = E_j$, where E_j is the unit vector of $\mathbb{R}^d$ in the jth coordinate direction. The matrix having the $V_j(x)$ as columns is precisely $\Psi_\Lambda(x; \Lambda_0)$ and so $\Delta(x, a) = \det[V_1(x)\,\vdots\,V_2(x)\,\vdots\cdots\vdots\,V_d(x)]$. The Jacobi condition may be restated in terms of the linear independence of these V_j.

Example 1. $f(y, z) = [y(1 + z^2)]^{1/2}$ is strictly convex on $(0, \infty) \times \mathbb{R}$. (Example 4, §3.3.) The stationary functions for f on say $[0, 1]$ are C^2 by 7.14 and satisfy the *second* equation (of §6.3) $f(x) - y'(x)f_{y'}(x) = \text{const.}$, or after simplification, $(y/1 + y'(x)^2)^{1/2} = 1/\sqrt{c}$, say, so that

$$1 + y'(x)^2 = cy(x). \tag{68}$$

With the *geometric* substitution $y'(x) = \tan \theta(x)$, we obtain

$$cy(x) = \sec^2 \theta(x),$$

which, when differentiated becomes

$$cy'(x) = 2 \sec^2 \theta(x) \tan \theta(x)\theta'(x),$$

or (when $y'(x) = \tan \theta(x) \neq 0$),

$$\frac{d}{dx} \tan \theta(x) = \sec^2 \theta(x)\theta'(x) = \frac{c}{2} = c_1, \text{ say.}$$

Thus $y'(x) = \tan \theta(x) = c_1(x - c_0)$, for a constant c_0, or finally, using (68) once again, we obtain

$$2c_1 y(x) = 1 + y'(x)^2 = 1 + c_1^2(x - c_0)^2. \tag{69}$$

We shall consider those functions of this form capable of providing local (minimal) values for

$$F(\hat{y}) = \int_0^1 f[\hat{y}(x)]\, dx$$

on

$$\hat{\mathcal{D}} = \{\hat{y} \in C^1[0, 1]: \hat{y}(0) = 1;\ \hat{y}(1) = \tfrac{1}{2}\}.$$

To satisfy the condition $y(0) = 1$, we must take $c_1^2 c_0^2 = 2c_1 - 1$ in (69), and after some manipulation, we obtain for $c_1 = (\lambda^2 + 1)/2$, the central family of quadratic functions:

$$\psi(x; \lambda) = \frac{(\lambda^2 + 1)x^2}{4} + \lambda x + 1, \tag{70}$$

in which the parameter λ has been chosen to have $\psi'(0; \lambda) = \lambda$. Observe that

$$\Delta(x, 0) = \psi_\lambda(x; \lambda) \equiv \frac{\lambda x^2}{2} + x, = 0 \quad \text{when } x = -\frac{2}{\lambda}. \tag{71}$$

If one of the associated parabolas passes through the point (x, y), then from (70), we have after competing the square, $4y = [(\lambda x + 2)^2 + x^2]$. Thus *no* curve of the family enters the region in which $0 < 4y < x^2$, and exactly *one* curve of the family passes through each point on the boundary of this region where $4y = x^2$. However there will be *two* curves of the family passing through each point of the complementary region in which $4y > x^2$. In particular, for the point $(1, \tfrac{1}{2})$, there are the curves corresponding to the parameters $\lambda = -1$ and $\lambda = -3$ shown in Figure 9.13. From (71), $\psi_\lambda(x; -1) \neq 0$ on $(0, 1]$. Hence the Jacobi condition *is* satisfied, and we are assured by Theorem 9.24 that $y_1(x) = \psi(x; -1) = x^2/2 - x + 1$, provides a strong local minimum value for F on $\hat{\mathcal{D}}$. But also from (71), $\psi_\lambda(x; -3) = 0$ when $x = \tfrac{2}{3}$. Thus by Theorem 9.25, the curve $y_3(x) = 5x^2/2 - 3x + 1$ cannot provide even a weak local minimum (or maximum) value.

If we consider the corresponding problem in the parametric version of minimizing

$$G(\hat{Y}) = \int_0^{5/2} \hat{y}(t)^{1/2}|\hat{Y}'(t)|\, dt,$$

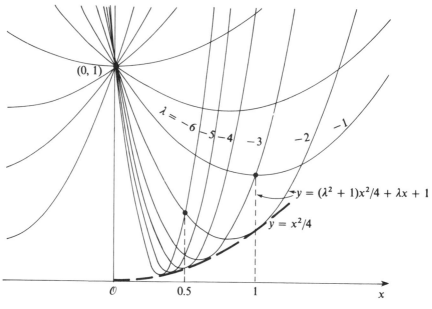

Figure 9.13

on
$$\hat{\mathscr{D}} = \{\hat{Y} = (\hat{x}, \hat{y}) \in C^1[0, 1])^2 \colon \hat{Y}(0) = (0, 1); \hat{Y}(\tfrac{5}{2}) = (1, \tfrac{1}{2}), \hat{y} \geq 0\},$$

then, as in the application of §7.5, we must admit the Goldschmidt curve $\hat{\Gamma}_0$ consisting of the horizontal segment along the x axis and the two vertical segments at its ends. By direct calculation in which we utilize the arc length as the parameter t, we obtain

$$G(\hat{Y}_0) = \int_0^1 y^{1/2}\, dy + \int_1^2 0\, dx + \int_0^{1/2} y^{1/2}\, dy = \tfrac{2}{3}(1 + 2^{-3/2}),$$

and this is *less* than

$$F(y_1) = \int_0^1 [y_1(x)(1 + y_1'(x)^2)]^{1/2}\, dx = \sqrt{2c_1} \int_0^1 y_1(x)\, dx$$

$$= \sqrt{2} \int_0^1 (x^2/2 - x + 1)\, dx = (\tfrac{2}{3})\sqrt{2}.$$

(In making this last computation we have used (69), with $2c_1 = 1 + \lambda^2$, where $\lambda = -1$.) We can approximate the Goldschmidt function $\hat{Y}_0$ by smooth functions, Y for which the integral $G(Y) = F(y)$ has values as near $G(\hat{Y}_0)$ as we wish (Proposition 7.6), and we conclude that y_1 gives only a strong *local* minimum value for F. While it is true that for this problem, the upper of two possible (stationary) parabolas through the same pair of points always provides a *local* minimum value, it may sometimes provide a minimum *less* than that for the Goldschmidt curve (Problem 9.26). A complete discussion is given in [P].

The situation illustrated in this problem occurs with some more general strongly convex functions of the form $p(x, y)\sqrt{1 + z^2}$ with $p > 0$, including the minimal surface of revolution function where $p(x, y) = y$. In particular, when there are two catenaries through the same points (as in Problem 7.15), then the upper satisfies the Jacobi condition (9.22) and will give a strong *local* minimal surface area which may or may not be less than that provided by the Goldschmidt solution, while the lower will not supply even a weak local extremal value. These arguments are facilitated by a geometric interpretation of the Jacobi conditions. See [P], [Bl].

§9.10. Concluding Remarks

In Chapter 3, we demonstrated that convexity of $f(\underline{x}, Y, Z)$ provides an elementary sufficiency proof for minimization even with variable end point conditions and constraints. In this chapter, we have obtained analogous results when $f(\underline{x}, \underline{Y}, Z)$ is convex, provided that the existence of a suitable field can be established.

Although this field theory approach appears to be the only one capable of supplying the results sought within the scope of the text, it involves subtleties which as we have seen, make each application to a specific problem extremely difficult. It should be used only as a last resort.

In this book, an effort has been made to present those results which can be established without use of the Lebesgue integral. However, we have not examined the special class of homogeneous integrand functions which yield integrals independent of the particular parametrization of the underlying trajectories. The study of the associated curve dependent integrals is presented in such works as [S], [G–F] and [Ak], and the relation between a resulting field theory and the Hamilton–Jacobi equations for the trajectories is investigated thoroughly in [C] and in [Ru]. In the latter work will also be found a corresponding development for multidimensional integral functions. Convexity plays a significant role in the examination of certain multidimensional problems in which the underlying domain of integration is permitted to vary. See [G–F] and [P–S].

The principal benefit derived from the introduction of the Lebesgue integral is the possibility of establishing existence of a minimizing function in a larger class of functions than those admitted in this book. These methods were first carried through successfully by Tonelli (c. 1922) and his results will be found in [Ak]. They depend on the fact that integrands, f, convex as in this chapter (§9.2), will produce integral functions, F, which are *semi-*

continuous in a weak sense on sets of a (Sobolev) space of functions which are *compact* in the *same* weak sense. Then, an extended version of Proposition 5.3 may be invoked to guarantee the existence of a minimum, at least where the integrand admits suitable estimates. Moreover, it may be possible to show that a minimizing function is a C^1 solution of the classical Euler–Lagrange equations of Chapter 6. In any case, a basis is obtained for systematic approximation to the minimizing value which provides theoretical justification for the numerical procedures of Ritz and Galerkin and their extensions. ([Mi], [Ak], [C–H], [G–F] and [We].)

Most normed linear spaces of functions used in this book are Banach spaces, and from the amalgamation of the algebraic theory of linear structures with the standard techniques of analysis has emerged the rich and powerful theory of functional analysis for such spaces. When its methods are invoked, then, for example, it is possible to establish the existence of Lagrangian multiplier "operators" for problems with Lagrangian constraints. This theory (of Liusternik) is presented in [E–T] and [I–T], where also will be found corresponding results for optimal control problems, and much, much, more. [V] should also be consulted.

Most of the general results alluded to above, remain valid when the elementary convexity of the integrand functions used in this book is replaced by a more geometric convexity of the form indicated in Problem 0.7. ([B–P].) Moreover, there is a theoretical construction which permits an "arbitrary" integrand f to be replaced by a new integrand f^{**} which is convex in this extended sense. To obtain this, it is necessary to examine a related problem for a conjugate integrand f^* which acts on the dual of an appropriate real Banach space. This theory, which is presented in [E–T], [I–T], and [R], uses some of the deeper results in functional analysis. It also extends into the realm of nonsmooth analysis where derivative sets are assigned pointwise to functions that are not differentiable at these points in any reasonable sense. See [Ro] and [Cla] for illuminating presentations.

Finally, we should mention a modern form of the homogeneous integrand problem, that of minimization of quadratic functionals by means of Hilbert space methods. (Hilbert spaces are Banach spaces in which the norm is obtained from a scalar or inner product exactly as in Euclidean space. Because of the additional geometric structure afforded by the inner product, and the corresponding Cauchy–Schwarz inequality, these spaces exhibit most features which might be expected from a Euclidean space of infinite dimension.) This approach is presented in the works of Mikhlin, [Mi], and its extension will be found in [V], and in [K–S]. Its methods are directly applicable to multidimensional problems, and the resulting Euler–Lagrange equations are, in general, elliptic partial differential equations. [Mo] gives an exhaustive treatment of multidimensional problems from a modern perspective.

Many other applications have been made of variational methods and a representative sample of suggestive titles will be found in the bibliography. In particular, [O–R] contains an extensive list of works in which variational methods have been applied to problems in mechanics and related areas, and [Ka–S] provides a similar list for problems in economics and management.

The foregoing remarks are merely suggestive and cannot do justice either to the comprehensive scope of this field, or to its contributors. Despite classical origins, the principles of the variational calculus remain a vital force in both mathematical and philosophical thought. Although the problems are now clearly delineated, methods for their solution are far from exhausted, and there remains a formidable gap between the known theoretical methods and their satisfactory application in specific instances. New techniques will be developed, and, it is to be hoped, that they will admit expression in even simpler forms than those now employed. The subject both needs and deserves an elegance of expression which is comparable to the idealism embodied in its concepts.

PROBLEMS

9.1. (a) Show directly that for $b > \pi$, the function

$$F(y) = \int_0^b [y'(t)^2 - y(t)^2]\, dt$$

is not minimized on

$$\mathcal{D} = \{ y \in C^1[0, b]: y(0) = y(b) = 0 \}$$

by $y_0 = \mathcal{O}$. Hint: It is easy to construct a function $\bar{y} \in \mathcal{D}$ such that $F(\bar{y}) < 0$.
   (b) For a single particle of mass $m = 2$, whose *rectilinear* motion is opposed by a linear spring with spring constant $k = 2$, verify that F is Hamilton's action integral from §8.1. What does this problem show about Hamilton's principle?
   (c) Make a limiting analysis as $x \nearrow b = \pi$ as in §9.1, to show that when $b = \pi$, $F(y) \geq 0$ still holds on $\mathcal{D}$, with equality *iff* $y(x) = \lambda \sin x$. (This proves Dido's conjecture of Problem 1.5.)

9.2. (a) Use the Weierstrass method of §9.1 to prove that $y = cx$ minimizes

$$F(y) = \int_0^1 \sqrt{1 + y'(x)^2}\, dx$$

on

$$\mathcal{D} = \{ y \in C^1[0, 1]: y(0) = 0,\ y(1) = c \}.$$

   (b) Use Hilbert's theorem (9.7) to reach the same conclusion.
   (c) What problem are we solving?

9.3. (a) Show that the stationary trajectories for

$$f(y, z) = \sqrt{1 + z^2}/y, \qquad y > 0,$$

are semicircles with centers on the x axis.

(b) Prove *directly* that those having the origin as a common (left) end point determine an *exact* field φ for f in a quarter plane.

(c) Use a theorem to reach the same conclusion as in (b).

9.4. (a) Examine Facts 3.11 and make appropriate vector valued extensions for [strictly] convex functions $f(\underline{x}, Y, Z)$.

(b) Prove your results as in Problems 3.2 and 3.3.

9.5. By the method of §9.1, consider the function

$$F(y) = \int_0^b [y'(x)^2 + 2y(x)y'(x) - 16y(x)^2]\, dx$$

on

$$\mathcal{D} = \{y \in C^1[0, b]: y(0) = y(b) = 0\}.$$

(a) Show that $y_0 = \mathcal{O}$ provides a strong minimum for F on $\mathcal{D}$, if $0 < b < \pi/4$.

(b) Show that F does not achieve a minimum on $\mathcal{D}$, if $b > \pi/4$.

9.6. Show that $y_0(x) = \sin 2x - 1$ provides a strong *maximum* for

$$F(y) = \int_0^{\pi/4} [4y(x)^2 - y'(x)^2 + 8y(x)]\, dx$$

on

$$\mathcal{D} = \{y \in C^1[0, \pi/4]: y(0) = -1, y(\pi/4) = 0\}.$$

9.7. Suppose that $f(\underline{x}, Y, Z)$ is [strictly] convex in a set $S = D \times D_0$ where D is the domain of an exact field (central field) for f, and D_0 is a convex domain in $\mathbb{R}^d$, as in 9.2.

(a) If $Y_0 \in (C^1[a, b])^d$ represents a field trajectory Γ_0 in D, show that Y_0 minimizes $F(\hat{Y}) = \int_a^b f[\hat{Y}(x)]\, dx$ [uniquely] on a certain set $\hat{\mathcal{D}}_0$.

(b) Prove that Y_0 in (a) is a [unique] weak *local* minimum point for F on $\hat{\mathcal{D}}_0$. Hint: Use Hilbert's formula (21).

9.8. (a) If $f(\underline{x}, Z)$ is [strictly] convex on $D \times \mathbb{R}^d$, use Hilbert's theorem to prove that each stationary function Y_0 for f minimizes F on an appropriate $\hat{\mathcal{D}}$ [uniquely]. Hint: There is a simple exact field for such f.

(b) Obtain the same result as in (a) *without* using the methods of this chapter.

9.9. (a) Construct an exact field φ for the brachistochrone problem which utilizes a family of geometrically similar "concentric" cycloids.

(b)* Define $\varphi(x, y)$ so obtained. (See §9.3.)

9.10. Verify the analysis leading to (23).

9.11. (a) Use Problem 6.42(b) to show that a central family for $f(s, y, z) = -y\sqrt{1 - z^2}$ is given by $\psi(s; \lambda) = \lambda^{-1} \sin(\lambda s)$ for $\lambda > 0$, where the center is at the origin.

(b) Graph three members of this family noting that $\psi'(0; \lambda) = 1$.

(c) Argue graphically that when $0 < y < s$, there is a *unique* λ with $0 < \lambda s < \pi$ for which $\lambda y = \sin(\lambda s)$, i.e., $y = \psi(s; \lambda)$. Hint: make the substitution $u = \lambda s$.

(d) Explain the geometrical significance of the restriction $0 < y < s$.

(e) Examine ψ_λ and conclude that this central family determines a *central field* for f in $D = \{(s, y) \in \mathbb{R}^2: 0 < y < s\}$.

(f)   Show that $f(\underline{s}, y, z)$ is strictly convex on $D \times \mathbb{R}$, and apply Hilbert's theorem to make *some* comparisons about curves in D.

(g)* Sketch a limiting argument in terms of the integral

$$F_\varepsilon(y) = -\int_\varepsilon^{l-\varepsilon} y(s)\sqrt{1 - y'(s)^2}\, ds,$$

for some $\varepsilon > 0$, which could lead to a proof that y_0 minimizes $-F$ on $\mathscr{D}^*$ of Problem 6.42. Hint: If $y_0 = y_0(s)$ defines the semicircle and $\hat{y} = \hat{y}(s)$ another curve both joining the point $(l, 0)$ to the origin and lying otherwise in D, consider the small segments σ_ε on which $s = \varepsilon$ and $s = l - \varepsilon$. What would you like to prove concerning these segments?

(h)* Show that Hilbert's integral I for the field

$$\varphi(s, y) = \psi'(s; \lambda)\Big|_{\lambda y = \sin \lambda s} = \sqrt{1 - \lambda^2 y^2}\,\Big|_{\lambda y = \sin \lambda s}$$

on each "vertical" segment σ_ε considered in (g) reduces to

$$I(\sigma_\varepsilon) = \int_a^b p(s, y)\, dy = \int_a^b \sqrt{r(y)^2 - y^2}\, dy$$

for small positive a and b, where $r(y) = \lambda^{-1}|_{\lambda y = \sin \lambda \varepsilon}$, or $\lambda^{-1}|_{\lambda y = \sin \lambda(l-\varepsilon)}$, are the related radii of the arcs defining the central family.

(i)   Argue geometrically that as $\varepsilon \searrow 0$, $|I(\sigma_\varepsilon)| \to 0$, while $F_\varepsilon(y) \to -A(y)$. Conclude that Dido's conjecture in the form $A(\hat{y}) \leq A(y_0)$ with equality *iff* $\hat{y} = y_0$, is true. (See Problem 1.5.)

(j)   Compare this solution with the approach taken in Problem 6.42. In what sense are both incomplete solutions of Dido's Problem?

9.12. The Zenodoros Problem. (See Problems 1.9, 6.43, and 8.27.)

(a)   Use Problem 8.27(d) to show that a central family for $f(x, y, z) = -(y(1 - z^2))^{1/2}$ is given by the parabolic functions $\psi(x; \lambda) = x - \lambda x^2$, for $\lambda > 0$.

(b)   Graph three members of this family noting that $\psi'(0; \lambda) = 1$.

(c)   Show by direct computation that through each point $(x, y) \in \mathbb{R}^2$ with $0 < y < x$, passes precisely one trajectory of this family generating the field $\varphi(x, y) = 2(y/x) - 1$.

(d)   Establish that this is a central field for f in $D = \{(x, y) \in \mathbb{R}^2: 0 < y < x\}$.

(e)   Argue that $f(\underline{x}, y, z)$ is strictly convex in $D \times (-1, 1)$, and apply Hilbert's comparison to related curves in D.

(f)   For $T > 0$, let Γ_0 be the graph of the function $y_0(x) = x - x^2/T = \psi(x; T^{-1})$ on $[0, T]$, and let $\hat{\Gamma}$ be the graph of a function $\hat{y} \in \hat{\mathscr{D}} = \{\hat{y} \in \hat{C}^1[0, T]: \hat{y}(0) = \hat{y}(T) = 0, \text{ with } (x, \hat{y}(x), \hat{y}'(x)) \in D \times (-1, 1).\}$ Finally, for $a > 0$, let σ_a be the segment joining these curves at $x = a$. Show that

$$\int_0^T f[\hat{y}(x)]\, dx \geq \int_a^T f[\hat{y}(x)]\, dx \geq \int_a^T f[y_0(x)]\, dx + I(\sigma_a),$$

for proper orientation of σ_a, where I is Hilbert's integral for φ.

(g) Prove that for a suitable function p,

$$|I(\sigma_a)| = \left|\int_{\sigma_a} p \, dy\right| \le M \int \frac{dy}{\sqrt{a-y}}$$

$$= 2M|\sqrt{a - \hat{y}(a)} - \sqrt{a - y_0(a)}| \to 0 \quad \text{as } a \searrow 0.$$

(h) Conclude that the function y_0 minimizes $F(\hat{y}) = \int_0^T f[\hat{y}(x)] \, dx$ on $\hat{\mathscr{D}}$.

(i)* Discuss the restriction, $0 < y < x$, in terms of the original variables of Problem 1.9, and show that this is equivalent to the requirement, $\pi \hat{y}^2(x) < 2\pi \int_0^x \hat{y}(\xi)\sqrt{1 + \hat{y}'(\xi)^2} \, d\xi$, which is *always* true. Hint: See Problem 1.7.

9.13*. In a domain D of $\mathbb{R}^{d+1}$, suppose that we are given *arbitrary* C^1 functions $h = h(x, Y)$ and $P = P(x, Y)$, for which the line integral

$$I(\Gamma) = \int_\Gamma (h \, dx + P \cdot dY)$$

has a value that depends only on the end points of C^1 curves $\Gamma \subseteq D$.

(a) Consider only those curves Γ which can be parametrized by $Y \in \mathscr{Y} = (C^1[a, b])^d$; argue that for fixed $A, B \in \mathbb{R}^d$,

$$F^*(Y) \stackrel{\text{def}}{=} \int_a^b [h(x, Y(x)) + P(x, Y(x)) \cdot Y'(x)] \, dx$$

is *constant* on

$$\mathscr{D} = \{Y \in \mathscr{Y} : Y(a) = A, \ Y(b) = B, (x, Y(x)) \in D\},$$

and hence $\delta F^*(Y; V) = 0$, $\forall V$ in a set $\mathscr{D}_0$.

(b) Show that the resulting *first* equation for a typical $Y \in \mathscr{D}$, is

$$P_x - h_Y + (\bar{P}_Y - P_Y) \cdot Y' = 0,$$

where the terms are evaluated at $(x, Y(x))$, and P_Y is the Jacobian matrix with rows indexed by Y, while $\bar{P}_Y$ is its transpose.

(c) Since (a, A), (b, B) could be any points in D with $a < b$, conclude that h, P *must* satisfy the exactness conditions (10). Hint: Reason geometrically.

(d) Next, let Φ be a *stationary* field in D for a *given* $f = f(x, Y, Z)$ (assumed C^1), and suppose that $Y \in \mathscr{D}$ is a solution of the associated field equation (§9.3). Show that $F(Y) = \int_a^b f[Y(x)] \, dx = F^*(Y)$ given in (a), if h is defined as in (9), where P is as yet unspecified.

(e) Use conditions (10) [obtained in (c)] to prove that when Y is as in (d) and $V(x) \stackrel{\text{def}}{=} P(x, Y(x)) - f_Z[Y(x)]$, then $V'(x) = -\Phi_Y(x, Y(x))V(x)$ with trivial solutions $V \equiv 0$. Now define P. Hint: See Proof of 9.4.

9.14*. To obtain a multidimensional version of Problem 9.13, proceed as follows: ($d > 1$)

(a) For arbitrary C^1 functions $h = h(X, u)$, $P = P(X, u)$, defined in a domain D of $\mathbb{R}^{d+1}$ require that the integral

$$F^*(u) \stackrel{\text{def}}{=} \int_R (h(X, u(X)) + P(X, u(X)) \cdot \nabla u(X)) \, dX$$

be *constant* on sets

$$\mathscr{D} = \{u \in C^1(\bar{R}) : u|_{\partial R} = \gamma \text{ fixed}\},$$

where R is a domain in $\mathbb{R}^d$, and conclude as in Problem 9.13, that this requires $\nabla \cdot P = h_u$ in R.

(b)* Conversely, when R is *simply connected*, apply Green's theorem in the form of equation (47) of §6.9, to show that when $\nabla \cdot P \equiv h_u$ in R, then we might expect that $I(S) \overset{\text{def}}{=} \int_S (-h, P) \cdot (n, N) \, d\sigma$ is invariant with all those hypersurfaces $S \subseteq \partial G$, where G is a Green's domain in R, and the "*edges*" of S are fixed. Here (n, N) is the outward normal to S from G. (Reason geometrically, concentrating on the case $d = 2$.)

(c) When S is the graph of a function $u \in C^1(R)$, show that $I(S) = F^*(u)$. Hint: argue that $d\sigma = \sqrt{1 + |\nabla u|^2} \, dX$ while (n, N) is proportional to $(-1, \nabla u)$. Again, concentrate on the case $d = 2$.

9.15. To obtain a multidimensional form of Theorem 9.7 which is valid for C^1 integrand functions $f = f(X, u, Z)$ as in §6.9, reproduce Hilbert's reasoning as follows: Let $\Phi = \Phi(X, u)$ be a C^2 function in a *simply connected* domain D of $\mathbb{R}^{d+1}$, for which $\nabla \cdot P \equiv h_u$ (as in Problem 9.14), when $P(X, u) \overset{\text{def}}{=} f_Z(X, u, \Phi(X, u))$, and $h(X, u) \overset{\text{def}}{=} f(X, u, \Phi(X, u)) - P(X, u) \cdot \Phi(X, u)$.

(a) Show that each solution $u_0 \in C^2(R)$ of the field equation, $\nabla u(X) = \Phi(X, u(X))$, is stationary for f (Theorem 6.13).

(b) Suppose that this u_0 is in $\mathscr{D}$ of Problem 9.14(a). Show that $F^*(u_0) = F(u_0) = \int_R f[u_0(X)] \, dX$.

(c) Use the constancy of F^* on $\mathscr{D}$ to prove that when also $u \in \mathscr{D}$:

$$F(u) - F(u_0) = \int_R \mathscr{E}(X, u(X), \Phi(X, u(X)), \nabla u(X)) \, dX,$$

where

$$\mathscr{E}(X, u, Z, W) \overset{\text{def}}{=} f(X, u, W) - f(X, u, Z) - f_Z(X, u, Z) \cdot (W - Z).$$

(d) Define an appropriate convexity for $f(X, u, Z)$, and formulate a multidimensional version of Hilbert's theorem.

(e) Solutions u of the above field equation are now difficult to produce. When $d = 2$, give a condition on the field Φ which is necessary for it to have *any* C^2 solutions.

(f)* How *might* an exact field for f be obtained from a family of stationary functions $\psi = \psi(T; (X, u))$?

9.16. (a) When $f = f(x, z) \in C^1(\mathbb{R}^2)$, and $\tau = \tau(x, y)$ is C^1, with $\nabla \tau \neq 0$, show that the family of lines which cut T, the zero level set of τ, *orthogonally* will determine an exact field for f which meets the transversal condition (25) of §9.4 in *any* domain D which they cover consistently.

(b) Find such a family for the parabolic function $\tau(x, y) = y - x^2$, and determine domains which they cover consistently.

(c) Make the construction of (b) for the function $\tau(x, y) = x^2 + y^2 - 1$.

(d) For $f(z) = z^2$, with τ as in (b), what related minimization problems can you solve?

(e) Repeat part (d) when $f(z) = \sqrt{1 + z^2}$.

9.17. Suppose that Φ is an exact field for f in a domain D of $\mathbb{R}^{d+1}$, where $f(\underline{x}, Y, Z)$ is [strictly] convex in $D \times \mathbb{R}^d$.

(a) If Φ satisfies the transversal condition (25) for a C^1 arcwise connected transversal $T \subseteq D$, which cuts one field trajectory Γ represented by $Y \in$

$(C^1[a, b])^d$ twice, verify directly that

$$\int_a^{t_1} f[Y(x)]\, dx = \int_a^{t_2} f[Y(x)]\, dx,$$

where $(t_j, Y(t_j)) \in T$, $j = 1, 2$, and $t_1 < t_2$. Hint: Use the invariance of I on the corresponding *sub*trajectories Γ_j of Γ.

(b) If for some fixed b, the set of points $(b, Y) \in D$ is convex, then it defines a transversal T. Show that the transversal condition (25) reduces to requiring that $f_Z(b, Y, \Phi(b, Y)) = 0$, $\forall\, (b, Y) \in T$. How could this transversal condition have been anticipated? (See §6.4.) If Φ meets this condition, which minimization problems can you solve?

(c) What condition should Φ satisfy on $T = \{(x, B): x_1 \le x \le x_2\}$ in order that Theorem 9.9 guarantee minimization for Y_0, if $Y_0(t) = B$, but $x_1 < t < x_2$?

(d)* Prove that the cycloid provides a minimum for Jakob Bernoulli's brachistochrone problem from §6.4. Hint: Use a field from Problem 9.9(a), in conjunction with the limiting analysis for the ordinary brachistochrone problem given in §9.4.

9.18. Verify that the functions defined by (38) satisfy (for constant μ) the equations of stationary for $\tilde{f}$ of (36), under the constraint (35).

9.19. Chaplygin's Problem.

(a) Using the formulation with *Lagrangian* constraint from Problem 1.4, show why it might be appropriate to consider the flight path of maximal area as being stationary for the modified function

$$\tilde{f}(t, Y, Y') = -xy' + \lambda(t, Y)[(x' - w)^2 + y'^2 - 1],$$

where $Y = (x, y)$, and $Y' = (x', y')$ is regarded as a variable. λ is an unknown function which we may suppose C^1.

(b) Demonstrate that when $\lambda > 0$ on a domain D of $\mathbb{R}^3$, then $\tilde{f}(t, Y, Y')$ is strictly convex on $D \times \mathbb{R}^2$.

(c) Under the additional constraint $(x' - w)^2 + y'^2 \equiv 1$, prove that the stationary functions for $\tilde{f}$ satisfy with $\mu = 2\lambda$ the equations

$$\mu(x' - w) = -y + c_2; \qquad \mu y' = x - c_1$$

for trajectory constants c_1, c_2.

(d) Show that $\mu^2 = (x - c_1)^2 + (y - c_2)^2$, and differentiate *along the trajectory* to get

$$\mu' = wy' \quad \text{so that} \quad \mu = wy + r,$$

for a trajectory constant r.

(e) Conclude that under the constraint in (c), each stationary trajectory must project onto an arc of a y ellipse whose ratio of minor to major axes is the wind-constant $\sqrt{1 - w^2}$.

(In particular, an associated closed flight path through the origin would be along the ellipse of the type in (e) fixed by the flight time and the initial flight path angle. It is considerably more difficult to reach this conclusion with the isoperimetric formulation of the problem. See [Sm] for details.)

(f) Explain what would be necessary to prove that flight along such an elliptical path would in fact *maximize* the enclosed area.

9.20. **Isoperimetric Problem.**
   (a) Take the Lagrangian formulation suggested by Problem 9.19, with $w = 0$. Show that with the "steering angle" parametrization $x'(t) = \cos \alpha(t)$, $y'(t) = \sin \alpha(t)$, then $\mu \alpha'(t) = 1$, where now μ is a trajectory constant which we may assume positive.
   (b) Conclude that if $Y(0) = 0$, the equations of motion are given in *complex* form by
$$v(x(t) + iy(t)) = vz(t) = -ie^{i\gamma}(e^{ivt} - 1),$$
   where $v = 1/\mu$ and γ is a real trajectory constant; $i = \sqrt{-1}$.
   (c)* Prove that each point $(t, x, y) \in \mathbb{R}^3$ with $t > 0, (x, y) \ne 0$, may be joined to the origin by a *unique* trajectory of this type.
   (d)* Prove that the resulting central family provides a central field for this problem. Hint: See the computation in §9.5, Example 3.

9.21. Let $\Delta(x, \tilde{a}) = \det \Psi_\Lambda(x, \tilde{a})$ as in Proposition 9.20.
   (a) For $\tilde{a}$ and $\tilde{a}_0 \in I_\varepsilon = [a - \varepsilon, b + \varepsilon]$, verify that with $V(x) \stackrel{\text{def}}{=} \Psi_\Lambda(x, \tilde{a}_0) - \Psi_\Lambda(x, \tilde{a})$, then $T(x) \stackrel{\text{def}}{=} (V(x), V'(x))$ is a solution of (60) with $T(\tilde{a}) = (\Psi_\Lambda(\tilde{a}, \tilde{a}_0), \Psi'_\Lambda(\tilde{a}, \tilde{a}_0) - I_d)$.
   (b) Argue that as $\tilde{a} \to \tilde{a}_0, |T(\tilde{a})| \to 0$.
   (c) Verify that for $x \in I_\varepsilon, \exists\, M > 0$ for which $|T(x)| \le e^M |T(\tilde{a})|$. Hint: Proposition 9.17.
   (d) Conclude that as $\tilde{a} \to \tilde{a}_0$, each component of the matrix $\Psi_\Lambda(x, \tilde{a})$ approaches the corresponding component of the matrix $\Psi_\Lambda(x, \tilde{a}_0)$.
   (e) Why does (d) imply that $\Delta(x, \tilde{a}) \to \Delta(x, \tilde{a}_0)$, as $\tilde{a} \to \tilde{a}_0$?

9.22. **The Jacobi Equation.**
   (a) Verify that equation (65) is the Euler–Lagrange equation for the function q of (66), first when $d = 1$, and then for $d > 1$.
   (b) With $h(\varepsilon) = F(Y_0 + \varepsilon \hat{V})$, show by formal differentiation of $h'(\varepsilon)$ that
$$h''(0) = Q(\hat{V}) = \int_a^b q[\hat{V}(x)]\, dx,$$
   again first when $d = 1$, then for $d > 1$. Explain why $Q(\hat{V})$ might be denoted $\delta^2 F(Y_0; \hat{V})$.

9.23. **Legendre's Approach.**
   (a) Show that when $d = 1$, with the substitutions $p(x) = f_{yy}[y_0(x)]$, $s(x) = f_{yz}[y_0(x)]$, $r(x) = f_{zz}[y_0(x)]$, equation (65) reduces to
$$(rv')' + (s' - p)v = 0.$$
   (b) Suppose that $r > 0$ and the Jacobi equation in (a) has a *positive* solution v_1 on $[a, b]$. Verify that the function $u = u_1 \stackrel{\text{def}}{=} -r(v'_1/v_1) - s$ satisfies the *Riccati* equation $u' = -p + (s + u)^2/r$.
   (c) With r, u_1 as in (b), prove that for q of (66),
$$Q(v) = \int_a^b q[v(x)]\, dx = \int_a^b r[v' + (s + u_1)v/r]^2(x)\, dx,$$
   when $v \in C^1[a, b]$, and $v(a) = v(b) = 0$.
   (d) Conclude that for such v, $Q(v) > 0$ unless $v \equiv 0$. Hint: If $v'(x) = -a(x)v(x)$ then $v(x) = \text{const.}\, e^{-A(x)}$, where $A(x) = \int a(x)\, dx$.

(e)   Argue, (with Legendre), that if $r(x) = f_{zz}[y_0(x)] > 0$ on $[a, b]$, and a solution u_1 of the Riccati equation in (b) is available, then the stationary
function y_0 will minimize $F(y) = \int_a^b f[y(x)]\, dx$ on a typical set $\mathscr{D}$, in *each*
direction v.

(f)   Why would the conclusion in (e) *not* establish y_0 as a weak local minimum
point for F on $\mathscr{D}$? What might strengthen Legendre's approach into a
proof for this minimality?

(g)* Attempt a formal extension of Legendre's argument to the case where
$d = 2$.

9.24. (a)   For the Sturm–Liouville function $\tilde{f}(x, y, z) \equiv \tau(x)z^2 + [q(x) - \lambda\rho(x)]y^2$,
with q, ρ, $\tau \in C[a, b]$ (as in §7.3), verify that the solutions of the Jacobi
equation (65) will be stationary for $\tilde{f}$.

(b)* When $\tau > 0$ on $[a, b]$, relate Picard's argument of Problem 7.19 in terms
of a nonvanishing stationary (eigen)function y_1, to that of Problem 9.23(b)
in terms of a nonvanishing solution v_1 of Jacobi's equation.

(c)   Explain how Picard's inequality for F of Problem 7.19(f) follows from
that of Legendre for Q in Problem 9.23(d).

9.25. Suppose that $f = f(x, Y, Z)$ is C^2, that $f_{ZZ}[Y_0(x)]$ is positive definite $\forall\, x \in [a, b]$
(as in 9.2), and that Γ_0, the graph of Y_0, is a trajectory of *some* exact field for f.
Conclude that "most" of Γ_0 is the trajectory of a *central* field for f. Hint:
Combine 9.7, 9.25, and 9.21.

9.26. For Example 1 in §9.9,
(a)   Verify that when $\lambda = -3$, the function from (70), $y_3 = \psi(\cdot; -3) \in C^1[0, 1]$,
gives an integral $F(y_3) > F(y_1) = (\tfrac{2}{3})\sqrt{2}$.
The two parabolas defined by (70) with $\lambda = -2$ and $\lambda = -6$ intersect at
$\beta = (\tfrac{1}{2}, \tfrac{5}{16})$.
(b)   Show that the corresponding integral for the Goldschmidt curve joining
$(0, 1)$ to β is greater than $F(y_2)$, (on $[0, \tfrac{1}{2}]$), where $y_2 = \psi(\cdot; -2)$.
(c)   Let $y_6 = \psi(\cdot; -6)$ on $[0, \tfrac{1}{2}]$. Is Jacobi's condition satisfied for y_6?

9.27*. (Newton's drag profile minimizes.) In Problem 7.27 we saw that the only
candidate for minimizing $F(\hat{y})$ on $\mathscr{D}^*$ is the cornered curve y_0 for which
$y_0(x) = h$, $x \le a$, and $y_0'(a+) = -1$. Moreover, for $a \le x \le 1$, y_0 is described
by (14′) of §3.4(c), and so has a stationary extension above the line $y = h$
(whose graph is indicated in Figure 3.6), as well as one below the x-axis. It is
given parametrically in Problem 3.40(b) where $c = a$ and $y_1 = h$.

(a)   When $y_1 = h$, the *family* of curves for $c > 0$ in Problem 3.39(b) cover
consistently a simply-connected domain $D \supseteq \{0 < x \le 1, 0 \le y \le h\}$. Explain, and graph several curves from this family.

(b)   Explain why the slopes φ of these curves define a field in D that is stationary, hence exact. Also explain why $\varphi(x, h) = -1$ and why $\varphi \le 0$ in D.
Conclude that when $y = h$, the field functions in (9) (here denoted by p and
$\tilde{h}$) are, respectively, $x/2$ and x.

(c)   Let $\hat{y} \in \hat{\mathscr{D}}^*$, and suppose that like y_0, $\hat{y}(x) = h$, for $0 < x \le c$ say. Let σ be
the segment from (c, h) to (a, h) parametrized in this direction), $\hat{\Gamma}$, the graph
of $\hat{y}$ for $x \ge c$, and Γ_0, the graph of y_0 for $x \ge a$. Then show that $I(\hat{\Gamma}) =
I(\Gamma_0) + I(\sigma)$, where I is Hilbert's invariant integral for φ. Use previous

information to conclude that

$$I(\sigma) = \int_c^a x \, dx = \int_0^a x \, dx - \int_0^c x \, dx$$

so that as in (21), $F(\hat{y}) \geq F(y_0)$ as desired. Hint: Recall that $\varepsilon \geq 0$ for the appropriate $\hat{y}'$, φ.

(d) If $\hat{y} \in \hat{\mathcal{D}}^*$ is not constant on some initial x-interval, we can modify the previous construction by introducing the vertical segment σ_c from $(c, \hat{y}(c))$ to (c, h) for small $c < a$ and letting $\hat{\Gamma}_c$ and Γ_c be the graphs of y and y_0 respectively for $x \geq c$. Show that $I(\hat{\Gamma}_c) = I(\Gamma_c) + I(\sigma_c)$ and that

$$|I(\sigma_c)| \leq \int_0^h \frac{2c|\varphi(c, y)|}{(1 + \varphi^2(c, y))^2} \, dy \leq ch \quad \left(\text{since } \frac{|u|}{1 + u^2} \leq \frac{1}{2} \text{ when } u \in R\right).$$

Let $c \searrow 0$ and conclude that as before $F(\hat{y}) \geq F(y_0)$.

PART THREE

OPTIMAL CONTROL

AN OBSERVATION

"Since the fabric of the universe is most perfect, and is the work of a most wise Creator, nothing whatsoever takes place in the universe in which some form of maximum and minimum does not appear."

LEOHARD EULER, *1744*

CHAPTER 10*

Control Problems and Sufficiency Considerations

The discipline now identified as optimal control emerged during the decade 1940–1950, from the efforts by engineers to design electromechanical apparatus which was efficiently self-correcting, relative to some targeted objective. Such efficiency is clearly desirable in, say, the tracking of an aircraft near a busy airport or in the consumption of its fuel, and other economically desirable objectives suggest themselves. The underlying mathematical problems were attacked systematically in the next decade by Bellman [Be], by Hestenes [He], and by a Russian group under Pontjragin [Po]. Their results were quickly adapted to characterize optimal processes in other fields (including economics itself) and the feasibility of optimal control is now a standard consideration in contemporary design strategy.

In examining the associated idealized problems, it is natural to employ the techniques of the variational calculus to obtain models for what can occur. We have already attacked two such problems successfully by such methods—the production problem of §3.4(d), and the fuel consumption problem in §3.5. Indeed, most (deterministic) problems in optimal control admit formulation as one of steering a system so as to minimize a performance integral over an interval (in time), in the presence of Lagrangian constraints, with certain additional target conditions and control restrictions (§10.1). (We shall not consider problems involving multidimensional integrals.) Control constraints are usually a reflection of physical limitations, and although their presence imposes severe complications on the theoretical derivation of necessary conditions (§11.1), it seems far less inimical to sufficiency considerations. (We avoid attempts at presenting general existence theory which for optimal control problems is truly formidable; see [Ce] and [Cla]). In this chapter we concentrate on developing effective sufficiency methods which can usually be attempted, and which, when successful, will lead to a solution—in many cases—*the* unique solution to the problem.

We demonstrate (in §10.2) that representative problems can be solved by reformulation utilizing elementary convexity as explored in this book (especially in Chapters 3 and 7). In §10.3 we develop an underlying sufficiency theory for a class of such problems (on a fixed interval) expressed in terms of the Pontrjagin function h which arises naturally in this context. The theory is applied to provide a complete solution to an important linear state-quadratic performance problem without control constraints. When subsequently admitting such constraints in §10.4, we encounter the fundamental control minimization principle which characterizes h along the optimal trajectory; we show that it too may provide sufficiency criteria, even for a nonconvex performance integrand.

The material in this chapter and the next is essentially independent of that in Chapters 8 and 9, but it demands the setting from Chapter 7.

§10.1. Mathematical Formulation and Terminology

For our purposes, a control problem arises whenever the state of a system at time t, as described by a vector $Y \in \mathbb{R}^d$, evolves according to a prescribed law given usually in the form of a first-order vector differential equation,

$$\hat{Y}'(t) = G(t, \hat{Y}(t), \hat{U}(t)), \tag{1}$$

under assignment of a (vector valued) control function $\hat{U}$ with $\hat{U}(t) \in \mathbb{R}^k$. G is given and assumed C^1 and $\hat{Y}$ is $\hat{C}^1$ (as in Chapter 7), but $\hat{U}$ may be only piecewise continuous ($\hat{C}$) on each time interval. Controls which have the same values except at common points of discontinuity will be considered identical.

This becomes a problem in optimal control when we wish to find those controls $\hat{U}_0$ which produce states $\hat{Y}_0$ that optimize a performance criterion assessed by an integral of the form

$$F(\hat{Y}, \hat{U}) = \int_a^b f(t, \hat{Y}(t), \hat{U}(t)) \, dt, \tag{2}$$

where f is a given real valued function (usually supposed C^1), and a and b might be permitted to vary to allow the state vectors $\hat{Y}$ to meet various initial/terminal conditions, $\hat{Y}(a)$, $\hat{Y}(b)$. In addition, the controls or state vectors might be subjected to other restrictions, such as $|\hat{U}(t)| \leq 1$, or $U(t) \in \mathcal{U}$, or $\hat{y} > 0$, etc. In general, we will assume that the initial state is prescribed, but allow various terminal conditions, and replace $[a, b]$ by $[0, T]$, where the target time T may be permitted to vary.

The problem is said to be *autonomous* when both the state function G and the performance function f (as well as any other interval constraints) do not depend explicitly on t. It is *linear* when both f and G are linear in Y and U, and it is *time-optimal* when the performance function $f \equiv 1$ on $[0, T]$.

Observe that system dynamics involving higher-order state derivatives can in principle be replaced by an equivalent first-order law of evolution of the form (1) but with a higher state dimension. However, the special character of the admissible controls $\hat{U}$ make it inadvisable in general to regard the pair $(\hat{Y}, \hat{U})$ as a larger "state" vector, especially since $\hat{Y}'$ is present explicitly, while $\hat{U}'$ is probably nonexistent.

Nor is it appropriate in general to integrate the state equation incorporating state boundary conditions and thereby convert a Lagrangian constraint into an isoperimetric one, since the solution to the resulting problem, even when available, might not satisfy the Lagrangian constraints.

The reduction of problems in a given discipline to the form considered above is a matter which usually requires significant insight. However, what we wish to stress in this chapter is that an important key to obtaining solutions to each problem is suitable reformulation. In particular, as we will show by representative examples, reformulation emphasizing aspects of convexity to determine sufficient conditions may again provide access to complete solutions.

By contrast, because of the restrictions on the controls, it is far more difficult to characterize necessary conditions for these same problems, since the optimal controls usually lie (at least partially) on the boundary of the control region, $\mathcal{U}$. Those controls which lie entirely on the boundary extremes are called *bang–bang* controls after the model of a light switch which is effective only when fully "on" or "off." Some rather deep theorems provide conditions under which an optimal control must be bang–bang—or at least can be replaced by a bang–bang control which is also optimal. We shall obtain a simple result of this type in Corollary 11.14. For the more general case, see [M–S], [Ne].

In order to simplify the presentation, the dimension of the function spaces as well as carats indicating piecewise continuity, differentiability, etc., will be suppressed when possible. Thus $\hat{Y} \in (\hat{C}^1[0, T])^d$ will be replaced by $Y \in \hat{C}^1(0, T)$, and differentiation with respect to t will be denoted by a dot. Finally, differential equations such as (1) will be understood to hold only when t is a point of continuity of U.

To better appreciate the significance of convexity in this exposition, the following observations may be of benefit:

On a *fixed interval* $[0, T]$, the typical problem in optimal control is seen to be that of minimizing

$$F(Y, U) = \int_0^T f(t, Y(t), U(t)) \, dt$$

on a set $\mathscr{D}$ incorporating initial and terminal conditions (usually on Y alone), subject to Lagrangian constraints of two forms:

(a) the state laws symbolized by $[Y, U] = 0$;
(b) the state-control restrictions symbolized by $\langle Y, U \rangle \le 0$;

where for each $(Y, U) \in \mathcal{D}$, each of the expressions $[\ ]$ and $\langle \ \rangle$ is an (integrable) *real valued* function of t on $[0, T]$.

Now, if for some (continuous) Lagrangian multiplier functions ρ and μ, $(Y_0, U_0) \in \mathcal{D}$ minimizes

$$\tilde{F}(Y, U) = F(Y, U) + \int_0^T (\rho(t)[Y, U] + \mu(t)\langle Y, U \rangle) \, dt$$

on $\mathcal{D}$ [uniquely], it follows that (Y_0, U_0) will minimize $F(Y, U)$ on $\mathcal{D}$ [uniquely] under (a) and (b) provided that for $t \in [0, T]$:

(a$'$) $[Y_0, U_0] = 0$; and

(b$'$) $\mu > 0$ with $\mu\langle Y_0, U_0 \rangle = 0$;   (3)

(except possibly at a finite set of values of t).

Indeed, then $\langle Y, U \rangle \leq 0 \Rightarrow \mu\langle Y, U \rangle \leq 0 = \mu\langle Y_0, U_0 \rangle$, so that

$$F(Y, U) \geq \tilde{F}(Y, U) \geq \tilde{F}(Y_0, U_0) = F(Y_0, U_0).$$

For the vector valued versions of (a) and (b) we simply add additional terms to f, each with its own multiplier function as in §2.3, resulting in a new integrand of the form

$$\tilde{f} = f + P \cdot [\ ] + M \cdot \langle \ \rangle,$$

which can then be subjected to a similar analysis.

Of course, if the new integrand $\tilde{f}$ is convex in the sense of this book, then minimization of $\tilde{F}$ can, in general, be obtained from a $(Y_0, U_0) \in \mathcal{D}$ which satisfies the Euler–Lagrange equations for $\tilde{f}$ together with the corresponding Weierstrass–Erdmann conditions of §7.5 at any corner points.

Also, sufficient strong convexity will guarantee uniqueness of the minimization, and this requires only convexity of each term in $\tilde{f}$ plus strong convexity of one of these terms. Moreover, as in §7.4, these arguments remain valid even when the terms of $\tilde{f}[Y(t), U(t)]$ are only piecewise continuous, as can be seen by the usual partitioning of the integrals.

When the target time T is not fixed, it may be possible to solve the problem as if it were, and then optimize over T as was done with the performance problem in §3.5. If this is not possible—and it *cannot* be for time optimal problems in which it is this T itself being minimized—then there may be a transformation which replaces the problem by a (convex) one over a fixed interval in some other independent variable. If this fails, then there are certain other sufficiency theorems including some of the field theory type of Chapter 9, but they are usually more difficult to implement.

§10.2. Sample Problems

It is difficult to identify a canonical set of problems in optimal control because of the relatively recent origin of this field. Moreover, only a few of those which have been formulated yield satisfactory explicit solutions, and those presented here may be more representative of their solvability than of the entire field. Nevertheless, their solutions do exhibit some of the characteristics to be expected, principally in the discontinuities (with or without bang–bang behavior) of the optimal controls, and in the changes in character of the optimal solutions with the relative size or "geometry" of the given conditions. The latter behavior has already been encountered in investigating the minimal surface of revolution (§1.4(a) and §7.5) but its presence clearly complicates both the analysis and the presentation of results.

Whenever possible, we solve these problems by methods that involve preliminary skirmishing, with direct attack once convexity seems at hand, as opposed to the usual approach of forcing the problem into the procrustean format of general control theory. In some of these problems, the optimal control is continuous, but in the Bolza problem of part (b), it is neither continuous nor bang–bang.

Problems in parts (a), (b), and (c) are presented without considering possible applications, but the remaining problems do have physical origins. The first extends the rocket analysis of §3.5 to a corresponding problem for which the optimal control *is* bang–bang. The next (whose statement is taken from [Ber]) shows that both continuous and bang–bang controls can be optimal for different time intervals. The last, that of an oscillator is selected because various natural questions concerning its optimal behavior(s) require different methods for satisfactory answers. The time-optimal problem for both it and the simpler nonoscillatory "docking" case (discussed in §11.2) are to be found in most expositions of optimal control.

The final problem utilizes a preliminary minimum principle instead of convexity to obtain the optimal time of transfer for certain simple state equations including those for a spinning asymetric body in space.

(a) Some Easy Problems

Suppose that we wish to minimize

$$F(y, u) = \int_0^1 y^2(t)\, dt$$

on

$$\mathscr{D} = \{(y, u) \in \hat{C}^1[0, T] \times \hat{C}[0, 1], \text{ with } y(0) = 0, y(1) = 1\},$$

subject to $\dot{y}(t) = u(t)$, with $u(t) \le 1$.

We note that these last two conditions can be replaced by the single Lagrangian inequality,

$$\dot{y}(t) - 1 \le 0 \tag{4}$$

(thereby eliminating u from the problem); thus we examine the modified integrand

$$\tilde{f}(t, y, \dot{y}) = y^2 + \mu(t)(\dot{y} - 1)$$

(considering $\dot{y}$ as a variable), which for any continuous μ, is strongly convex on $[0, 1] \times \mathbb{R}^2$. (Why?) It follows that each y_0 in

$$\mathscr{D}^* = \{y \in \hat{C}^1[0, 1]: y(0) = 0, y(1) = 1\}$$

which satisfies the Euler–Lagrange equation

$$\frac{d}{dt}\tilde{f}_{\dot{y}} = \tilde{f}_y \quad \text{or} \quad \dot{\mu} = 2y \quad \text{on } (0, 1)$$

with

$$\mu(\dot{y} - 1) = 0, \qquad \mu > 0,$$

will minimize $F^*(y) = \int_0^1 y^2(t)\, dt$ on $\mathscr{D}^*$ under (4).

By inspection, $y_0(t) = t$ satisfies all conditions if we take $\mu(t) = 1 + t^2$ (since $\dot{\mu}(t) = 2y_0(t) = 2t$). Thus, this y_0, with $u_0 = \dot{y}_0 = 1$, provides the unique minimum for this problem.

It also provides the unique maximum since in fact for this problem, if $y \in \mathscr{D}^*$ and $\dot{y} \le 1$, then

$$1 = y(1) = \int_0^1 \dot{y}(t)\, dt \le \int_0^1 1\, dt = 1,$$

where equality is possible only if $\dot{y}(t) \equiv 1$ or $y(t) = t = y_0(t)$. Therefore y_0 is the *only* competing function, and naturally it excels in all respects. This solution is independent of the optimization question being posed, and in particular, it is independent of the integral $\int_0^1 y^2(t)\, dt$; this fact will have more significance in our discussion of Theorem 11.6.

Note also that the above simple device of setting $\dot{Y} = U$, transforms most problems in variational calculus considered in the previous chapters into those which have the appearance of problems in optimal control. This supplies a convenient source of counterexamples.

For example, it is easily verified by the techniques of Chapter 3, that the convex function

$$\int_0^1 [y^2(t) + \dot{y}^2(t)]\, dt$$

is minimized uniquely on

$$\mathscr{D} = \{y \in \hat{C}^1[0, 1]: y(0) = 1, y(1) = e\}$$

by

$$y_0(t) = e^t.$$

Hence

$$u_0(t) = \dot{y}_0(t) = e^t$$

provides the unique optimal control for minimizing

$$F(y, u) = \int_0^1 [y^2(t) + u^2(t)]\, dt \quad \text{on} \quad \mathscr{D} \times \hat{C}[0, 1]$$

under the state law $\dot{y} = u$, and, if desired, the control restriction

$$|u(t)| \le 4.$$

(This is a simple example of the linear state-quadratic performance problem to be investigated in §10.3.)

Moreover, some related problems can be solved by brute force once the state equation is used properly. For example, the minimization of

$$F(y, u) = \int_0^1 (u^2 - y^3)\, dt \quad \text{when } u \in \hat{C}[0, 1],\ -1 \le u(t) \le 0,$$

on

$$\mathscr{D} = \{y \in \hat{C}^1[0, 1]\colon y(0) = y(1) = 1,\ y(t) \ge 0\}^1$$

under the state equation $\dot{y} = yu$, appears to be difficult primarily because of the $-y^3$ term. However, if we integrate this term by parts, we see that when $y \in \mathscr{D}$:

$$F(y, u) = \int_0^1 (u^2 + 3(t - 1)y^2 \dot{y})\, dt - (t - 1)y^3(t)\Big|_0^1$$

$$= \int_0^1 (u^2 + 3(t - 1)y^3 u)\, dt - 1.$$

Now, when $y \ge 0 \ge u$, the new integrand is ≥ 0 with equality when $u(t) = u_0(t) = 0$, so that $\dot{y}_0(t) = 0$ or $y_0(t) = y_0(1) = 1$. (This problem resembles a more significant one of Zeidan that yields to the same attack. See [K–P] and related works. However, it also resembles the Wirtinger problem of §9.5 which does not give itself up so easily!)

(b) A Bolza Problem

The methods employing convexity may extend also to a problem such as the following of Bolza type in which the performance integral is augmented by a function of the endpoint values.

$$\text{Minimize} \quad F(y) = y^2(2) + \int_0^2 y^2(t)\, dt$$

on $\mathscr{D} = \{y \in \hat{C}^1[0, 2]\colon y(0) = 1\}$

under the state law $\dot{y}(t) = u(t)$ with the control restriction $|u(t)| \le 1$.

Here it is simplest to replace the last two requirements by $\dot{y}(t)^2 - 1 \le 0$, and consider the modified function

$$\tilde{F}(y) = F(y) + \int_0^2 \mu(t)(\dot{y}(t)^2 - 1)\, dt,$$

---

[1] The requirement that $y(t) \ge 0$ is superfluous for this problem. From the state equation, we see that $\dot{y} \ge -y$ when $u \ge -1$, on any interval $[0, t]$ in which $y \ge 0$. But then $y(t) \ge y(0)e^{-t}$, so this interval cannot terminate.

which is convex on $\mathscr{D}$ when $\mu \geq 0$. (Why?) We want $y = y_0 \in \mathscr{D}$ such that

$$\delta\tilde{F}(y; v) = 2y(2)v(2) + \int_0^2 [2y(t)v(t) + 2\mu(t)\dot{y}(t)\dot{v}(t)]\, dt$$

$$= 0, \quad \forall\, v \in \mathscr{D}_0 = \{v \in \hat{C}^1[0, 2], v(0) = 0\},$$

with $\mu(t)(\dot{y}_0(t)^2 - 1) \equiv 0$. (Why?)

Integrating by parts, incorporating the fact that $v(0) = 0$, we obtain

$$\tfrac{1}{2}\delta\tilde{F}(y; v) = [y(2) + \mu(2)\dot{y}(2)]v(2) + \int_0^2 [y(t) - (\mu\dot{y})^{\cdot}]v(t)\, dt$$

Thus we want $\mu\dot{y}$ to be continuous with $(\mu\dot{y})^{\cdot} = y$ except at corner points, and

$$\mu(\dot{y}^2 - 1) \equiv 0, \quad \text{with} \quad y(0) = 1 \quad \text{and} \quad y(2) + \mu(2)\dot{y}(2) = 0.$$

Suppose $\mu(0) > 0$, so that either $\dot{y}(0) = +1$ or $\dot{y}(0) = -1$. In the former case $\mu\dot{y}$ increases (why?) so that μ can never vanish and $\dot{y}(t) = +1$; therefore, $y(t) = t + 1$, but this violates the terminal condition. In the latter case, $\dot{y}(t) = -1$, and both $y(t) = 1 - t$ and $\mu(t) = (1 - t)^2/2 + c$ decrease until μ vanishes and permits a corner point at $t = t_1$, say. For example, if $c = 0$, then $t_1 = 1$, $y(t_1) = 0$, and we could take $\mu(t) = y(t) = 0$, for $t > 1$, since this satisfies all requirements. Thus, one possible solution is to take

$$y_0(t) = 1 - t \quad \text{and} \quad \mu(t) = (1 - t)^2/2, \qquad t \leq 1,$$
$$= 0 \qquad\qquad\qquad\quad = 0, \qquad\qquad\quad t \geq 1,$$

but are these the only choices? Observe that

$$\tilde{F}(y_0 + v) - \tilde{F}(y_0) - \delta\tilde{F}(y_0; v) = v^2(2) + \int_0^2 [v^2(t) + \mu(t)\dot{v}(t)^2]\, dt$$

$$= 0 \qquad \text{iff } v \equiv 0.$$

Thus $\tilde{F}(y_0 + v) = \tilde{F}(y_0)$ iff $v \equiv 0$, so that the solution y_0 is unique, and

$$u_0 = \dot{y}_0 = \begin{cases} -1 & \text{on } (0, 1), \\ 0 & \text{on } (1, 2), \end{cases}$$

is the unique optimal control.

Here the optimal control u_0, although discontinuous, is not of the bang–bang type, since $u_0 = 0$ is not on the boundary of the control region $\mathscr{U} = [-1, 1]$. (However, it would be of this type on the smaller control region $\mathscr{U} = [0, 1]$.)

(c) Optimal Time of Transit

The state $X(t) = (x(t), x_1(t))$ of a system at time t is governed by the linear equation

$$\dot{X}(t) = (1 + u(t), u_1(t)),$$

under the control $U(t) = (u(t), u_1(t))$ with the constraint $|U(t)| = 1$.

In order to find the minimum time T required to transfer this system from its initial state at the origin $(X(0) = \mathcal{O})$ to a prescribed state $(X(T) = B = (b, b_1))$, we can transform this problem to one on a fixed interval as follows:

Under admissible motions we have

$$1 = u^2 + u_1^2 = (\dot{x} - 1)^2 + \dot{x}_1^2,$$

or $2\dot{x} = \dot{x}^2 + \dot{x}_1^2 \geq 0$, which implies that since $x(0) = 0$, $x(T) = b$,

$$b = \int_0^T \dot{x}(t)\, dt \geq 0, \tag{5}$$

and only such b can be permitted. Moreover, $b = 0 \Rightarrow \dot{x} \equiv 0$ on $(0, T)$, (why?), so that $\dot{x}_1 \equiv 0$, as well. Therefore $b_1 = x_1(T) = x_1(0) = 0$, and the problem is trivial.

Thus for $b > 0$, we shall assume that $\dot{x} > 0$, and consider x as the new *independent* variable, while $y = t(x)$ and $y_1(x) = x_1(t(x))$ will be the *new* state variables, governed by the *new* state equation

$$y_1'^2 = 2y' - 1 \tag{6}$$

on the *fixed* interval $[0, b]$. (Why?) Since

$$\begin{cases} \dot{x}y' = 1, \\ y_1' = \dot{x}_1 y', \end{cases}$$

we shall permit discontinuities in y_1' provisionally. Then we have the simple convex problem of minimizing

$$T = \int_0^b y'(x)\, dx = \frac{1}{2}\int_0^b (y_1'^2 + 1)\, dx = F(y_1) \tag{7}$$

on

$$\mathcal{D} = \{y_1 \in \hat{C}^1[0, b]: y_1(0) = 0,\ y_1(b) = b_1\},$$

where the Lagrangian constraint (6) can now be considered after y_1 has been determined. By the methods of §3.3, it is easy to show that the unique minimum is

$$T_0 = F(Y_0) = \int_0^b y_0'(x)\, dx = y_0(b) = \frac{b_1^2 + b^2}{2b^2} = \beta, \quad \text{say.}$$

It follows that β is the desired minimum time T_0 required to reach state B under the mild restriction that $\dot{x} > 0$; it is obtained uniquely with the linear trajectory $X_0(t) = (t/\beta,\ b_1 t/b\beta)$.

Remarks. In solving this problem, we were fortunate in that it is convex when expressed in terms of new variables on a *fixed* interval. Although it is usually possible to reduce a time optimal problem (or other problem involving a varying time interval) to one for a fixed interval (see the proof of Theorem 11.10), the resulting integrand is seldom convex. Some special forms of time-optimal problems will be discussed in §11.2.

(d) A Rocket Propulsion Problem

In §3.5 we examined a problem of minimum fuel consumption for a vertically rising rocket, neglecting dynamic effects of the fuel mass loss in so doing. A seemingly better model for this same rocket obeys the dynamic law

$$m\ddot{y} = -p\dot{m} - mg, \tag{8}$$

where p is a given positive constant, and $m = m(t)$ is the mass of the rocket and fuel at time t. Then $-\dot{m}$, the rate of fuel consumption, is nonnegative, controllable, and limited by the design of the engine; its effect in providing thrust is clearly visible in (8).

Since $m(t) \geq m_R$, the mass of the rocket alone, we can divide (8) by m and obtain

$$\ddot{y} = u - g, \tag{9}$$

where $u = -p\dot{m}/m \geq 0$, is a thrust control with, say, $u(t) \leq \beta$; g is the gravitational acceleration which we still assume constant. Note that (9) is identical to the dynamic law used in the earlier analysis, but our new model admits more realistic applications.

For example, if the initial mass of the fueled rocket is M_0, what is the maximum altitude which can be reached by this rocket with the consumption of a fixed mass M of fuel during the first stage of ascent, and how should the fuel be burned to achieve it?

For each program $m(t)$ of fuel consumption over the fixed time interval $[0, T]$, with $m(T) = M_0 - M$, there is a control $u \in \hat{C}[0, T]$, with

$$\int_0^T u(t)\, dt = p \log(M_0/(M_0 - M)) \quad \text{(Why?)}$$
$$= k > 0, \quad \text{say}, \tag{10a}$$

Under this isoperimetric constraint, we wish to find that control u which, by (8), (9), and (10a) maximizes

$$y(T) = \int_0^T (T - t)\ddot{y}(t)\, dt$$

$$= \int_0^T (T - t)(u(t) - g)\, dt = \int_0^T (T - t)u(t)\, dt - gT^2/2, \tag{10b}$$

on

$$\mathscr{D}_T = [u \in \hat{C}[0, T]: 0 \leq u(t) \leq \beta].$$

To use convexity, we consider instead the problem of finding that $u_0 \in \mathscr{D}_T$ which under (9) and (10a) *minimizes*

$$F(u) = \int_0^T (t - T)u(t)\, dt.$$

If we ignore the control constraints, $0 \le u(t) \le \beta$, and examine for constant λ, the modified integrand

$$\tilde{f}(t, u) = (t - T)u + \lambda u$$

(which is convex), for that u_0 which makes $0 = \tilde{f}_u[u_0(t)] = t - T + \lambda$, we see that this is not feasible. (Why?) Hence the inequality constraints on u appear to be an essential feature of this problem.

To take them into account most effectively observe that the pair of inequalities $0 \le u \le \beta$ is equivalent to the single *quadratic inequality* $u(u - \beta) \le 0$, since the possibility of $u < 0 < u - \beta$ is untenable.

Then according to the approach taken in §10.1, we should consider the modified integrand

$$\tilde{f}(t, u) = (t - T + \lambda)u + \mu(t)(u^2 - \beta u)$$

which is strongly convex on $[0, T] \times \mathbb{R}$ when $\mu \in C[0, T]$ is positive.

If we can find a λ, μ, and $u_0 \in \hat{C}[0, T]$ such that

$$0 = \tilde{f}_u[u_0(t)] = (t - T + \lambda) + \mu(t)(2u_0(t) - \beta), \tag{11a}$$

while

$$\mu(t)u_0(t)[u_0(t) - \beta] = 0 \quad \text{with } \mu > 0, \tag{11b}$$

and (10a) is satisfied, we have a unique solution. (Why?) Observe that by (11a), $\mu(t) = 0$ only when $t = \tau \overset{\text{def}}{=} T - \lambda$ for $\lambda < T$, so that for any other t, (11b) requires $u_0(t) = 0$ or β. But then $\mu > 0$ in (11a) requires that

$$u_0(t) = \begin{cases} \beta, & t \in (0, \tau), \\ 0, & t \in (\tau, T). \end{cases}$$

Finally, it remains to select λ, if possible, to make

$$k = \int_0^T u_0(t)\, dt = \int_0^\tau \beta\, dt = \beta\tau = \beta(T - \lambda);$$

since this simple equation has the unique solution

$$\lambda = T - k/\beta < T,$$

we are assured that u_0 *is* optimal for the problem. Moreover, although $\mu(\tau) = 0$, μ is otherwise positive on $(0,T)$ so that by subdivision of the integrals, u_0, is unique. Note that it is discontinuous and switches in value from β to 0 at $\tau = k/\beta$.

The corresponding maximum altitude at time T is from (10b) given by

$$y_{\max}(T) = \int_0^T (T - t)(u_0(t) - g)\, dt$$

$$= \beta \int_0^\tau (T - t)\, dt - gT^2/2$$

$$= \beta(T\tau - \tau^2/2) - gT^2/2.$$

Finally, if we maximize this parabolic expression in T, we obtain the absolute maximum amplitude of

$$Y_{max} = (\beta - 2g)k^2/2\beta g$$

attained at time $T = T_0 = k/g$.

Observe that although $u_0 = \beta$ is constant during the burn interval $(0, \tau)$, $\dot{m}_0$, the optimal rate of fuel consumption, is not. Instead, from (9), follows $(d/dt) \log m_0 = -\beta/p$, so that on $(0, \tau) = (0, \beta/k)$,

$$m_0(t) = M_0 e^{-\beta t/p}$$

or

$$\dot{m}_0(t) = -(\beta/p)M_0 e^{-\beta t/p}.$$

Note that here, the optimal control u_0 must be bang–bang for the physical control region $\mathcal{U} = [0, \beta]$. If formulated as a control problem this yields an $\tilde{F}(y, u)$ which is (only) convex and in fact u can be arbitrary on $[\tau, T]$, resulting in a *singular control* problem [B–J].

(e) A Resource Allocation Problem

At time t, a fraction $u(t)$ of a quantity being produced at a rate $y(t)$ is allocated for investment to improve productive capacity, the rest being sold for profit. It is desired to choose an investment fraction u_0 and a production rate y_0 to maximize the total profit over a *fixed* time interval $[0, T]$.

At time t, goods for profit are produced at the rate $(1 - u(t))y(t)$ (why?) and thus we wish to *minimize*

$$F(y, u) = -\int_0^T (1 - u(t))y(t)\, dt$$

on

$$\hat{\mathcal{D}} = \{(y, u) \in \hat{C}^1[0, T] \times \hat{C}[0, T]: y(0) = a_1 \geq 0,\ y(t) \geq 0\},$$

subject to the control restriction $0 \leq u(t) \leq 1$, and an investment constraint. The choice

$$\dot{y}(t) = u(t)y(t),$$

allows the production rate to increase directly with the amount available for investment.

This is clearly of the form which identifies it as a problem in optimal control. However, the product $u(t)y(t)$ which appears in the integrand makes convexity arguments difficult. Fortunately, in this case, we can use the constraint to replace the product.

Moreover, since $y \geq 0$, we can replace the control restriction $0 \leq u \leq 1$ by the Lagrangian inequalities $0 \leq \dot{y}(t) \leq y(t)$, and we obtain the simpler problem of minimizing

$$F(y) = \int_0^T [\dot{y}(t) - y(t)]\, dt$$

on

$$\mathscr{D} = \{y \in \hat{C}^1[0, T]: y(0) = a_1, y \geq 0\}$$

under the pair of Lagrangian inequalities

$$\dot{y}(t) - y(t) \leq 0, \qquad -\dot{y}(t) \leq 0$$

expressed in the most useful form.

According to our general analysis in §7.4, we should introduce Lagrangian multiplier functions, $\lambda(t)$, $\mu(t)$, and try instead to minimize the modified integral

$$\tilde{F}(y) \overset{\text{def}}{=} F(y) + \int_0^T [\lambda(t)(\dot{y}(t) - y(t)) - \mu(t)\dot{y}(t)] \, dt$$

$$= \int_0^T [(1 + \lambda(t) - \mu(t))\dot{y}(t) - (1 + \lambda(t))y(t)] \, dt.$$

Now, the associated integrand

$$\tilde{f}(t, y, z) = (1 + \lambda(t) - \mu(t))z - (1 + \lambda(t))y$$

is convex, and if we can find $y_0 \in \mathscr{D}$, $\lambda \geq 0$, $\mu \geq 0$, for which y_0 minimizes $\tilde{F}$ on $\mathscr{D}$ with

$$\lambda(t)(\dot{y}_0(t) - y_0(t)) \equiv -\mu(t)\dot{y}_0(t) \equiv 0,$$

then we will have a solution to the problem. For minimization, we require only the Euler–Lagrange equation

$$\frac{d}{dt}\tilde{f}_z = \tilde{f}_y \quad \text{or} \quad \frac{d}{dt}(1 + \lambda - \mu) = -(1 + \lambda), \tag{12}$$

with continuity of $1 + \lambda - \mu$ at the corner points, together with the natural boundary condition $\tilde{f}_z[y_0(T)] = (1 + \lambda - \mu)(T) = 0$.

At T: $\mu(T) - 1 = \lambda(T)$, and if $\lambda \geq 0$,

then: $\mu(T) \geq 1$; hence *near* T, we must have $\dot{y}_0(t) \equiv 0$ to obtain

$\mu(t)\dot{y}_0(t) \equiv 0$;

but then $y_0(t) = c_0$ near T, so that $\lambda(t)(\dot{y}_0(t) - y_0(t)) \equiv 0$, with $c_0 \neq 0$, implies that $\lambda(t) \equiv 0$ near T, and $\mu(T) = 1$. Thus, near T, the Euler–Lagrange equation (12) reduces to $-\dot{\mu} = -1$, so that $\mu(t) = 1 - (T - t)$, in an interval which terminates at a switching time $\tau < T$. It is possible that $\tau \leq 0$. Otherwise, for $\tau < T$, we expect $\lambda \neq 0$, so that $\dot{y}_0 = y_0$, and thus $y_0(t) = c_0e^t$, where we can take $c_0 = y_0(0) = a_1$, if all of the other conditions can be met.

Since $\dot{y}_0 \neq 0$, it follows that $\mu \equiv 0$, and (12) again simplifies to

$$(1 + \lambda)^{\cdot} = -(1 + \lambda),$$

with the exponential solution

$$1 + \lambda(t) = ce^{-t},$$

where the corner condition gives $ce^{-\tau} = T - \tau$ (why?), or $c = (T - \tau)e^\tau$. Requiring continuity of μ at τ (and hence of λ) gives $0 = 1 - (T - \tau)$ provided that the real switching time $\tau = T - 1$, when $T \geq 1$. Finally, when $T \leq 1$, we take $\mu(t) = 1 - (T - t)$, and $\lambda(t) = 0$, which gives an optimal solution, $y_0(t) = a_1$.

When $T > 1$, there is a single switching time $\tau = T - 1$, and

$$y_0(t) = \begin{cases} a_1 e^t, & t \leq \tau, \\ a_1 e^\tau, & \tau \leq t \leq T, \end{cases}$$

provides an optimal solution since the auxiliary functions

$$\mu(t) = \begin{cases} 0, & t \leq \tau, \\ 1 - (T - t), & \tau \leq t \leq T, \end{cases}$$

$$\lambda(t) = \begin{cases} -1 + (T - \tau)e^{\tau - t}, & t \leq \tau, \\ 0, & \tau \leq t \leq T, \end{cases}$$

are nonnegative and continuous as required. Observe that in this latter case, the optimal control function is bang–bang;

$$u_0(t) = \frac{\dot{y}_0(t)}{y_0(t)} = \begin{cases} 1, & t < \tau, \\ 0, & \tau < t \leq T, \end{cases}$$

is discontinuous with values entirely on the boundary extremes of the control region $\mathcal{U} = [0, 1]$. It dictates that initially *all* of the output should be used to improve plant production capability, after which *all* material produced should be sold for profit. In a real situation, the work force might object to this "ideal" solution when T is large!

It remains to consider uniqueness of the optimal solution. What we have actually established is that if $y \in \hat{C}^1[0, T]$, with $y(0) = y_0(0) = a_1$, then with $v = y - y_0$:

$$\tilde{F}(y) - \tilde{F}(y_0) = \int_0^T [(1 + \lambda - \mu)\dot{v} - (1 + \lambda)v](t) \, dt$$

$$= [(1 + \lambda - \mu)v](t) \Big|_0^T = 0. \quad \text{(Why?)}$$

Hence, $\tilde{F}(y) = \tilde{F}(y_0) = F(y_0)$, and under the required inequalities $0 \leq \dot{y} \leq y$, since λ and μ are continuous and nonnegative, it follows that

$$F(y) = F(y_0) \quad \text{iff} \quad \lambda(\dot{y} - y)(t) \equiv -\mu\dot{y}(t) \equiv 0.$$

However, when $T \leq 1$, $\mu \neq 0$, so that $\dot{y} \equiv 0$ or $y(t) = y(0) = y_0(t)$; and similarly for $T > 1$, we conclude that

$$\begin{cases} \dot{y} = y, & 0 \leq t \leq \tau, \\ \dot{y} = 0, & \tau \leq t \leq T, \end{cases}$$

which again means that $y(t) \equiv y_0(t)$. (Why?)

Thus we have obtained the unique solution y_0, to the given problem and hence the unique optimal control u_0. A related application which arises in the fishing industry is explored in Problem 10.29.

(f) Excitation of an Oscillator

If a particle at rest at the origin is forced into rectilinear motion whose position y at time t is governed by the equation

$$\ddot{y} + y = u, \tag{13}$$

where the forcing term $u \in \hat{C}$ is bounded, say, $|u(t)| \leq 1$, then the resulting motion is in general oscillatory in character.

Indeed by variation of parameters [C–L], the motion with

$$y(0) = \dot{y}(0) = 0$$

is given uniquely by

$$y(t) = \int_0^t \sin(t - \tau)u(\tau)\, d\tau, \tag{14}$$

as can be verified by using Leibniz' formula (A.14) to differentiate the integral at each point of continuity of u.

In particular, if $u \equiv +1$, then the substitution $s_1 = t - \tau$ gives

$$y(t) = \int_0^t \sin s_1\, ds_1 = 1 - \cos t,$$

and the oscillatory behavior is evident.

However, if at some time σ, we switch from $u = +1$ to $u = -1$, then the subsequent oscillations will be about $y = -1$, but the amplitude will vary. For example, $\sigma = \pi$ produces for $t \geq \pi$, $y(t) = -1 - 3 \cos t$; but then, we can switch again to get a still larger amplitude, etc.

Thus we might ask for the largest $y(T)$ which can be obtained at a given time T under the above restrictions, and which control(s) u will be optimal in producing it. But from (14), we see that

$$y(T) \leq \int_0^T |\sin(T - \tau)|\, d\tau = y_0(T),$$

and $y_0(T)$ is obtained uniquely under the optimal control $u_0(t) = \operatorname{sgn} \sin(T - t)$.

For example, when $T = 2\pi$, $y_0(2\pi) = 4$, and it is obtained solely from the control

$$u_0 = \begin{cases} -1 & \text{on } (0, \pi), \\ +1 & \text{on } (\pi, 2\pi). \end{cases}$$

However, if we consider instead the equally reasonable problem of maximizing at time T, the energy measured by $E \overset{\text{def}}{=} (y^2 + \dot{y}^2)/2$, then although under (13),

$$E(T) = \int_0^T \dot{E}(t)\,dt = \int_0^T (y + \ddot{y})\dot{y}\,dt = \int_0^T \dot{y}u\,dt, \qquad (15)$$

we cannot use the above approach as successfully since $|\dot{y}u|$ is not as easy to analyze. From (14) follows

$$\dot{y}(t) = \int_0^t \cos(t - \tau)u(\tau)\,d\tau, \qquad (16)$$

and since $\dot{E} = \dot{y}u$ (why?), we can obtain after some manipulation left to Problem 10.6 the useful formula

$$2E(T) = \left(\int_0^T u(t)\cos t\,dt\right)^2 + \left(\int_0^T u(t)\sin t\,dt\right)^2. \qquad (17)$$

Moreover, for $t \le T \le \pi/2$, both $\sin t$ and $\cos t$ are nonnegative, and we see that when $|u| \le 1$,

$$E(T) \le \tfrac{1}{2}\left[\left(\int_0^T \cos t\,dt\right)^2 + \left(\int_0^T \sin t\,dt\right)^2\right] = 1 - \cos T = E_0(T)$$

and thus maximal energy $E_0(T)$ is achievable with either of the controls $u_0 \equiv +1$ or $u_0 \equiv -1$. When $T > \pi/2$, this maximal energy problem is significantly more difficult than its predecessors. (A related time-optimal problem is solved completely in [Y].) We shall return to it in §11.1 after we have developed a more general approach for attacking it.

For the present, we shall convert it into the standard form by introducing the new variable

$$\dot{y} = y_1,$$

so that   (18)

$$\dot{y}_1 = u - y.$$

Thus, with $Y = (y, y_1)$, we wish to minimize

$$F(Y, u) = -\int_0^T y_1 u\,dt$$

on

$$\mathscr{D} = \{(Y, u) \in \hat{C}^1[0, T] \times \hat{C}[0, T], \text{ with } Y(0) = \mathcal{O} \text{ and } |u(t)| \le 1\}$$

under the new linear state equation (18), which takes the matrix form

$$\dot{Y} = \begin{bmatrix} 0 & 1 \\ -1 & 0 \end{bmatrix} Y + u \begin{vmatrix} 0 \\ 1 \end{vmatrix}.$$

We see that this problem is autonomous, and that the terminal state $Y(T)$ is unspecified.

(g) Time-Optimal Solutions by Steepest Descent

Suppose $k = d$ and the state equation has the simple form

$$\dot{Y}(t) = \psi(t, Y(t)) + U, \tag{19}$$

with the control restriction $|U| \le 1$. Then it may be possible to solve the time-optimal problem for which $Y(0) = A$ is given and $Y(T) = \mathcal{O}$ is required, by arguing that we would like to force $|Y(t)|$ to descend from its value $|A| \ne 0$ to 0 as quickly as possible.

Since except at corner points, for $Y(t) \ne \mathcal{O}$,

$$|Y| \frac{d}{dt} |Y| = \frac{1}{2} \frac{d}{dt} (|Y|^2) = Y \cdot \dot{Y},$$

we obtain from (19), that with $V = Y/|Y|$,

$$\frac{d}{dt} |Y| = V \cdot \psi(t, Y) + V \cdot U$$
$$\ge V \cdot \psi(t, Y) - 1 \quad \text{if } |U| \le 1, \tag{20}$$

with equality precisely when, $U = -Y/|Y|$, if we employ the Cauchy inequality (§0.1) to infer that $V \cdot U \ge -|V||U| \ge -1$, when $|U| \le 1$ for the unit vector $V = Y/|Y|$.

Therefore, if we could solve the state equation (19) with $U = -Y/|Y|$, and meet the appropriate boundary conditions at some T_0, then the solution $Y_0(t)$ would be the (unique) optimal trajectory, T_0 would be the minimal time of transfer, and $U_0(t) = -Y_0(t)/|Y_0(t)|$, $t < T_0$, would supply the (unique) optimal control.

However, if we could just solve

$$\frac{d}{dt} |Y| = \frac{Y}{|Y|} \cdot \psi(t, Y) - 1, \tag{21}$$

with $|Y(0)| = |A|$, $|Y(T_0)| = 0$, then by the same argument, T_0 will at least be the (unique) minimum time of transfer. Now we can surely do this if, say,

$$Y \cdot \psi(t, Y) = 0 \tag{22}$$

since then (21) is just $(d/dt)|Y| = -1$ with the trivial solution

$$|Y(t)| = |Y(0)| - t = |A| - t \quad \text{so that} \quad T_0 = |A|. \tag{23}$$

Moreover, the optimal control $U_0(t)$ is that which opposes the (unit) state vector $Y(t)/|Y(t)|$ at each subsequent instant. (This provides an illustration of control *synthesis*: a state dependent prescription for the optimal control at each instant.)

Condition (22) is realized in the problem where a spinning fully asymmetrical body (say, a satellite) in space is to be brought to (spin) rest by the application of a (vector) torque U.

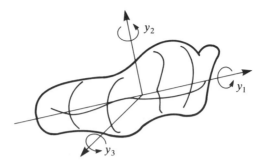

Figure 10.1

Indeed, if the body has distinct moments of inertia I_j about three body fixed principal axes through its center of mass as illustrated in Figure 10.1, then the associated angular momentum vector $Y = (y_1, y_2, y_3)$ is governed by Euler's equations. (See [Ce].) The first requires gyroscopic coupling due to the assumed asymmetry $I_2 \neq I_3$; it is

$$\dot{y}_1 = \frac{I_2 - I_3}{I_2 I_3} = y_2 y_3 + u_1,$$

and the rest are obtained by cyclic permutation of the indices (1, 2, 3).

Since the inertial coefficients are constant, it is easy to see that (22) requires only

$$\frac{I_2 - I_3}{I_2 I_3} + \frac{I_3 - I_1}{I_3 I_1} + \frac{I_1 - I_2}{I_1 I_2} = 0$$

and the sum of the first two terms does cancel the last.

It follows from (23) that if all control torques $|U| \leq 1$ are available, then the minimal time of transfer from a spin state A is $T_0 = |A|$, and it is given by that control $U_0(t) = - Y(t)/|Y(t)|$ which opposes the spin state direction at each subsequent instant. For further discussion of this problem when one axis of symmetry $(I_2 = I_3)$ is allowed, see [A–F].

In addition to (22), other conditions may lead to a solution by this method in related problems where it is desired to drive a system from a given state A to the origin. Some of these are explored in Problems 10.7–10.9, while time-optimal problems for different control regions are investigated in §11.2. Observe, however, that the fundamental device utilized in the above analysis, was that of choosing at each instant t an admissible control $U_0(t)$ which *minimizes* a certain expression related to the given functions. The development of a suitable generalization (the Pontjragin principle) is taken up in §10.4.

§10.3. Sufficient Conditions Through Convexity

In the preceding section we have seen several examples of control problems which (perhaps after suitable transformation) yield to the convexity methods of this text. We shall now subject the general control problem *on a fixed interval* as formulated in §10.1 to the same methods, by imposing on it whatever additional convexity is required to assure theoretical success.

We begin with the simple problem in which there are no control restrictions ($\mathcal{U} = \mathbb{R}^k$), and derive conditions sufficient to provide a complete solution to the problem in which the state equation is linear while the performance integrand is quadratic in the state-control variables.

As the basic analysis in §2.3, revealed, for the *fixed* interval $[0, T]$, if we can minimize

$$\tilde{F}(Y, U) = \int_0^T \tilde{f}[Y(t), U(t)]\, dt$$

on some set $\mathcal{D}$ specifying initial/target conditions on Y, where for some Lagrangian multiplier function, $P \in \hat{C}^1[0, T]$,

$$\tilde{f}(t; Y, U; Z) \overset{\text{def}}{=} f(t, Y, U) + P(t)\cdot(G(t, Y, U) - Z), \tag{24}$$

then in the absence of further state/control constraints, we minimize

$$F(Y, U) = \int_0^T f[Y(t), U(t)]\, dt$$

on $\mathcal{D}$, under the state equation

$$\dot{Y}(t) = G[Y(t), U(t)], \qquad t \in (0, T).$$

Observe that (24) can be rewritten

$$\tilde{f}(t; Y, U; Z) = h(t, Y, U) - P(t)\cdot Z,$$

where

$$h(t, Y, U) \overset{\text{def}}{=} f(t, Y, U) + P(t)\cdot G(t, Y, U). \tag{25}$$

Moreover, $\tilde{f}(t; Y, U; Z)$ is [strongly] convex, precisely when $h(t, Y, U)$ is [strictly] convex, and if the state function G is linear in Y and U, then only the [strict] convexity of $f(t, Y, U)$ is required.

Assuming sufficient differentiability, from (25) we get

$$\tilde{f}_Y = -P \quad \text{and} \quad \tilde{f}_U = \mathcal{O},$$

and the Weierstrass–Erdmann conditions on $\tilde{f}$ (§7.5) are fulfilled by the requirement that P be continuous. The corresponding Euler–Lagrange equations (§6.7) are, in condensed form,

$$\frac{d}{dt}\tilde{f}_{\dot{Y}} = \tilde{f}_Y, \quad \text{or} \quad -\dot{P} = h_Y(t, Y, U), \tag{26a}$$

$$\frac{d}{dt}\tilde{f}_{\dot{U}} = \tilde{f}_U, \quad \text{or} \quad \mathcal{O} = h_U(t, Y, U), \tag{26b}$$

and they must be considered together with the state equation

$$\dot{Y} = G(t, Y, U) \qquad (26c)$$

as constituting the differential system governing the problem (except at corner points of Y and/or points of discontinuity in U). Equation (26a) is called *the adjoint equation*, and P is called the adjoint function. P can be given interpretations related to particular disciplines. For example, see [K–S].

With this preparation, we can get our first major result.

(10.1) Theorem (Sufficiency). *Suppose that D is an open set in $\mathbb{R}^d$, $\mathscr{U} \subseteq \mathbb{R}^k$, and $h(t, Y, U)$ is [strictly] convex on $[0, T] \times D \times \mathscr{U}$. Then each solution Y_0, $P \in \hat{C}^1[0, T]$, $U_0 \in \hat{C}[0, T]$, of the system (26a, b, c) minimizes*

$$F(Y, U) = \int_0^T f[Y(t), U(t)] \, dt$$

[uniquely] under the Lagrangian inequality

on:
$$P(t) \cdot (G[Y(t), U(t)] - \dot{Y}(t)) \leq 0 \qquad (27)$$

(i) $\mathscr{D} = \{(Y, U) \in \hat{C}^1[0, T] \times \hat{C}[0, T]: \; (Y(t) \cdot U(t)) \in D \times \mathscr{U}\}$, if $P(0) = P(T) = \mathcal{O}$;

(ii) $\mathscr{D}_T = \{(Y, U) \in \mathscr{D}: Y(0) = Y_0(0)\}$, if $P(T) = \mathcal{O}$;

(iii) $\mathscr{D}_0 = \{(Y, U) \in \mathscr{D}: Y(0) = Y_0(0), Y(T) = Y_0(T)\}$;

or on

(iv) $\mathscr{D}_\Phi = \{(Y, U) \in \mathscr{D}: Y(0) = Y_0(0), \Phi(Y(T)) = \mathcal{O}\}$, if $\bar{P}(T) = \Lambda\Phi_Y(Y_0(T))$,

for some $\Lambda \in \mathbb{R}^l$ that makes $\Lambda \cdot \Phi$ convex. In the last case Φ is an l-vector valued function on an open set containing $Y_0(T)$ such that $\Phi(Y_0(T)) = \mathcal{O}$, and the columns of Φ_Y are indexed by Y.

PROOF. For the stated conditions, if (Y, U) and (Y_0, U_0) are in $\mathscr{D}$, then

$F(Y, U) - F(Y_0, U_0)$

$\geq \tilde{F}(Y, U) - \tilde{F}(Y_0, U_0)$

$= \displaystyle\int_0^T [h[Y(t), U(t)] - h[Y_0(t), U_0(t)] - P(t) \cdot (\dot{Y}(t) - \dot{Y}_0(t))] \, dt$

$\geq \displaystyle\int_0^T [h_Y[Y_0(t), U_0(t)] \cdot (Y(t) - Y_0(t))$

$\qquad + h_U[Y_0(t), U_0(t)] \cdot (U(t) - U_0(t)) - P(t) \cdot (\dot{Y}(t) - \dot{Y}_0(t))] \, dt$

(by convexity)

$= -\displaystyle\int_0^T \frac{d}{dt}(P(t) \cdot (Y(t) - Y_0(t))) \, dt = -P(t) \cdot (Y(t) - Y_0(t)) \Big|_0^T$

(by (26a, b))

$= 0$ under (i), (ii), and (iii);

[with equality $\Rightarrow$ equality in integrands $\Rightarrow Y(t) = Y_0(t)$, $U(t) = U_0(t)$, $t \in [0, T]$, if we recall the convention concerning points of discontinuity in U and U_0.]

For part (iv), Λ acts as Lagrangian multipliers for the additional constraints $\Phi(Y(T)) = 0$. As in §3.4(d), we consider minimization of the augmented function of Bolza type

$$F^*(Y, U) \stackrel{\text{def}}{=} \tilde{F}(Y, U) + \Lambda \cdot \Phi(Y(T)),$$

for which we have by the previous inequality and the hypothesized convexity of $\Phi \cdot \Lambda$, the comparison

$$F^*(Y, U) - F^*(Y_0, U_0)$$

$$\geq -P(t) \cdot (Y(t) - Y_0(t)) \Big|_0^T + \Lambda \Phi_Y(Y_0(T)) \cdot (Y(T) - Y_0(T))$$

$$= [-\bar{P}(T) + \Lambda \Phi_Y(Y_0(T))](Y(T) - Y_0(T)) = 0$$

[with equality only if $Y(t) = Y_0(t)$, $U(t) = U_0(t)$, $t \in [0, T]$]. $\qquad \square$

Remarks. In condition (iii), both the initial and final states are prescribed; in both (ii) and (iv), the initial state is prescribed, but in (ii), there is essentially no restriction on the final state, while in (iv) (which has significance only when $d > 1$) the final state is restricted to lie in the zero level set of Φ. In each of the latter cases, and in (i), we require that the Lagrangian multiplier function P satisfy boundary conditions which compensate for the freedom permitted in Y. If $d > 1$, we only need to compensate in each component of Y, and we could establish similar results when there is a componentwise mixture of initial/target conditions of the above types. It is also easy to compensate for a target inequality such as $y(T) \leq y_0(T)$ by requiring that $p(T) \geq 0$.

In the usual case, where only the actual solutions Y of the state equation are of interest, we obtain minimization under the stated conditions among those Y which satisfy $Y(t) = G[Y(T), U(t)]$. However, our formulation of Theorem 10.1, permits consideration of certain state differential inequalities provided that corresponding multipliers $P(t)$ can be found with components p_j of the correct signs. With further modifications of the integrand $\tilde{f}$ by addition of terms such as $\mu(t)\psi(t, Y, U)$, it is straightforward to formulate corresponding theorems which permit such state-control inequalities as $\psi(t, Y(t), U(t)) \leq 0$. (See Problem 10.18.)

Linear State-Quadratic Performance Problem

A special but important nonautonomous control problem is that governed by a linear state equation of the form

$$\dot{Y}(t) = \mathbb{A}(t)Y(t) + \mathbb{B}(t)U(t) \tag{28}$$

for appropriate matrix functions $\mathbb{A}$ and $\mathbb{B}$, together with a "quadratic" performance function of the type associated with energy assessments:

$$f(\underline{t}, Y, U) = \tfrac{1}{2}(\bar{U}\mathbb{Q}(\underline{t})U + \bar{Y}\mathbb{S}(\underline{t})Y).$$

For symmetric *positive definite* matrices $\mathbb{Q}$ and $\mathbb{S}$ with elements in $C[0, T]$, $f(\underline{t}, Y, U)$ will be strictly convex on $[0, T] \times \mathbb{R}^{d+k}$. (Why?)
Then

$$h(\underline{t}, Y, U) = f(\underline{t}, Y, U) + P(\underline{t}) \cdot (\mathbb{A}(\underline{t})Y + \mathbb{B}(\underline{t})U)$$

will be strictly convex on $[0, T] \times \mathbb{R}^{d+k}$ for *any* P in $\hat{C}^1$, and the equations (26a) and (26b) are, respectively,

$$-\dot{\bar{P}}(t) = \bar{Y}(t)\mathbb{S}(t) + \bar{P}(t)\mathbb{A}(t), \tag{28a}$$

$$\mathcal{O} = \bar{U}(t)\mathbb{Q}(t) + \bar{P}(t)\mathbb{B}(t), \tag{28b}$$

where $\bar{P}$, $\bar{Y}$, and $\bar{U}$ are row vectors to be determined to satisfy these equations together with (28) (the state equation) and with certain initial/target conditions on Y, P.

Since $\mathbb{Q}$ is invertible (§0.13), we can rewrite (28b) as

$$\bar{U} = -\bar{P}\mathbb{B}\mathbb{Q}^{-1}, \tag{29}$$

and substitute into (28) to obtain the first-order system which we abbreviate

$$\begin{aligned}
\dot{Y} &= \mathbb{A}Y - \mathbb{B}(\bar{\mathbb{Q}})^{-1}\bar{\mathbb{B}}P, \\
\dot{P} &= -\bar{\mathbb{S}}Y - \bar{\mathbb{A}}P,
\end{aligned} \tag{30}$$

using bars to denote transposes.

Each solution (Y_0, P_0) of the linear system (30) that meets the boundary conditions of Theorem 10.1(i), (ii), or (iii), supplies through (29) a solution of the system (28) and (28a, b). As we know from the strict convexity of $h(\underline{t}, Y, U)$, this will be the *unique* solution to our optimal control problem.

From the general analysis of linear systems as given in §A.5 (in particular, (A.19)) we can assert that system (30) has a unique solution on any interval $[0, T]$ for any given initial conditions $(Y(0), P(0))$. However, we cannot prescribe both Y and P at either endpoint, and this awkward feature of optimal control systems complicates the analysis. Fortunately, here, we can give simple arguments that guarantee existence of solutions meeting the split boundary conditions of Theorem 10.1, and so, the unique solution of the control problem. In case (i), $Y_0(t) = P(t) = \mathcal{O}$ and $U_0(t) = \mathcal{O}$ minimizes uniquely; (but see Problem 10.19).

Upon introducing the enlarged vector $V = (Y, P) \in \mathbb{R}^{2d}$, the system (30) written in condensed form $\dot{V} = \mathbb{K}V$ for an appropriate matrix $\mathbb{K}$, has for each $j = 1, 2, \ldots, 2d$ a unique solution V_j on $[0, T]$ with $V_j(0) = E_j$, the unit vector in the jth coordinate direction in $\mathbb{R}^{2d}$. Moreover, each solution V of the condensed system may be expressed as a (unique) linear combination $V = \sum_{j=i}^{2d} c_j V_j$ of these base solutions, with coefficients $C = (c_1, c_2, \ldots, c_{2d}) =$

$V(0)$. Note that then

$$V(t) = \mathbb{V}_t V(0) \quad \text{where} \quad \mathbb{V}_t = [V_1(t) : V_2(t) : \cdots : V_{2d}(t)].$$

In particular

$$V(T) = \begin{vmatrix} Y(T) \\ P(T) \end{vmatrix} = \mathbb{V}_T \begin{vmatrix} Y(O) \\ P(O) \end{vmatrix} = \begin{bmatrix} \mathbb{V}_1 & \mathbb{V}_2 \\ \mathbb{V}_3 & \mathbb{V}_4 \end{bmatrix} \begin{vmatrix} Y(O) \\ P(O) \end{vmatrix},$$

say, where $\mathbb{V}_T$ has been partitioned into four $d \times d$ matrices.

Now in case (ii), where $Y(O)$ and $P(T) = O$ are prescribed, we need to solve the smaller system

$$O = P(T) = \mathbb{V}_3 Y(O) + \mathbb{V}_4 P(O), \tag{31}$$

which is possible provided that $\mathbb{V}_4$ is invertible. However, for the admissible choice $Y(O) = O$, each solution $P(O)$ of the reduced system $O = \mathbb{V}_4 P(O)$ provides the *unique* solution of the associated control problem which, by inspection, is clearly $Y_0(t) = P(t) = O$. Thus, $P(O) = O$, and $\mathbb{V}_4$ is invertible.

Similarly, in case (iii), where $Y(O)$ and $Y(T)$ are prescribed, we must be able to choose $P(0)$ to ensure that $Y(T) = \mathbb{V}_1 Y(O) + \mathbb{V}_2 P(O)$, which will be possible if $\mathbb{V}_2$ is invertible. Again, the admissible choice $Y(O) = Y(T) = O$ gives $P(O) = O$ as the only solution of $\mathbb{V}_2 P(O) = O$, and thus the invertibility of $\mathbb{V}_2$ is assured by the same appeal to uniqueness of solution of the control problem.

Case (iv) is the most difficult to examine, and to remain within the framework of linear analysis, we *assume* that $\Phi(Y) = \mathbb{M} Y + L$ for a given vector $L \in \mathbb{R}^l$ and a given *constant* matrix $\mathbb{M}$ of *maximal rank* $l \, (\leq d)$. Then $\Phi_Y(Y) = \mathbb{M}$, so that we must find a solution (Y_0, P) of the system (30) which meets the prescribed state conditions $Y_0(0)$, with $Y_0(T) \in \mathscr{A}$, the affine space in $\mathbb{R}^d$ for which $\mathbb{M} Y = -L$, together with the adjoint condition, that $\overline{P}(T) = \Lambda \mathbb{M}$ for *some* $\Lambda \in \mathbb{R}^l$. Thus we must find a Λ for which

$$\begin{aligned} Y(T) &= \mathbb{V}_1 Y(O) + \mathbb{V}_2 P(O) = Y \in \mathscr{A}, \\ P(T) &= \mathbb{V}_3 Y(O) + \mathbb{V}_4 P(O) = \overline{\Lambda \mathbb{M}}. \end{aligned} \tag{32}$$

From the previous analysis, both $\mathbb{V}_2$ and $\mathbb{V}_4$ are invertible, so that if a solution exists, we would have

$$P(0) = \mathbb{V}_2^{-1}(Y - \mathbb{V}_1 Y(O)) = \mathbb{V}_4^{-1}(\overline{\Lambda \mathbb{M}} - \mathbb{V}_3 Y(O)), \tag{33a}$$

or

$$Y = \mathbb{V}_1 Y(O) + \mathbb{V}_2 \mathbb{V}_4^{-1}(\overline{\mathbb{M} \Lambda} - \mathbb{V}_3 Y(O)). \tag{33b}$$

But then, $\mathbb{M} Y = -L$ gives the following equation for $\overline{\Lambda}$:

$$\mathbb{M} \mathbb{V}_2 \mathbb{V}_4^{-1} \overline{M \Lambda} = -L + \overline{M}[\mathbb{V}_2 \mathbb{V}_4^{-1} \mathbb{V}_3 - \mathbb{V}_1] Y(0), \tag{33c}$$

which may be solved, provided that the matrix $\mathbb{M}_0 = \mathbb{M} \mathbb{V}_2 \mathbb{V}_4^{-1} \overline{M}$ is invertible. For the admissible case $Y(O) = O$, $L = O$, we note that $Y(T) = O \in \mathscr{A}$, and thus as before, $Y_0(t) \equiv P(t) \equiv O$ will give the *unique* solution to the optimal control problem, while each solution $\overline{\Lambda}$ of $\mathbb{M}_0 \overline{\Lambda} = O$ gives, by

(33b), a Y, which will be an admissible $Y(T)$ for the system (32), and hence provide a solution to the control problem. We conclude that $P(T) = 0 = \overline{M}\overline{\Lambda}$, or since M is of maximal rank $l \leq d$, that $\overline{\Lambda} = 0$; i.e., M_0 is invertible.

Since M_0 is invertible, we can always solve (33c) for $\overline{\Lambda}$, and find $Y \in \mathscr{A}$ from (33b), resulting in a compatible $P(0)$ from (33a) which satisfies the system (32) with $Y(T) = Y$. The resulting $Y_0(t)$, $U_0(t)$ provide the unique solution to the control problem.

For other linear systems and split-boundary conditions, such arguments may not be possible—and even when they prevail there is in general no closed method for obtaining the desired solutions to such two-point boundary value problems. For a survey of the required numerical "shooting" methods, see [Ke]. We note however, that in principle the fundamental matrix $\mathbb{V}_T$ can be approximated quite accurately for explicit linear systems which are not too large, and thus used to obtain $P(0)$ as an explicit solution to a set of simultaneous linear equations, from which follows

$$\begin{vmatrix} Y_0(t) \\ P(t) \end{vmatrix} = [\mathbb{V}_t] \begin{vmatrix} Y(0) \\ P(0) \end{vmatrix}.$$

The unique optimal control $U_0(t)$ is then given by (29) and it is always continuous.

In cases (ii), (iii), and (iv), in which $Y(0)$ is prescribed, we can obtain the same results under slightly weakened hypotheses. Let's summarize our findings

(10.2) Theorem. *Let $\mathbb{A}$, $\mathbb{B}$, $\mathbb{Q}$, and $\mathbb{S}$ be given matrix functions with elements in $C[0, T]$, of sizes $d \times d$, $d \times k$, $k \times k$, and $d \times d$, respectively, where at each t, $\mathbb{Q}(t)$ is symmetric positive definite while $\mathbb{S}(t)$ is symmetric positive semi-definite. Then there exists a unique optimal control $U_0 \in C^1[0, T]$ and trajectory $Y_0 \in C^1[0, T]$ that minimizes*

$$F(Y, U) = \int_0^T [\overline{U}(t)\mathbb{Q}(t)U(t) + \overline{Y}(t)\mathbb{S}(t)Y(t)]\, dt$$

under the state equation $\dot{Y}(t) = \mathbb{A}(t)Y(t) + \mathbb{B}(t)U(t)$, on:

(ii) $\mathscr{D}_T = \{(Y, U) \in \hat{C}^1[0, T] \times \hat{C}[0, T]$, *with* $Y(0)$ *prescribed*$\}$;
(iii) $\mathscr{D}_0 = \{(Y, U) \in \mathscr{D}_T$ *with* $Y(T)$ *prescribed*$\}$;
(iv) $\mathscr{D}_M = \{(Y, U) \in \mathscr{D}_T$ *with* $MY(T) = -L\}$.

where $L \in \mathbb{R}^l$ is given and M is a given matrix of maximal rank $l \leq d$.

PROOF. Only the weakened hypothesis on $\mathbb{S}$ requires comment. The integrand

$$f(t, Y, U) = \overline{U}\mathbb{Q}(t)U + \overline{Y}\mathbb{S}(t)Y$$

remains convex, and is in fact *semi*-strongly convex in that

$$f(t, Y, U) = f(t, Y_0, U_0) \Rightarrow U = U_0 \quad \text{(why?)}.$$

Thus by arguments for uniqueness of the solution, we now can conclude only that equality of performance implies that $U(t) = U_0(t)$. But the competing state functions $Y(t)$ must satisfy the linear equation

$$\dot{Y}(t) = \mathbb{A}(t)Y(t) + \mathbb{B}(t)U_0(t) \quad \text{with } Y(O) = Y_0(O).$$

Therefore $Y(t) \equiv Y_0(t)$ on $[0, T]$, so that full uniqueness still obtains, and we can claim invertibility for the various (sub)matrices required in the foregoing analysis.   □

(10.3) **Remarks.** The methods of linear analysis extend to provide corresponding results for this problem when the state equation takes the form

$$\dot{Y}(t) = \mathbb{A}(t)Y(t) + \mathbb{B}(t)U(t) + E(t) \quad \text{for given } E \in C[0, T].$$

(See [C–L].) As we have noted, the approach indicated in each case is numerically feasible, and in simple examples with small constant matrices, can in fact be carried out explicitly. See Problems 10.14, and 10.15 This is one example of an optimal control problem in which control synthesis is achievable, in that if the system is in state Y at time τ, then there is an associated optimal control $U_0(t, \tau)$ say which can be used to bring the state to its assigned target area at time T.

Such control synthesis although clearly desirable cannot always be expected, especially if there are not enough conditions present to guarantee uniqueness of the optimal control.

Thus, we have theoretical access to a complete solution for certain linear state-quadratic performance problems under the tacit assumption that there are no restrictions on the admissible controls (other than having piecewise continuity). Otherwise, U_0 as defined through (29) might violate these assumptions. It may be possible to accomodate control restrictions such as $|U(t)| \le 1$ by adding a Lagrangian function such as $\mu(t)(|U|^2 - 1)$ to the integrand (25), and proceeding as in the previous examples. Here, we would try to find $\mu \ge 0$ so that $\mu(t)(|U_0(t)|^2 - 1) \equiv 0$, and we have already experienced the complications such conditions impose on the ensuing analysis. In the next section, we shall give a new approach to handling control constraints of an even more general form.

§10.4. Separate Convexity and the Minimum Principle

As we have seen, an optimal control problem (with convexity present) is characterized naturally by the auxiliary function

$$h(t, Y, U) = f(t, Y, U) + P(t) \cdot G(t, Y, U)$$

expressed in terms of the performance function f, the state transformation function G, and the Lagrangian multiplier (vector) function P awaiting further specification. In this section, we uncover an alternative to the convexity of h with respect to U, in the form of the minimum principle of Pontjragin, which dominates the remainder of the book. We use this principle to solve a problem with a nonconvex performance integrand.

In Theorem 10.1, we required convexity of $h(\underline{t}, Y, U)$ on an appropriate set, which entails separate convexity in Y and U, and it is this which we now wish to examine.

(10.4) **Proposition.** *Suppose that for some* $U_0 \in \hat{C}[0, T]$ *and* $P \in \hat{C}^1[0, T]$, $h(\underline{t}, Y, U_0(\underline{t}))$ *is convex on* $[0, T] \times D \times \mathcal{U}$, *where D is open in* $\mathbb{R}^d$ *and* $\mathcal{U} \subseteq \mathbb{R}^k$, *then each* $Y_0 \in \hat{C}^1[0, T]$, *which satisfies the adjoint equation*

$$\dot{P}(t) = -h_Y(t, Y(t), U_0(t)), \tag{34}$$

makes $h_0(t) \overset{\text{def}}{=} h(t, Y_0(t), U_0(t))$ *continuous on* $[0, T]$; *and constant in the autonomous case if additionally,*

$$\dot{Y}_0(t) = G(Y_0(t), U_0(t)) \qquad on \; (0, T).$$

PROOF. With $\tilde{f}(t, Y, U, Z) = h(t, Y, U) - P(t) \cdot Z$, it follows that for $Y \in \hat{C}^1[0, T]$,

$$\tilde{F}(Y, U_0) - \tilde{F}(Y_0, U_0) \geq \int_0^T \{h_Y[Y_0(t), U_0(t)] - P(t) \cdot (\dot{Y}(t) - \dot{Y}_0(t)\} \, dt$$

$$= P(t) \cdot (Y(t) - Y_0(t)) \Big|_0^T = 0,$$

if $Y(0) = Y_0(0)$ and $Y(T) = Y_0(T)$. Hence Y_0 satisfies the *second* Euler–Lagrange equation for $\tilde{f}$, which in the integral form given in §7.5, is that

$$h_0(t) = \tilde{f}[Y_0(t), U_0(t)] - \tilde{f}_Z[Y_0(t), U_0(t)] \cdot \dot{Y}_0(t)$$

$$= \int_0^t \tilde{f}_t[Y_0(\tau), U_0(\tau)] \, d\tau + c_0,$$

for some constant c_0. Thus h_0 is continuous, and it is constant in the autonomous case where

$$\dot{Y}_0 = G(Y_0, U_0) \quad \text{and} \quad h(t, Y, U) = f(Y, U) + P(t) \cdot G(Y, U),$$

since then

$$\tilde{f}_t[Y_0(t), U_0(t)] = \dot{P}(t) \cdot G(Y_0(t), U_0(t)) - \dot{P}(t) \cdot \dot{Y}_0(t) = 0,$$

(except at corner points). $\square$

In particular, under the hypotheses of Theorem 10.1, we may expect the optimal h_0 to be continuous, possibly constant, and this can assist in the detailed analysis of particular candidates for optimality, since discontinuit-

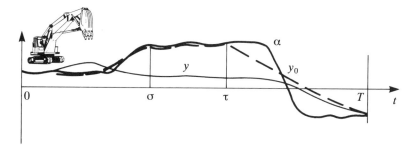

Figure 10.2

ies in U_0 are limited both in location and values to those for which $h(t, Y_0(t), U_0(t-)) = h(t, Y_0(t), U_0(t+))$.

Example 1 (Optimal Highway Design). A construction company wants to build a highway in terrain that has local altitude of $\alpha = \alpha(t)$, over a horizontal distance of T meters as indicated in Figure 10.2. Assume that land preparation costs (excavation and fill) for a highway of altitude $y = y(t)$ at $t \in [0, T]$ is proportional to $(y(t) - \alpha(t))^2$ per meter, and that safety concerns dictate that the grade $|\dot{y}|$ not exceed $a = 1/10$.

Then the company would try to find $y \in \hat{\mathcal{Y}} = \hat{C}^1[0, T]$ that minimizes

$$F(y, u) \stackrel{\text{def}}{=} \int_0^T (y - \alpha)^2(t)\, dt, \quad \text{with} \quad \dot{y}(t) = u(t) \in \mathcal{U} = [-a, a]. \quad (35)$$

Here, $f(t, y, u) = (y - \alpha(t))^2$, so the problem is not autonomous and for a suitable $p \in \hat{\mathcal{Y}}$,

$$h(\underline{t}, y, u) = (y - \alpha(\underline{t}))^2 + p(\underline{t})u$$

is convex. Therefore the choices of an optimal grade strategy $u_0 = \dot{y}_0$ are limited to those for which

$$h_0 = (y_0 - \alpha)^2 + pu_0 \quad \text{is continuous;}$$

in particular, pu_0 must be continuous which means that u_0 can have a discontinuity only at points where p vanishes. Of course, we suppose that p satisfies the adjoint equation (26a)

$$-\dot{p} = h_y = 2(y_0 - \alpha), \quad (36)$$

and we can use Theorem 10.1(i) if we take

$$p(0) = p(T) = 0, \quad (36')$$

since y_0 is unspecified at either endpoint. However, we would also need to invoke (26b), $0 = h_u = p$, which means that $y_0(t) \equiv \alpha(t)$, thereby eliminating our ability to limit the grade-size. We will return to this problem after we examine an alternative to (26b).

In the context of the previous proposition, suppose that for each non-corner point $t \in [0, T]$

$$0 = h_U[Y_0(t), U_0(t)]. \tag{37}$$

Then under convexity of $h(t, Y_0(t), U)$ on $\mathcal{U}$, it follows by §0.8, that this function of U is *minimized* at $U_0(t)$, or that

$$\min_{U \in \mathcal{U}} h(t, Y_0(t), U) = h_0(t) = h(t, Y_0(t), U_0(t)); \tag{38}$$

moreover, under continuity of h_0, this must hold for all t, since then the equivalent inequality

$$h(t, Y_0(t), U) \geq h_0(t), \qquad \forall\, U \in \mathcal{U}.$$

persists. (Why?)

This last inequality shows that the control U_0 is *instantaneously* optimal in minimizing $h(t, Y_0(t), U)$ at each t, and (38) affords our first glimpse of this principle credited to Pontrjagin (c. 1955),[1] (which is formulated solely as a *minimum principle* in this book.) Observe that if it holds for all U near $U_0(t)$ (in particular, if $\mathcal{U}$ is open) and $h(t, Y_0(t), U)$ is differentiable with respect to U, then by ordinary calculus we recover (37). However, it can hold even when these conditions are not met, as for example, when $f(t, Y, U) = -|U|$. Moreover, if $\mathcal{U}$ is *convex* and h is continuously differentiable in the components of U, then by the mean value theorem,

$$(38) \Rightarrow h_U[Y_0(t), U_0(t)] \cdot (U - U_0(t) \geq 0, \qquad U \in \mathcal{U}. \tag{39}$$

This result of Loewen shows that Theorem 10.1 holds as stated when the minimal inequality replaces equation (26b). (See Problem 10.28.)

Now, let's see how this helps the highway design project of Example 1. First, we note that $\mathcal{U} = [-a, a]$ is convex, and that for any $y_0 \in \widehat{\mathcal{Y}}$,

$$h(t, y_0(t), u) = (y_0 - \alpha)^2(t) + p(t)u \tag{40}$$

is minimized on $\mathcal{U}$ by $u = u_0(t) = \pm a$ if $p(t) \lessgtr 0$. (A minimizing u_0 is undetermined at points where p vanishes.) Therefore according to this analysis, the optimal highway consists of segments of maximal grade where $\dot{y}_0 = \pm a$, joining sections that follow the given terrain (over which $p = 0$). Theorem 10.1 as modified by (39) shows that this descriptive choice *is* optimal, and in fact it is unique because $h(\underline{t}, y, \underline{u})$ is strictly convex. (See also, the next proposition.) For the terrain of Figure 10.2, y_0 is as indicated, where the interval endpoints τ and σ can be approximated numerically. (See Problem 10.20.) Observe that through the minimum principle we have obtained a useful qualitative description of the optimal highway without actually solving the problem.

---

[1] In fact, the first version of the principle appeared in 1950 in a then little-known work of Hestenes (see [He]).

In using (39) to modify Theorem 10.1, we still required joint convexity of $h(\underline{t}, Y, U)$. The minimum principle permits alternatives. For instance, if

$$h^*(t, Y) = \min_{U \in \mathcal{U}} h(t, Y, U), \tag{41}$$

is defined, then we have the following result first considered by Arrow and Kurtz. (For some generalizations, see [S–S].)

(10.5) Proposition. *In Theorem 10.1, suppose that $h^*(\underline{t}, Y)$ is [strictly] convex on D. Then the conclusions remain valid [with uniqueness in Y only] when instead of (26b) and the convexity of h, the minimum principle (38) holds; i.e., $h^*(t, Y_0(t)) = h_0(t) = h(t, Y_0(t), U_0(t))$.*

PROOF*. When (Y, U) and (Y_0, U_0) are in $\mathcal{D}$ and we suppress the argument t in the integrands, we have as before that

$$F(Y, U) - F(Y_0, U_0) \geq \tilde{F}(Y, U) - \tilde{F}(Y_0, U_0)$$

$$= \int_0^T [h(t, Y, U) - h(t, Y_0, U_0)] \, dt - \int_0^T P \cdot (\dot{Y} - \dot{Y}_0) \, dt$$

$$\geq \int_0^T [h^*(t, Y) - h^*(t, Y_0) \, dt - \int_0^T P \cdot (\dot{Y} - \dot{Y}_0) \, dt,$$

(by (41) and (38))

$$\geq \int_0^T h_Y^*(t, Y_0) \cdot (Y - Y_0) \, dt - \int_0^T P \cdot (\dot{Y} - \dot{Y}_0) \, dt,$$

(by the hypothesized convexity of $h^*(\underline{t}, Y)$).

It is not obvious, but we show below that

$$h_Y^*(t, Y_0) = h_Y(t, Y_0, U_0) \quad (= -\dot{P} \text{ from (26a)}). \tag{42}$$

Therefore, the right side of the last inequality becomes $-P \cdot (Y - Y_0)(t)|_0^T$, which vanishes in cases (i), (ii), or (iii) of Theorem 10.1, and establishes our result [since equality implies that $Y = Y_0$ under strict convexity of $h^*(\underline{t}, Y)$]. For case (iv), we can introduce the Bolza function F^* and modify the argument as before to reach the same conclusions.

It remains to establish (42), and in doing so, we can hold $t \in [0, T]$ fixed. Then, for all Y sufficiently near $Y_0 = Y_0(t)$,

$$Y_\varepsilon \overset{\text{def}}{=} (1 - \varepsilon) Y_0 + \varepsilon Y, \qquad 0 < \varepsilon \leq 1,$$

is also near Y_0 so that with $U_0 = U_0(t)$:

$$h(t, Y_\varepsilon, U_0) - h(t, Y_0, U_0) \geq h^*(t, Y_\varepsilon) - h^*(t, Y_0)$$

$$\geq h_Y^*(t, Y_0) \cdot (Y_\varepsilon - Y_0) = \varepsilon h_Y^*(t, Y_0) \cdot (Y - Y_0).$$

If we now divide by $\varepsilon > 0$ and let $\varepsilon \searrow 0$, we find by the chain-rule that

$$h_Y(t, Y_0, U_0) \cdot (Y - Y_0) \geq h_Y^*(t, Y_0) \cdot (Y - Y_0)$$

and this can hold for *all* Y near Y_0, *iff* the derivative vectors on each side are identical.   □

(10.6) **Remarks.** Since the form of h* depends on the unknown adjoint function P, it is not easy to predict that h* will exhibit the desired convexity. However, it clearly does so when

$$h(t, Y, U) = h'(t, Y) + h''(t, U) \tag{43}$$

and $h'(t, Y)$ is [strictly] convex on D.

Example 2. To minimize

$$F(y, u) = \int_0^1 [y^2 - y - (1 - u)^2](t)\, dt$$

on

$$\mathcal{D} = \{y \in \hat{C}^1[0, 1]; u \in \hat{C}[0, 1]: y(0) = 0\}$$

under the state equation $\dot{y} = u$, where $u(t) \in \mathcal{U} = [0, 2]$, we note that the problem is autonomous. Moreover, if $p \in \hat{C}^1[0, 1]$, then

$$h(t, y, u) \overset{\text{def}}{=} y^2 - y - (1 - u)^2 + p(t)u$$

$$= h'(y) + h''(t, u), \tag{44}$$

where $h'(y) = y^2 - y$ is strictly convex, but

$$h''(t, u) = -(1 - u)^2 + p(t)u$$

is not convex.

In fact $h''(t, u)$ is a concave parabolic function in u, so that for $0 \le u \le 2$, it is minimized at $u = u_0$ where $u_0 = 0$ or $u_0 = 2$. Thus, by Proposition 10.4, if $y_0 \in \hat{C}^1[0, 1]$, and $\dot{y}_0 = u$, then

$$h_0 = y_0^2 - y_0 - (1 - u_0)^2 + pu_0 = c_0, \qquad \text{a constant,}$$

when p satisfies the adjoint equation

$$-\dot{p} = h_y = 2y_0 - 1 \quad \text{(with } p(1) = 0). \tag{45}$$

Since $(1 - u_0)^2 = 1$ when $u_0 = 0$ or 2, we see that $y_0^2 - y_0 + pu_0 = c_0 + 1$ is also constant, and conclude that switching from $u_0 = 0$ to $u_0 = 2$ can occur only at points where p vanishes. Finally, we know that if it exists, the minimizing y_0 with $y_0(0) = 0$, is unique. Hence we can first try the simplest possibilities.

Case 1. $u_0(t) \equiv 0$. Then $\dot{y}_0 = u_0 = 0$ so that $y_0(t) = y_0(0) = 0$, and by (45), $\dot{p} = 1$ with $p(1) = 0$, so that $p(t) = t - 1 \le 0$. However, from (44), we see that for $t < 1$,

$$h(t, y_0, u) = -(1 - u)^2 + (t - 1)u$$

is minimized only when $u = 2$ and has the minimum value of $-1 + (t - 1)2 \ne h_0(t)$, since $h_0(t)$ is constant. This solution does not satisfy the minimum principle (38). It may be the optimal solution but we cannot use our proposition to verify its optimality (or that of the other simple case $u_0(t) = 2$. (See Problem 10.22.)

Case 2. We must look for a solution with at least one switching point σ, say, where p vanishes. Since $p(1) = 0$, it is simplest to try $p(t) = 0$ for $t \geq \sigma$. Then $2y_0(t) - 1 = -\dot{p} = 0$ or $y_0(t) = \frac{1}{2}$, so that $u_0(t) = \dot{y}_0(t) = 0$ for $t \geq \sigma$. But then for $t \leq \sigma$ we should try $u_0(t) = 2$ so that $y_0(t) = 2t$, and continuity of y_0 at σ requires that $\sigma = \frac{1}{4}$. Finally, for $t \leq \frac{1}{4}$: $\dot{p}(t) = -2y_0(t) + 1 = -4t + 1$ so that $p(t) = -2t^2 + t + c$, and $p(1/4) = 0 \Rightarrow c = -\frac{1}{8}$ so that $p(t) = -2(t - \frac{1}{4})^2 \leq 0$. Now for $t \leq \frac{1}{4}$, $h(t, 2t, u)$ is minimized when $u = u_0 = 2$ (since $p \leq 0$) and it has the minimum value $h_0(t) = -\frac{5}{4}$. Similarly for $t \geq \frac{1}{4}$: $h(t, \frac{1}{2}, u) = (\frac{1}{2})^2 - \frac{1}{2} - (1 - u)^2$ is minimized when either $u = 2$ or $u = u_0 = 0$ and it has the minimum value $h_0(t) = -\frac{5}{4}$. Therefore for all $t \in [0, 1]$ this y_0 satisfies the minimum principle. Consequently, the unique solution to our problem is given by

$$y_0(t) = 2t; \quad \text{and} \quad u_0(t) = 2, \qquad t < \tfrac{1}{4},$$

$$= \tfrac{1}{2}; \qquad\qquad\qquad = 0, \qquad t > \tfrac{1}{4},$$

and the optimal control is bang–bang.

One final observation may be of value. As we have seen, in analysis of optimal control problems, the Lagrangian multiplier P acquires equal status with the state vector Y. In some sense we are optimizing with respect to Y and P through the control U, and it is sometimes useful to reflect this fact by writing the auxiliary function as

$$H(t, Y, U, P) = f(t, Y, U) + P \cdot G(t, Y, U).$$

Then note that formally, equations (26a) and (26c) become

$$\dot{P} = -H_Y, \qquad \dot{Y} = H_P, \tag{46}$$

which resemble those canonical equations of §8.4 for a kind of "Hamiltonian" H. Although the presence of U prevents this from providing a complete analogy, it is possible to incorporate some features of Hamiltonian mechanics in analyzing problems of optimal control. Indeed, it is this partial analogy which seems to be responsible for the original motivation of the Pontrjagin (maximal) principle; (particularly illuminating discussions are given in [Y] and in [Lo]). Further sufficiency theorems incorporating convexity-type inequalities can be formulated in terms of hypothesized solutions to an associated Hamilton–Jacobi equation, but these too, cannot be readily implemented in the solution of actual problems. However, there are related sufficiency theorems utilizing suitable field theory-like extensions of the Weierstrass method of §9.1 which can be carried through. (See [Le].) A sufficiency result free of convexity assumptions is given in Problem 10.25.

PROBLEMS

10.1.   Minimize $F(y, u) = \int_0^1 u^2(t)\, dt$ under $\dot{y} = -2y + u$ with $y(0) = 1$, $y(1) = 0$. Is your solution unique?

10.2.   Minimize $F(y, u) = \int_0^1 (2 - 5t)u(t)\, dt$, with $y(0) = 0$, $y(1) = e^2$, under:
(a) $\dot{y}(t) = 2y(t) + 5e^{2t}u(t)$, $|u(t)| \le 1$. Is your solution unique?
(b*) $\dot{y}(t) = 2y(t) + 4e^{2t}u(t)$, $|u(t)| \le 1$. Hint: Let $v(t) = e^{-2t}y(t)$.

10.3.   To *maximize* $F(y, u) = \int_0^1 y^2(t)\, dt$ under $\dot{y} = u$ with $y(0) = 0$, $|u(t)| \le 1$, show that $|y(t)| \le |t|$, and hence $F(y, u) \le ?$ When does equality hold?

10.4.   In minimizing $F(y, u) = \int_0^1 (u(t) + 1)\, dt$, under $\dot{y} = -u$ with $y(0) = y(1) = 0$, show that each control $u \in \hat{C}[0, 1]$ with $\int_0^1 u(t)\, dt = 0$ is optimal.

10.5*. An old cylindrical concrete tank of height 30 feet contains water of depth y used for drinking; it leaks at a rate proportional to y. The initial depth is 12 feet and over the next 100 days we can supply the tank with water from a controllable but limited source. Suppose that the resulting state equation is approximated by $\dot{y} = -0.1y + u$, ft/day, where $u(t) \in [0, 1]$ represents the net inflow limited by $\int_0^{100} u(t)\, dt = 60$, say. How should we control u to *maximize* the mean depth $100^{-1} \int_0^{100} y(t)\, dt$ during the time interval $[0, 100]$? Hints: Let $y_1 = e^{\cdot lt} y$, and write the control inequality as $u(u - 1) \le 0$.

10.6.   Obtain the energy formula (17) of §10.2(f) by writing the square of a single integral as a double integral.

10.7.   (a) (Solution by steepest descent.) If (22) holds, then (20) becomes $|Y|^{\cdot} = V \cdot U$. Use the Cauchy inequality (§0.1) to get $|Y|^{\cdot} \ge -|U|$ (with equality...). By integration, conclude that $\int_0^T |U|\, dt \ge |A|$ under given hypotheses (with equality...). Formulate an optimal control problem for the spinning body in which the fuel consumed (by say torque producing jets) in bringing the body to spin rest at time T, as measured by the above integral, is minimized. Which controls U_0 will be optimal; i.e., how will they be directed?
(b) Combine the inequality in part (a) with the Schwarz inequality for integrals [Rud] to obtain

$$|A|^2 \le T\int_0^T |U|^2\, dt \quad \text{(with equality only if } |U(t)| = \text{const.).}$$

Conclude that the energy integral on the right is minimized for fixed $T \ge |A|$ by the controls

$$U_0(t) = -\frac{|A|}{T} \frac{Y_0(t)}{|Y_0(t)|}.$$

10.8.   Under the same conditions as in Problem 10.7, show that for $k > 0$, the controls $U_0(t) = -mY_0(t)/|Y_0(t)|$ with $m = \min(k, 1)$, minimize $\int_0^T [k + |U|^2(t)]\, dt$. Determine the associated minimum value(s) and optimal time(s) of transfer. Argue that this control prescription is unique.

10.9.   For the spinning body in §10.2(g), a measure of resistance to motion can be admitted if aY is *subtracted* from the right side of the state equation(s), where a is a positive scalar constant. Carry out the time optimal analysis for the resulting problem. Hint: Equation (20) reduces to $|Y|^{\cdot} = -a|Y| + V \cdot U$. Now let $v(t) = e^{at}|Y(t)|$.

10.10. To *maximize* $\dot{y}(2)$, if $\ddot{y} = u$ with $|u| \le 1$, where

$$y(0) = \dot{y}(0) = 0; \qquad y(2) = 1:$$

(a) Explain why it would suffice to find (y_0, u_0) meeting the above conditions which minimizes

$$\tilde{F}(y, u) = \int_0^2 [-u + \rho(\ddot{y} - u) + \mu(u^2 - 1)](t)\, dt$$

with ρ, μ continuous and $\mu \ge 0$ on $[0, 2]$, such that

$$\mu(t)(u_0^2(t) - 1) \equiv 0. \tag{47}$$

(b) Examine the convexity of the associated integrand $\tilde{f}(t, u, \ddot{y})$.
(c) Consider $\Delta = \tilde{F}(y, u) - \tilde{F}(y_0, u_0)$, and integrate the $\rho(\ddot{y} - \ddot{y}_0)$ term by parts to show that if y also meets the above conditions, then

$$\Delta \ge \rho(2)(\dot{y} - \dot{y}_0)(2) + \int_0^2 [\ddot{\rho}(y - y_0)](t)\, dt$$

when

$$-(1 + \rho) + 2\mu u_0 \equiv 0. \tag{48}$$

(d) Conclude that $\Delta \ge 0$, if also $\rho(t) = c(2 - t)$ for any constant c.

10.11. (e) Observe that $u_0 = \pm 1$ satisfies (47) and with (d), (48), provided that

$$2\mu = (1 + \rho)u_0 = (1 + c(2 - t))u_0 \ge 0.$$

Argue that $u_0 = \operatorname{sgn}(1 + c(2 - t))$ suffices.
(f) We must determine c so that

$$y_0(2) = 1 = \int_0^2 \dot{y}_0(t)\, dt = \int_0^2 (2 - t)u_0(t)\, dt,$$

or after substitution from (e):

$$1 = \int_0^2 \tau \operatorname{sgn}(1 + c\tau)\, d\tau.$$

Show that this requires $c = -1/\sqrt{3}$.
(g) Find the optimal $\dot{y}_0(2)$, and explain why y_0 is unique. Hint: Why is the associated optimal control u_0 unique?
(h) Note that if $y_0(2) = 2$, then any $c \ge -\frac{1}{2}$ suffices in (f). Does this contradict uniqueness?

10.12. To minimize $F(y, u) = \int_0^\pi [(y - e^{-t})^2 + u^2]\, dt$ under

$$\dot{y}(t) = g(t, u) \stackrel{\text{def}}{=} (1 + \sin t)u(t) \quad \text{with} \quad y(0) = 0.$$

(a) Reformulate this as a related problem for

$$\tilde{F}(y, u) = F(y, u) + \int_0^\pi p(t)[g(t, u) - \dot{y}(t)]\, dt;$$

(b) Explain how convexity might be used to characterize a minimum for $\tilde{F}$ when $p(\pi) = 0$.
(c) Obtain the Euler–Lagrange equations for the integrand of (b);
(d)* Try to solve these equations.
(e) Should the solution be unique? Explain.

10.13. Minimize $F(Y, u) = \int_0^2 u^2(t)\, dt$ where

$$Y(0) = (y, y_1)(0) = \mathcal{O} \quad \text{and} \quad \dot{Y} = (y_1, u),$$

using multipliers $P = (p, p_1)$ with $P(2) = \mathcal{O}$. Is optimal control u_0 unique? bang–bang?

10.14. Minimize $F(Y, u) = \int_0^2 u^2(t)\, dt$, where

$$\dot{Y} = (\dot{y}, \dot{y}_1) = (y_1, u - y_1), \qquad Y(0) = \mathcal{O},$$

and $y + 5y_1 = 15$ at time $t = 2$. Hint: Theorem 10.2 with $\mathbb{M} = [1, 5]$.

10.15*. (a) To minimize

$$F(u) = \int_0^{100} \frac{dx}{x + (25 + x/4)u(x)}$$

on $\hat{C}[0, 100]$, with

$$G(u) = \int_0^{100} \frac{u(x)\, dx}{x + (25 + x/4)u(x)} = 1,$$

where $0 \le u(x) \le 1$, show that it suffices to find u_0 which minimizes

$$\tilde{F}(u) = F(u) + \lambda G(u) + \int_0^{100} \mu(x)(u(u - 1))(x)\, dx$$

for an appropriate constant λ, and continuous $\mu \ge 0$, such that $\mu u_0(u_0 - 1) \equiv 0$.

(b) For the appropriate $\tilde{f}$, set $\tilde{f}_u = 0$ and note that $\mu(x) = 0$ only when $x = \zeta = 25/\lambda_1$, where $\lambda_1 = \lambda - 4^{-1}$. Otherwise, we must take $u_0(x) = 1$ for $x < \zeta$, $u_0(x) = 0$, $x > \zeta$, in order to have $\mu \ge 0$. Show that $G(u_0) = 1 \Rightarrow \zeta = 20(e^{5/4} - 1)$, so that $\lambda = 25/\zeta + 1/4$ is known.

(c)* For convexity, show that $\tilde{f}_{uu} \ge 0$ when $u \ge 0$. (This is immediate when $x < \zeta$; for $x > \zeta$, it is essential to use the fact that $\mu(x) = (x\lambda_1 - 25)/x^2$.)

(d) Observe that μ obtained in (b) is continuous, and conclude (through (3.10)) that u_0 is indeed the unique optimal "control" for this problem (which is given a physical origin in Problem 11.26).

10.16. Under the conditions of Problem 10.13, find $P(2)$ to minimize

$$F^*(Y, u) = |Y(2) - (5, 2)|^2 + F(Y, u).$$

10.17. (a) Transform the problem of minimizing

$$\int_0^{2\pi} (y^2 + u^2)(t)\, dt \quad \text{under} \quad \ddot{y} + \dot{y} = u$$

with say $y(0) = \dot{y}(0) = 0$, $y(2\pi) = 1$, into one for which Theorem 10.1 applies. Does Theorem 10.2 apply? Explain.

(b) Find an optimal trajectory for this problem. Is it unique? Hint: Eliminate u and transform to a Bolza problem with $(\ddot{y} + \dot{y})(2\pi) = 0$.

10.18. (a) Explain how the hypotheses of Theorem 10.1 should be modified to admit a single state-control inequality such as

$$\psi(t, Y(t), U(t)) \le 0, \quad \text{where } \psi \text{ is } C^1.$$

(b) How should the proof be modified?
(c) How will the conclusions change?
(d) Use your result to reexamine Problem 10.10.
(e) Make a similar analysis of Proposition 10.5.

10.19. Consider Theorem 10.2 in case (i) of Theorem 10.1. Does a solution (Y_0, U_0) exist? Is it unique?

10.20. For Example 1, in §10.4:
(a) Explain why σ should be chosen to make $\int_0^\sigma (y_0 - \alpha)\, dt = 0$, and how you might locate σ by using the graph of α in Figure 10.2.
(b) How should τ be chosen?

10.21. (a) In minimizing

$$F(y, u) = \int_0^T |u|(t)\, dt \quad \text{under} \quad \dot{y} = u,\ |u(t)| \le 1,$$

with given $y(0)$ and $y(T) = \mathcal{O}$, show that we should choose

$$u_0(t) = \begin{cases} 0 & |p(t)| < 1, \\ -\operatorname{sgn} p(t), & |p(t)| > 1, \end{cases}$$

for an appropriate adjoint function p. Use this fact to solve the problem when $y(0) = 1$, and $T = 1$.
(b) What happens if $T = 2$? if $T = \frac{1}{2}$?

10.22. (a) In Example 2 of §10.4, verify that $u_0(t) = 2$ does not satisfy the minimum principle for h.
(b) Solve the corresponding problem when y is added to the integrand of F.

10.23. Minimize $\int_0^1 (u^2 - 2u)(t)\, dt$, under the *isoperimetric* inequality $\int_0^1 t|u|(t)\, dt \le \frac{1}{6}$. Hint: Introduce y with $\dot{y}(t) = t|u|(t)$, $y(0) = 0$.

10.24. Minimize $\int_0^1 [\dot{y} - (2t - 1)]^2\, dt$, subject to $y(0) = y(1) = 0$, with $y \ge 0$, by introducing $\dot{y} = u$. Hint: Problem 10.18.

10.25. An elementary sufficiency criterion can be obtained as follows:
(a) Set $\tilde{h} = h - \dot{P} \cdot Y$ and as in Theorem 10.1, show that

$$\tilde{F}(Y, U) - \tilde{F}(Y_0, U_0) = \int_0^T [\tilde{h}(t, Y(t), U(t)) - \tilde{h}(t, Y_0(t), U_0(t))]\, dt$$

$$- P(t) \cdot (Y(t) - Y_0(t)) \Big|_0^T.$$

(b) Conclude that if $\tilde{h}(t, Y, U) \ge \tilde{h}(t, Y_0(t), U_0(t))$ when $(t, Y, U) \in [0, T] \times D \times \mathcal{U}$, then (Y_0, U_0) minimizes $F(Y, U)$ on $\mathcal{D}_\Phi$, if $P(T) \cdot (Y - Y_0(T)) \le 0$ when $\Phi(Y) = \mathcal{O}$, under the given Lagrangian inequality.
(c) Here, there are no convexity assumptions. How would you modify the hypotheses to obtain a unique Y_0? optimal control?

10.26. (a) Use Theorem 10.2 to minimize

$$\int_0^2 |U(t)|\, dt \quad \text{under} \quad \dot{Y} = U \in \mathbb{R}^2 \quad \text{where} \quad U = (u, u_1),$$

with $Y(0) = \mathcal{O}$ and $(e^y - y_1)(2) = 0$. (Take $|P| = 1$.)

(b) This problem has a simple graphical interpretation. Use the result of Problem 10.25 to contrast what happens when instead $(e^y - y_1)(2) = 2$.

10.27. (Optimal oil production.) An offshore oil field estimated to contain K barrels is to be pumped out by w identical rigs over a large number of days, T. Each oil rig costs R dollars to install and has a maximum production rate of m barrels per day of oil that will sell at a fixed price of D dollars per barrel. At discount rate r, the company should control both w and $v = v(t)$, the fraction of its total productive capacity in use at time t, to maximize

$$D \int_0^T e^{-rt} v(t) mw \, dt - Rw \quad \text{where} \quad \dot{y} = v(mw), \; y(0) = 0,$$

with $0 \le v(t) \le 1$ and $y(T) \le K$. (See [S–S] and Problem 6.44.)
(a) Show that this leads to a nonconvex (and nonstandard) minimization problem in optimal control.
(b) Explain why it is enough to introduce $W = mw$, $B = R/mD$, and minimize

$$F(y, W) = BW - \int_0^T e^{-rt} \dot{y}(t) \, dt, \quad \text{where} \quad \dot{y}(0) = 0, \; y(T) \le K,$$

and $0 \le \dot{y}(t) \le W$.
(c)* Verify that (y_0, W_0) with $0 \le \dot{y}_0 \le W_0$, $y_0(0) = 0$, $y_0(T) = K$, minimizes provided that we can find $\mu = \mu(t) \le 0$, $v = v(t) \le 0$, and $p > 0$, for which

$$B = \int_0^T \mu(t) \, dt, \quad \mu(\dot{y}_0 - W_0) \equiv v\dot{y}_0 \equiv 0 \quad \text{and} \quad (\mu - v)(t) = e^{-rt} - p.$$

Hint: See §10.2(e).
(d) Show that one optimal strategy is to take $\dot{y}_0 \equiv W_0$, $t \le \tau$; $\dot{y}_0 = 0$, $t > \tau$, where $W_0 = K/\tau$; find τ implicitly when $Br < 1$. (Assume $T > \tau$, and ignore the fact that $w_0 = W_0/m$ should be an integer.) Is this the only optimal strategy?

10.28. To establish (39): if $U_0 = U_0(t)$ and U are in $\mathcal{U}$, then for $0 < \varepsilon \le 1$ note that $U_\varepsilon \stackrel{\text{def}}{=} (1 - \varepsilon)U_0 + \varepsilon U$ is also in $\mathcal{U}$. Show that for some $\tilde{\varepsilon} < \varepsilon$,

$$0 \le h(t, Y_0, U_\varepsilon) - h(t, Y_0, U_0) = \varepsilon h_U(t, Y_0, U_{\tilde{\varepsilon}}) \cdot (U - U_0),$$

where $Y_0 = Y_0(t)$. Now divide by ε and let $\varepsilon \to 0$. Where is the convexity of $\mathcal{U}$ used?

10.29*. (A fish harvesting problem.) A large population of fish is to be taken commercially over the next 4 years. During this period, its size y (in millions) at time t can be described approximately by the equation

$$\dot{y} = y - y^2 - uy, \qquad 0 \le t \le 4,$$

where $u = u(t)$ represents the fraction of the existing population to be harvested at time t, and the quadratic term accounts for possible natural loss due to overcrowding, fighting, etc. (If there is no fishing, this population could stabilize at one million.) The fishing industry wants to control u to maximize its yield $\int_0^4 (uy) \, dt$, where $y(0) = a$ is assumed known and positive, and $0 \le u(t) \le 1$.

(a) Show that this leads to a nonconvex minimization problem in optimal control, and identify the source(s) of nonconvexity.

(b) Following the model in §10.2(e) show that it suffices to minimize

$$F(y) = \int_0^4 (y^2 - y + \dot{y})(t)\, dt$$

under the Lagrangian *inequalities*

$$y^2 - y + \dot{y} \le 0, \qquad -y^2 - \dot{y} \le 0.$$

Conclude that it suffices to find $y_0 > 0$ with $y_0(0) = a$ which minimizes

$$\tilde{F}(y) = \int_0^4 [(1 + \mu - v)(y^2 + \dot{y}) - (1 + \mu)y]\, dt,$$

for appropriate continuous multiplier functions $\mu \ge 0$, $v \ge 0$, if

$$\mu(y_0^2 - y_0 + \dot{y}_0) \equiv v(y_0^2 + \dot{y}_0) \equiv 0 \qquad (\Rightarrow \mu v y_0 \equiv 0 \Rightarrow \mu v \equiv 0).$$

(c) Note that the integrand $\tilde{f}(t, y, \dot{y})$ is strongly convex when $p \overset{\text{def}}{=} (1 + \mu - v) > 0$. Explain why it then suffices to find $y_0 > 0$, $p > 0$, satisfying $\dot{p} = 2py - (1 + \mu)$, with $y_0(0)$ given, $p(4) = 0$, and μ, v as in (b).

(d)* For $y(0) = \frac{1}{4}$, conclude that

$$y_0(t) \overset{\text{def}}{=} \begin{cases} (1 + 3e^{-t})^{-1}, & 0 \le t \le \log 3, \\ 1/2, & \log 3 \le t \le 2, \\ 1/t, & 2 \le t \le 4, \end{cases}$$

provides the *unique* minimum sought (so that $u_0 \overset{\text{def}}{=} 1 - y_0 - (\dot{y}_0/y_0)$ is the unique optimal control). Hint: Calculate $y_0^2 - y_0 + \dot{y}_0$ and $y_0^2 + \dot{y}_0$ in each subinterval, and conclude that $\mu = 0$ except on $[0, \log 3]$, while $v = 0$ except on $[2, 4]$. Then use the reduced equation for p to find these functions on the remaining intervals. Check that they are nonnegative, while $p > 0$, with $p(4) = 0$.

(e)** Obtain the solution in (d) directly by solving the equations for (nonnegative) p, μ, v, y_0 from the previous parts. Outline: Note that $\mu v \equiv 0$ with $p(4) = 0 \Rightarrow$ *near* 4, $v > 0$, $\mu = 0$, so that $\dot{y}_0 = -y_0^2$ which can be integrated. Then solve $\dot{p}$ equation for $(1 - v)$ and determine τ_2 at which $v(\tau_2) = 0$. Argue that in next interval $\mu = v = 0$ so that $y_0 = \frac{1}{2}$. Finally, permit $\mu > 0$, and conclude that in such intervals (where $v = 0$), $\dot{y}_0 = -y_0^2 + y_0$, which is an integrable equation (for $x = 1/y_0$) with a unique solution having $y_0(0) = \frac{1}{4}$. Then solve $\dot{p}$ equation for $(1 + \mu)$ and find $\tau_1 < \tau_2$ at which $\mu(\tau_2) = 0$, while $y_0(\tau_1) = \frac{1}{2}$. Assemble results.

(f)* Analyze the simpler case when $y(0) = \frac{1}{2}$.

CHAPTER 11

Necessary Conditions for Optimality

What conditions are necessary for optimal performance in our problems? In Chapter 10 we saw that if a control problem can be formulated on a fixed interval and its defining functions are suitably convex, then the methods of variational calculus can be adapted to suggest sufficient conditions for an optimal control. In particular, the minimum principle of §10.3 and §10.4 can guarantee optimality of a solution to the problem. In §11.1 we will discover that this principle is necessary for optimality whether or not convexity is present, even when the underlying interval is not fixed (Theorem 11.10). Then in §11.2, we examine the simple but important class of linear time-optimal problems for which the time interval itself is being minimized and the adjoint equation (a necessary condition) can be used to suggest sufficient conditions for optimality. Finally, in §11.3, we extend our control-theory approach to more general problems involving Lagrangian inequality constraints, and in Theorem 11.20 we obtain a Lagrangian multiplier rule of the Kuhn–Tucker type.

The material in this chapter (and in §A.7) is significantly more difficult than that of previous chapters, even though it represents only a small selection of the results to be found in [Po], [Ber], [He], and [M–S], among others.

§11.1. Necessity of the Minimum Principle

In this section we wish to show that the sufficient conditions obtained in §10.4 in the presence of convexity, are in general necessary for an optimal solution whether or not the integrand exhibits convexity. Recall that this was also the case for the fixed interval problem in the variational calculus (Chapters 3 and 6).

Specifically, we shall prove the necessity of the minimum principle, which asserts that optimality of a control U_0 guarantees its pointwise minimization of some related function h, in that

$$h(t, Y_0(t), U) \geq h(t, Y_0(t), U_0(t)), \qquad \forall\, U \in \mathcal{U}, \tag{1}$$

where $\mathcal{U}$ is the control set, and Y_0 is an associated optimal state function.

Moreover, we expect that for performance and state functions f and G, h might take the form

$$h(t, Y, U) = f(t, Y, U) + P(t) \cdot G(t, Y, U) \tag{2}$$

for some function P to be determined, but which probably satisfies the adjoint equation

$$\dot{P} = -h_Y(\cdot, Y_0, U_0), \tag{3}$$

along the optimal trajectory, with related target conditions.

Now, under classical assumptions that $\mathcal{U}$ is open and h is differentiable in U, (1) implies that $h_U(\cdot, Y_0, U_0) \equiv \mathcal{O}$. Then the minimum principle is just the Weierstrass necessary condition (7.15), for the modified integrand

$$\tilde{f}(t, Y, U, \dot{Y}) = h(t, Y, U) - P(t) \cdot \dot{Y}, \tag{4}$$

if we think temporarily of U as $\dot{U}$. Indeed, the relevant combination for $W = (U, V) \in \mathbb{R}^{k+d}$, is the excess function given by

$$\tilde{f}(\cdot, Y_0, U, V) - \tilde{f}(\cdot, Y_0, U_0, \dot{Y}_0) - \tilde{f}_{(U,\dot{Y})}(\cdot, Y_0, U_0, \dot{Y}_0) \cdot (W - (U_0, \dot{Y}_0))$$

$$= h(\cdot, Y_0, U) - h(\cdot, Y_0, U_0) - P \cdot (V - \dot{Y}_0) + P \cdot (V - \dot{Y}_0)$$

$$= h(\cdot, Y_0, U) - h(\cdot, Y_0, U_0) \quad \text{(assuming that } h_U(\cdot, Y_0, U_0) = \mathcal{O})^1$$

and this is nonnegative when (1) holds.

Finally, since from (4),

$$\tilde{f}(t, Y, U, \dot{Y}) = f(t, Y, U) + P(t)(G(t, Y, U) - \dot{Y}), \tag{5}$$

we see that the minimum principle involves the existence of Lagrangian multiplier functions P, for which the modified integrand $\tilde{f}$ satisfies the Weierstrass condition. This too, is something we might expect, but we have not established it previously with Lagrangian constraints of the form $\dot{Y} = G(t, Y, U)$; moreover, we know that $\mathcal{U}$ might not be open and that h_U need not exist.

For these reasons, the necessary conditions for optimality of a control must be established independently, and we should not anticipate success by elementary methods.

We begin by considering the effects of certain permissible control modifications on an *autonomous* state equation and obtain a precise description of

---

[1] If $\mathcal{U}$ is convex, then (39) of Chapter 10 can be used to eliminate this assumption.

the resultant states at later times. We use this in part (b) to establish the minimum principle for autonomous problems on a fixed interval (albeit with a slightly different h), and apply our results to the oscillator energy problem from §10.2(f). Finally in part (c) we transform the general problem into the case covered in part (b) and give illustrative applications which indicate the techniques employed when using these conditions, as well as some of the difficulties encountered in doing so.

We will sometimes refer to a state $Y \in \hat{C}^1[0,T]$ as a state trajectory. We may also assign a vector-valued function or a matrix-valued function any smoothness property common to its elements.

(a) Effects of Control Variations

Suppose that some control $U_0 \in \hat{C}[0, T]$ produces a solution $Y_0 \in \hat{C}^1[0, T]$ of an (autonomous) state equation

$$\dot{Y} = G(Y, U_0) \quad \text{on } (0, T), \tag{6}$$

where both G and G_Y are continuous on $D \times \mathcal{U}$ for a domain D of $\mathbb{R}^d$ and a *bounded* control set $\mathcal{U} \subseteq \mathbb{R}^k$. T is finite.

We wish to characterize analytically, the state target values $Y(T)$ which arise through solutions of (6) with $Y(0) = Y_0(0)$, when the control U_0 is replaced by *constant* values $V \in \mathcal{U}$ over small subintervals $[\tau', \tau]$ of continuity, as illustrated in Figure 11.1.

Intuitively, we expect that small intervals of replacement should produce small target effects, but we require a far more precise mathematical description. The results are presented in Propositions 11.2, 11.3, and 11.4, and if desired, they can be accepted without proof on first reading.

We consider first the effects of a *single* interval of replacement $[\tau', \tau]$, where $\tau' = \tau - \varepsilon c$ for some given $c > 0$, and a positive ε so small that U_0 is continuous on $[\tau', \tau]$. The result, denoted U_ε, is again an admissible control and it produces a unique associated state trajectory Y_ε that satisfies the

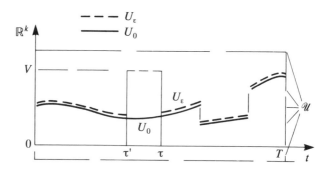

Figure 11.1

equation $\dot{Y} = G(Y, U_\varepsilon)$ on $(0, T)$, with $Y(0) = Y_0(0)$. Indeed, when $t \leq \tau'$, we just take $Y_\varepsilon(t) = Y_0(t)$. On $[\tau', \tau]$, the small interval of replacement, we can solve the equation $\dot{Y}(t) = G(Y(t), V)$ with $Y_\varepsilon(\tau') = Y_0(\tau')$, by the methods of §A.5, and obtain

$$Y_\varepsilon(\tau) = Y_0(\tau) + \int_{\tau'}^{\tau} (G(Y_\varepsilon(t), V) - G(Y_0(t), U_0(t))) \, dt. \tag{7}$$

For small ε, $Y_\varepsilon(\tau)$ will be sufficiently near $Y_0(\tau)$ to permit the embedding methods of Theorem A.18, to be applied to obtain the solution Y_ε on the successive subintervals of continuity of U_0 over $[\tau, T]$. In particular, the resulting target value $Y_\varepsilon(T)$ is known, and it can be compared with $Y_0(T)$.

(11.1) **Lemma.** *There are G-dependent positive constants M and δ_0 such that on $[\tau', \tau]$,*

$$\sigma(t) = |Y_\varepsilon(t) - Y_0(t)| \leq Mc\varepsilon, \quad when \ c\varepsilon \leq \delta_0.$$

PROOF. Replacing τ with t in (7), we get

$$Y_\varepsilon(t) - Y_0(t) = \int_{\tau'}^{t} [G(Y_0(s), V) - G(Y_0(s), U_0(s))] \, ds$$

$$+ \int_{\tau'}^{t} [G(Y_\varepsilon(s), V) - G(Y_0(s), V)] \, ds. \tag{8}$$

The first integrand on the right is continuous, hence bounded by M_0 (say) on the interval $[0, T]$, while the second may be estimated by some

$$\gamma |Y_\varepsilon(s) - Y_0(s)|, \quad \text{exactly as in (A.16).}$$

It follows that for $t \in [\tau', \tau]$:

$$\sigma(t) \leq M_0 c\varepsilon + \gamma \int_{\tau'}^{t} \sigma(s) \, ds = \psi(t), \quad \text{say,} \tag{9}$$

so that

$$\dot{\psi}(t) = \gamma\sigma(t) \leq \gamma\psi(t) \quad \text{or} \quad (e^{-\gamma t}\psi(t))^{\cdot} \leq 0.$$

Hence

$$\sigma(t) \leq \psi(t) \leq e^{\gamma(t-\tau')}\psi(\tau')$$

$$\leq e^{\gamma c\varepsilon} M_0 c\varepsilon \leq Mc\varepsilon, \tag{10}$$

if say $c\varepsilon \leq 1/\gamma = \delta_0$, and we replace eM_0 with M. $\square$

Henceforward, we suppose $c\varepsilon \leq \delta_0$.

If we examine (8) when $t = \tau$, we see that

$$Y_\varepsilon(\tau) = Y_0(\tau) + \varepsilon c\Delta G(\tau, V) + \varepsilon_3(\varepsilon), \tag{11}$$

where for small ε, we have approximated the first integrand by

$$\Delta G(\tau, V) \stackrel{\text{def}}{=} G(Y_0(\tau), V) - G(Y_0(\tau), U_0(\tau)), \tag{12}$$

and used the preceding lemma to estimate the second integrand uniformly by some $\gamma\varepsilon$. The final expression in (11) collects the error terms associated with these approximations into some $\mathfrak{z}(\varepsilon)$ that approaches 0 with ε; ($\varepsilon c = \tau' - \tau$ is just the length of $[\tau', \tau]$).

Next, we note that on the remaining interval $[\tau, T]$ both Y_ε and Y_0 satisfy the *same* state equation

$$\dot{Y} = G(Y, U_0)$$

but with initial value $Y(\tau) = A$, for *different* A.

By the arguments used in Theorem A.20 we know that as a function of A, the resulting solution $Y(t, A)$, say, is differentiable with respect to the components of A, and when evaluated at $A_0 = Y_0(\tau)$, $\mathbb{W}_\tau \overset{\text{def}}{=} Y_A(\cdot, A_0)$ is the unique solution to the linear system

$$\dot{\mathbb{W}} = G_Y(Y, U)\mathbb{W} \quad \text{on } (\tau, T) \tag{13}$$

with $\mathbb{W}_\tau(\tau) = \mathbb{I}_d$ (the $d \times d$ identity matrix), when the columns of G_Y are indexed by Y. It follows that since $A = Y_\varepsilon(\tau)$ is given by (11), then on $[\tau, T]$

$$\frac{d}{d\varepsilon} Y_\varepsilon(\cdot)\Big|_{\varepsilon=0} = Y_A(\cdot, A_0)\frac{dA}{d\varepsilon}\Big|_{\varepsilon=0} = c\mathbb{W}_\tau(\cdot)\Delta G(\tau, V);$$

i.e.,

$$Y_\varepsilon(t) = Y_0(t) + \varepsilon c \mathbb{W}_\tau(t)\Delta G(\tau, V) + \varepsilon\mathfrak{z}(t, \varepsilon), \quad \tau \le t \le T, \tag{14}$$

for some function $\mathfrak{z}(t, \varepsilon)$ that approaches 0, as $\varepsilon \to 0$. When $t = T$, let $\mathbb{W}_\tau = \mathbb{W}_\tau(T)$ and $\mathfrak{z}(\varepsilon) = \mathfrak{z}(T, \varepsilon)$, so that

$$Y_\varepsilon(T) = Y_0(T) + \varepsilon c \mathbb{W}_\tau\Delta G(\tau, V) + \varepsilon\mathfrak{z}(\varepsilon). \tag{14'}$$

At this point, we digress to obtain a simple result.

(11.2) Proposition. *If $P \in \hat{C}^1[0, T]$ is a solution of the adjoint equation*

$$\dot{P} = -\bar{G}_Y(Y_0, U_0)P \quad \text{on } [0, T], \tag{15}$$

then $\bar{P}(\tau) = \bar{P}(T)\mathbb{W}_\tau$.

PROOF. Since under transposition, $\dot{\bar{P}} = -\bar{P}\bar{G}_Y(Y_0, U_0)$, we have by (13) that $(\bar{P}\mathbb{W}_\tau)^\cdot = \bar{P}\dot{\mathbb{W}}_\tau + \dot{\bar{P}}\mathbb{W}_\tau = 0$, while $\mathbb{W}_\tau(\tau) = \mathbb{I}_d$. Thus $\bar{P}(\tau) = \bar{P}(\tau)\mathbb{W}_\tau(\tau) = \bar{P}(T)\mathbb{W}_\tau(T) = \bar{P}(T)\mathbb{W}_\tau$. ☐

Now we return to our principal development. From (14), it follows that if we next replace U_0 on a "later" disjoint interval of continuity $[\tau_1', \tau_1]$ by some constant value $V_1 \in \mathcal{U}$, then at $\tau_1' = \tau_1 - \varepsilon c_1$ the new Y_ε agrees with the old so that we can repeat the construction to obtain for $t > \tau_1$:

$$Y_\varepsilon(t) = Y_0(t) + \varepsilon(c\mathbb{W}_\tau(t)\Delta G(\tau, V) + c_1 \mathbb{W}_{\tau_1}(t)\Delta G(\tau_1, V_1)) + \varepsilon\mathfrak{z}(t, \varepsilon).$$

In particular, this would apply when t is the target time T, and we can evidently repeat the process at a finite set of τ_i. This observation establishes the following:

(11.3) **Proposition.** *If U_0 is replaced by constant values $V_i \in \mathcal{U}$ on disjoint intervals of continuity $[\tau_i - \varepsilon c_i, \tau_i]$, $i = 1, 2, \ldots, n$, then the new control U_ε produces a state trajectory Y_ε with target value*

$$Y_\varepsilon(T) = Y_0(T) + \varepsilon \sum_{i=1}^{n} c_i \mathcal{V}_i + \varepsilon_3(\varepsilon), \tag{16}$$

where $\varepsilon > 0$, $c_i \geq 0$, and

$$\mathcal{V}_i = \mathbb{W}_{\tau_i} \Delta G(\tau_i, V_i), \qquad i = 1, 2, \ldots, n, \tag{16'}$$

with terms defined in (12) and (13), while $_3(\varepsilon) \to 0$, with ε. □

The next argument is rather deep analytically, although it is not difficult to appreciate its geometrical plausibility. We suppose as we may that $Y_0(T) = \mathcal{O}$, and observe that for *fixed $\mathcal{V}_i \in \mathbb{R}^d$, $i = 1, 2, \ldots, d$,* as the $c_i \geq 0$ are varied, the set

$$K \overset{\text{def}}{=} \left\{ X = \sum_{i=1}^{d} c_i \mathcal{V}_i \colon c_1 \geq 0, i = 1, 2, \ldots, d \right\}$$

forms an infinite pyramid in $\mathbb{R}^d$ with apex at $\mathcal{O}$, and "edges" in the directions of the $\mathcal{V}_i$ as indicated in Figure 11.2. This pyramid need not have an interior, but when it does, we can establish the following:

(11.4) **Proposition.** *If K contains a point X_0 in its interior, then for some admissible control U_ε, the corresponding $Y_\varepsilon(T) = \rho X_0$, if $\rho > 0$ is sufficiently small.*

PROOF*. To have an interior, the edge vectors $\mathcal{V}_i$ of K must be linearly independent, and we note that for some small ρ, ρX_0 is also in the interior of

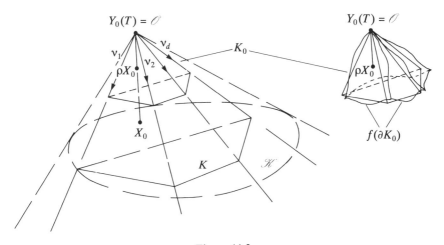

Figure 11.2

the *compact* subpyramid

$$K_0 = \left\{ X = \varepsilon \sum_{i=1}^{d} c_i \mathscr{V}_i : 0 \le \varepsilon \le \varepsilon_0, 0 \le c_i \le 1, \sum_{i=1}^{d} c_i = 1 \right\},$$

also shown in Figure 11.2.

Of course, we know that for ε_0 sufficiently small, we can get all of the target values from (16) (with $n = d$); they take the form

$$Y_\varepsilon(T) = \varepsilon \sum_{i=1}^{d} c_i \mathscr{V}_i + \varepsilon_3(\varepsilon),$$

but the correction term cannot be ignored. Instead, we consider the function f which assigns to each $X \in K_0$, the target value from (16) with the *same ε, c_i*, so that $f(X) = X + \varepsilon_3(\varepsilon)$. Since for ε_0 sufficiently small, and $0 \le c_i \le 1$, we can take a single function $_3$, it follows that this $f: K_0 \to \mathbb{R}^d$ is continuous. Moreover, for $X \in \partial K_0$, the boundary of K_0,

$$|f(X) - X| = \varepsilon |_3(\varepsilon)| \le \varepsilon_0 < |X - \rho X_0|,$$

if ε_0 is sufficiently small and $\rho > 0$ is even smaller. Then examining Figure 11.2 again, we can "see" that f does not move the boundary of K_0 far enough to remove ρX_0 from $f(K_0)$; i.e., *some $X \in K_0$*, has the target value $f(X) = \rho X_0$ as required.

(To complete this argument rigorously we must appeal to a form of Brouwer's Fixed Point Theorem as given, say, in [L–M]; this circumstance gives analytical depth to the proposition, and to subsequent results dependent upon it.) $\square$

(11.5) **Remark.** This proposition also holds if K is replaced by the *target cone*, $\mathscr{K}$ also indicated in Figure 11.2, where

$$\mathscr{K} = \left\{ X = \sum_{i=1}^{n} c_i \mathscr{V}_i : c_i \ge 0, i \le n = 1, 2, \dots \right\}, \tag{17}$$

and the $\mathscr{V}_i$ are *any* of the vectors defined by ((12), (13), and (16)) for *some τ_i, V_i*. For if X_0 is in the interior of $\mathscr{K}$, then there must be d linearly independent $\mathscr{V}_i$ forming a subpyramid K containing X_0 in *its* interior. With a slight shift if necessary, we can suppose these V_i to be associated with distinct τ_i, so that the construction in the proposition is valid for K, and so for $\mathscr{K}$.

(b) Autonomous Fixed Interval Problems

To understand the relevance of the next result, note that the autonomous optimal control problem from §10.1 admits (Mayer) formulation as that of finding a $\tilde{Y} = (y, Y) \in \hat{C}^1[0, T]$ with *minimal $y(T)$* among those which satisfy an augmented state equation

$$\dot{\tilde{Y}} = \tilde{G}(Y, U) \quad \text{on } (0, T), \tag{18}$$

with prescribed $\tilde{Y}(0)$, and, say, $Y(T)$, if the target is fixed.

Indeed, if $\tilde{G} = (f, G)$ so that $\dot{y} = f(Y, U)$ is the "first" state equation in (18), and we set $y(0) = 0$, then $y(T) = \int_0^T f(Y(t)), U(t)) \, dt = F(Y, U)$. Hence minimizing $y(T)$ under (18), is equivalent to minimizing $F(Y, U)$ under the usual state equation $\dot{Y} = G(Y, U)$ with prescribed $Y(0)$ and, say, $Y(T)$. However, as we shall now show, this Mayer problem admits attack by the geometrical methods just developed. The tilde over G will be suppressed but d will be replaced by $d + 1$ to facilitate application to the control problem as originally formulated.

(11.6) **Theorem** (Pontrjagin). *Let D be a domain of $\mathbb{R}^{d+1}$ and $\mathcal{U}$ be a bounded set in $\mathbb{R}^k$. Suppose that on a fixed interval $[0, T]$, $(\tilde{Y}_0, U_0)$ minimizes $y(T)$ among those with $\tilde{Y} = (y, Y)$ in*

$$\mathcal{D}_\Phi = \left\{ \begin{array}{c} (\tilde{Y}, U) \in \hat{C}^1[0, T] \times \hat{C}[0, T] : (\tilde{Y}(t), U(t)) \in D \times \mathcal{U} \\ \tilde{Y}(0) = \tilde{Y}_0(0), \Phi(Y(T)) = \mathcal{O} \end{array} \right\},$$

under the state equation
$$\dot{\tilde{Y}} = G(\tilde{Y}, U) \quad \text{on } (0, T).$$

(Here G and G_Y are continuous while either $\Phi \equiv \mathcal{O}$ or Φ is a C^1 l-vector valued function having $\Phi_Y(Y_0(T))$ of maximal rank $l \le d$.) Then, there exists a solution $\tilde{P} = (p, P)$ of the adjoint equation

$$\dot{\tilde{P}} = -\overline{G}_{\tilde{Y}}(\tilde{Y}_0, U_0)\tilde{P} \quad \text{on } (0, T), \quad \text{with} \quad \overline{\tilde{P}}(T) = \Lambda\Phi_Y(Y_0(T)), \tag{19}$$

for some $\Lambda \in \mathbb{R}^l$, such that the minimal inequality

$$\tilde{P}(\tau) \cdot G(\tilde{Y}_0(\tau), V) \ge \tilde{P}(\tau) \cdot G(\tilde{Y}_0(\tau), U_0(\tau)), \quad \forall \, V \in \mathcal{U}, \tag{19'}$$

holds at each point τ of continuity of U_0.

Remark. In fact, $\tilde{P} \cdot G(\tilde{Y}_0, U_0)$ is constant so that (19') holds at all t. (See Remarks 11.11.)

PROOF. Assume that $\tilde{Y}_0(T) = \tilde{\mathcal{O}}$, and consider first the simpler cases where the target is either fixed ($\Phi(Y) \equiv Y$) or free ($\Phi(Y) \equiv \mathcal{O}$). In these cases, the target cone $\mathcal{K}$ of 11.5, cannot contain in its interior the "downward" pointing vector $X_0 = (-1, 0, \ldots, 0)$. [For then by Proposition 11.4 (as extended), there would be a modified control U_ε, with associated state target value $\tilde{Y}_\varepsilon(T) = (-\rho, 0, 0, \ldots, 0)$ for some $\rho > 0$. But then $Y(T) = Y_0(T) = \mathcal{O}$, while $y(T) = -\rho < 0 = y_0(T)$ contradicting minimality.] Thus, there is a unit vector $\tilde{N}$ orthogonal to a d-dimensional subspace $\mathcal{T}$ for which

$$\tilde{N} \cdot X \ge 0, \quad \forall \, X \in \mathcal{K}, \quad \text{with} \quad \tilde{N} = (n, N) \text{ and } n \ge 0. \tag{20}$$

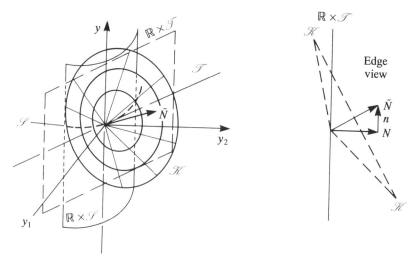

Figure 11.3

(When $\Phi \equiv 0$, we can take $\tilde{N} = (1, 0, \ldots, 0)$) ([N]). In case $d = 2$, (20) guarantees that each vector in the cone is within an angle of $\pi/2$ of the fixed upward-slanted vector $\tilde{N}$, as illustrated in Figure 11.3.

(11.7) For the more general Φ, the requirement that $\mathbb{M} = \Phi_Y(Y_0(T))\,(= \Phi_Y(\mathcal{O}))$ have maximal rank $l \le d$ means that locally (through implicit function theory [Ed]), $\mathcal{S}$, the $\mathcal{O}$-level set of Φ through $\mathcal{O}$ can be represented as a (smooth) $r\,(= d - l)$-dimensional submanifold in $\mathbb{R}^d$. Moreover, $\mathcal{S}$ has an associated r-dimensional tangent space $\mathcal{T}$ which is orthogonal to the subspace spanned by the l linearly independent (column) vectors of $\mathbb{M}$. A similar argument to that given above shows that $\mathcal{K}$ cannot contain in its interior any *downward directed tangent* vector $X_0 \in \mathbb{R} \times \mathcal{T}$.[1] We conclude that there is again a vector $\tilde{N} = (n, N)$ with N orthogonal to $\mathbb{R} \times \mathcal{T}$ for which (20) holds. But then $\overline{N} = \Lambda\Phi_Y(Y_0(T))$ for some $\Lambda \in \mathbb{R}^l$.

In Figure 11.3, we illustrate the latter situation for $d = 2$, where the target surface is defined by the single equation $\varphi(y_1, y_2) = 0$. Maximal rank at $\mathcal{O}$ means that $\nabla\varphi(\mathcal{O}) \ne \mathcal{O}$ and as we know say, by the discussion at the end of §5.6, this vector is orthogonal to the $r\,(= 1)$-dimensional subspace $\mathcal{T}$ (a line) tangent to $\mathcal{S}$, the 0-level set at $\mathcal{O}$. Since the interior of $\mathcal{K}$ cannot contain vectors in the "lower" half of the plane $\mathbb{R} \times \mathcal{T}$, then (20) holds for some $\tilde{N} = (n, N)$, where N is orthogonal to $\mathcal{T}$ and thus is given by some $\lambda\nabla\varphi(\mathcal{O})$.

---

[1] For if so, then $\mathcal{K}$ also contains a small downward directed target vector $\tilde{Y}_\varepsilon(T)$ which lies in $\mathbb{R} \times \mathcal{S}$, and this is clearly inadmissible. (Why?)

It is now very easy to complete the derivation of the minimal inequality (19′). First, we observe that the linear homogeneous adjoint equation (19) always has a unique solution over $[0, T]$ on successive subintervals of continuity of U_0 with prescribed target value $\tilde{P}(T) = \tilde{N}$ as obtained above. But, then by Proposition 11.2, we know that at each point of continuity $\tau > 0$ of U_0, and for any $V \in \mathcal{U}$,

$$\tilde{P}(\tau)\Delta G(\tau, V) = P(T)\mathbb{W}_\tau \Delta G(\tau, V)$$

$$= \tilde{N} \cdot \mathcal{V} \quad \text{(by (16′))}$$

$$\geq 0, \quad \text{since } \mathcal{V} \in \mathcal{K},$$

so that the result follows from definition (12).   □

Now let's see what these results tell us about our original autonomous fixed interval problem. Note that in its Mayer formulation, when $\tilde{Y} = \tilde{G}(Y, U)$ only, then with $\tilde{P} = (p, P)$, we must have from (19) that

$$\dot{p} = -\tilde{G}_y \cdot \tilde{P} = 0,$$

so that $p(t) = \text{const.} = p(T) = n$, the first component of $\tilde{N}$, which by our arguments *cannot* be slanted downward. Thus $p(t) = \lambda_0 \geq 0$, and this means that either $\lambda_0 = 0$, which although atypical can occur;[1] or by division and redefinition of $\tilde{P}$, and Λ as required, we can have $\lambda_0 = 1$. Moreover, we always have $(\lambda_0, P(T)) = \tilde{N} \neq \tilde{0} \in \mathbb{R}^{d+1}$, so that for the free target case ($\Phi \equiv \mathcal{O}$), where $P(T) = \mathcal{O}$, we must take $\lambda_0 = 1$. Let's summarize our findings.

(11.8) **Theorem.** *Suppose that we have an autonomous control problem on the fixed interval $[0, T]$ for performance and state functions f and G, respectively, with a target that is either fixed or free. If (Y_0, U_0) is a minimizing solution then there is a Pontjragin function*

$$h(t, Y, U) \overset{\text{def}}{=} \lambda_0 f(Y, U) + P(t) \cdot G(Y, U) \tag{21a}$$

that obeys the minimum principle

$$h(t, Y_0(t), U) \geq h_0(t) \overset{\text{def}}{=} h(t, Y_0(t), U_0(t)), \qquad V, U \in \mathcal{U}, \tag{21b}$$

and makes h_0 constant.
    P is a $\hat{C}^1$ solution of the adjoint equation

$$\dot{P} = -h_Y(\cdot, Y_0, U_0) \quad \text{on } [0, T], \tag{21c}$$

for which $(\lambda_0, P(T)) \neq \mathcal{O}$. λ_0 is 0 or 1, with $\lambda_0 = 1$ and $P(T) = \mathcal{O}$ in the free target case.

PROOF. The constancy of h_0 is a consequence of more general results to be established in part (c). (See Remarks 11.11.) Therefore (21b) which holds at

---

[1] See the first example in §10.2(a). Such problems are referred to as being *abnormal* because their optimal criteria do not depend on the performance measure being optimized.

each point t of continuity of U_0 by (19′) also holds at a point of discontinuity, in that it is satisfied by either limiting value of $h(t, Y_0(t), U)$.   □

(11.9) Remarks. 1. If fixed (or more general) target values are defined by requiring that $\Phi(Y(T)) = 0$ for some C^1 l-vector valued function Φ for which the matrix $\Phi_Y(Y_0(T))$ has maximal rank $l \leq d$, then by (19),

$$\bar{P}(T) = \Lambda\Phi_Y(Y_0(T)) \quad \text{for some} \quad \Lambda \in \mathbb{R}^l. \tag{21d}$$

Maximal rank is essential for this conclusion. See Example 1 below.

2. The Lagrange multiplier theorem (5.16) admits similar formulation for $\lambda_i = p_i$, with

$$\tilde{J} = \lambda_0 J + \sum_{i=1}^{N} p_i G_i.$$

Then alternative (a) permits $\lambda_0 = 0$, while if (a) is not satisfied, we must take $\lambda_0 \neq 0$, and, if desired, $\lambda_0 = 1$.

3. Observe that under classical smoothness conditions, we have just established a Lagrange multiplier rule for Lagrangian constraints of the form $\dot{Y} = G(Y, U)$. It guarantees the Weierstrass condition, but for the modified integrand $\tilde{f} = \lambda_0 f + P \cdot (G - \dot{Y})$ (where now U is to be thought of as $\dot{U}$). The more general situation is discussed in §11.3.

Example 1. As an illustration, recall the linear-quadratic problem at the end of §10.2(a) where we found that $y_0(t) = u_0(t) = e^t$ minimizes

$$F(y, u) = \int_0^1 [y^2(t) + u^2(t)]\, dt$$

uniquely on

$$\mathscr{D} = \{(y, u) \in \hat{C}^1[0, 1] : y(0) = 1, y(1) = e\}$$

under $\dot{y} = u$, with $|u| \leq 4$,

Thus, for some $p \in \hat{C}^1[0, 1]$ and $\lambda_0 \, (= 0 \text{ or } 1)$,

$$h(t, y, u) = \lambda_0(y^2 + u^2) + p(t)u$$

obeys the minimum principle for $u_0(t) = e^t$; hence,

$$\forall\, |u| \leq 4: \quad h(t, y_0(t), u) \geq h(t, y_0(t), e^t) = h_0(t).$$

Since $|e^t| < 4$ for $t \in [0, 1]$, it follows that

$$h_u(t, y_0(t), e^t) = 0 \quad \text{or} \quad 2\lambda_0 u_0(t) + p(t) = 0,$$

so that

$$p(t) = -2\lambda_0 u_0(t) = -2\lambda_0 e^t \quad \text{and} \quad p(1) = -2\lambda_0.$$

Since $(\lambda_0, p(1)) \neq (0, 0)$, we must take $\lambda_0 = 1$, even though this is a fixed target problem. If we use $\varphi(y) = y - e$ to fix the target, we can take $\lambda = -2$ and satisfy (21d). However, if we use $\varphi(y) = (y - e)^2$ to fix the target, then $\varphi_y(y_0(1)) = 2(y_0(1) - e) = 0$, and we cannot find a suitable λ because the maximal rank condition is not satisfied.

Note that
$$h_0(t) = (e^t)^2 + (e^t)^2 - 2e^t e^t = 0.$$

Example 2 (Oscillator Energy Problem). At the end of §10.2(f), we considered how to maximize the energy of an oscillator that can be reached from a state of rest in a specified time T. If we set
$$\mathcal{U} = [-1, 1], \qquad Y = (y, y_1),$$
$$f(Y, u) = -y_1 u,$$
and
$$G(Y, u) = (y_1, u - y),$$

then Theorem 11.8 is applicable to each
$$(Y_0, u_0) \in \mathcal{D} = \{(Y, u) \in \hat{C}^1[0, T] \times \hat{C}[0, T]: Y(0) = 0, u(t) \in \mathcal{U}\}$$

that minimizes
$$F(Y, u) = \int_0^T -y_1(t)u(t)\, dt \quad \text{on } \mathcal{D},$$

under the state equation $\dot{Y} = G(Y, u)$.

For this free-target problem, we must take $\lambda_0 = 1$ and, for each
$$P = (p, p_1) \in \hat{C}^1[0, T],$$
we have
$$h(t, Y, u) = -y_1 u + p(t)y_1 + p_1(t)(u - y)$$
$$= p(t)y_1 + p_1(t)y - u(y_1 - p_1(t))$$
$$= l(t, Y) - u(y_1 - p_1(t)), \quad \text{say.} \tag{22}$$

In our earlier analysis, we found that
$$Y_0(t) = \int_0^t \begin{Bmatrix} \sin(t - \tau) \\ \cos(t - \tau) \end{Bmatrix} u_0(\tau)\, d\tau, \tag{23}$$

if we recall equations (14) and (16) in §10.2(f).

Similarly, we know that P satisfies the adjoint equation(s)
$$\dot{p} = -h_y[Y_0, u_0] = p_1,$$
$$\dot{p}_1 = -h_{y_1}[Y_0, u_0] = u_0 - p, \tag{24}$$

which are seen to be identical to those satisfied by Y_0. Since $P(T) = \mathcal{O}$, we conclude that
$$P(t) = -\int_t^T \begin{Bmatrix} \sin(t - \tau) \\ \cos(t - \tau) \end{Bmatrix} u_0(\tau)\, d\tau. \tag{25}$$

To determine possible optimal controls, we invoke the minimum principle. When we examine the dependence on u of
$$h(t, Y_0(t), u) = l(t, Y_0(t)) - u(y_1(t) - p_1(t)),$$

we see that in order that $u_0(t)$ minimize this expression when $|u| \leq 1$, we should choose

$$u_0(t) = \operatorname{sgn}(y_1(t) - p_1(t)). \quad \text{(Why?)}$$

Thus, from (23) and (25), we find that

$$u_0(t) = \operatorname{sgn}\left(\int_0^T \cos(t - \tau) u_0(\tau)\, d\tau\right)$$

$$= \operatorname{sgn}(\cos(t - \alpha)), \tag{26}$$

where $\cos \alpha = c/\sqrt{c^2 + s^2}$, $\sin \alpha = s/\sqrt{c^2 + s^2}$, if

$$\left. \begin{matrix} c \\ s \end{matrix} \right\} = \int_0^T \left\{ \begin{matrix} \cos \tau \\ \sin \tau \end{matrix} \right\} u_0(\tau)\, d\tau$$

are not both zero. If we exclude the nonoptimal case $u_0 \equiv 0$, then we must find α if possible that makes these equations for u_0 compatible. We see that the optimal control is of the bang–bang type; in fact, it is constant, $+1$ or -1, on successive intervals of length π, exactly as our preliminary analysis in §10.2(f) suggested.

For example, when $T = 2\pi$, we can ask whether for $\alpha = \pi/2$

$$u_0 = \begin{cases} +1 & \text{on } (0, \pi), \\ -1 & \text{on } (\pi, 2\pi), \end{cases}$$

could be optimal?

For this we need

$$\int_0^{2\pi} \cos(t - \tau) u_0(\tau)\, d\tau = \int_0^{\pi} \cos(t - \tau)\, d\tau - \int_{\pi}^{2\pi} \cos(t - \tau)\, d\tau$$

$$= 2 \sin t + 2 \sin t = 4 \sin t,$$

to have the sign of $u_0(t)$ on intervals of constancy. It does, and hence this control could be optimal when $T = 2\pi$, or when $T = \pi$. (Why?)

On the other hand, this same u_0 cannot be optimal when $T = 3\pi/2$ since

$$\int_0^{3\pi/2} \cos(t - \tau) u_0(\tau)\, d\tau = 2 \sin t - \int_{\pi}^{3\pi/2} \cos(t - \tau)\, d\tau$$

$$= 3 \sin t + \cos t$$

which equals -1 when $t = \pi$, and so by continuity, is negative in a neighborhood of $t = \pi$, while $u_0(t) = +1$, for $t \in (0, \pi)$.

From Theorem 11.8, we also know that $h_0(t) = h(t, Y_0(t), U_0(t))$ is constant. In particular, $h_0(0) = h_0(T)$, and since $Y_0(0) = P(T) = \mathcal{O}$, from (22), we see that

$$u_0(0+) p_1(0) = -u_0(T-) y_1(T),$$

or from (23) and (25), that

$$u_0(0+) \int_0^T (\cos \tau) u_0(\tau) \, d\tau = u_0(T-) \int_0^T \cos(T - \tau) u_0(\tau) \, d\tau. \qquad (27)$$

This equation provides a simple test which possible optimal controls must meet. For example, when $T = \pi/2$, the control

$$u = \begin{cases} +1 & \text{on } (0, \pi/6), \\ -1 & \text{on } (\pi/6, \pi/2), \end{cases}$$

cannot be optimal since $\int_0^{\pi/2} \cos \tau u(\tau) \, d\tau = 0$, while the corresponding integral on the right side of (27) is nonvanishing. On the other hand, for $T = \pi$, this condition becomes

$$u_0(0+) \int_0^\pi \cos \tau u_0(\tau) \, d\tau = -u_0(\pi-) \int_0^\pi \cos \tau u_0(\tau) \, d\tau,$$

so that either the integral vanishes (which it does when u_0 is constant) or $u_0(0+) = -u_0(\pi-)$, and this is true of all other candidates permitted by $u_0(t) = \text{sgn} \cos(t - \alpha)$. Thus all of these $u_0(t)$ are still candidates for optimality (and for the latter, $h_0(t) \neq 0$).

However, for $T = 5\pi/4$, we can have $u_0(0+) = \pm u_0(5\pi/4-)$, (why?), so that the corresponding integrals in (27) must be checked for both types of possible u_0. Supposing that initially $u_0 = +1$, then after tedious computations outlined in Problem 11.2, we find that only one candidate of each type survives the test, and upon comparing the associated energies, that only one of these, namely

$$u_0 = \begin{cases} +1 & \text{on } (0, \sigma), \\ -1 & \text{on } (\sigma, 5\pi/4), \end{cases} \quad \text{where} \quad \tan \sigma = -(1 + \sqrt{2}), \qquad (27')$$

could be optimal.

Unfortunately, without either a sufficiency theorem, or an existence theorem, we still cannot guarantee the optimality of this sole survivor.

(c) General Control Problems

In Theorem 11.8, the interval $[0, T]$ is fixed, which precludes direct application to say, time optimal problems. However, it is straightforward to use the theorem (and subsequent remarks) to obtain an extension to the non-autonomous case where T also varies, and thereby strengthen the previous results. It is less straightforward to present the new results.

(11.10) **Theorem** (Necessity of the Minimum Principle). Hypotheses. *Let D be a bounded domain of* $\mathbb{R}^d$ *and* $\mathcal{U}$ *be a bounded set in* $\mathbb{R}^k$. *Suppose that*

$$F(Y, U, T) \stackrel{\text{def}}{=} \int_0^T f(t, Y(t), U(t)) \, dt$$

is minimized by (Y_0, U_0, T_0) *on*

$$\mathcal{D}_\Phi \stackrel{\text{def}}{=} \left\{ \begin{array}{l} (Y, U) \in \hat{C}^1[0, T] \times \hat{C}[0, T], \text{ for } T > 0; \\ \text{with } Y(0) = Y_0(0), \; \Phi(T, Y(T)) = \mathcal{O} \in \mathbb{R}^l; \\ \text{and } (Y(t), U(t)) \in D \times \mathcal{U} \end{array} \right\}$$

under the state equation

$$\dot{Y}(t) = G(t, Y(t), U(t)) \quad \text{on } (0, T). \tag{28a}$$

Assume that $f, f_{(t, Y)}$ *and* $G, G_{(t, Y)}$ *are continuous on* $[0, b] \times D \times \mathcal{U}$ *for some* $b > T_0$, *while either* $\Phi \equiv \mathcal{O}$, *or* Φ *is a* C^1 *function in a neighborhood of* $B_0 \stackrel{\text{def}}{=}$ $(T_0, Y_0(T_0)) \in \mathbb{R}^{d+1}$ *such that* $\Phi_{(t, Y)}(B_0)$ *has maximal rank* $l \le d + 1$.

Conclusions. *Then there exist* $(\lambda_0, \Lambda) \ne \mathcal{O}$ *in* $\mathbb{R}^{l+1}$ *with* $\lambda_0 = 0$ *or* 1, *and a* $P \in \hat{C}^1[0, T_0]$, *with* $\bar{P}(T_0) = \Lambda\Phi_Y(B_0)$ *such that*

$$\dot{P} = -h_Y[Y_0, U_0] \quad \text{on } (0, T_0), \tag{28b}$$

where

$$h(t, Y, U) \stackrel{\text{def}}{=} \lambda_0 f(t, Y, U) + P(t) \cdot G(t, Y, U),$$

obeys the minimum principle that $\forall \, U \in \mathcal{U}$

$$h(t, Y_0(t), U) \ge h_0(t) \stackrel{\text{def}}{=} h(t, Y_0(t), U_0(t)), \qquad t \in [0, T_0]. \tag{29}$$

Moreover, $h_0 \in \hat{C}^1[0, T_0]$, *with* $h_0(T_0) = -\Lambda \cdot \Phi_t(B_0)$, *while*

$$-\dot{h}_0 = \lambda_0 f_t[Y_0, U_0] + P \cdot G_t[Y_0, U_0], \tag{30}$$

and

$$(\lambda_0, h_0(T_0), P(T_0)) \ne \mathcal{O} \in \mathbb{R}^{d+2}. \tag{30'}$$

PROOF. Observe that when $\Phi = (\varphi, \tilde{\Phi})$ with $\varphi(t, Y) = t - T_0$, and the problem is specialized to that of the autonomous fixed interval case ($T = T_0$) considered previously, then the initial conclusions are just a statement of the results already obtained at each point of continuity of U_0. Hence we can assume that this part of the theorem is valid under these restrictions, with proper attention to dimensionality.

We will obtain the new results by reduction to this case on $[0, 1]$, through the parametric substitutions

$$t = t(\tau); \qquad X(\tau) = Y(t(\tau)), \qquad V(\tau) = U(t(\tau)).$$

Here t is regarded as a new state variable governed by the law $t'(\tau) = w$, where w is to be considered as a *new* nonnegative control in $\hat{C}[0, 1]$. Through the chain rule, (28a) is replaced by

$$X'(\tau) = G(t(\tau), X(\tau), V(\tau))w(\tau) \tag{31}$$

and it is seen that if $t_0(\tau) = \tau T_0$ (so that $w_0 = T_0$), with corresponding substitutions for X_0, V_0, then $(t_0, X_0; w_0, V_0)$ minimizes

$$F(t, X; w, V) \overset{\text{def}}{=} \int_0^1 f(t(\tau), X(\tau), V(\tau))w(\tau)\, d\tau \qquad (= F(Y, U, T), \text{ if } T = t(1))$$

on

$$\mathcal{D}_\Phi = \left\{ \begin{aligned} &(t, X; w, V) \in \hat{C}^1[0, 1] \times \hat{C}[0, 1], \text{ with } t(0) = 0, X(0) = Y_0(0) \in \mathbb{R}^d, \\ &\Phi(t(1), X(1)) = 0 \in \mathbb{R}^l, \text{ and } (w, V)(\tau) \in [0, b] \times \mathcal{U} \end{aligned} \right\},$$

under the state laws $t'(\tau) = w$, and (31). Observe that the transformed problem is autonomous as desired, since its defining functions do not depend explicitly on τ.

Hence by Theorem 11.8 and Remarks 11.9, with appropriate substitutions, $\exists \lambda_0 = 0$ or 1, $\Lambda \in \mathbb{R}^l$, and functions $(q_0, Q) \in \hat{C}^1[0, 1]$ with

$$(\lambda_0; q_0(1), Q(1)) \neq 0, \tag{32a}$$

such that

$$q_0(1) = \Lambda \cdot \Phi_t(T_0, X_0(1)); \qquad Q(1) = \Lambda \Phi_X(T_0, X_0(1))$$

(since $t_0(1) = T_0$); and

$$q_0'(\cdot) = -\hat{h}_t(\cdot, t_0, X_0, V_0)T_0, \qquad Q'(\cdot) = -\hat{h}_Y(\cdot, t_0, X_0, V_0)T_0, \tag{32b}$$

where

$$\hat{h}(\tau, t, X, V)w \overset{\text{def}}{=} [\lambda_0 f(t, X, V) + Q(\tau) \cdot G(t, X, V) + q_0(\tau) \cdot 1]w,$$

obeys the minimum principle on $[0, 1]$ for the control *pair* (w, V), relative to the trajectory $t_0(\tau) = \tau T_0$ and $X_0(\tau) = Y_0(\tau T_0)$, at each point τ of continuity of V_0.

In particular, if we take $V = V_0(\tau) = U(\tau T_0)$, and consider both $w(\tau) = T \geq T_0$ and $w(\tau) = T \leq T_0$ in the resulting inequality, it follows that at each point of continuity, $\hat{h}(\tau, \tau T_0, X_0(\tau), V_0(\tau)) = 0$, (why?), or, with

$$t = \tau T_0 \quad \text{and} \quad P(t) \overset{\text{def}}{=} Q(t/T_0), \tag{32c}$$

that $h_0(t) = -q_0(t/T_0)$ is *continuous* on $[0, T_0]$, where h_0 is as defined in (29).

Similarly, if we take $w(\tau) = T_0$ and consider the minimum inequality for V compared with $V_0(\tau)$, then we see that *after cancellation*, and resubstitutions, h as defined in (29) obeys the desired minimum principle relative to this same h_0. (Problem 11.3.)

Finally, from (32b) and (32c) follow the equations (28b) and (30); the fact that $(\lambda_0, h_0(T_0), P(T_0)) \neq 0$ is an immediate consequence of (32a) and this in turn would be violated were $(\lambda_0, \Lambda) = 0$. (Why?) □

(11.11) **Remarks.** 1. Equation (30) may be written

$$\dot{h}_0(t) = -H_t(t, Y_0(t), U_0(t), P(t)), \tag{33}$$

if h is replaced by $H(t, Y, U, P)$ as in the final part of §10.4. When the problem is autonomous so that $f_t \equiv 0$, $G_t \equiv 0$, then (30) implies that h_0 is constant on

$[0, T_0]$. If, in addition, $\Phi_t \equiv \mathcal{O}$, then we can conclude that $h_0 = h_0(T_0) = 0$, so that $(\lambda_0, P(T_0)) \neq \mathcal{O}$. However, for the fixed interval case which requires say $\Phi = (\varphi, \tilde{\Phi})$ with $\varphi(t, Y) = t - T_0$, then $\Phi_t = (1, \tilde{\Phi}_t) \neq \mathcal{O}$, and we know that for such problems h_0 need not vanish (Example 2). Observe that all conclusions about the behavior of h_0 were obtained under the assumption that the various integrands are defined on some time interval $(0, b) \supseteq (0, T_0]$. For certain nonautonomous fixed interval problems, this need not be valid. If, for example, $f(t, Y, U) = \sqrt{T_0^2 - t^2} \tilde{f}(Y, U)$, then more general methods are required to obtain analogous results. See [He].

2. This theorem required boundedness of the control set $\mathcal{U}$. When $\mathcal{U}$ can in addition be described by Lagrangian inequalities of the form $\Psi(U) \leq \mathcal{O} \in \mathbb{R}^m$, then under sufficient smoothness assumptions, we can replace the minimum principle as stated by $\tilde{h}_U(t, Y_0(t), U_0(t)) = \mathcal{O}$, where $\tilde{h} = h + M \cdot \Psi$ for suitable multiplier functions M on $[0, T_0]$. Recall that a similar sufficiency formulation was made in §10.1 and subsequently used to attack several of the problems in §10.2. This extension, as a necessary condition for optimality, will be established in §11.3.

3. In this theorem, the optimal target time T_0 could be any positive number, and this might not be realistic in certain applications. Problems in which T_0 is required to lie in some given interval are explored in [S–S].

Example 3. Suppose that for $Y = (y, y_1)$ and given $B = (b, b_1) \in \mathbb{R}^2$, we wish to minimize

$$F(Y, u, T) = \int_0^T (1 + y_1^2(t)) \, dt$$

on

$$\mathscr{D} = \left\{ \begin{array}{c} (Y, u) \in \hat{C}^1[0, T] \times \hat{C}[0, T) \text{ with } T > 0: \\ Y(0) = \mathcal{O}, Y(T) = B, |u(t)| \leq 1 \end{array} \right\},$$

under the state law

$$\dot{Y} = (\dot{y}, \dot{y}_1) = (y_1, u) \quad \text{on } (0, T). \tag{34a}$$

(A possible physical setting for this problem is presented in the discussion of the docking problem in the next section.)

Here, we have an autonomous fixed-target problem with a free terminal time and we can take $\Phi(Y) = Y - B$. If we suppose that (Y_0, u_0, T_0) supplies an optimal solution, then we may apply Theorem 11.10; accordingly, we introduce $\lambda_0 \, (= 0 \text{ or } 1)$ with $P = (p, p_1) \in \hat{C}^1[0, T_0]$, and form

$$h(t, Y, u) = \lambda_0(1 + y_1^2) + p(t)y_1 + p_1(t)u.$$

Then on $(0, T_0)$, P satisfies the adjoint equations, abbreviated as follows:

$$-\dot{p} = h_y = 0 \Rightarrow p(t) = c, \quad \text{const.,}$$
$$-\dot{p}_1 = h_{y_1} = 2\lambda_0 y_1 + p = 2\lambda_0 y_1 + c. \tag{34b}$$

Also, after simplification ($p(t) = c$)

$$h(t, Y, u) = \lambda_0(1 + y_1^2) + cy_1 + p_1(t)u, \tag{35}$$

satisfies the minimum principle which here reduces to the requirement that

$$p_1(t)u \geq p_1(t)u_0(t), \qquad \forall\, |u| \leq 1,$$

and this demands that

$$u_0(t) = -\operatorname{sgn} p_1(t). \tag{35'}$$

Moreover, $h_0(t) = h(t, Y_0(t), u_0(t))$ is $\hat{C}^1$, with $\dot{h}_0 = 0$ by autonomy; and $h_0(T) = 0$, since $\Phi_t = \mathcal{O}$, so that $h_0(t) \equiv 0$. (See Remarks 11.11.)

Now, if $\lambda_0 = 0$, then, since $y_1(0) = 0$,

$$0 = h_0(0) = p_1(0)u_0(0+),$$

and by (34b), $\dot{p}_1 = -c$ or $p_1(t) = c_1 - ct$ which can change sign at most once, at $\tau = c_1/c$. Thus by (35'), $u_0(t)$ is constant, either $+1$ or -1 in each interval excluding τ. But then, $p_1(0) = 0 \Rightarrow c_1 = 0 \Rightarrow \tau = 0$. Hence, in this case, no switching is permitted and we must take $u_0(t) \equiv +1$ or $u_0(t) \equiv -1$. However, this implies through the easily integrated state equations (34a), that $T_0 = |b_1|$, while

$$\text{either} \quad b = b_1^2/2 \text{ and } b_1 > 0,$$
$$\text{or} \quad b = -b_1^2/2 \text{ and } b_1 < 0. \tag{36}$$

For other targets B, we must consider $\lambda_0 = 1$. Then (34b) gives $-\dot{p}_1 = 2y_1 + c$ ($\Rightarrow \dot{p}_1$ is continuous) which combined with (34a) yields $-\ddot{p}_1 = 2\dot{y}_1 = 2u_0$, or finally by (35'),

$$\ddot{p}_1 = 2 \operatorname{sgn} p_1. \tag{37}$$

On each interval in which p_1 is of constant sign σ, this equation has only parabolic solutions of the form $p_1(t) = \sigma t^2 + at + a_1$. Moreover, the parabolas permitted cannot vanish twice unless p_1 vanishes in the intervening interval. (Why?) Since $\dot{p}_1$ is continuous, it follows that the only possible p_1 are those shown in Figure 11.4 where a possibly degenerate zero interval $[\tau, \tau']$ is flanked by tangential semiparabolas opening in either direction.

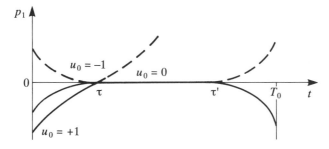

Figure 11.4

By (35′), such p_1 correspond to optimal controls $u_0 = 0$ on $[\tau, \tau']$, with $u_0 = +1$ or $u_0 = -1$ in each complementary interval. The period $[\tau, \tau']$ corresponds to one in which by (34a), $y_1(t)$ is constant while $y(t)$ varies linearly. In the complementary interval $(0, \tau)$,

$$y(t) = \sigma t^2/2, \quad \text{for } \sigma = \pm 1 \quad \text{or} \quad y = \sigma y_1^2/2. \tag{38}$$
$$y_1(t) = \sigma t,$$

At a switching time τ where u_0 changes from $+1$ to 0 or -1, we must have $p_1(\tau) = 0$ and $\dot{p}_1(\tau) \geq 0$ (Why?) Thus from (34b) and (35):

$$h_0(\tau) = 0 = 1 + y_1^2(\tau) + cy_1(\tau),$$
$$-\dot{p}_1(\tau) = 2y_1(\tau) + c \leq 0. \tag{39}$$

Moreover, having p_1 negative on $(0, \tau)$ corresponds to taking $\sigma = +1$ in (38) so that $y_1(\tau) = \tau > 0$. It follows that

$$0 \leq 1 + y_1^2(\tau) - 2y_1^2(\tau) \Rightarrow y_1^2(\tau) \leq 1. \tag{39′}$$

Thus such changes can occur only when $0 < y_1(\tau) \leq 1$, and a similar analysis shows that in the opposite case, where u_0 changes from -1 to 0 or $+1$, $0 < -y_1(\tau) \leq 1$. It follows that switching occurs according to this prescription or not at all. Moreover, $y_1 = \pm 1$ define the only lines of constancy for an optimal trajectory in the Y-plane, and $y(t) = \pm t + c_0$ only along such lines.

We can use convexity to argue that for the given time T_0, there can be at most one optimal control. For if we introduce μ as in §10.1, we see that the modified integrand

$$\tilde{f}(t, Y, u, \dot{Y}) = 1 + y_1^2 + p(t)(y_1 - \dot{y}) + p_1(t)(u - \dot{y}_1) + \mu(t)(u^2 - 1)$$

is for $\mu \geq 0$, convex on $[0, T_0] \times \mathbb{R}^3$, and strictly convex in y_1.

The Euler–Lagrange equations for $\tilde{f}$ in y and y_1 are just (34b), while that for u is

$$\tilde{f}_u = 0 = p_1 + 2\mu u_0,$$

which is automatically satisfied on intervals where $u_0 = 0$, and we can take $\mu = 0$ there. Otherwise, from (35′),

$$2\mu = -p_1/u_0 = |p_1| > 0,$$

and thus, $\mu(t)(u_0^2(t) - 1) \equiv 0$.

It follows that if $(Y_0, u_0) \in \mathscr{D}_{T_0}$, then it minimizes $\int_0^{T_0}(1 + y_1^2(t))\, dt$ on $\mathscr{D}_{T_0}$ among those satisfying (34a), and *uniquely* in y_1. But $\dot{y}_1 = u_0$ so that u_0 (hence Y_0) must be uniquely determined.

Assembling these facts, leads us to form the state diagram of Figure 11.5, where the primary parabolas corresponding to $u_0 \equiv +1$ and $u_0 \equiv -1$, respectively, are shown in bold lines.

The switching lines $y_1 = \pm 1$ are dashed, and the remaining parabolas are those which arrive at other targets with $u_0 = +1$ or -1. From the

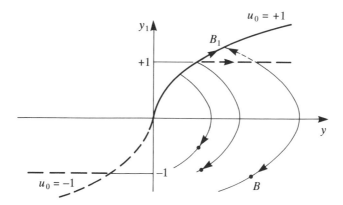

Figure 11.5

uniqueness of the optimal control-trajectory, and a careful analysis utilizing the symmetry of those parabolas of different families passing through the same point, it follows that the typical optimal trajectories to various target regions are as indicated by the arrows.

For example, to reach B_1 by a path utilizing the line $y_1 = +1$, would result in an integral for $\int_0^{T_0} y_1^2(t)$ which is at least as large as that along $y_1 = y^2/2$ and the time $\int_0^{T_0} 1\, dt$ would clearly be larger. Thus the dotted alternative is excluded.

For further discussion of this and related examples, see [M–S]. The simpler case in which $\int_0^T y_1^2(t)\, dt$ only is minimized is taken up in Problem 11.1(h), where it is shown that the necessary conditions preclude optimality for all targets B except those on the primary parabolas. The corresponding time-optimal problem is investigated in the next section.

§11.2. Linear Time-Optimal Problems

In this section, we confront some problems in which the target time T is being minimized and which are not replaced by equivalent problems on some fixed interval. To avoid further complications we will look only at time-optimal problems governed by linear state equations and consider in detail only the autonomous case, although certain results have nonautonomous (and nonlinear) extensions.

Even within this narrow range, the problems require special additional conditions of a geometrical nature in order to guarantee the optimality or uniqueness of a control which meets the necessary conditions of Theorem 11.10. The controls which emerge are in general bang–bang and piecewise constant; with further restrictions, it is possible to limit the number of switches in value which an optimal control can exhibit. These results are applied to obtain a complete solution to a docking problem, and other applications are discussed.

Problem Statement

We wish to find the minimal time T required to bring a system from a given state in $\mathbb{R}^d$ at time $t = 0$ to the origin under the linear autonomous state equation

$$\dot{Y}(t) = \mathbb{A} Y(t) + \mathbb{B} U(t) \quad \text{on } (0, T), \tag{40}$$

where $\mathbb{A}$ and $\mathbb{B}$ are constant matrices and $U(t) \in \mathscr{U} \subseteq \mathbb{R}^k$.

The state equations in all of our previous examples in Part III can be put in the form (40), so that it still permits wide application.

Suppose that (Y_0, U_0, T_0) gives a minimizing solution to this problem. To apply Theorem 11.10 we should take $f = 1$, $\lambda_0 = 0$ or 1, and form

$$h(t, Y, U) = \lambda_0 1 + P(t) \cdot (\mathbb{A} Y + \mathbb{B} U).$$

Then the adjoint equation (28b) takes the homogeneous form $-\dot{P} = \bar{P} \mathbb{A}$, which has the general solution

$$\bar{P}(t) = \bar{P}(0) e^{-t \mathbb{A}} = V e^{-t \mathbb{A}} \tag{40'}$$

for an appropriate square matrix function $e^{-t\mathbb{A}}$ and row matrix V.[1] From (29) we see that for each $U \in \mathscr{U}$:

$$h(t, Y_0(t), U) = \lambda_0 + V e^{-t \mathbb{A}} (\mathbb{A} Y_0(t) + \mathbb{B} U)$$

$$\geq h_0(t) = \lambda_0 + V e^{-t \mathbb{A}} (\mathbb{A} Y_0(t) + \mathbb{B} U_0(t)),$$

or

$$V e^{-t \mathbb{A}} \mathbb{B} U \geq V e^{-t \mathbb{A}} \mathbb{B} U_0(t), \qquad 0 \leq t \leq T_0. \tag{41}$$

Also $h_0 = 0$ (why?), so that by $(30'), (\lambda_0, V) \neq \mathcal{O}$.

For this linear autonomous case, there are additional conditions under which the minimum principle in the form (41) is effectively sufficient.

(11.12) Theorem. *Suppose that the origin is an interior point of the control set $\mathscr{U}$ in $\mathbb{R}^k$ and that the matrix $\mathbb{M} = (\mathbb{B} : \mathbb{A}\mathbb{B} : \mathbb{A}^2\mathbb{B} : \ldots : \mathbb{A}^{d-1}\mathbb{B})$ has rank d. If for some vector $V \in \mathbb{R}^d \sim \mathcal{O}$, $V e^{-t \mathbb{A}} \mathbb{B} U \geq V e^{-t \mathbb{A}} \mathbb{B} U_0(t)$, $t \in [0, T_0]$, $\forall\, U \in \mathscr{U}$, then*

---

[1] Formally

$$e^{-t\mathbb{A}} \overset{\text{def}}{=} \mathbb{I} - t\mathbb{A} + \frac{t^2}{2!}\mathbb{A}^2 - \frac{t^3}{3!}\mathbb{A}^3 + \cdots,$$

where the indicated series of matrices converges uniformly in t on each compact interval. Also $e^{-t\mathbb{A}}$ and $e^{t\mathbb{A}}$ are inverse matrices [C–L].

(Y_0, U_0, T_0) *minimizes $\int_0^T dt$ under* (40) *on*

$$\mathscr{D}_0 = \{(Y, U, T) \in \hat{C}^1[0, T] \times \hat{C}[0, T] \times (0, \infty);$$

$$Y(0) = Y_0(0); Y(T) = \mathcal{O}; U(t) \in \mathscr{U}\}.$$

PROOF. With the same matrix $e^{-t\mathbb{A}}$ used above, the state equation may be rewritten $(d/dt)(e^{-t\mathbb{A}} Y(t)) = e^{-t\mathbb{A}}\mathbb{B}U(t)$ so that upon integration over $[0, T]$, with $Y(T) = \mathcal{O}$ (and $e^{-0\mathbb{A}} = \mathbb{I}_d$), we obtain

$$\mathcal{O} = Y(0) + \int_0^T e^{-t\mathbb{A}}\mathbb{B}U(t) \, dt, \tag{42a}$$

and similarly, when $Y(0) = Y_0(0)$:

$$-Y_0(0) = \int_0^{T_0} e^{-t\mathbb{A}}\mathbb{B}U_0(t) \, dt = \int_0^T e^{-t\mathbb{A}}\mathbb{B}U(t) \, dt. \tag{42b}$$

Now suppose that this admissible control U, produces a state trajectory Y with $Y(0) = Y_0(0)$ and $Y(T) = 0$ in the time $T < T_0$. To reach a contradiction we utilize our hypotheses as follows:

Since $U = \mathcal{O} \in \mathscr{U}$, we see from (41) that

$$Ve^{-t\mathbb{A}}\mathbb{B}U_0(t) \leq 0, \qquad \forall \, t; \tag{43}$$

hence from (42b), we have upon premultiplication by the constant (row) vector V, that

$$-V \cdot Y_0(0) = \left\{ \int_0^T + \int_T^{T_0} \right\} (Ve^{-t\mathbb{A}}\mathbb{B}U_0(t)) \, dt$$

$$\leq \int_0^T Ve^{-t\mathbb{A}}\mathbb{B}U(t) \, dt = -V \cdot Y_0(0). \quad \text{(Why?)}$$

Then equality holds throughout so that in particular,

$$0 = \int_T^{T_0} Ve^{-t\mathbb{A}}\mathbb{B}U_0(t) \, dt,$$

and by (43),

$$Ve^{-t\mathbb{A}}\mathbb{B}U_0(t) \equiv 0 \quad \text{on } [T, T_0].$$

Using (41) again, we see that $U \in \mathscr{U} \Rightarrow Ve^{-t\mathbb{A}}\mathbb{B}U \geq 0$, and where also $-U \in \mathscr{U}$, we get $Ve^{-t\mathbb{A}}\mathbb{B}U \equiv 0$ on $[T, T_0]$. Since $\mathcal{O}$ is an interior point of $\mathscr{U}$, this must hold for all U near $\mathcal{O}$ so that the row vector $Ve^{-t\mathbb{A}}\mathbb{B} \equiv \mathcal{O}$ on $[T, T_0]$. By translation, with $t_1 = (T + T_0)/2$,

$$\Phi(t) \overset{\text{def}}{=} Ve^{-t_1\mathbb{A}}e^{-t\mathbb{A}}\mathbb{B},$$

$$= V_1 e^{-t\mathbb{A}}\mathbb{B} \equiv \mathcal{O}, \quad \text{near } t = 0.$$

In particular, $\Phi(0) = \Phi'(0) = \cdots = \Phi^{(d-1)}(0) = 0$, and since

$$\left[\frac{d}{dt}e^{-t\mathbb{A}} = -\mathbb{A}e^{-t\mathbb{A}}\right]_{t=0} = -\mathbb{A},$$

upon successive differentiation, we see that $V_1 = Ve^{-t_1\mathbb{A}}$ is orthogonal to the columns of $\mathbb{B}$, $\mathbb{A}\mathbb{B}$, $\mathbb{A}^2\mathbb{B}, \ldots, \mathbb{A}^{d-1}\mathbb{B}$, and hence to the columns of $\mathbb{M}$, hypothesized to be of rank d. It follows that $V_1 = \mathcal{O} \Rightarrow V = V_1 e^{t_1\mathbb{A}} = \mathcal{O}$, and this is the desired contradiction. $\square$

(11.13) Remarks. The condition on $\mathbb{M}$ will be discussed after Corollary 11.14. The same techniques are effective for certain nonlinear state equations of the simple form $\dot{Y}(t) = \mathbb{A}Y(t) + G(U(t))$, when $G(\mathcal{O}) = \mathcal{O}$. Details and applications are given in Problems 11.4 and 11.5.

Example 1. In $\mathbb{R}^2$, consider the state equation $\dot{Y} = U$ where $\mathbb{A} = \mathcal{O}$ and $\mathbb{B} = \mathbb{I}$, so that $\mathbb{M} = [\mathbb{B}:\mathcal{O}: \; :\mathcal{O}] = [\mathbb{I}:\mathcal{O}: \; :\mathcal{O}]$ has rank 2. If we take $\mathcal{U} = \{U \in \mathbb{R}^2 : |u_i| \le 1, \; i = 1, \; 2\}$ then for each nonzero $V = (v_1, v_2)$, and $U \in \mathcal{U}$:

$$Ve^{-t\mathbb{A}}\mathbb{B}U = V \cdot U = v_1 u_1 + v_2 u_2 \ge -|v_1| - |v_2|.$$

For instance, when $V = (1, 0)$, *each* control of the form $U_0(t) = (-1, u_2(t))$, with its associated state $Y_0(t) = (T_0 - t, -\int_t^{T_0} u_2(\tau)\, d\tau)$ on $[0, T_0]$, is optimal among these producing state trajectories Y, with $Y(T_0) = \mathcal{O}$, and $Y(0) = Y_0(0)$.

In particular, $U_0(t) = (-1, 0)$ is optimal among those on $[0, 2]$ for which $Y(0) = (2, 0)$, but it is not unique in this respect. (Problem 11.6.) Under slightly more restrictive conditions, we can obtain a much more attractive result.

(11.14) Corollary. *Suppose that $\mathcal{U}$ of Theorem 11.12 is a closed box (or a compact convex polyhedron) in $\mathbb{R}^k$ with vertices U_j. If the vectors $\mathbb{B}E$, $\mathbb{A}\mathbb{B}E, \ldots,$ $\mathbb{A}^{d-1}\mathbb{B}E$ are linearly independent for each edge vector E joining adjacent vertices, then the optimal control $U_0(t)$ is unique, and piecewise vertex-valued on $\mathcal{U}$.*

PROOF*. For each $t \in [0, T_0]$, $Ve^{-t\mathbb{A}}\mathbb{B}U$ is linear in U and continuous on $\mathcal{U}$. Hence by Proposition 5.3 it is minimized at some $U_0(t) \in \mathcal{U}$, and for the hypothesized shapes of $\mathcal{U}$, the minimum value must occur at some vertex U_j. (See Problem 11.7.) If this same minimum value occurs at another vertex, it must do so at an adjacent vertex U_i so that by subtraction, $Ve^{-t\mathbb{A}}\mathbb{B}E = 0$ for the edge vector $E = U_i - U_j$.

Suppose this nonuniqueness occurs for more than a finite set of values of $t = t_n$. Then since the number of edges is finite we can assume that all t_n are associated with the same edge E, i.e., that $\varphi(t_n) \overset{\text{def}}{=} Ve^{-t_n\mathbb{A}}\mathbb{B}E = 0$, $\forall n = 1, 2, \ldots$. From this it follows that the function $\varphi(t) = Ve^{-t\mathbb{A}}\mathbb{B}E$, which is real analytic in t, vanishes identically [Rud]. But if $\varphi(t) = Ve^{-t\mathbb{A}}\mathbb{B}E \equiv 0$, then by differentiating successively and evaluating at $t = 0$, we find as before that $V\mathbb{B}E = V\mathbb{A}\mathbb{B}E = \cdots = V\mathbb{A}^{d-1}\mathbb{B}E = 0$. Thus, if $V = (v_1, \ldots, v_d)$, then $\sum_{i=1}^{d} v_i \mathbb{A}^{i-1}\mathbb{B}E = 0$, and by the hypothesized linear independence, we must have $V = \mathcal{O}$, a contradiction. Therefore, for all but a *finite* set of $t \in [0, T_0]$; $U_0(t) = U_j$, for a *unique* vertex U_j.

If $t_j \in J$, a *maximal* open subinterval of $(0, T_0)$ that excludes these exceptional values and $U(t_j) = U_j$, then for $U_i \neq U_j$, we see from (41) that $Ve^{-t_j\mathbb{A}}\mathbb{B}U_i > Ve^{-t_j\mathbb{A}}\mathbb{B}U_j$; by continuity, this strict inequality holds for *all* t near t_j, so that $U_0(t) = U_j$ for all such t. (Why?) It follows that in the open interval J, $U_0(t)$ is locally constant, hence *constant* $(= U_j)$, since possible points of discontinuity have been precluded. This provides the desired description of an optimizing control U_0.

If $U(t)$ is another optimizing control on $[0, T]$, then obviously $T = T_0$, so that by (42b) we have

$$\int_0^{T_0} Ve^{-t\mathbb{A}}\mathbb{B}(U(t) - U_0(t))\, dt = 0;$$

by (41), the integrand is nonnegative. From the standard argument we conclude that $Ve^{-t\mathbb{A}}\mathbb{B}U(t) = Ve^{-t\mathbb{A}}\mathbb{B}U_0(t)$, but as we have just established, this is possible only for that U_0 described above, i.e., $U(t) = U_0(t)$. $\square$

Remarks. This result shows that for such problems, an optimal control should be of the bang–bang type since it is mostly vertex-valued. If we replace $\mathcal{U}$ by $\{U : |U| \leq 1\}$ then such distinguished vertices are not present, and we have already considered related problems in §10.2(g).

The matrix $\mathbb{M}$ of Theorem 11.12 is of rank d if the edge-vector condition of this corollary is satisfied by *any* $E \neq \mathcal{O}$, but not conversely. (See the previous example.) Similar results hold in the nonautonomous linear-state case where the given matrices $\mathbb{A}$ and $\mathbb{B}$ are continuous functions of t, but the required condition generalizing that for E is less geometrical and more awkward to verify. (See [Ber].) Through an even less accessible condition, Diliberto [TO] uses the methods of §11.1 to obtain a bang–bang principle for control sets $\mathcal{U}$ which are *nonconvex* polytopes.

Example 2 (A Free-Space Docking Problem). An ion propulsion vehicle in free space equipped with forward and reverse thrust engines has given position and velocity at time 0, and it is to be brought to rest in minimum time at a target position (taken to be the origin.)

If we assume rectilinear motion only and neglect gravitational and fuel mass-loss effects, then its position $y(t)$ at time t is governed by Newton's second law: $\ddot{y}(t) = u(t)$, where $mu(t)$ is the net thrust produced by the engines, and m is the mass of the vehicle (assumed constant). By normalization we can suppose $|u(t)| \leq 1$.

If we set $Y(t) = (y(t), y_1(t))$ where $y_1(t)$ is the velocity at time t, then the state vector $Y = (y, y_1) \in \mathbb{R}^2$ obeys the linear law

$$\dot{Y}(t) = (y_1(t), u(t)) = \mathbb{A}Y(t) + \mathbb{B}u(t),$$

where

$$\mathbb{A} = \begin{bmatrix} 0 & 1 \\ 0 & 0 \end{bmatrix} \quad \text{and} \quad \mathbb{B} = \begin{vmatrix} 0 \\ 1 \end{vmatrix} \quad \text{so that} \quad \mathbb{A}\mathbb{B} = \begin{vmatrix} 1 \\ 0 \end{vmatrix},$$

and if $\mathcal{U} = [-1, 1]$, we can apply the preceding results. Here $E = \pm 2 \in \mathbb{R}^1$ provide the only edge vectors, and the vectors $\{\mathbb{B}E, \mathbb{A}\mathbb{B}E\} = \pm 2\{(0, 1), (1, 0)\}$ are linearly independent in $\mathbb{R}^2$. Thus by Corollary 11.14 we expect an optimal control $u_0(t)$ on $[0, T_0]$ to be bang–bang, in fact to have only "vertex" values of $+1$ and -1 except at a finite set of points. If present, it is unique.

Since $\mathbb{A}^2 = \mathbb{O}$ (why?), then $e^{-t\mathbb{A}} = \mathbb{I} - t\mathbb{A}$, only, $= \begin{bmatrix} 1 & -t \\ 0 & 1 \end{bmatrix}.$[1] By Theorem 11.12, any $u_0 \in \hat{C}[0, T_0]$ is time-optimal if we can find some $V = (v, v_1) \neq \mathcal{O}$, such that $\forall |u| \leq 1$:

$$Ve^{-t\mathbb{A}}\mathbb{B}u = V \cdot (-t, 1)u \geq V \cdot (-t, 1)u_0(t)$$

or

$$(-tv + v_1)u \geq (-tv + v_1)u_0(t) \quad \text{on } [0, T_0], \tag{44}$$

The linear (switching) function $\sigma(t) = v_1 - tv$ can vanish at most once on any interval, and otherwise, its sign determines $u_0(t) = +1$ or -1. (Why?)

All that remains is to show that by suitable choice of a switching time τ, we can indeed reach the origin from the given initial state $A \in \mathbb{R}^2$, using only those controls of the form

$$u_0(t) = \begin{cases} a, & t < \tau, \\ -a, & t > \tau, \end{cases}$$

where for each A, $a = \pm 1$. The time T_0 required to do so is then minimal, and the associated control is unique.

Under the simple dynamics permitted here it is clear by time reversal that this is equivalent to asking whether we can travel from the origin to a given state A by such controls.

Obviously, we can do so for all states in $\mathbb{R}^2$ where $u_0(t) \equiv +1$, i.e., for all states A which lie along the upper parabolic arc with parametric equations

$$\begin{cases} y(t) = t^2/2, \\ y_1(t) = t, \end{cases} \quad (\text{so } y = y_1^2/2, y_1 > 0),$$

or along the lower parabolic arc $y = -y_1^2/2$, $y_1 < 0$, obtained when $u_0(t) = -1$, both shown in Figure 11.6. The other parabolic arcs shown in this figure are those obtained by reversing the thrust direction at various points along

---

[1] See the footnote on page 398.

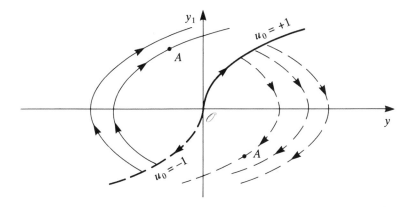

Figure 11.6

the principal pair. On geometrical grounds, it is evident that each point $A \neq \mathcal{O}$ can be joined to the origin by one (and only one) such combined curve. Moreover using analytic geometry, we can obtain the (unique) optimal time T_0 associated with each A. The details are given in Problem 11.8. The unique optimal control will be the *negative* of that used to reach A.

As this example shows, it is of considerable value to know how many switching times an optimal control can have. In general, this is a difficult problem. However, suppose that $\mathcal{U}$ is a closed box of the form

$$\mathcal{U} = \{U \in \mathbb{R}^k: -m_j \leq u_j \leq m_j, \text{ with } m_j > 0, j = 1, 2, \ldots, k\},$$

whose vertices are just the points where each $u_j = \pm m_j, j = 1, 2, \ldots, k$.
    Then the edge condition of Corollary 11.14 is equivalent to requiring that for each *column* B_j of $\mathbb{B}$, the vectors $B_j, \mathbb{A}B_j, \ldots, \mathbb{A}^{d-1}B_j$ are linearly independent (Problem 11.9). Moreover, with $\bar{P}(t) = Ve^{-t\mathbb{A}}$ as in (40'),

$$\bar{P}(t)\mathbb{B}U = \sum_{j=1}^{k} (\bar{P}(t) \cdot B_j)u_j = \sum_{j=1}^{k} \sigma_j(t)u_j, \tag{45}$$

where $\sigma_j(t) = \bar{P}(t) \cdot B_j$ is the *switching function* for the jth component of $U_0(t)$, so-called because as (45) shows, if $\sigma_j(t) \neq 0$ then

$$\sigma_j(t)u_j \geq -m_j \operatorname{sgn} \sigma_j(t) = \sigma_j(t)(U_0)_j(t),$$

which means that the jth component of U_0 is

$$\begin{cases} -m_j, & \text{when } \sigma_j(t) > 0, \\ +m_j, & \text{when } \sigma_j(t) < 0. \end{cases}$$

When the eigenvalues of $\mathbb{A}$ are all real, there can be at most $d - 1$ zeros of each σ_j (which is continuous), and so at most $d - 1$ switches in each component of the optimal control U_0. Some details are given in Problem 11.10, and generalizations are presented in [Le] and in [Y]. In particular, $\mathcal{U}$ as above can be replaced by *any* closed rectangular box. (But unless it has $\mathcal{O}$ as an interior point, we cannot use Theorem 11.12, to guarantee optimality, and without some other sufficiency argument—or an independent existence proof—we cannot assert that we have found an optimal control.) Even for the simple case considered here, existence theory is difficult. (See [M–S].) Moreover unless $\mathbb{A}$ has eigenvalues with only nonpositive real parts, there need not be a solution. (Problem 11.14.)

The eigenvalues of $\mathbb{A}$ are imaginary in the case of the corresponding time-optimal problem for the oscillator of §10.2(f) governed by the state equation

$$(\dot{y}, \dot{y}_1) = (y_1, u - y), \qquad |u| \le 1.$$

Indeed, the matrix

$$\mathbb{A} = \begin{bmatrix} 0 & 1 \\ -1 & 0 \end{bmatrix}$$

has eigenvalues λ where

$$|\mathbb{A} - \lambda\mathbb{I}| = \begin{vmatrix} -\lambda & 1 \\ -1 & -\lambda \end{vmatrix} = 0,$$

or $\lambda^2 + 1 = 0$, so that $\lambda = \pm i$.

Here again

$$\mathbb{B} = \begin{vmatrix} 0 \\ 1 \end{vmatrix} \quad \text{so that} \quad \mathbb{M} = [\mathbb{B} : \mathbb{A}\mathbb{B}] = \begin{bmatrix} 0 & 1 \\ 1 & 0 \end{bmatrix}$$

has rank 2, and thus each minimizing control for h on $[0 : T_0]$ will be the unique optimal control. However, the state-plane analysis is somewhat more complicated. (See [Po].) For this problem, sign changes in $u_0(t)$ occur over successive intervals of length π. Thus, the number of switches is not even bounded, but will vary with the location of the initial state, and this is characteristic of problems with nonreal eigenvalues. (Problem 11.10.)

§11.3. General Lagrangian Constraints

In addition to endpoint constraints, a problem in the variational calculus (or in optimal control), may also be subjected to a Lagrangian constraint of the form $G(t, Y(t), \dot{Y}(t)) = \mathcal{O}$, or more generally, to a Lagrangian *inequality* such as

$$G(t, Y(t), \dot{Y}(t)) \le \mathcal{O} \tag{46}$$

interpreted componentwise; (in control problems, $\dot{Y}$ is replaced by U).

When the $\dot{Y}$ terms do not appear explicitly, then such problems can often be attacked through implicit function theory, as was indicated in the proof of

Theorem 6.10. (See Problem 11.15.) Otherwise, the problem is considerably more difficult.

In this section we will use the minimum principle of Theorem 11.10 in conjunction with some more facts about cones in $\mathbb{R}^d$ (collected in §A.7) and implicit function theory to obtain existence of multiplier functions associated with minimizing in the presence of Lagrangian (in)equalities such as (46). From our introductory discussions in §2.3, §7.4, and §10.1, we know that such multipliers can be effective in producing sufficient conditions on fixed interval problems. We now want to consider their necessity in a more general setting.

We begin in part (a) by using Theorem 11.10 to extend the Pontjragin principle to optimal control problems in which the control set $\mathscr{U}$ can be described by C^1 inequalities. Then in part (b), we use these results to produce multipliers for the general problem in the variational calculus governed by Lagrangian (in)equalities of the type (46). Finally, in part (c), we indicate further extensions that could apply to analogous problems in optimal control. This material represents a small selection of the results that can be found in [He].

In addition to the results from §A.7, we will need the following:

(11.15) **Lemma.** *If* $\mathbb{G}$ *is a real* $m \times n$ *matrix with* m *linearly independent rows and* n *columns*[1]*, then* $\mathbb{G}\overline{\mathbb{G}}$ *is invertible.*

PROOF. Suppose that for some $X \in \mathbb{R}^m$, we have $X\mathbb{G}\overline{\mathbb{G}} = \mathcal{O}$. Then $X\mathbb{G}\overline{\mathbb{G}}\overline{X} = (X\mathbb{G})\cdot(X\mathbb{G}) = 0$; i.e., $X\mathbb{G} = \mathcal{O}$. But since $\mathbb{G}$ is of maximal rank m, this is possible *iff* $X = \mathcal{O}$, and the invertibility of $\mathbb{G}\overline{\mathbb{G}}$ follows as in (0.13).   □

(a) Control Sets Described by Lagrangian Inequalities

We suppose that all terms are as defined in Theorem 11.10. Then from Theorem A.28, we can easily get the following:

(11.16) **Corollary.** *In Theorem* 11.10, *suppose that the control set* $\mathscr{U}$ *can be described by Lagrangian inequalities* $\Psi(U) \leq \mathcal{O} \in \mathbb{R}^m$, *where* Ψ *is* C^1 *and* $\Psi_U(U_0(t))$ *has maximal row rank* m *while* f_U *and* G_U *are continuous. Then there exists a (unique)* $M = (\mu_1, \mu_2, \ldots, \mu_m) \in \hat{C}[0, T_0]$, *with* $M \geq \mathcal{O}$, *for which* $\tilde{h}_U(\cdot, Y_0, U_0) = \mathcal{O}$ *on* $[0, T_0]$, *where*

$$\tilde{h}(t, Y, U) = h(t, Y, U) + M(t) \cdot \Psi(U). \tag{47}$$

Moreover

$$\mu_i(t)\psi_i(U_0(t)) \equiv 0, \qquad i = 1, 2, \ldots, m.$$

PROOF. The minimum principle (29) implies that for each t, $h(t, Y_0(t), U)$ is minimized at $U_0(t)$ under the constraint $\Psi(U) \leq \mathcal{O}$. By Theorem A.28, a

---

[1] In this case we will say that $\mathbb{G}$ has maximal row rank m ($\leq n$).

unique $M \geq \mathcal{O}$ and $\tilde{h}$ exist as required to satisfy

$$\tilde{h}_U(\cdot, Y_0, U_0) = h_U(\cdot, Y_0, U_0) + M(\cdot)\Psi_U(U_0) = \mathcal{O}$$

(where the rows of Ψ_U are indexed by U), with $\mu_i(t)\psi_i(U_0(t)) \equiv 0$, $i = 1, 2, \ldots$, m. But since $\Psi_U(U_0(\cdot))$ is hypothesized to have maximal row rank then $\mathbb{Q}_0(\cdot)$, its product with its transpose, is invertible (by Lemma 11.15) and the inverse is necessarily continuous with U_0; hence, $M(\cdot) = -h_U(\cdot, Y_0, U_0)\overline{\Psi}_U(U_0)\mathbb{Q}_0^{-1}(\cdot)$. However, the matrices on the right are continuous except at the discontinuities of U_0, so that $M \in \hat{C}[0, T_0]$. $\square$

Note that when a component ψ_i is *inactive* at $U_0(t)$ in that $\psi_i(U_0(t)) < 0$, then $\mu_i(t) = 0$. In §10.1 and in several of the applications in §10.2, we have seen how this extension affects the search for optimal solutions.

(b)* Variational Problems with Lagrangian Constraints

In this part, we use the optimal control results just obtained, to establish the existence of appropriate multipliers for problems in the variational calculus with general Lagrangian constraints as mentioned in §6.7. To do so we require some rather technical extensions of standard implicit function theory. (These can be avoided in a simpler case taken up in Problem 11.16.)

In the next theorem and its corollary, all sets are in Euclidean spaces, and all mappings are vector valued.

(11.17) Theorem. *Suppose that $\mathcal{F} = \mathcal{F}(T, V)$ is C^1 and that $\mathcal{F}_V(T, V)$ is invertible on a nonempty set S_0 on which $\mathcal{F}(T, V) = \mathcal{O}$. If $S_0 \supseteq \{(T, \mathcal{O}): T \in \mathcal{T}_0\}$ for some set $\mathcal{T}_0$, then there exist $\varepsilon > 0$ and a (locally) unique C^1 function F defined on a neighborhood $\mathcal{T}$ of $\mathcal{T}_0$ such that $\mathcal{F}(T, F(T)) \equiv \mathcal{O}$, with $|F(T)| < \varepsilon, \forall T \in \mathcal{T}_0$.*

PROOF. The local existence of F is a consequence of ordinary implicit function theory [Ed], which guarantees that near each $T_0 \in \mathcal{T}_0$, the $\mathcal{O}$-level set of $\mathcal{F}$ near $(T_0, \mathcal{O})$ can be represented by the graph of a unique C^1 function, $F(\cdot, T_0)$, say. When a compact part of $\mathcal{T}_0$ is covered by a finite number of such neighborhoods, the relevant locally defined functions must agree in any common region of definition (why?) so that F is well defined in a neighborhood of this part, and it is locally unique. By considering extensions to larger compact sets, it follows that an F is defined and (locally) unique as required. $\square$

Note: An analogous argument was used in proving Theorem 9.24, and Figure 9.12 may help to visualize the construction.

(11.18) Corollary. *If $\mathcal{G} = \mathcal{G}(X, U)$ is C^2 and $\mathcal{G}_U(X, U)$ has maximal row rank on S_0, the nonempty $\mathcal{O}$-level set of $\mathcal{G}$, then there exists a C^1 function G defined* *on a neighborhood of S_0 such that*

$$\mathcal{G}(X, G(X, U)) \equiv \mathcal{O}; \quad \text{and when} \quad (X, U) \in S_0 \colon G(X, U) = U.$$

PROOF. Set $T = (X, U)$, suppose the columns of $\mathcal{G}_U$ indexed by U, and consider $\mathcal{F}(T, V) \overset{\text{def}}{=} \mathcal{G}(X, U + \mathcal{G}_U(T)V)$, which is C^1 (since $\mathcal{G}$ is C^2). Also $\mathcal{F}_V(T, V) = \mathcal{G}_U(T)\bar{\mathcal{G}}_U(T)$ is invertible on S_0 (by Lemma 11.15) and $\mathcal{F}(T, \mathcal{O}) = \mathcal{G}(X, U) = \mathcal{O}$ on S_0, so that we can apply the theorem to obtain an F defined in a neighborhood of S_0. Then

$$G(X, U) \overset{\text{def}}{=} U + \mathcal{G}_U(X, U)F(X, U)$$

has the desired properties, since

$$\mathcal{G}(X, G(X, U)) = \mathcal{G}(X, U + \mathcal{G}_U(X, U)F(X, U)) = \mathcal{F}(T, F(T)) \equiv \mathcal{O};$$

$$(X, U) \in S_0 \Rightarrow \mathcal{G}(X, U) = \mathcal{F}(T, \mathcal{O}) = \mathcal{O},$$

but then $\mathcal{F}(T, F(T)) = \mathcal{O} \Rightarrow F(T) = \mathcal{O}$, by local uniqueness of F; i.e., $G(X, U) = U$ as desired. □

By using the previous results, we can transform a problem in the variational calculus with Lagrangian constraints into a standard problem in control theory. (It may be helpful to first read Problem 11.16.) To facilitate later comparison, we will employ notation which differs slightly from that used in previous chapters to formulate the problem.

(11.19) *Suppose that $Y_0 \in \hat{C}^1[0, T_0]$ minimizes*

$$\mathcal{F}(Y, T) \overset{\text{def}}{=} \int_0^T f(t, Y(t), \dot{Y}(t)) \, dt$$

on

$$\mathcal{D} = \left\{ \begin{matrix} Y \in \hat{C}^1[0, T] \colon T > 0, \text{ with } Y(0) = Y_0(0), \\ (Y(t), \dot{Y}(t)) \in D, \text{ and } \Phi(T, Y(T)) = \mathcal{O} \in \mathbb{R}^l \end{matrix} \right\},$$

under the Lagrangian constraint

$$\mathcal{G}(t, Y(t), \dot{Y}(t)) \equiv \mathcal{O} \quad \text{on } (0, T).$$

Here D is a domain in $\mathbb{R}^{2d}$ while $f = f(t, Y, Z)$, f_Y and f_Z are assumed continuous, and $\Phi = \Phi(t, Y)$ is C^1 while $\Phi_{(t, Y)}(T_0, Y_0(T_0))$ has maximal row rank $l \le d + 1$. Similarly, we suppose that $\mathcal{G} = \mathcal{G}(t, Y, U)$ is C^2, and that $\mathcal{G}_U(t, Y, U)$ has maximal row rank $m \le d$ whenever $\mathcal{G}(t, Y, U) = \mathcal{O}$, in a neighborhood S_0 of $\mathcal{C}_0$, the trajectory of Y_0 in $\mathbb{R}^{2d+1}$. According to Corollary 11.17, with $X = (t, Y)$, there exists a C^1 function $G(t, Y, U)$ defined in a neighborhood of S_0, such that

$$\mathcal{G}(t, Y, G(t, Y, U)) \equiv \mathcal{O} \in \mathbb{R}^m, \quad \text{while} \quad G(t, Y, U) = U \quad \text{on } S_0. \tag{48}$$

Next, let

$$f(t, Y, U) \overset{\text{def}}{=} f(t, Y, G(t, Y, U)). \tag{49}$$

We claim that with $U_0 = \dot{Y}_0$, the triple (Y_0, U_0, T_0) minimizes

$$F(Y, U, T) = \int_0^T f(t, Y(t), U(t))\, dt,$$

on

$$\tilde{\mathcal{D}} = \left\{ \begin{array}{l} (Y, U) \in \hat{C}^1[0, T] \times \hat{C}[0, T];\ T > 0,\ \text{with } Y(0) = Y_0(0); \\ U(t) \in \mathcal{U}(t), \qquad \Phi(T, Y(T)) = \mathcal{O} \in \mathbb{R}^l, \end{array} \right\}$$

under the state law

$$\dot{Y} = G(t, Y, U), \tag{50}$$

where

$\mathcal{U}(t)$ is a bounded neighborhood of $\dot{Y}_0(t)$ such that

$(Y_0(t), U(t)) \in D, \quad \text{when} \quad U(t) \in \mathcal{U}(t).$

Indeed, if $(Y, U, T) \in \tilde{\mathcal{D}}$, then defining $\dot{Y} = U$, gives $(Y, T) \in \mathcal{D}$ and if $\dot{Y} = G(t, Y, U)$ then

$$F(Y, U, T) = \int_0^T f(t, Y, \dot{Y})\, dt = \mathcal{F}(Y, T) \geq \mathcal{F}(Y_0, T_0),$$

since by (48),

$$\mathcal{G}(t, Y, \dot{Y}) = \mathcal{G}(t, Y, G(t, Y, U)) \equiv \mathcal{O}.$$

However, before we can simply invoke Theorem 11.10, we must address the complications which arise from the time-varying control set $\mathcal{U}(t)$. Note first that since $\dot{Y}_0$ is $\hat{C}$, we can find a fixed finite set of subintervals of $[0, T_0]$ on each of which $\mathcal{U}(t) = \mathcal{U}_j$ does not vary with t. Then if we reexamine the proof of those results about propagation of vectors V_i (Propositions 11.3–11.5) we see that the arguments are not really affected by the finite set of $\mathcal{U}_j$ along $[0, T]$, since all we require is that the actual V_i be chosen within the relevant $\mathcal{U}_j$. Moreover, the arguments about target values utilized specific V_i, not their local bounds. Thus we can conclude that Theorem 11.6, and (hence) Theorem 11.10 remain vaid.

In particular, from the latter, we know that there exist $P \in \hat{C}^1[0, T_0]$ and $\lambda_0 (= 0 \text{ or } 1)$ for which

$$h(t, Y, U) \overset{\text{def}}{=} \lambda_0 f(t, Y, U) + P(t) \cdot G(t, Y, U) \tag{51}$$

satisfies the minimum principle for admissible U; i.e., for $t \in [0, T_0]$:

$$h(t, Y_0(t), U) \geq h_0(t) \overset{\text{def}}{=} h(t, Y_0(t), U_0(t)),$$

whenever $U \in \mathcal{U}(t)$. Thus if we set

$$\hbar(t, Y, U) = \lambda_0 f(t, Y, U) + P(t) \cdot U,$$

so that

$$\hbar(t, Y, G(t, Y, U)) = h(t, Y, U), \tag{51'}$$

we can say that $\hbar(t, Y_0(t), U) \geq \hbar(t, Y_0(t), \dot{Y}_0(t))$, $\forall\, U \in \mathcal{U}(t)$ such that $\mathcal{G}(t, Y_0(t), U) = \mathcal{O}$, since by (48), this implies that $G(t, Y_0(t), U) = U$.

Moreover, by hypothesis and construction, at each point of continuity t, $h_U(t, Y_0(t), U)$ will be continuous in U near $\dot{Y}_0(t)$. Thus from Theorem A.28 and our assumption that $\mathscr{G}_U(t, Y_0(t) \cdot)$ has maximal row rank at $\dot{Y}_0(t)$, (and so in a U neighborhood), it follows that there exist *unique* multipliers $M(t)$ such that

$$\tilde{h}_U(t, Y_0(t), \dot{Y}_0(t)) \equiv \mathcal{O} \tag{52}$$

where

$$\tilde{h}(t, Y, U) \stackrel{\text{def}}{=} h(t, Y, U) + M(t) \cdot \mathscr{G}(t, Y, U). \tag{53}$$

Thus from (51) and (51′),

$$\lambda_0 f_U + P + M \mathscr{G}_U = 0, \tag{54}$$

when evaluated at $(t, Y_0(t), \dot{Y}_0(t))$.

However, again, since $\mathscr{G}_U$ has maximal row rank, we conclude from Lemma 11.15 that $\mathcal{Q}$, its product with its transpose, has an inverse which is continuous with $\dot{Y}_0$; i.e.,

$$M = -(\lambda_0 f_U + P)\bar{\mathscr{G}}_U \mathcal{Q}^{-1} \tag{55}$$

is continuous with $\dot{Y}_0$.

Also from (54),

$$P = -\tilde{f}_U[Y_0(\cdot)]$$

if we introduce

$$\tilde{f} = \lambda_0 f + M \cdot \mathscr{G}, = \tilde{h} - P \cdot U. \tag{56}$$

Next, from (51′) and (53) we have

$$h(t, Y, U) = \tilde{h}(t, Y, G(t, Y, U)),$$

since

$$\mathscr{G}(t, Y, G(t, Y, U)) \equiv \mathcal{O},$$

and from our original optimal control characterization (Theorem 11.10) it follows that

$$-\dot{P}(t) = [h_Y = \tilde{h}_Y + \tilde{h}_U G_Y]|_{(t, Y_0, \dot{Y}_0)}$$
$$= \tilde{h}_Y(t, Y_0, \dot{Y}_0) = \tilde{f}_Y[Y_0(t)] \quad \text{(by (52))}.$$

Combining this with (56) and replacing U by $\dot{Y}$, we see that except at corner points,

$$\frac{d}{dt}\tilde{f}_Y[Y_0(t)] = \tilde{f}_Y[Y_0(t)];$$

i.e., Y_0 is stationary for $\tilde{f}$ in such intervals.

At corner points, $\tilde{f}_Y[Y_0(\cdot)]$ is continuous, (with P), while from the continuity of $h_0(t) = \tilde{f}[Y_0(t)] - \tilde{f}_Y[Y_0(t)] \cdot \dot{Y}_0(t)$ follows the second Weierstrass–Erdmann condition of §7.5.

From equation (30) for $\tilde{h}_0$ in Theorem 11.10, we can show that Y_0 also satisfies the second Euler–Lagrange equation for $\tilde{f}$. Similarly, the minimum principle gives the Weierstrass necessary condition on $\tilde{f}$ of Theorem 7.15. (Problem 11.20.)

It remains only to obtain the appropriate transversality conditions on $\tilde{f}$. By Theorem 11.10, there is a $\Lambda \in \mathbb{R}^l$ such that $(\lambda_0, \Lambda) \neq \mathcal{O}$ for which

$$\tilde{f}_{\dot{Y}}[Y_0(T_0)] = -P(T_0) = -\Lambda \Phi_Y(T_0, Y_0(T_0)),$$

while   (57)

$$\tilde{f}[Y_0(T_0)] - \tilde{f}_{\dot{Y}}[Y_0(T_0)] \cdot \dot{Y}_0(T_0) = h_0(T_0) = -\Lambda \cdot \Phi_t(T_0, Y_0(T_0)).$$

We have established the following major result:

(11.20) Theorem. *Suppose that (Y_0, T_0) is minimizing for the problem formulated in (11.19). Then there exist λ_0 ($= 0$ or 1) and Lagrangian multipliers $M \in \hat{C}[0, T_0]$ such that Y_0 is stationary for $\tilde{f} = \lambda_0 f + M \cdot \mathscr{G}$ except possibly at its corner points. At these corner points, both $\tilde{f}_{\dot{Y}}[Y_0(\cdot)]$ and $(\tilde{f} - (\tilde{f}_{\dot{Y}}) \cdot \dot{Y})[Y_0(\cdot)]$ are continuous. Moreover, for $\tilde{f}$, Y_0 satisfies the Weierstrass condition (7.15), together with the second Euler–Lagrange equation, and it meets the transversal conditions (57), for an appropriate Λ, with $(\lambda_0, \Lambda) \neq \mathcal{O}$.* $\quad\square$

(c) Extensions

(11.21) Various extensions of Theorem 11.20 are possible. For example, suppose that instead of the fixed left end condition $Y(0) = Y_0(0)$, the competing Y are required to satisfy a condition of the form $\tilde{\Phi}(a, Y(a)) \equiv \mathcal{O}$ for some C^1 vector valued function $\tilde{\Phi}$ whose Jacobian matrix at $(a_0, Y_0(a_0))$ has maximal rank $\leq d$. The f defined as above on $[a_0, T_0]$ will satisfy transversality conditions analogous to (57) at $(a_0, Y_0(a_0))$ for an appropriate vector $\tilde{\Lambda}$. It is also possible to consider minimization of a Bolza type functional in which $\mathscr{F}$ is replaced by

$$\mathscr{F} + \varphi_0(T, Y(T)), \quad \text{where } \varphi_0 \text{ is } C^1.$$

Then the theorem holds as stated if $\Lambda \cdot \Phi$ is replaced by $\lambda_0 \varphi_0 + \Lambda \cdot \Phi$ in the transversality conditions (57). (Details are indicated in Problem 11.22*.)

Corresponding problems with additional isoperimetric constraints may be handled by replacing these with Lagrangian constraints in a higher-dimensional vector space as in Proposition 9.11. The resulting multipliers associated with such constraints will be constant as expected. A simple example in discussed in Problem 11.23.

(11.22) Lagrangian inequalities of the form $\mathscr{G}(t, Y(t), \dot{Y}(t)) \leq \mathcal{O}$, can be treated with the device of slack variables introduced by Valentine [CCV], and used in Problem 7.22*.

To accomplish this, we let $v_i = w_i^2$, $i = 1, 2, \ldots, m$, and for the new function

$$\mathscr{G}^*(t, Y, Z, W) = \mathscr{G}(t, Y, Z) + V \tag{58}$$

we consider the problem of minimizing $\mathscr{F}$ as before on

$$\mathscr{D}^* = \{(Y, W, T) : (Y, T) \in \mathscr{D}, W \in \hat{C}[0, T]\}$$

under the Lagrangian constraint

$$\mathscr{G}^*(t, Y(t), \dot{Y}(t), W(t)) \equiv \mathcal{O}. \tag{59}$$

Observe that $\nabla g_i^* = (\nabla g_i, 2w_i E_i)$, where E_i is the unit vector in the ith direction in $\mathbb{R}^m$, so that the new gradients are still linearly independent.

We consider W as though it were $\dot{W}$ in a problem in which W is not present explicitly, but could be obtained from $\dot{W}$ by integration, with say, $W(0) = \mathcal{O}$.

If (Y_0, T_0) minimizes $\mathscr{F}$ on $\mathscr{D}$, under $\mathscr{G}[Y(t)] \leq \mathcal{O}$, then defining $V_0(t) = -\mathscr{G}[Y_0(t)]$, gives a W_0 for which (Y_0, W_0, T_0) minimizes $\mathscr{F}$ on $\mathscr{D}^*$ under (59). For any $(Y, W, T) \in \mathscr{D}^*$ which satisfies (58), equation (59) gives a $(Y, T) \in \mathscr{D}$ satisfying $\mathscr{G}[Y(t)] \leq \mathcal{O}$, since $V \geq \mathcal{O}$.

Thus Theorem 11.20 is applicable, and we have multipliers $M \in \hat{C}[0, T]$ for which all conditions satisfied by $\tilde{f}$ are now satisfied by

$$f^* \stackrel{\text{def}}{=} \lambda_0 f + M \cdot \mathscr{G}^* = \tilde{f} + M \cdot V, \tag{60}$$

but with respect to both Y and W (as $\dot{W}$), where $\tilde{f} = \lambda_0 f + M \cdot \mathscr{G}$, as before, is independent of W.

In particular, the first Weierstrass–Erdmann condition relative to W is that $f_{\dot{W}}^*$ be constant when evaluated along the minimizing trajectory. However, since $v_i = w_i^2$, this means that $2\mu_i w_i = \text{const.}$, when so evaluated; moreover, either $\mu_i \equiv 0$, or $g_i[Y_0(t)]$ vanishes at some point along Y_0, for otherwise this constraint would remain inactive throughout. Thus, upon proper evaluation we must have

$$\mu_i w_i \equiv 0 \quad \text{or} \quad \mu_i w_i^2 \equiv 0, \quad \text{so that}$$

$$\mu_i(t) g_i[Y_0(t)] \equiv 0 \quad \text{on } [0, T_0]. \tag{61}$$

But, then the second equation for f^*, the Weierstrass condition for f^*, and the second transversality condition (57) for f^* reduce to those for $\tilde{f}$ alone; it follows that for this new problem, Theorem 11.20, holds in its entirety and in addition we have (61). Moreover, by a careful repetition of the transformation to the optimal control problem, we can conclude that $M \geq \mathcal{O}$. (See Problem 11.24.)

(11.23) Finally, there is an optimal control analogue for each of the above extensions which can be obtained by applying the same techniques. Let's consider only one such problem, that for which (Y_0, U_0, T_0) minimizes

$$F(Y, U, T) = \int_0^T f(t, Y(t), U(t)) \, dt$$

on

$$\mathscr{D}_\Phi = \left\{ \begin{array}{l} (Y, U, T): (Y, U) \in \hat{C}^1[0, T] \times \hat{C}[0, T]; \ T > 0; \\ \quad Y(0) = Y_0(0), \quad \Phi(T, Y(T)) = \mathcal{O}, \end{array} \right\}$$

subject to the Lagrangian inequality

$$\mathscr{G}(t, Y, U) \leq \mathcal{O}, \quad 0 \leq t \leq T, \tag{62a}$$

together with the state law

$$\dot{Y} = G(t, Y, U), \qquad 0 \le t \le T. \tag{62b}$$

Observe that (62a) permits description of state-dependent control sets $\mathcal{U}$. Here, it is supposed that $\mathcal{G}$ and G are C^2 (although slightly weaker hypotheses suffice), where $\mathcal{G}_U(t, Y_0(t), U_0(t))$ is of maximal rank $m \le k$ while f, f_Y, and f_U are continuous. The additional assumptions are those made in Theorem 11.10.

We can reduce this problem to the one covered in (11.22) if we consider U as $\dot{U}$ (from which an "honest" U could be obtained by integration with say $U(0) = \mathcal{O}$), and the resulting (Y, U) as an enlarged state vector $\in \mathbb{R}^{d+k}$. Then equations (62) can be combined into the single Lagrangian inequality

$$\mathcal{G}^*(t, Y, \dot{Y}, U) \overset{\text{def}}{=} \left\{ \begin{matrix} \mathcal{G}(t, Y, U) \\ G(t, Y, U) - \dot{Y} \end{matrix} \right\} \le \mathcal{O}$$

and since

$$\mathcal{G}^*_{(\dot{Y}, U)} = \left[\begin{array}{c|c} \mathcal{O} & \mathcal{G}_U \\ \hline -I_d & G_U \end{array} \right]$$

has maximal rank $m^* = m + d \le d + k$ along the optimal trajectory, the previous results are applicable. Accordingly, there exist multipliers $M^* = (M, P)$, say (which are continuous except at the discontinuities of U), and a λ_0 $(= 0$ or $1)$ for which the modified integrand

$$\tilde{f} \overset{\text{def}}{=} \lambda_0 f + P \cdot (G - \dot{Y}) + M \cdot \mathcal{G}$$

satisfies the usual necessary conditions along the optimal trajectory, together with the appropriate transversality conditions at T_0. In addition, from (61), it follows that

$$M \ge \mathcal{O} \quad \text{with} \quad \mu_i g_i(t, Y_0(t), U_0(t)) \equiv 0, \qquad i = 1, 2, \ldots, m. \tag{63}$$

In particular, we expect that $\tilde{f}_U$ is constant and hence $\mathcal{O}$ (since $\Phi_U \equiv 0$) along the optimal trajectory, while both $-P = \tilde{f}_{\dot{Y}}$ and $\tilde{f} - \tilde{f}_{\dot{Y}} \cdot \dot{Y} - \tilde{f}_U \cdot U$ are continuous. Observe that in view of (63), the latter reduces to $h_0 = \lambda_0 f[Y_0, U_0] + P(\cdot) \cdot G[Y_0, U_0]$. Next, $P(T_0) = \Lambda \Phi_{(T, Y)}(T_0, Y_0(T_0))$ for an appropriate Λ, with $(\lambda_0, \Lambda) \ne \mathcal{O}$. In addition, we expect that except at corner points (Y_0, U_0) is stationary for $\tilde{f}$; i.e.,

$$-\dot{P} = \frac{d}{dt} \tilde{f}_{\dot{Y}} = \tilde{f}_Y = \tilde{h}_Y, \quad \text{say,}$$

if

$$\tilde{h} \overset{\text{def}}{=} h + M \cdot \mathcal{G} = \tilde{f} + P \cdot \dot{Y}, \tag{64}$$

where

$$h \overset{\text{def}}{=} \lambda_0 f + P \cdot G, \quad \text{as usual;}$$

moreover,

$$\tilde{h}_U = \tilde{f}_U = \mathcal{O}, \quad \text{by the above argument.}$$

Finally, we have from the Weierstrass necessary condition for $\tilde{f}$ that with $V = (W, U)$ say, near $(\dot{Y}_0(t), U_0(t))$,

$$\tilde{h}(t, Y_0(t), V) - \tilde{h}(t, Y_0(t), \dot{Y}_0(t), U_0(t))$$

$$= \tilde{f}(t, Y_0(t), V) - \tilde{f}(t, Y_0(t), \dot{Y}_0(t), U_0(t)) + P(t) \cdot (W - \dot{Y}_0(t))$$

$$\geq 0 \quad \text{(since } P = -\tilde{f}_Y \text{ and } \tilde{f}_U = 0 \text{ along the optimal trajectory.)}$$

As two special cases of the latter inequality, we obtain (when $U = U_0(t)$) that

$$\tilde{h}(t, Y_0(t), W, U_0(t)) \geq h_0(t), \qquad \forall\, W \text{ near } \dot{Y}_0(t)$$

and (when $W = \dot{Y}_0(t)$) that

$$\tilde{h}(t, Y_0(t), \dot{Y}_0(t), U) \geq h_0(t), \qquad \forall\, U \text{ near } U_0(t).$$

We have now closed the circle of ideas begun in §10.1, by proving that the conditions shown there to be sufficient on a fixed interval with convexity are indeed necessary for more general problems. In [He] other "smooth" problems are considered wherein the various given functions are supposed C^2 as required. Related problems with $\hat{C}$ state variables are examined in [S–S], and numerical methods that implement some of these results are indicated in [G–L].

Associated nonsmooth problems are explored in [Ce], [Cla], [Lo], [I–T], [R], [Wa], [Z] among others. Such problems demand far more sophisticated tools, and constraints of the type considered here impose substantial difficulties for the development of a governing theory. Nevertheless, after proper interpretation, the results are essentially the same; viz., constrained optima guarantee existence of associated multipliers for which a modified performance function is optimized as though such constraints were not present. In this theory, the role of the multipliers receives functional analytic clarification, but the solution of any given problem, even when smooth, remains as difficult as before. Each problem must be analyzed by methods peculiar to itself. What is most needed is simple effective sufficiency criteria, especially for nonconvex problems on variable intervals, or a better understanding of the transformations which might be used to reduce such problems to those for which such criteria are already known. Peterson and Zalkind (see [Le]) supply a valuable comparative survey of earlier work, while recent results of Zeidan (see [Cla] and [Lo]) provide hope for further success.

PROBLEMS

11.1.   After suitable reformulation if required, apply Theorem 11.8 to Problem: (a) 10.1; (b) 10.3; (c) in §10.2(e); (d) 10.29; and (e) 10.10 (see Remark 11.9). After suitable reformulation, if necessary, apply Theorem 11.10 to Problem (f), 10.2(a); and (g) 10.14. Try to identify necessary conditions with those previously found to be sufficient. (h) In Example 3 of §11.1, show what happens when we replace F by $F(Y, u, T) = \int_0^T y_1^2(t)\, dt$. Hint: Equation (39′).

11.2.   (Oscillator energy problem when $T = 5\pi/4$.)

(a) Suppose $\sigma \in [\pi/4, T)$. To examine possible optimality of

$$u_\sigma = \begin{cases} +1 & \text{on } (0, \sigma), \\ -1 & \text{on } (\sigma, T), \end{cases}$$

compute

$$\int_0^T \cos \tau u_\sigma(\tau)\, d\tau \quad \text{and} \quad \int_0^T \sin \tau u_\sigma(\tau)\, d\tau$$

as functions of σ, and using $u_\sigma(0+) = -u_\sigma(T-)$, determine σ from equation (27).

(b) For the same choices of u_σ, use (17) of Chapter 10 to calculate $2E_\sigma(T) = a + b \sin(\sigma - \beta)$ and note that it is maximized by the σ found in part (a).

(c) Make a similar analysis for

$$\tilde{u}_\sigma = \begin{cases} -1 & \text{on } (\sigma, \sigma + \pi), \\ +1 & \text{otherwise}, \end{cases}$$

when $\sigma \in (0, \pi/4)$, and note that $\max \tilde{E}_\sigma(T) \le \max E_\sigma(T)$ from part (b). What can you conclude?

11.3.   Verify the assertions made in the proof of Theorem 11.10 concerning the properties of h_0 and h as obtained by applying the minimum principle to $\hat{h}w$.

11.4.   In Theorem 11.12 replace the linear state equation by $\dot{Y}(t) = \mathbb{A}Y(t) + G(U(t))$, where G is $\mathscr{C}^1$, with $G(-U) = -G(U)$, $\forall U$ near $\mathcal{O} \in \mathcal{U}$.

(a) Observe that $G(\mathcal{O}) = \mathcal{O}$, and verify that the inequalities in the proof remain valid when $G(U)$ replaces $\mathbb{B}U$ in each occurrence. Conclude that on $[T, T_0]$: $Ve^{-t\mathbb{A}}G(U) \equiv \mathcal{O}$, $\forall$ small U.

(b) Show that if $\mathbb{B} \overset{\text{def}}{=} G_U(\mathcal{O})$ is used to define $\mathbb{M}$, then the conclusions remain valid.

(c) State your result as a theorem.

11.5.   (a) Observe that the hypotheses on $\mathbb{M}$ in both Theorem 11.12 and the previous problem preclude having $\mathbb{B} = \mathcal{O}$. Could $\mathbb{A} = \mathcal{O}$? Explain.

(b) The new result in the previous problem applies to the state equation $\dot{y} = -2y + \sin u$, with say $|u| \le 1$. (Why?) Solve the associated time-optimal problem with $y(0) = 1$.

(c) Give a vector-valued nonlinear state equation to which the new theorem applies. Give one with a vector-valued control.

11.6.   Find alternate time-optimal controls for the problem in Example 1 of §11.2. Verify that the edge condition in Corollary 11.14 is *not* satisfied for this problem.

11.7.   (a) When $\mathscr{P}$ is a compact convex polyhedron in $\mathbb{R}^d$ with vertices $U_j, j = 1, 2, \ldots, n$ say, then for each *fixed* nonzero $V \in \mathbb{R}^d$, show that the function $l(U) = V \cdot U$ is linear and continuous (on $\mathscr{P}$). Conclude that it assumes its *minimum* value m at some point U_0 on $\mathscr{P}$.

(b) Suppose that $U_0 = \mathcal{O}$, and note that $U = -\rho V \notin \mathscr{P}$ for *any* $\rho > 0$, so that $U_0 = \mathcal{O}$ must be a boundary point of $\mathscr{P}$. Also $l(U) = l(U_0) = m$, whenever $(U - U_0) \perp V$, which means that l also assumes its minimum value at any such U. Since all such U must lie on the boundary of $\mathscr{P}$ (why?), conclude that we can suppose U_0 is a vertex of $\mathscr{P}$.

(c) If this same minimum occurs at another vertex, why must it do so at an *adjacent* vertex U_1 (one for which the segment joining U_0 and U_1 is also on the boundary of each face containing these vertices)?

11.8. In Example 2 of §11.2:

(a) Verify that $u \equiv +1$ provides a unique parabolic trajectory of the state equation through each point $A = (a, a_1)$ in the upper region shown in Figure 11.6, while $u \equiv -1$ accomplishes the same for the lower region.

(b) In the upper region, show that the parabola through A meets the primary trajectory for $u \equiv -1$ at the point $\tilde{A} = (\tilde{a}, \tilde{a}_1)$ where $\tilde{a}_1 = -\sqrt{a_1^2/2 - a}$, and find the corresponding point when A is in the lower region.

(c) Show that the time of travel $\alpha = T(A)$ from $\mathcal{O}$ to A along the combined trajectories is given by $T(A) = \sqrt{2a_1^2 - 4a} + a_1$; or $\sqrt{2a_1^2 + 4a} - a_1$, according to whether A is in the upper or lower region.

(d) Using such trajectories, time optimality can be established independently of the results of §11.2 by showing that when $t \in [0, \alpha]$, for each trajectory $Y(t)$ of the state equation joining $\mathcal{O}$ to A, we have $(d/dt)T(Y(t)) \le 1$, where $T(Y(t))$ is the associated time of travel from $\mathcal{O}$ to $Y(t)$ given in part (c).

(e)* Obtain the last inequality by proving that

$$\frac{d}{dt}T(Y(t)) - 1 \le \begin{cases} (u - 1)\left(\dfrac{2y_1}{\sqrt{2y_1^2 - 4y}} + 1\right), & \text{in the upper region,} \\[3mm] (u + 1)\left(\dfrac{2y_1}{\sqrt{2y_1^2 + 4y}} - 1\right), & \text{in the lower region.} \end{cases}$$

Hint: $(d/dt)T(Y(t)) = T_y(Y(t))\dot{y}_1(t) + T_{y_1}(Y(t))u(t)$.

11.9. (a) When $\mathcal{U}$ is a closed coordinate box centered at the origin with sides of length $2m_i$, $i = 1, 2, \ldots, k$, explain why the edge vectors E joining adjacent vertices are essentially the coordinate vectors in each direction.

(b) What does the edge condition in Corollary 11.14 become for such $\mathcal{U}$?

11.10*. (Analysis of switching function σ of equation (45).)

(a) Suppose that the eigenvalues λ_i of $\mathbb{A}$ are real and distinct, and B is a column of $\mathbb{B}$. Show that for vectors C_i,

$$\sigma(t) = Ve^{-t\mathbb{A}}B = V \cdot \left(\sum_{i=1}^{d} C_i e^{-\lambda_i t}\right) = \sum_{i=1}^{d} \beta_i e^{-\lambda_i t}, \quad \text{say.}$$

Hint: $e^{-t\mathbb{A}}B$ is a solution of $\dot{X} = -\mathbb{A}X$, where $\mathbb{A}$ is similar to a diagonal matrix. (See [C–L].)

(b) Prove by induction that this continuous switching function σ associated with B is either $\equiv 0$, or it cannot vanish at more than $d - 1$ points $t \in [0, \infty)$. Why does this limit its number of sign changes?

(c) If the eigenvalues are real but not necessarily distinct, then from [C–L],

$$\sigma(t) = \sum_{i=1}^{e} q_i(t)e^{-\lambda_i t}$$

for appropriate $e < d$ and polynomials q_i associated with the *distinct* λ_i, $i = 1, 2, \ldots, e$. Argue that the conclusion in part (b) applies to this case as well.

(d) If nonreal eigenvalues occur, then they must do so in conjugate pairs $\lambda = \mu \pm iv$ for real μ, v, giving terms in $\sigma(t)$ such as $e^{-\mu t} \cos(vt + \alpha)$. Why would this eliminate bounds on the number of sign changes?

11.11. A system is governed by the state law $\ddot{y} = u$, with $|u| \leq 1$. Analyze the time-optimal problem of bringing this system from state $A = (y(0), \dot{y}(0), \ddot{y}(0))$ to $\mathcal{O} = (0, 0, 0)$. Characterize possible optimal controls u_0, including bang–bang behavior, number of switches, etc. Is an optimal control unique (if it exists)?

11.12. Perform an analysis similar to that in the previous problem for the system governed by the state equation $\dot{Y} = (\dot{y}, \dot{y}_1) = (y_1 + u, -y + u_1)$, where $\mathcal{U} = \{(u, u_1): |u| \leq 1, |u_1| \leq 1\}$.

11.13. Perform an analysis similar to that in Problem 11.11 for the system governed by the state equation $\dot{Y} = (\dot{y}, \dot{y}_1) = (y_1 + u, u)$, where $|u| \leq 1$.

11.14. (Time-optimal controls need not exist.)
   (a) Show that for $|u| \leq 1$, the equation $\dot{y} = y + u$ has solutions from $y(0)$ to $y(T) = 0$ only if $|y(0)| < 1$. Hint: Multiply by e^{-t} and integrate to obtain the estimate $|y(0)| \leq 1 - e^{-t}$. Note that here the eigenvalue of $\mathbb{A}$ is positive.
   (b) For $\dot{y} = -y + u$, show that the corresponding argument is not conclusive.

11.15. (Lagrangian constraints $G(Y) = \mathcal{O}$ or $G(x, Y) \equiv \mathcal{O}$.)
   (a) Extend the argument used in proving Theorem 6.10 to the case where g is replaced by $G = (g_1, g_2, \ldots, g_m)$ $(m < d)$, where G is C^2 and at each $x \in [a, b]$, the gradients $\nabla g_i(x)$, $i = 1, 2, \ldots, m$, are linearly independent. Observe that now the continuous vector function, $\Lambda = (\lambda_1, \ldots, \lambda_m)$ is selected to make

$$\frac{d}{dx} f_{y'}(x) - f_y(x) = \Lambda(x) G_y(Y_0(x))$$

   for *some* locally defined $y \in \mathbb{R}^m$ such that $y = \Psi(\overline{Y})$ "solves" $G(y, \overline{Y}) = 0$.
   (b) Extend the argument in part (a) to cover the case when $G = G(x, Y)$ for a C^2 function G with $G_Y[Y_0(x)]$ of maximal row rank $m < d$ at each x.
   (c)* Explain how you could use slack variables to consider Lagrangian inequalities such as $G(Y(x)) \leq \mathcal{O}$ or $G(x, Y(x)) \leq \mathcal{O}$. Hint: See (11.22) in §11.3.

11.16*. (Simple differential constraints: $\mathcal{G}(t, Y(t), \dot{Y}(t)) = \mathcal{O}$.)
   (a) Suppose that for some choice of variables $Y = (X, V) \in \mathbb{R}^{d-k} \times \mathbb{R}^k$, the constraint function $\mathcal{G}$ takes the form $\mathcal{G}(t, Y, \dot{Y}) = G(t, Y, \dot{V}) - \dot{X}$. Then show that the problem of minimizing

$$\mathcal{F}(Y, T) = \int_0^T f(t, Y(t), \dot{Y}(t)) \, dt \quad \text{under} \quad \mathcal{G}[Y(t)] = \mathcal{O},$$

   with say $Y(0)$, $Y(T)$ prescribed, can be transformed into a control problem of minimizing

$$F(Y, U, T) = \int_0^T f(t, Y(t), U(t)) \, dt,$$

where $f(t, Y, U) \stackrel{\text{def}}{=} f(t, Y, G(t, Y, U), U)$ under the new state law $\dot{Y} = (G(\cdot, Y, U), U)$ with the control set $\mathscr{U} = \mathbb{R}^k$.

(b) Apply the minimum principle (Theorem 11.10) to a possible minimizing (Y_0, U_0, T_0), to obtain an associated Lagrangian multiplier function (P, Q), say, and a $\lambda_0 = 0$ or 1 for which (since $\mathscr{U}$ is *open*): $\lambda_0 f_U + G_U P + Q = \mathcal{O}$, when evaluated at (Y_0, U_0).

(c) Establish that (P, Q) satisfies the equations

$$(P, Q)^{\cdot} = -(h_X, h_V) \quad \text{for} \quad h = \lambda_0 f + P \cdot G + Q \cdot U.$$

(d) Introduce $M = P + \lambda_0 f_{\dot{X}}[Y_0]$ and conclude that the above equations combine to form the Euler–Lagrange equations for

$$f \stackrel{\text{def}}{=} \lambda_0 f + M \cdot (G - \dot{X}) = \lambda_0 f + M \cdot \mathcal{G}.$$

(e) Discuss the related transversal conditions, and formulate your result as a theorem.

11.17. (a) Obtain conditions which are necessary for $Y = (y, y_1) \in C^1[0, 1]$ to minimize $F(Y) = \int_0^1 \dot{y}^2 \, dt$ under $\dot{y}_1^2 - \dot{y} = 0$ with $Y(0) = \mathcal{O}$, $y(1) = 1$, $y_1(1) = 2$.

(b) Show that these conditions will also be sufficient provided that an appropriate multiplier, $\mu \geq 0$, can be found.

(c) Explain how to find an optimal solution from these conditions.

11.18. (a) Apply an analysis similar to that used in the previous problem to minimize $F(Y) = \int_0^1 |\dot{Y}|^2 \, dt$ under $\dot{y}_1^3 + \dot{y} = 0$ with $Y(0) = \mathcal{O}$ and $y(1) = 1$.

(b) What is the minimum value?

11.19. (a) Obtain necessary conditions for the time-optimal problem of a system whose state $Y = (y, y_1)$ at time t is governed by the *nonlinear* state equation $\dot{Y} = (u, u^2)$ where $|u| \leq 1$ with $Y(0)$, $y_1(T)$ prescribed.

(b) Can you find a sufficiency argument?

11.20. In the arguments leading to Theorem 11.20 verify that as defined $\tilde{f}$ does satisfy the second Euler–Lagrange equation and the Weierstrass necessary condition.

11.21. (Bolza functionals, I)

(a) Suppose that for some C^1 function φ, (Y_0, U_0, T_0) minimizes

$$F^*(Y, U, T) = \varphi(Y(T)) + \int_0^T f(t, Y(t), U(t)) \, dt,$$

where $U \in \mathscr{U}$, on

$$\mathscr{D}^* = \left\{ \begin{array}{c} (Y, U) \in \hat{C}^1[0, T], \; T > 0, \; Y(0) = A, \\ \Phi(T, Y(T)) = \mathcal{O} \end{array} \right\}$$

subject to $\dot{Y} = G(t, Y, U)$ on $(0, T)$.

   Introduce a *new* state variable y and control $u = \dot{y}$, form $\tilde{Y} = (Y, y)$, $\tilde{U} = (U, u)$, and show that $(\tilde{Y}_0, \tilde{U}_0, T_0)$ minimizes

$$\tilde{F}(\tilde{Y}, \tilde{U}, T) = \int_0^T [f(t, Y(t), U(t)) + u(t)] \, dt$$

on a related $\widetilde{\mathscr{D}}$, under the above augmented state equations, provided that $y_0(0) = 0$, $y(T_0) = \varphi(Y(T_0))$, $u \in \mathbb{R}$.

(b) Apply the minimum principle from Theorem 11.10 to

$$\tilde{h} = \lambda_0(f + u) + P \cdot G + pu,$$

noting that then

$$\tilde{h}_u(\cdot, \tilde{Y}_0, \tilde{U}_0) = \lambda_0 + p = 0.$$

(c) Examine the associated transversality conditions for the augmented $\tilde{\Phi}(\tilde{Y}) = (\Phi(Y), \varphi(Y) - y)$, and verify that the conclusions of Theorem 11.10 are valid as stated if $\Lambda\Phi$ is replaced by $\lambda_0\varphi + \Lambda\Phi$.

(d) Argue that the same is true when $\Phi = \Phi(t, Y)$ and $\varphi = \varphi(t, Y)$. Hint: Introduce t as a new state variable as in the proof of Theorem 11.10.

11.22*. (Bolza Functionals, II)

Assume that Theorem 11.10 remains valid for Bolza functionals of the type considered in Problem 11.21, when $\Lambda\Phi$ is replaced by $\lambda_0\varphi + \Lambda\Phi$. Reexamine the arguments leading to Theorem 11.20, and conclude that it, too, is valid for

$$\mathscr{F}^*(Y, T) = \varphi_0(T, Y(T)) + \int_0^T f(t, Y(t), \dot{Y}(t))\, dt,$$

under the same replacement.

11.23. Transform the problem of minimizing

$$\mathscr{F}(Y, T) = \int_0^T f(t, Y(t), \dot{Y}(t))\, dt \quad \text{under} \quad \mathscr{G}(t, Y(t), \dot{Y}(t)) = 0$$

on some set $\mathscr{D}$ subject to the single isoperimetric constraint $\int_0^T g(t, Y(t), \dot{Y}(t))\, dt = l$, by introducing a new variable y satisfying $\dot{y} = g[Y(\cdot)]$, with $y(0) = 0$.

11.24*. Repeat the arguments leading to Theorem 11.20 when $\mathscr{G}$ is replaced by $\mathscr{G}^*(t, Y, U, W)$ defined in (58). W is to be considered as a new control in the state equation $\dot{Y} = G^*(t, Y, U, W)$ where G^* is an appropriate function defined implicitly by $\mathscr{G}^* = 0$. Conclude that for a suitably defined h^*, $h^*(t, Y_0(t), U) \geq h^*(t, Y_0(t), \dot{Y}_0(t))$, $\forall U$ such that $\mathscr{G}(t, Y_0(t), U) \leq 0$, and hence that the associated multiplier function $M \geq 0$.

11.25. Obtain necessary conditions for the time-optimal problem of a system whose state $Y = (y, y_1)$ at time t is governed by $\dot{Y} = (y_1, u)$, with $Y(0)$, $Y(T)$, prescribed, when $u^2 \leq y_1^4$.

11.26. (a) A cup filled with tea at $100\,°C$ is to be cooled to a temperature $0\,°C$ by adding a *fixed* amount of milk. An equation for the temperature y in the cup permitting heat loss due to overflow can be approximated by $\dot{y} = -y - 25u - \frac{1}{4}uy$, with $y(0) = 100$, $y(T) = 0$. Obtain necessary conditions for a time-optimal solution under an external liquid flow u controlled by $0 \leq u(t) \leq 1$, with $\int_0^T u(t)\, dt = 1$, and show that $u_0 = 0$, $0 \leq t \leq 0.69$.

(b) Note that $\dot{y} < 0$, permits $x = y$ to be used as the independent variable in an equivalent *fixed* interval problem. Formulate the new problem and compare with Problem 10.15. What can you conclude concerning uniqueness and sufficiency of the control u_0 obtained in part (a)?

Appendix

In this part, we present complete proofs of several standard results from advanced calculus and basic analysis which are utilized in the text. Frequent appeal is made to the proven facts that a continuous function on a compact set is *uniformly* continuous (5.2), is bounded, and assumes maximum and minimum values (5.3). However, these results can be invoked as needed, since compactness of closed and bounded sets in $\mathbb{R}^d$ is established at the outset.

Section A.5 is devoted to establishing the existence and uniqueness of solutions to systems of differential equations as required for Chapters 9 and 11. The next section provides a brief discussion of the Rayleigh ratio as connected with the higher eigenvalues of the Sturm–Liouville problem from §7.3. The appendix concludes with an examination of tangent cones in $\mathbb{R}^d$ that is needed (only) in §11.3.

§A.0. Compact Sets in $\mathbb{R}^d$

To establish the compactness of the bounded interval $[a, b] \subseteq \mathbb{R}$ (as defined in §5.2), it suffices to prove that the unit interval $I = [0, 1]$ is compact. Indeed, a scale change and a translation permits us to regard $[a, b]$ as the unit interval in another coordinate system, and convergence is clearly preserved under these elementary operations.

(A.0) Lemma. $I = [0, 1]$ *is compact.*

PROOF. Each $x \in I$ has the *binary* expansion $x = \sum_{k=1}^{\infty} b_k 2^{-k}$, where $b_k = 0$ or 1, $k = 1, 2, \ldots$, and in case of ambiguity, we choose that expansion which terminates; e.g., we take $2^{-(m-1)}$ instead of $\sum_{k=m}^{\infty} 2^{-k}$, $m = 1, 2, \ldots$.

Now, for each $n = 1, 2, \ldots$, suppose that $x_n \in I$ has the binary expansion $x_n = \sum_{k=1}^{\infty} b_{nk} 2^{-k}$.

We will select a subsequence $\bar{x}_1, \bar{x}_2, \ldots$ of these x_n which converges to a limit point $\bar{x} \in I$.

If there is an infinite set of the x_n with *first* binary coefficient $b_{1n} = 0$, then choose $\bar{x}_1$ to be that x_n from this set with *least* index n_1. Otherwise, there is an infinite set of the x_n with $b_{1n} = 1$, and $\bar{x}_1$ should be taken to be that from among them with least index. In either case let $\bar{b}_1$ be the first binary coefficient of $\bar{x}_1$.

Next, let $\bar{x}_2$ be that x_n of least index $n_2 > n_1$ for which $b_{1n} = \bar{b}_1$ and $b_{2n} = 0$, if there is an infinite set of such x_n; otherwise, for which $b_{1n} = \bar{b}_1$ and $b_{2n} = 1$. Let $\bar{b}_2$ be the *second* binary coefficient of $\bar{x}_2$.

This selection process can be continued indefinitely, since at each stage we consider only those remaining x_n which form an infinite set. There results a *subsequence* $\bar{x}_m = x_{n_m}$, and an associated sequence of binary coefficients $\bar{b}_m$, $m = 1, 2, \ldots$, determining the *number* $\bar{x} \stackrel{\text{def}}{=} \sum_{k=1}^{\infty} \bar{b}_k 2^{-k} \in I$.

But by construction, $\bar{x}$ and $\bar{x}_m$ will have the *same* initial m binary coefficients $\bar{b}_1, \bar{b}_2, \ldots, \bar{b}_m$, and the remaining binary coefficients cannot exceed 1. Hence, by an easy estimate,

$$|\bar{x} - \bar{x}_m| \leq \sum_{k=m}^{\infty} 2 \cdot 2^{-k} = 4 \cdot 2^{-m}, \quad \text{which} \to 0 \quad \text{as } m \to \infty. \qquad \square$$

Remarks. The selection process used in proving the lemma may be visualized by successively bisecting (sub)intervals of I, and retaining at each stage, the *left*most interval which contains an infinite set of the x_n. The $\bar{b}_k$ index these selections, and, more importantly, define the limit point $\bar{x}$ (as the sum of a series of positive terms). We are assuming here (and elsewhere) that the set of real number is complete in that it contains a sum for each convergent series of its elements. For further discussion of this somewhat subtle point, see [Ed].

Compactness is the basis for many surprising results in analysis. For example, suppose f is differentiable at each point of a compact interval, $[a, b]$ and $f(x_n) = 0$ for distinct points $x_n \in [a, b]$, $n = 1, 2, \ldots$. Then a subsequence $\{x_{n_k}\}$ has a limit point $x_0 \in [a, b]$ at which *both*

$$f(x_0) = \lim_{k \to \infty} f(x_{n_k}) = 0 \quad \text{and} \quad f'(x_0) = \lim_{k \to \infty} \frac{f(x_{n_k}) - f(x_0)}{x_{n_k} - x_0} = 0.$$

Compactness is essential for this conclusion: $f(x) = \sin x$ and $f'(x) = \cos x$ never vanish simultaneously even though $f(n\pi) = 0$ for $n = 1, 2, \ldots$.

If we consider *successively* each component of a sequence $X_n \in B = \{X \in \mathbb{R}^d : a_j \leq x_j \leq b_j, j = 1, 2, \ldots, d\}$, $n = 1, 2, \ldots$, then it is straightforward to establish the compactness of the *closed* box B. [The proven compactness of $[a_1, b_1]$ guarantees a convergent sequence of the *first* components of the X_n with a limit point $\bar{x}_1 \in [a_1, b_1]$. Then from *this subsequence* of the second components, we may extract a convergent subsequence with the limit point $\bar{x}_2 \in [a_2, b_2]$. Continuing *successively*, we obtain a dth sub-sub-subsequence $\{X_n\}_{n=1}^{\infty}$, say, for which the dth components converge to a limit point $\bar{x}_d \in [a_d, b_d]$, while each of the other components converges to its previous limit. (Why?) It follows that $\lim_n \bar{X}_n = \bar{X} = (\bar{x}_1, \bar{x}_2, \ldots, \bar{x}_d) \in B$ by application of the triangle inequality.]

(A.1) Theorem. *Each closed and bounded set $K \subseteq \mathbb{R}^d$ is compact.*

PROOF. Since K is bounded, $K \subseteq B$ for some (large) closed box B. If $X_n \in K \subseteq B$, $n = 1, 2, \ldots$, the compactness of B guarantees a convergent subsequence $\{X_{n_k}\}$, say, with limit point $X_0 \in B$. But K is closed and by definition contains the limits of each of its convergent sequences. Hence $X_0 \in K$.   □

§A.1. The Intermediate and Mean Value Theorems

In the text, we are often required to show that a given function f assumes a prescribed value in an interval I. In this section we establish two standard results of this type.

First, an observation: When f is *continuous* at $x_0 \in I$ and $f(x_0) > 0$, then $f(x) > 0$, $\forall\, x$ in a neighborhood of x_0.

[Indeed, by the *reverse* triangle inequality we have

$$f(x) \geq f(x_0) - |f(x) - f(x_0)| > \frac{f(x_0)}{2}, \quad \text{when } |f(x) - f(x_0)| < \frac{f(x_0)}{2},$$

and this holds $\forall\, x$ in a neighborhood of x_0.]

A similar analysis shows that when $f(x_0) > c$, then $f > c$ in a neighborhood of x_0, and the corresponding result holds when $f(x_0) < c$.

(A.2) Intermediate Value Theorem. *If $f \in C[a, b]$ then f assumes each value c between $f(a)$ and $f(b)$.*

PROOF. For definiteness, suppose that $f(a) > c > f(b)$. Then by continuity (as above) $f > c$ on (a, x) for some x sufficiently near a. We wish to find the *largest* such interval. To do so rigorously, we let $P = \{x \in (a, b): f > c$ on $(a, x)\}$ and consider the (open) set $I = \bigcup_{x \in P}(a, x)$, which with x_1, and $x_2 > x_1$ also contains the interval $[x_1, x_2]$. (Why?) Hence I is an open *interval* of the form (a, x_0) for some $x_0 \in (a, b]$ on which $f > c$. Were $f(x_0) < c$ we could use continuity as above to conclude that $f < c$, on *some* (x, x_0), and this is a contradiction. Similarly, were $f(x_0) > c$, we could conclude that $f > c$ on a larger interval $(a, \tilde{x})$ with $\tilde{x} > x_0$, but then $\tilde{x} \in P$, and $\tilde{x} \in (a, x_0]$, which is another contradiction. Thus $f(x_0) = c$ as desired.   □

f may of course assume the value c at many distinct points, unless f is *strictly increasing* on $[a, b]$ (or *strictly decreasing* on $[a, b]$). In particular, each sign change of f on the interval must be accompanied by an interior point at which f vanishes.

If f is also differentiable on the compact interval $[a, b]$, then we can gain further insight about its behavior. For example, unless f changes sign only finitely often on the interval, we can use our earlier compactness argument to produce a point at which *both* f and f' vanish.

Similarly, we have the following:

(A.3) Mean Value Theorem. *If f is differentiable on an interval I, then when $a < b$ and $[a, b] \subseteq I : m \overset{\text{def}}{=} [f(b) - f(a)]/(b - a) = f'(x^*)$ for some $x^* \in (a, b)$.*

PROOF. On the *compact* interval $[a, b]$, by (5.3), the *continuous* function $g(x) = f(x) - m(x - a)$ assumes maximum and minimum values. Since $g(a) = f(a) = g(b)$, at least one of these must occur at an *interior* point x^*. Then, by the usual argument, $g'(x^*) = 0 = f'(x^*) - m$, as desired.   □

(A.4) Corollary. *If $f' \equiv 0$ on an interval I, then f is constant on I.*

PROOF. For fixed $a \in I$, we may apply the previous result to each $x = b \in I$, if $a < x$. Then for an associated x^* we must have $f(x) - f(a) = (x - a)f'(x^*) = 0$, or $f(x) = f(a)$, for any $a < x$.   □

(A.5) Corollary. *If f is differentiable on an interval I and $f'(x) \neq 0$ in I, then when $[a, b] \subseteq I$, f assumes each value between $f(a)$ and $f(b)$ precisely once on $[a, b]$.*

PROOF. From A.3, we could not have $f(a) = f(b)$, for *any* distinct points a, $b \in I$. But f is continuous so that A.2 applies.   □

It is straightforward to obtain a multidimensional version of the mean value theorem for a C^1 function on a *convex* set (one which contains the segment joining each pair of its points.)

(A.6) Proposition. *If S is a convex open set in $\mathbb{R}^d$, and $f \in C^1(S)$, then when X_0, $X_1 \in S$, $f(X_1) - f(X_0) = \nabla f(X^*) \cdot (X_1 - X_0)$, for some point X^* on the line segment joining X_0 and X_1.*

PROOF. A typical point on the line segment is given by $X_u = uX_1 + (1 - u)X_0$, $(0 \leq u \leq 1)$, and each such point is in S. Thus

$$g(u) \overset{\text{def}}{=} f(X_u) = f(uX_1 + (1 - u)X_0)$$

satisfies the conditions of the mean value theorem on $I = [0, 1]$, so that $f(X_1) - f(X_0) = g(1) - g(0) = g'(u^*)(1 - 0)$ for some $u^* \in (0, 1)$. But since f is C^1, it follows that $g'(u^*) = \nabla f(X_{u^*}) \cdot (X_1 - X_0)$ and setting $X_{u^*} = X^*$ establishes the proposition.   □

(A.7) **Corollary.** *If $f \in C^1(S(X_0))$, where $S(X_0)$ is an open spherical neighborhood of X_0, then f is differentiable at X_0. (See 0.10.)*

PROOF. $S(X_0)$ is convex. (Why?) With $X_1 = X$ we have from A.6, that when $X \neq X_0$, and $U = (X - X_0)/|X - X_0|$, then

$$f(X) = f(X_0) + \nabla f(X_0) \cdot (X - X_0) + |X - X_0|(\nabla f(X^*) - \nabla f(X_0)) \cdot U,$$

for a *unit* vector U, and an X^* "between" X and X_0. From the Cauchy inequality it follows that the correction term $\mathfrak{z}(X - X_0) \overset{\text{def}}{=} (\nabla f(X^*) - \nabla f(X_0)) \cdot U$ may be estimated by

$$|\mathfrak{z}(X - X_0)| \leq |\nabla f(X^*) - \nabla f(X_0)| \leq \sum_{j=1}^{d} |f_{x_j}(X^*) - f_{x_j}(X_0)|;$$

and $\mathfrak{z}(X - X_0) \to \mathcal{O}$ as $X \to X_0$, since $|X^* - X_0| \leq |X - X_0|$. Thus f is differentiable at X_0. $\qquad\square$

§A.2. The Fundamental Theorem of Calculus

In this appendix (as elsewhere in the text) we assume as known that a real valued function f which is piecewise continuous on a compact interval $[a, b]$ generates a unique real number denoted $\int_a^b f(x)\, dx$ called its (Riemann) integral which may be regarded as the *signed* area "under the curve" of its graph. When these conditions are met on each compact *subinterval* $[a, b] \subset (\alpha, \beta)$, then it may also be possible to define the *improper* Riemann integral

$$\int_\alpha^\beta f(x)\, dx = \lim \left(\int_a^b f(x)\, dx \right), \quad \text{as } a \searrow \alpha \text{ and } b \nearrow \beta.$$

As defined, the integral is a linear function of its integrand in the sense of §2.2. The following results from advanced calculus are consistent with an areal interpretation of the integral, and we shall assume as known those basic additivity and monotonicity properties and estimates for the integral which can be inferred from this interpretation [Fl].

(A.8) **Theorem** (Fundamental Theorem of Calculus). *If f is a (Riemann) integrable function on $[a, b]$, and $F(x) \overset{\text{def}}{=} \int_a^x f(t)\, dt$, then $F'(x_0) = f(x_0)$ at each point $x_0 \in (a, b)$ where f is continuous. If $f \in C^1[a, b]$, then*

$$\int_a^b f'(x)\, dx = f(b) - f(a).$$

(*f* is also integrable over $[a, x] \subseteq [a, b]$ so that F is defined.)

PROOF. For each $x \in [a, b]$ with $x \neq x_0$, we may express the difference quotient

$$\frac{F(x) - F(x_0)}{x - x_0} = f(x_0) + \frac{1}{x - x_0} \int_{x_0}^x [f(t) - f(x_0)]\, dt. \tag{1}$$

Since f is continuous at x_0 by hypothesis, for each $\varepsilon > 0$, $\exists\, \delta > 0$ such that

$$|t - x_0| \leq |x - x_0| < \delta \Rightarrow |f(t) - f(x_0)| < \varepsilon.$$

Thus, for $0 < x - x_0 < \delta$, the term with the integral can be estimated by

$$\frac{1}{|x - x_0|}\left|\int_{x_0}^{x} [f(t) - f(x_0)]\, dt\right| \leq \frac{1}{|x - x_0|}\int_{x_0}^{x} |f(t) - f(x_0)|\, dt$$

$$\leq \frac{\varepsilon}{|x - x_0|}\int_{x_0}^{x} dt = \varepsilon;$$

reversing the limits on the integral shows that the same estimate is valid when $0 < x_0 - x < \delta$. Hence as $x \to x_0$, the integral term in (1) approaches zero and we conclude that

$$F'(x_0) = \lim_{x \to x_0} \frac{F(x) - F(x_0)}{x - x_0} = f(x_0).$$

Finally, when $f \in C^1[a, b]$, then by the result just established, $F(x) \overset{\text{def}}{=} \int_a^x f'(t)\, dt$ is defined on $[a, b]$ with derivative $F'(x) = f'(x)$, $x \in (a, b)$. Hence by A.4, $f(b) - f(a) = F(b) - F(a) = \int_a^b f'(x)\, dx$. $\qquad\square$

Remark. When x_0 is a or b, the initial derivation—restricted to those x for which $F(x)$ is defined—shows that $f(x_0)$ is the corresponding *one-sided* derivative of F at x_0. The first part also holds when the real valued function $F(x)$ is defined by an *improper* Riemann integral $\int_a^x f(t)\, dt$.

(A.9) **Lemma.** *If $0 \leq p \in C(a, b)$ and $\int_a^b p(x)\, dx = 0$, then $p(x) = 0$, $\forall\, x \in (a, b)$.*

(The existence of the possibly improper integral is presupposed by the hypotheses.)

PROOF. For $x \in (a, b)$, $0 \leq P(x) \overset{\text{def}}{=} \int_a^x p(t)\, dt \leq \int_a^b p(t)\, dt = 0$, since $p \geq 0$. Hence $P(x) \equiv 0$, so that by Theorem A.8, $p(x) = P'(x) = 0$. $\qquad\square$

(A.10) **Proposition.** *If f and g are in $C(a, b)$ and $f \leq g$, then $\int_a^b f(x)\, dx \leq \int_a^b g(x)\, dx$ provided that both integrals exist and are finite; with equality iff $f(x) = g(x)$, $\forall\, x \in (a, b)$.*

PROOF. $0 \leq p \overset{\text{def}}{=} g - f$ is also in the linear space $C(a, b)$. Thus as in Lemma A.9.

$$0 \leq \int_a^b (g(x) - f(x))\, dx = \int_a^b g(x)\, dx - \int_a^b f(x)\, dx,$$

with equality of the integrals on the right iff $g(x) - f(x) = 0$, $\forall\, x \in (a, b)$, or $f \equiv g$ on (a, b). $\qquad\square$

Remark. The conclusions of A.9 and A.10 hold for piecewise continuous functions p, and f, g, respectively, except at the points of discontinuity.

§A.3. Partial Integrals: Leibniz' Formula

When a function $f(x, y)$ can be integrated with respect to say the first variable (x), then the resulting integral defines a function of the *second* variable (y) whose properties we wish to ascertain.

(A.11) Lemma. *If f is continuous on the compact rectangle $[a, b] \times [u, v] \subseteq \mathbb{R}^2$, then*

$$g(y) \overset{\text{def}}{=} \int_a^b f(x, y)\, dx \qquad (2)$$

determines a continuous function on $[u, v]$.

PROOF. For each $y \in [u, v]$, $f(\cdot, y)$ is continuous on $[a, b]$ and hence the integral in (2) is defined. Since f is *uniformly* continuous (Lemma 5.2) it follows that for each $\varepsilon > 0$, $\exists\, \delta > 0$ such that $|f(x, y) - f(x, \tilde{y})| < \varepsilon$ when $y, \tilde{y} \in [u, v]$ and $|y - \tilde{y}| < \delta$. Thus for such $y, \tilde{y}$:

$$|g(y) - g(\tilde{y})| = \left| \int_a^b [f(x, y) - f(x, \tilde{y})]\, dx \right|$$

$$\leq \int_a^b |f(x, y) - f(x, \tilde{y})|\, dx$$

$$\leq \varepsilon \int_a^b dx = \varepsilon(b - a),$$

and this implies that g is (uniformly) continuous on $[u, v]$. $\qquad\square$

Since g of (2) is continuous, we can form the *iterated integral*

$$\int_u^v dy \int_a^b f(x, y)\, dx = \int_u^v g(y)\, dy$$

and by symmetry, the other iterated integral

$$\int_a^b dx \int_u^v f(x, y)\, dy$$

which has the *same* value:

(A.12) Theorem. *If f is continuous on the compact rectangle $[a, b] \times [u, v] \subseteq \mathbb{R}^2$, then*

$$\int_a^b dx \int_u^v f(x, y)\, dy = \int_u^v dy \int_a^b f(x, y)\, dx. \qquad (3)$$

PROOF. By the above remarks, both iterated integrals exist and it remains only to establish their equality.

We shall regard u as fixed and permit v to vary, say $v \in [u, v_0]$. Then from Theorem A.8,

$$\frac{d}{dv} \int_u^v dy \int_a^b f(x, y) \, dx = \int_a^b f(x, v) \, dx = g(v),$$

(since by Lemma A.11, the inner integral function is continuous).

The theorem will follow if we can show that the *left* side of (3), $F(v)$, say, has the *same* derivative, since both sides vanish at $v = u$. (Here we use the well-known fact that functions with the same derivative on an interval differ by a constant (See A.4).)

For small $t \neq 0$, with v and $v + t$ in $[u, v_0]$, we may express the difference quotient as follows:

$$\frac{F(v + t) - F(v)}{t} = \int_a^b dx \frac{1}{t} \int_v^{v+t} [f(x, y) - f(x, v)] \, dy + \int_a^b f(x, v) \, dx. \quad (4)$$

Now, given $\varepsilon > 0$, $\exists \, \delta_0$ such that $|f(x, y) - f(x, v)| < \varepsilon$ when $|y - v| < \delta_0$ and $y, v \in [u, v_0]$, $\forall \, x \in [a, b]$ by the aforementioned *uniform* continuity of f on $[a, b] \times [u, v_0]$ (Lemma 5.2). Thus when $0 < t < \delta_0$, the iterated integral in (4) can be estimated by

$$\int_a^b dx \frac{1}{t} \int_v^{v+t} |f(x, y) - f(x, v)| \, dy \leq \int_a^b dx \frac{\varepsilon}{t} \int_v^{v+t} dy = \varepsilon(b - a),$$

and the same estimate holds for $0 < |t| < \delta_0$. Hence, recalling the definition of F, we see that in the limit:

$$\frac{d}{dv} \int_a^b dx \int_u^v f(x, y) \, dy = \lim_{t \to 0} \frac{F(v + t) - F(v)}{t} = F'(v) = \int_a^b f(x, v) \, dx$$

as desired.   □

Interchanging the order of integration as permitted by this theorem may be interpreted as integrating "under" the integral sign. It may also be possible to differentiate under the integral sign.

(A.13) **Theorem.** *If $f = f(x, y)$ and its partial derivative f_y are continuous on $[a, b] \times [u, v]$ then*

$$g(y) \overset{\text{def}}{=} \int_a^b f(x, y) \, dx \quad (5)$$

is C^1 in the interval (u, v) with the derivative

$$g'(y) = \frac{d}{dy} \int_a^b f(x, y) \, dx = \int_a^b f_y(x, y) \, dx. \quad (6)$$

PROOF. $h(y) \overset{\text{def}}{=} \int_a^b f_y(x, y) \, dx$ is continuous on $[u, v]$ (by Lemma A.11) and hence by the theorem just proven,

$$\int_u^v h(y) \, dy = \int_u^v dy \int_a^b f_y(x, y) \, dx = \int_a^b dx \int_u^v f_y(x, y) \, dy.$$

However, the last inner integration may be carried out by A.8, and the last iterated integral becomes

$$\int_a^b [f(x, v) - f(x, u)] \, dx,$$

so that from (5) follows

$$\int_u^v h(y) \, dy = g(v) - g(u). \tag{7}$$

For *fixed* u, by Theorem A.8, we may differentiate (7) with respect to v to get $g'(v) = h(v)$ and when $v = y$, this is the desired result. □

Finally, from Theorem A.13 and the chain rule we obtain Leibniz' formula:

(A.14) Theorem (Leibniz). *If $f = f(x, y)$ and f_y are continuous on $[a, b] \times [u, v]$ and $h \in C^1(u, v)$ with $a \leq h(y) \leq b$, $\forall \, y \in (u, v)$, then*

$$\frac{d}{dy} \int_a^{h(y)} f(x, y) \, dx = \int_a^{h(y)} f_y(x, y) \, dx + h'(y) f(h(y), y). \tag{8}$$

PROOF. Set $g(y, z) = \int_a^z f(x, y) \, dx$ so that $g_y(y, z) = \int_a^z f_y(x, y) \, dx$ by A.13 while $g_z(y, z) = f(z, y)$ by A.8. Hence from the chain rule,

$$\frac{d}{dy} g(y, h(y)) = g_y(y, h(y)) + g_z(y, h(y)) h'(y),$$

which, after obvious substitutions, is (8). □

§A.4. An Open Mapping Theorem

As we observed in §5.7, the full strength of the inverse function theorem (5.13) is not required to establish the existence of Lagrangian multipliers. The following suffices:

(A.15) Theorem. *For a neighborhood $S = S(X_0) \subseteq \mathbb{R}^d$, let $F = (f_1, \ldots, f_d) \in (C^1(S))^d$ have an invertible Jacobian matrix $F'(X_0)$ (with elements $(\partial f_i / \partial x_j)(X_0)$, $i, j = 1, 2, \ldots, d$, and rows indexed by i). Then $\exists \, m > 0$ such that for each small $t > 0$, $Y_1 \in S_{mt}(F(X_0)) \Rightarrow Y_1 = F(X_1)$, for some $X_1 \in S_t(X_0)$.*

Remarks. The inverse function theorem guarantees, in addition, that the point X_1 is unique, and that the resulting inverse function is itself C^1. We will utilize the continuity of the partials of the f_j only at X_0.

PROOF. $\Delta(X) \overset{\text{def}}{=} \det[F'(X)]$ consists of sums and products of the elements of $F'(X)$ and so will inherit their hypothesized continuity. Moreover, $\Delta(X_0) \neq 0$, since $F'(X_0)$ is invertible, and thus, by continuity as in §A.1, $\Delta(X) \neq 0$ in an open neighborhood

$$S_r(X_0) = \{X \in \mathbb{R}^d \colon |X - X_0| < r\},$$

so that $F'(X)$ is also invertible in this neighborhood. The continuity of the partial derivatives (at X_0) in conjunction with the mean value theorem ensures that each component f_j is differentiable at X_0 as in A.7, and we may infer that F is differentiable at X_0 in that

$$F(X) = F(X_0) + F'(X_0)(X - X_0) + |X - X_0|\mathfrak{Z}(X - X_0), \qquad (9)$$

for an appropriate *vector* valued function $\mathfrak{Z}$ having limit $\mathcal{O}$ as $X - X_0 \to \mathcal{O}$. ($F(X)$ and $X - X_0$ are *column* vectors). When $X \neq X_0$, we may rewrite (9) as

$$\frac{F(X) - F(X_0)}{|X - X_0|} = F'(X_0)U + \mathfrak{Z}(X - X_0), \qquad (10)$$

for the *unit* vector $U = (X - X_0)/(|X - X_0|)$.

Now, $F'(X_0)U \neq \mathcal{O}$, since $F'(X_0)$ is invertible (§0.13), and hence the real valued function $\rho(U) \stackrel{\text{def}}{=} |F'(X_0)U|$ is *positive* and continuous on the *compact* surface of the unit sphere, $\{U \subseteq \mathbb{R}^d \colon |U| = 1\}$ (§A.0). Thus, by Proposition 5.3, ρ assumes the *necessarily positive* minimum value, $4m$, say, so that from (10), and the reverse triangle inequality

$$\frac{|F(X) - F(X_0)|}{|X - X_0|} \geq |F'(X_0)U| - |\mathfrak{Z}(X - X_0)| > 4m - m = 3m(>2m), \quad (11)$$

on a sphere $S_t(X_0)$, with $t < r$, so small that $|\mathfrak{Z}(X - X_0)| < m$.

We claim that with m, t, as above and $Y_0 = F(X_0)$, each point $Y_1 \in S_{mt}(Y_0)$ is obtained as $Y_1 = F(X_1)$ for *some* point $X_1 \in S_t(X_0)$. To establish this assertion (which completes the proof of the theorem), we appeal once again to Proposition 5.3 as follows:

On the *compact* sphere $\{X \colon |X - X_0| \leq t\}$, the continuous function $\mu(X) \stackrel{\text{def}}{=} |F(X) - Y_1|$ assumes its minimum value at a point, X_1, say. Moreover, if $|X_1 - X_0| = t$, we would have from the reverse triangle inequality:

$$|F(X_1) - Y_1| \geq |F(X_1) - F(X_0)| - |Y_0 - Y_1|;$$

or, by (11),

$$\mu(X_1) > 2m|X_1 - X_0| - mt = 2mt - mt = mt,$$

which contradicts our knowledge that the minimum value,

$$\mu(X_1) \leq \mu(X_0) = |F(X_0) - Y_1| = |Y_0 - Y_1| < mt.$$

Thus the minimum value occurs at X_1 with $|X_1 - X_0| < t$, and X_1 must also give the minimum value of

$$\mu^2(X) = |F(X) - Y_1|^2 = (F(X) - Y_1) \cdot (F(X) - Y_1).$$

Hence $\nabla\mu^2(X_1) = 2(F(X_1) - Y_1))F'(X_1) = \mathcal{O}$ (where now $F(X_1) - Y_1$ is a *row* vector). But $F'(X_1)$ is invertible, since $|X_1 - X_0| < r$; therefore $F(X_1) - Y_1 = \mathcal{O}$, or $Y_1 = F(X_1)$ as asserted.

Finally, observe that while m is fixed, t ($<r$) may be taken to be as small as we wish.   □

Remark. A further application of the mean value theorem can be used to prove that the point X_1 is uniquely determined in $S_t(X_0)$ for a given $t > 0$, and this establishes the existence of a *local* inverse function, F^{-1}, in $S_{mt}(Y_0)$. The elements of the *inverse* matrix to $F'(X)$ constitute the Jacobian matrix for F^{-1}, and their continuity will follow from Cramer's rule. For this analysis, which would establish the inverse function theorem, and a derivation of its corollary, the implicit function theorem, see [Ed].

§A.5. Families of Solutions to a System of Differential Equations

In §9.7 we obtained central families as solutions of a first-order system of differential equations of the form

$$T'(x) = G(x, T(x)), \qquad x \in [a, b],$$

$$\text{with } T(a) = L \text{ prescribed.} \tag{12}$$

Here, for a fixed $N = 1, 2, \ldots$, $G = G(x, T)$ is a given N-vector valued function which is continuous on a domain D of $\mathbb{R}^{N+1}$.

We suppose that $T_0 \in \mathcal{T} = (C^1[a, b])^N$ is a *known* solution to (12), with $T_0(a) = L_0$, and that for some $\delta_0 > 0$, D contains the *compact* strip $K_0 = \{(x, T): |T - T_0(x)| \le \delta_0, x \in [a, b]\}$, as shown in Figure A.1.

For each L with $|L - L_0| < \delta_0$, we seek a C^1 solution $T \in \mathcal{T}$ to (12) with $T(a) = L$ whose graph is contained in K_0. We further desire that these solu-

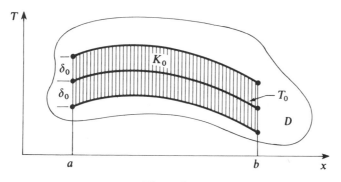

Figure A.1

tions, denoted $T(\cdot\,; L)$, have a Jacobian matrix T_L with continuous elements, and to obtain this we shall require that G have a Jacobian matrix G_T with continuous elements (and columns indexed by T).

(A.16) Lemma. *If the elements of G_T are continuous on K_0, then for (x, T_0) and (x, T_1) in K_0:*

$$G(x, T_1) - G(x, T_0) = \left(\int_0^1 G_T(x, T_u)\, du \right)(T_1 - T_0), \tag{13}$$

where $T_u \overset{\text{def}}{=} uT_1 + (1 - u)T_0,\ u \in [0, 1]$; so that

$$|G(x, T_1) - G(x, T_0)| \le \gamma |T_1 - T_0|, \tag{14}$$

for a constant $\gamma > 0$.

PROOF. For each $u \in [0, 1]$, $(x, T_u) \in K_0$ (Why?), and by the chain rule $(d/du)G(x, T_u) = G_T(x, T_u)(T_1 - T_0)$ for the constant vector $T_1 - T_0$. For a *typical component*, g, of G, we have

$$g(x, T_1) - g(x, T_0) = \int_0^1 \frac{d}{du} g(x, T_u)\, du = \left(\int_0^1 g_T(x, T_u)\, du \right) \cdot (T_1 - T_0),$$

and this gives (13).

Moreover, the continuous function $|g_T(x, T_u)|$ is bounded on the *compact* strip K_0 by 5.3, and hence from standard estimates:

$$|g(x, T_1) - g(x, T_0)| \le \left(\int_0^1 |g_T(x, T_u)|\, du \right) |T_1 - T_0|$$

$$\le \gamma_g |T_1 - T_0|, \quad \text{say.}$$

(14) follows for a sufficiently large γ. $\square$

(A.17) Proposition. *With G as above, equation (12) has at most one solution $T \in \mathcal{T}$ with prescribed $T(a) = L_0$.*

PROOF. Let T_0 be the known solution, and take K_0, δ_0, γ as above. Observe that upon integration of (12) with $T_0(a) = L_0$, we obtain

$$T_0(x) = L_0 + \int_a^x G(s, T_0(s))\, ds, \qquad x \in [a, b]. \tag{15}$$

Each solution $T \in \mathcal{T}$ of (12) with $T(a) = L_0$ also satisfies (15), and from continuity $(x, T(x)) \in K_0$, for a *maximal* interval $[a, \tilde{b}]$, with $\tilde{b} \le b$. Consequently, on $[a, \tilde{b}]$ we may use (14) to estimate $V(x) \overset{\text{def}}{=} T(x) - T_0(x)$, as follows:

$$v(x) \overset{\text{def}}{=} |V(x)| = \left| \int_a^x (G(s, T(s)) - G(s, T_0(s)))\, ds \right|$$

$$\le \gamma \int_a^x |V(s)|\, ds = \gamma \int_a^x v(s)\, ds = \sigma(x), \quad \text{say.}$$

When A.8 is applied, we get

$$\sigma'(x) = \gamma v(x) \leq \gamma \sigma(x) \quad \text{or} \quad (e^{-\gamma x} \sigma(x))' \leq 0,$$

which, when integrated, gives

$$|V(x)| = v(x) \leq \sigma(x) \leq e^{\gamma(x-a)} \sigma(a) = 0. \quad \text{(Why?)}$$

Thus $T(x) = T_0(x)$ on $[a, \tilde{b}]$, and in particular: $T(\tilde{b}) = T_0(\tilde{b})$. Were $\tilde{b} < b$, we would conclude again by continuity that $(x, T(x)) \in K_0$ for a larger interval than $[a, \tilde{b}]$ contradicting its supposed maximality. Hence $\tilde{b} = b$ and $T = T_0$.

□

This proposition also shows that each solution $T(\cdot; L)$ considered above will be fixed uniquely by its initial value at $x = a$. We will now prove that such solutions exist.

(A.18) **Theorem.** *Let $T_0 \in \mathcal{T}$ be a solution of* (12) *with $T_0(a) = L_0$, and G, K_0, δ_0, γ be as above. For each L with $|L - L_0| = \delta < \delta_0 e^{-\gamma(b-a)}$, there exists a unique solution $T \in \mathcal{T}$ of* (12), *with $T(a) = L$, and for this solution*

$$\|T - T_0\| = \max_{x \in [a,b]} |T(x) - T_0(x)| < \delta e^{\gamma(b-a)}.$$

PROOF. We shall obtain T as a (continuous) solution to the integral equation

$$T(x) = L + \int_a^x G(s, T(s)) \, ds. \tag{16}$$

It then follows that $T(a) = L$, and from the fundamental theorem of calculus that $T \in \mathcal{T}$, with $T'(x) = G(x, T(x))$, $x \in [a, b]$. The uniqueness is then a consequence of A.17.

Let $r = b - a$.

The solution to (16) is obtained by successive approximation from (15) as follows

$$T_0(x) = L_0 + \int_a^x G(s, T_0(s)) \, ds;$$

$$\tag{17}$$

$$T_{n+1}(x) = L + \int_a^x G(s, T_n(s)) \, ds, \quad n = 0, 1, 2\ldots.$$

Observe that $T_1(x) = T_0(x) + L - L_0$, so that

$$\|T_1 - T_0\| = |L - L_0| = \delta < \delta_0, \quad \text{i.e., } (x, T_1(x)) \in K_0. \tag{18}$$

Next, by (17), (14), and (18):

$$|T_2(x) - T_1(x)| = \left| \int_a^x [G(s, T_1(s)) - G(s, T_0(s))] \, ds \right|$$

$$\leq \gamma \int_a^x |T_1(s) - T_0(s)| \, ds \leq \gamma(x - a)\delta \leq \gamma r \delta,$$

so that by the triangle inequality (§5.1, Example 6):

$$\|T_2 - T_0\| \le \|T_1 - T_0\| + \|T_2 - T_1\| \le \delta(1 + \gamma r) \le \delta e^{\gamma r} < \delta_0.$$

Thus $(x, T_2(x)) \in K_0$, and we may continue inductively to conclude that for $n = 0, 1, 2 \ldots$

$$|T_{n+1}(x) - T_n(x)| \le \frac{[\gamma(x - a)]^n}{n!} \delta, \qquad (19)$$

while

$$\|T_{n+1} - T_0\| \le \delta e^{\gamma r} < \delta_0, \quad \text{so that } (x, T_{n+1}(x)) \in K_0.$$

From (19), we see that when $n \ge 1$,

$$T_n(x) = T_0(x) + \sum_{k=0}^{n-1} [T_{k+1}(x) - T_k(x)],$$

is the nth partial sum of an absolutely (and uniformly) convergent series; hence $T(x) = \lim_{n \to \infty} T_n(x)$ exits, and $\|T_n - T\| \to 0$ as $n \to \infty$. Also

$$|T(x) - T_0(x)| = \lim_{n \to \infty} |T_n(x) - T_0(x)| \le \delta e^{\gamma r};$$

i.e.,

$$\|T - T_0\| \le \delta e^{\gamma(b-a)} < \delta_0, \quad \text{so that } (x, T(x)) \in K_0. \qquad (20)$$

Finally, T, as the *uniform* limit of continuous functions T_n, is itself continuous (see [Ed]), and using (17) and (14), we have from the triangle inequality

$$\left| T(x) - L - \int_a^x G(s, T(s)) \, ds \right|$$

$$\le |T(x) - T_{n+1}(x)| + \left| \int_a^x [G(s, T_n(s)) - G(s, T(s))] \, ds \right|$$

$$\le |T(x) - T_{n+1}(x)| + \gamma(b - a)\|T_n - T\|. \quad \text{(Why?)}$$

In the limit as $n \to \infty$; the right side vanishes and we conclude that $T(x) = L + \int_a^x G(s, T(s)) \, ds$ as required. $\qquad \square$

(A.19) **Remark.** A slight modification of the construction used in this theorem yields a proof of *local* existence for a solution T to (12) with $T(a) = L$, even when the solution $T_0(x)$ is *not* available. Indeed, if we set $T_0(x) \equiv L$, and define $T_n(x)$ recursively by (17) then it is seen that

$$|T_1(x) - T_0(x)| \le \int_a^x |G(s, L)| \, ds < \delta_0, \quad \text{when } |x - a| \le r,$$

for a *sufficiently small* r. The remaining estimates will be as before, and the functions $T_n(x)$ will converge to a continuous solution $T(x)$ of (16), but only in the neighborhood $\{|x - a| < r\}$. However, Proposition A.17 still applies and we can conclude that the solution to (12), with $T(a) = L$, is uniquely determined in this neighborhood. (It is possible to give growth conditions on G, which permit the extension of this solution to the entire interval $[a, b]$. See [C–L] and [Ak].)

The *continuous dependence* of the solution $T = T(\cdot\,; L)$ (established by Theorem A.18) on its initial data, L, is apparent. The estimate (20); viz.,

$$\|T - T_0\| \le \delta e^{\gamma(b-a)} = e^{\gamma(b-a)}|L - L_0|,$$

shows this when L is near L_0, and by the uniqueness previously established, we may obtain a similar estimate for L sufficiently near $\tilde{L}_0$, when $|L_0 - \tilde{L}_0| < \delta$.

To establish the *differentiable dependence* on L requires more effort, and we shall do so for a typical component l, of L. If E is the associated unit vector in $\mathbb{R}^N$, and we set $T(x, h) = T(x; L_0 + hE)$, for small h in $\mathbb{R}$, then we must guarantee the existence of the directional derivative.

$$V(x) \stackrel{\text{def}}{=} \lim_{h \to 0} V(x, h), \tag{21}$$

where

$$V(x, h) = \frac{T(x, h) - T_0(x)}{h} \quad (h \ne 0). \tag{22}$$

It is not difficult to see by formal operations that if it exists (and is continuous) then V should satisfy the *linear* integral equation

$$V(x) = E + \int_a^x G_T(t, T_0(t))V(t)\, dt. \tag{23}$$

By means similar to those employed in proving A.18, we may show that for a given unit vector E, equation (23) has a unique continuous solution V. (The estimate corresponding to (14) is trivial, and the solution discussed in A.19 is valid for the entire interval.)

(A.20) **Theorem.** *The solutions $T = T(\cdot\,; L)$ of (12) have a Jacobian matrix $T_L = T_L(\cdot\,; L)$ with jointly continuous components, $\partial t_i/\partial l_j$, $i, j = 1, 2, \ldots, N$.*

PROOF. Following the above discussion, we shall first establish (21), assuming the existence of V as a solution to (23). Thus with $T(x) = T(x, h) = T(x; L)$, for $L = L_0 + hE$, we have from (22) and (16),

$$V(x, h) - V(x) = \int_a^x \{[G(s, T(s)) - G(s, T_0(s))]h^{-1} - G_T(s, T_0(s))V(s)\}\, ds.$$

Introducing

$$T_u(s) = uT(s) + (1 - u)T_0(s) \quad \text{for } u \in [0, 1], \tag{24}$$

and using an analogous form of (13), we get after some manipulation:

$$V(x, h) - V(x)$$

$$= \int_a^x \left[\left(\int_0^1 G_T(s, T_u(s)) \, du \right) V(s, h) - G_T(s, T_0(s)) V(s) \right] ds$$

$$= \int_a^x \left(\int_0^1 [G_T(s, T_u(s))] \, du \right) (V(s, h) - V(s)) \, ds$$

$$+ \int_a^x \left(\int_0^1 [G_T(s, T_u(s)) - G_T(s, T_0(s))] \, du \right) V(s) \, ds, \qquad (25)$$

since $G_T(s, T_0(s))$ is constant with respect to the u integration.

From A.18, we know that when $|h| = |L - L_0| = \delta < \delta_0 e^{-\gamma(b-a)}$, then $\|T - T_0\| < \delta\kappa_0 = |h|\kappa_0$, where $\kappa_0 = e^{\gamma(b-a)}$, and by (24),

$$\|T_u - T_0\| = |u| \|T(\cdot, h) - T_0\| < |h|\kappa_0 \quad \text{for } u \in [0, 1]. \qquad (26)$$

From the *uniform* continuity of the components of G_T on the compact set K_0 (5.2), then for a typical *row* g_T of G_T and each $\varepsilon > 0$, we can conclude by (26) that $|g_T(s, T_u(s)) - g_T(s, T_0(s))| \le \varepsilon$, when $|h|$ is sufficiently small.

Thus we may estimate (25) in the manner used to obtain (14), and see that for sufficiently small h, and an *appropriately large* γ,

$$v_h(x) \stackrel{\text{def}}{=} |V(x, h) - V(x)| \le \gamma \left(\int_a^x v_h(s) \, ds + \varepsilon \|V\| \right) = \sigma(x), \quad \text{say}.$$

Then, as in the proof of A.17, $\sigma'(x) = \gamma v_h(x) \le \gamma\sigma(x)$, or $(e^{-\gamma x}\sigma(x))' \le 0$, so that $v_h(x) \le \sigma(x) \le e^{\gamma(x-a)}\sigma(a) \le e^{\gamma(b-a)}\varepsilon\gamma\|V\|$. Since ε may be made as small as we wish, we see that as $h \to 0$, $v_h(x) \to 0$ and in fact $\|V(\cdot, h) - V\| \to 0$.

Recalling the definitions (22) and (21) of $V(x, h)$ and $V(x)$ respectively, we have proved that for an appropriate E, the partial derivative $T_l = T_l(x; L_0)$ exists and provides a *continuous* solution to equation (23)

$$T_l(x; L_0) = E + \int_a^x G_T(s, T_0(s)) T_l(s; L_0) \, ds.$$

Similarly, for each L near L_0, with the *same E*,

$$T_l(x; L) = E + \int_a^x G_T(s, T(s; L)) T_l(s; L) \, ds,$$

and by a repetition of the arguments used above in estimating (25), it can be shown that for each $\varepsilon > 0$, there is a $\delta > 0$ so small that $|L - L_0| < \delta$ makes $\|T(\cdot; L) - T_0\| \le \varepsilon_0(\le \delta_0)$, and consequently, $\|T_l(\cdot; L) - T_l(\cdot; L_0)\| \le \varepsilon$.

Therefore, for this δ, ε:

$$|T_l(x; L) - T_l(x_0; L_0)| \leq |T_l(x; L) - T_l(x; L_0)| + |T_l(x; L_0) - T_l(x_0; L_0)|$$

$$\leq \varepsilon + |T_l(x; L_0) - T_l(x_0; L_0)| \leq 2\varepsilon,$$

say, when also $|x - x_0|$ is small. Thus, in particular, the components of T_L are (jointly) continuous as required. $\qquad\qquad\square$

§A.6. The Rayleigh Ratio

In the discussion of the nonuniform string in §8.9, it was shown that the natural frequencies of motion are directly related to the eigenvalues of a Sturm–Liouville problem of the type presented in §7.3. Moreover, both in Problem 7.19 and in §8.9(a), we have indicated the role played by the associated Rayleigh ratio in characterizing the fundamental frequency, or *least* eigenvalue, λ. In this section, we will examine briefly, how this same ratio is related to the higher eigenvalues. Although we consider only the problem studied in §7.3, the methods employed admit far reaching generalizations which form the basis for most modern results in this subject. (See [W–S].)

We first establish that the eigenfunction associated with a given eigenvalue is essentially unique, and that the eigenfunctions for distinct eigenvalues exhibit a natural orthogonality. We then use this orthogonality to define (inductively) the higher eigenvalues as successive minima of the Rayleigh ratio, and obtain an equivalent characterization by means of the Courant–Weyl maximin principle. We then use comparison to establish the existence of arbitrarily large eigenvalues and this provides useful summation formulas for coefficients related to eigenfunction expansions. Finally, we apply these formulae to give a proof for a Wirtinger inequality.

For given functions ρ, q, $\tau \in C[a, b]$, with ρ, τ, positive, and $q \geq 0$, we recall that the problem of minimizing the *Rayleigh ratio*

$$R(y) = \frac{\int_a^b [\tau y'^2 + q y^2](x)\, dx}{\int_a^b [\rho y^2](x)\, dx} \tag{27}$$

on

$$\mathscr{D} = \{y \in \mathscr{Y}_0 : y \neq \mathcal{O}\}, \quad \text{where} \quad \mathscr{Y}_0 = \{y \in C^1[a, b] : y(a) = y(b) = 0\},$$

could be solved by a *positive* solution $y_1 \in \mathscr{Y}_0$ of the Euler–Lagrange equation

$$(\tau y')' = (q - \lambda \rho) y \quad \text{on } (a, b), \tag{28}$$

when the eigenvalue $\lambda = \lambda_1 = R(y_1)$ (Problem 7.19). In fact, with the normalization $G(y_1) = 1$, where

$$G(y) \overset{\text{def}}{=} \int_a^b [\rho y^2](x)\, dx, \tag{29}$$

y_1 is within a sign, the *unique* eigenfunction in $\mathscr{Y}_0$ for this eigenvalue λ_1. [Otherwise, when $\lambda = \lambda_1$, the equation (28) would have two *linearly independent* solutions, both in $\mathscr{Y}_0$. It would then follow that *all* solutions to this second-order linear equation on $[a, b]$ would be in $\mathscr{Y}_0$, contradicting the fact that a solution y can be constructed by the method of §A.5 with $y(a) \neq 0$. (See [C–L].)]

This argument is equally applicable to any eigenvalue λ, so that the associated eigenfunction y is uniquely determined within a sign by the normalization $G(y) = 1$. In what follows we suppose the eigenfunctions to be so normalized. (We also suppress the argument x in the integrands.)

(A.21) **Lemma.** *The eigenfunctions* y, $\tilde{y}$ *associated with distinct eigenvalues* λ, $\tilde{\lambda}$, *respectively, are* ρ-*orthogonal in that* $\int_a^b \rho y \tilde{y}\, dx = 0$.

PROOF.

$$(\lambda - \tilde{\lambda}) \int_a^b \rho y \tilde{y}\, dx = \int_a^b [(\lambda \rho y)\tilde{y} - (\tilde{\lambda}\rho\tilde{y})y]\, dx$$

$$= \int_a^b [(\tau\tilde{y}')'y - (\tau y')'\tilde{y}]\, dx = ((\tau\tilde{y}')y - (\tau y')\tilde{y})\Big|_a^b = 0,$$

where we have used (28) and the analogous equation for $\tilde{\lambda}$, $\tilde{y}$, to produce the central term, and subsequent integration by parts to obtain the boundary terms which for y, $\tilde{y} \in \mathscr{Y}_0$, must vanish; other terms which arise in this process cancel. Since $\lambda \neq \tilde{\lambda}$, the orthogonality follows. □

Thus, in seeking another eigenvalue, it would be reasonable to consider minimizing R on $\mathscr{D}$ subject to the constraint

$$G_1(y) \equiv \int_a^b \rho y y_1\, dx = 0, \tag{30}$$

or equivalently, to consider minimizing

$$F(y) \equiv \int_a^b (\tau(y')^2 + q y^2)\, dx \tag{31}$$

on the *subspace* $\mathscr{Y}_0$ subject to the constraints $G(y) = 1$, $G_1(y) = 0$. By Theorem 6.4b with Remark 5.17, it follows that if y_2 accomplishes this minimization then $\exists$ Lagrangian multipliers μ and $-\lambda_2$ such that y_2 is a C^1 solution of the equation

$$(\tau y')' = (q - \lambda_2 \rho)y + \mu \rho y_1. \tag{32}$$

(The constraints eliminate the alternative conclusion 6.4a, since

$$\begin{vmatrix} \delta G(y; y) & \delta G(y; y_1) \\ \delta G_1(y; y) & \delta G_1(y; y_1) \end{vmatrix} = \begin{vmatrix} 2G(y) & 2G_1(y) \\ G_1(y) & G(y_1) \end{vmatrix} = \begin{vmatrix} 2 & 0 \\ 0 & 1 \end{vmatrix} = 2.)$$

Here $\mu = 0$. [Indeed, with *two* integrations by parts repeatedly utilizing the constraint $G_1(y) = 0$, we have from (29) that

$$\mu = \mu G(y_1) = \mu \int_a^b \rho y_1^2 \, dx = \int_a^b [(\tau y')'y_1 - q y y_1] \, dx$$

$$= -\int_a^b [\tau y' y_1' + q y y_1] \, dx$$

$$= \int_a^b [(\tau y_1')' - q y_1] y \, dx = -\int_a^b \lambda_1 \rho y_1 y \, dx = 0,$$

since y_1 is a solution of (28) for $\lambda = \lambda_1$. (The boundary terms vanish as usual with y and y_1.)]

It follows that y_2 is an eigenfunction to the eigenvalue λ_2, and as in Problem 7.19(a), $\lambda_2 = R(y_2) \geq R(y_1) = \lambda_1$. (Why?) Moreover, $\lambda_2 > \lambda_1$, for otherwise y_2 would be an eigenfunction to λ_1 which is essentially different from y_1 (since $G_1(y_2) = 0$). Continuing in this manner, we can define $\lambda_3 = \min\{R(y): y \in \mathcal{D}, G_j(y) = 0, j = 1, 2\}$ where

$$G_j(y) \overset{\text{def}}{=} \int_a^b \rho y y_j \, dx,$$

and conclude that if a minimizing function y_3 exists, then it is necessarily an eigenfunction to the eigenvalue λ_3 ($> \lambda_2 > \lambda_1$). Finally, assuming that for $j < n$, eigenfunctions y_j have been found to eigenvalues λ_j (and normalized to have $G(y_j) = 1$), we set

$$\lambda_n = \min\{R(y): y \in \mathcal{D}; G_j(y) = 0, j < n\}. \tag{33}$$

In order to continue this construction we must at each stage produce a minimizing (eigen)function, or at least have a proof that one exists. Instead, investigators have sought methods which could be used to approximate these eigenfunctions and eigenvalues. For example, numerical methods can be employed to reduce (28) to an ordinary eigenvalue problem for a large $N \times N$ matrix. A corresponding approximation for the orthogonality conditions $G_j(y) = 0$, shows that the solution is to be sought in a *subspace* of $\mathbb{R}^N$, and matrix methods are available to attack these problems. There are also other methods available to establish the minimality. (See [G–F].)

However, if we assume that the relevant eigenfunctions do exist, then the associated eigenvalues can be approximated both from above by (33) and from below, as in the following "maximin" result.

(A.22) **Proposition** (Courant). *For $j = 1, 2, \ldots, n - 1$, let $\varphi_j \in C[a, b]$. Then*

$$\mu_n \overset{\text{def}}{=} \min\left\{R(y): y \in \mathcal{D}; \Phi_j(y) \overset{\text{def}}{=} \int_a^b y \varphi_j \, dx = 0, j < n\right\} \leq \lambda_n.$$

PROOF. We may select constants c_i not all zero so that the function $y = \sum_i^n c_i y_i$ meets the orthogonality conditions $\Phi_j(y) = \sum_{i=1}^n c_i \Phi_j(y_i) = 0$, $j = 1, 2, \ldots,$

$n - 1$, since this underdetermined system of homogeneous linear equations *always* has a nontrivial solution. Then, *for this* $y \in \mathcal{Y}_0$, we know that $\mu_n \leq R(y)$, since

$$G(y) = \int_a^b \rho y^2 \, dx = \sum_{i,j=1}^n c_i c_j \int_a^b \rho y_i y_j \, dx = \sum_{i=1}^n c_i^2 \neq 0, \qquad (34)$$

in view of Lemma A.21 and the assumed normalization $G(y_j) = G_j(y_j) = 1$, $j = 1, 2, \ldots, n$. Similarly from (31) and (28),

$$F(y) = \sum_{i,j=1}^n c_i c_j \int_a^b (\tau y_i' y_j' + q y_i y_j) \, dx$$

$$= -\sum_{i,j=1}^n c_i c_j \int_a^b [(\tau y_i')' - q y_i] y_j \, dx$$

$$= \sum_{i,j=1}^n c_i c_j \lambda_i \int_a^b \rho y_i y_j \, dx = \sum_{i=1}^n \lambda_i c_i^2,$$

or

$$F(y) \leq \lambda_n \sum_{i=1}^n c_i^2, \quad \text{since } \lambda_1 < \lambda_2 < \cdots < \lambda_n.$$

Thus $\mu_n \leq R(y) = F(y)/G(y) \leq \lambda_n$ by (34). $\qquad\qquad \square$

Remarks. Writing $\mu_n = \mu(\varphi_1, \varphi_2, \ldots, \varphi_{n-1})$, so that by (33):

$$\lambda_n = \mu(\rho y_1, \rho y_2, \ldots, \rho y_{n-1}),$$

we see that λ_n is the *maximum* (over all sets of continuous functions φ_1, $\varphi_2, \ldots, \varphi_{n-1}$) *of the minimum* for the Rayleigh ratio of functions orthogonal to these φ_j, for $j < n$. This explains the appellation "maximin" applied to the last result. (There are also corresponding minimax characterizations of the higher eigenvalues. See [W–S].) Moreover, as in §7.3, we may replace the φ_j by piecewise C^1 functions $\hat{\varphi}_j$—in particular, by piecewise *linear* functions— which are easier to produce and to control computationally. Thus, this proposition may be implemented and leads to the method of finite elements; it is also related to the approximation methods of Ritz and Galerkin. See [C–H] and [S–F].

Now, when $\rho = \rho_0$ and $\tau = \tau_0$ are (positive) constants, while $q = 0$, then (28) has for eigenvalues $\tilde{\lambda}_n > 0$, the elementary eigenfunctions

$$\tilde{y}_n(x) = a_n \cos \omega_n x + b_n \sin \omega_n x,$$

where $\omega_n = n\pi/l$,

$$\tilde{\lambda}_n = (\tau_0/\rho_0)(n\pi/l)^2, \qquad l = b - a;$$

and the coefficients a_n and b_n are selected to have $\tilde{y}_n \in \mathcal{D}$. (See §8.9(a).)

(A.23) **Corollary.** *When* $\rho_0 = \max \rho$, *and* $\tau_0 = \min \tau$, *then* $\tilde{\lambda}_n \leq \lambda_n$, $n = 1$, $2, \ldots$, *so that* $\lambda_n \to +\infty$, *as* $n \to \infty$.

PROOF. For this ρ_0, τ_0 and $q = 0$, let $\tilde{R}$ be the associated Rayleigh ratio which is defined since $\rho_0 > 0$. Observe that when $y \in \mathscr{D}$:

$$\tilde{R}(y) = \left(\tau_0 \int_a^b y'^2 \, dx\right) \Big/ \left(\rho_0 \int_a^b y^2 \, dx\right) \le \frac{F(y)}{G(y)} = R(y),$$

so that clearly $\tilde{\lambda}_1 = \min_{y \in \mathscr{D}} \tilde{R}(y) \le \min_{y \in \mathscr{D}} R(y) = \lambda_1$. To obtain the corresponding result for the higher eigenvalues, we set

$$\varphi_j = \rho_0 \tilde{y}_j, \quad j < n,$$

so that

$$\tilde{\lambda}_n = \min\{\tilde{R}(y): y \in \mathscr{D}; \Phi_j(y) = 0, j < n\} \le \mu_n \le \lambda_n,$$

by Proposition A.22. $\qquad\square$

(A.24) **Theorem.** *If $\lambda_n \to +\infty$ as $n \to \infty$, and $\varphi \in C[a, b]$, set*

$$c_n = G_n(\varphi) = \int_a^b \rho \varphi y_n \, dx, \qquad n = 1, 2, \dots.$$

Then:

(a) $\sum_{n=1}^{\infty} c_n^2 \le G(\varphi) = \int_a^b \rho \varphi^2 \, dx$ *(Bessel's inequality).*
(b) *If $\varphi \in \mathscr{Y}_0$, then $\sum_{n=1}^{\infty} c_n^2 = \int_a^b \rho \varphi^2 \, dx$ (Parseval's formula).*

PROOF. For $N \ge 1$, set $\varphi_N = \varphi - \sum_{n=1}^{N} c_n y_n$, and observe that for $j \le N$:

$$G_j(\varphi_N) = \int_a^b \rho \varphi_N y_j \, dx = c_j - \sum_{n=1}^{N} c_n G_j(y_n) = c_j - c_j = 0, \tag{35}$$

by the assumed *orthonormality* of the y_n. Hence

$$0 \le G(\varphi_N) = \int_a^b \rho \varphi_N \left(\varphi - \sum_{n=1}^{N} c_n y_n\right) dx = \int_a^b \rho \varphi_N \varphi \, dx,$$

or

$$G(\varphi_N) = \int_a^b \rho \varphi^2 \, dx - \sum_{n=1}^{N} c_n \int_a^b \rho \varphi y_n \, dx = G(\varphi) - \sum_{n=1}^{N} c_n^2. \tag{36}$$

Thus $\sum_{n=1}^{N} c_n^2 \le G(\varphi)$ and (a) follows as $N \to \infty$. (Observe that we have used *only* the orthonormality of the y_n.)

If, in addition, $\varphi \in \mathscr{Y}_0$, then we may suppose that $G(\varphi_N) \ne 0$, $\forall N$. [Otherwise we would have $\sum_{n=1}^{N} c_n^2 = G(\varphi)$, while $\varphi_N = 0$ (Why?), so that $c_j = G_j(\varphi) = \sum_{n=1}^{N} c_n G_j(y_n) = 0$ if $j > N$, and (b) would follow.] Then $\varphi_N \in \mathscr{Y}_0$ and by (33) and (35), $\lambda_N \le R(\varphi_N) = F(\varphi_N)/G(\varphi_N)$, or by (31),

$$0 \le G(\varphi_N) \le \lambda_N^{-1} \int_a^b (\tau \varphi_N'^2 + q \varphi_N^2) \, dx = \lambda_N^{-1}\left(F(\varphi) - \sum_{n=1}^{N} c_n^2 \lambda_n\right) \le \lambda_N^{-1} F(\varphi),$$

where we have made repeated use of the devices of integration by parts in conjunction with the vanishing of the boundary terms, the differential equation (28), and the orthogonality of the eigenfunctions. Hence by (36),

$$0 \leq G(\varphi) - \sum_{n=1}^{N} c_n^2 = G(\varphi_N) \leq \lambda_N^{-1} F(\varphi) \to 0 \quad \text{as } N \to \infty,$$

and (b) follows.   □

In particular, we note that when $a = 0$ and $b = 1$ then the eigenfunctions for the reduced problem of $\rho = \rho_0 = \tau = \tau_0 = 1$, and $q = 0$ are the normalized functions $u_n(x) = \sqrt{2} \sin n\pi x$ obtained in §8.9(a). It follows that to each $\varphi \in \mathscr{Y}_0 = \{y \in C^1[0, 1]: y(0) = y(1) = 0\}$, is associated a set of coefficients $c_n = \sqrt{2} \int_0^1 \varphi(x) \sin n\pi x \, dx$ such that $\int_0^1 \varphi^2(x) \, dx = \sum_{n=1}^{\infty} c_n^2$. This lends credence to the (proveable) conjecture that $\varphi(x) = \sum_{n=1}^{\infty} c_n \sin n\pi x$. Moreover, in the general setting we might expect to prove that functions $\varphi \in \mathscr{Y}_0$, admit the eigenfunction representation $\varphi = \sum_{n=1}^{\infty} c_n y_n$. Although with proper interpretation of convergence this, too is the case, we shall not pursue these series considerations, and instead refer the interested reader to [C–H].

However, the preceding theorem does afford a proof for a Wirtinger inequality which is analogous to that considered in §9.5. As above, we note that the trigonometric functions $y_n(x) = (1/\sqrt{\pi}) \sin(nx/2)$, constitute the normalized eigenfunctions on $[0, 2\pi]$ when $q = 0$ and $\rho \equiv \tau \equiv 1$, for the eigenvalues $\lambda_n = n^2/4$, $n = 1, 2, \dots$.

(A.25) **Corollary** (A Wirtinger Inequality). *If*

$$y \in \mathscr{Y}_0 = \{y \in C^1[0, 2\pi]: y(0) = y(2\pi) = 0\}$$

and

$$\int_0^{2\pi} y(x) \sin(x/2) \, dx = 0,$$

then

$$F(y) = \int_0^{2\pi} [(y')^2 - y^2] \, dx \geq 0,$$

with equality iff $y(x) = c_2 \sin x$.

PROOF. Since $y \in \mathscr{Y}_0$, we have for it the *Parseval* formula (A.24b):

$$\int_0^{2\pi} y^2 \, dx = \sum_{n=1}^{\infty} c_n^2, \quad \text{with } c_n = \frac{1}{\sqrt{\pi}} \int_0^{2\pi} y(x) \sin \frac{nx}{2} \, dx, \tag{37}$$

$n = 1, 2, \dots$, where by hypothesis, $c_1 = 0$. Similarly, a simple computation shows that the complementary functions $\bar{y}_n(x) = (1/\sqrt{\pi}) \cos(nx/2)$ are also *orthonormal* on $[0, 2\pi]$. Hence, for $\varphi = y'$, we have the *Bessel* inequality (A.24a):

$$\int_0^{2\pi} y'^2 \, dx \geq \sum_{n=1}^{\infty} \bar{c}_n^2, \tag{38}$$

where

$$\bar{c}_n = \frac{1}{\sqrt{\pi}} \int_0^{2\pi} y'(x) \cos\frac{nx}{2} dx$$

$$\frac{n}{2}\frac{1}{\sqrt{\pi}} \int_0^{2\pi} y(x) \sin\frac{nx}{2} dx = \frac{n}{2}c_n, \qquad n = 1, 2, \ldots,$$

since $y(0) = y(2\pi) = 0$. Thus combining (37) and (38) and recalling that $c_1 = 0$, we have

$$\int_0^{2\pi} (y'^2 - y^2)\, dx \geq \sum_{n=2}^{\infty} (\bar{c}_n^2 - c_n^2) = \sum_{n=2}^{\infty} \left[\left(\frac{n}{2}\right)^2 - 1\right]c_n^2 \geq 0,$$

with equality *iff* $c_n = 0$ when $n > 2$. But it then follows from (36) that $\varphi_2(x) = y(x) - c_2 \sin x$ satisfies $G(\varphi_2) = \int_0^{2\pi}(\varphi_2)^2\, dx = G(y) - (c_2)^2 = 0$, (by (37)) so that $\varphi_2 \equiv 0$; i.e., $y(x) = c_2 \sin x$.   □

Remark. There is a similar proof for the standard Wirtinger inequality used in Problem 1.6, involving the general theory of Fourier series. (See [H–L–P].) As we observed there, some restriction on the functions in $\mathcal{Y}_0$ is necessary for its validity. Fortunately, the integral condition $\int_0^{2\pi} y(x)\sin(x/2)\, dx = 0$ can always be achieved in the solution of the classical isoperimetric problem, without violating the additional requirements that $y(0) = y(2\pi) = 0$. [We suppose that the origin is placed on a typical bounding curve of length 2π, and "load" the curve nonuniformly with the mass function $\sin(s/2)$, where s is the arc length from the origin. Then, if the x axis is taken through the *centroid* of the loaded curve, it follows that $\int_0^{2\pi} y(s)\sin(s/2)\, ds = 0$.]

§A.7. Linear Functionals and Tangent Cones in $\mathbb{R}^d$

In this section, we combine results from linear algebra with our earlier efforts to describe tangency (§5.6) to obtain multipliers associated with minimization in $\mathbb{R}^d$ with inequality constraints.

Definition. If $A \in \mathbb{R}^d$, then $l(X) = A \cdot X$, $X \in \mathbb{R}^d$, determines the linear functional l on $\mathbb{R}^d$. When $A_i \in \mathbb{R}^d$, $i = 1, 2, \ldots, m$, then $L(X) = (l_1(X), l_2(X), \ldots, l_m(X))$ provides the vector valued counterpart, where $l_i(X) = A_i \cdot X$, $X \in \mathbb{R}^d$.

We say that the l_i are linearly independent when the A_i are linearly independent, $i = 1, 2, \ldots, m \leq d$. We write $A \geq \mathcal{O}$, when each component of A is nonnegative, and give a corresponding componentwise interpretation to other vector-valued inequalities. Let $M = (\mu_1, \mu_2, \ldots, \mu_m) \in \mathbb{R}^m$.

(A.26) Proposition. *Suppose the l_i as above are linearly independent, $i = 1, 2, \ldots, m \leq d$, and $l(X) = A \cdot X$.*

(a) *If* $L(X) = \mathcal{O} \Rightarrow l(X) = 0$, $\forall X \in \mathbb{R}^d$, *then* $l + M \cdot L \equiv 0$, *for a unique* $M \in \mathbb{R}^m$.

(b) *If* $L(X) \leq \mathcal{O} \Rightarrow l(X) \geq 0$, $\forall X \in \mathbb{R}^d$, *then* $l + M \cdot L \equiv 0$, *for a unique* $M \geq \mathcal{O}$ $(M \in \mathbb{R}^m)$.

PROOF. (a) If A is not in the m-dimensional subspace $\mathscr{S}$ spanned by the A_i, $i = 1, 2, \ldots, m$, then there is a vector X orthogonal to $\mathscr{S}$, but not to A. However, then $L(X) = \mathcal{O}$ while $l(X) \neq 0$, contradicting the hypothesis. It follows that $A = \sum_{i=1}^m (-\mu_i) A_i$ for some μ_i, or that

$$l(X) = A \cdot X = \sum_{i=1}^m (-\mu_i)(A_i \cdot X) = \sum_{i=1}^m (-\mu_i) l_i(X) = -M \cdot L(X).$$

Clearly, having $l(X) = -\tilde{M} \cdot L(X) \Rightarrow (M - \tilde{M}) \cdot L(X) \equiv \mathcal{O}$ which violates linear independence unless $M = \tilde{M}$.

(b) The inequality hypothesis implies that of part (a) since $L(-X) = \mathcal{O} \Leftrightarrow L(X) = \mathcal{O} \Rightarrow l(X) \geq 0$ and $-l(X) = l(-X) \geq 0$. Thus $L(X) = \mathcal{O} \Rightarrow l(X) = \mathcal{O}$, and the representation $l(X) = -M \cdot L(X)$ follows for a unique $M \in \mathbb{R}^m$. Were, say $\mu_1 < 0$, we could find an X orthogonal to the $(m-1)$-dimensional subspace spanned by A_i, $i \geq 2$, for which $A_1 \cdot X < 0$. But then $L(X) \leq \mathcal{O}$, while $l(X) = -\mu_1(A_1 \cdot X) < 0$, again contradicting the hypotheses. Thus $\mu_1 \geq 0$, and by extension, $M \geq \mathcal{O}$ as desired. □

We now wish to obtain "one-sided" characterizations for some of the concepts of tangency introduced in §5.6. By Theorem A.1, the surface of the unit sphere in $\mathbb{R}^d$ is compact. Suppose that $G = (g_i, \ldots, g_m)$ has (Frechét) differentiable components at $X_0 \in \mathbb{R}^d$ and $G(X_0) = \mathcal{O}$. Assume that X_0 is not an isolated point of the *sublevel* set $S = \{X \in \mathbb{R}^d : G(X) \leq \mathcal{O}\}$. There are sequences $X_n \in S$, $n = 1, 2, \ldots, (X_n \neq X_0)$, with $X_n \to X_0$ as $n \to \infty$, for which the corresponding *unit* vectors $(X_n - X_0)/|X_n - X_0|$ have convergent subsequences with unit vectors T as limits. Then

$$\mathscr{K}_0 = \{X = cT : c \geq 0 \text{ and } T \text{ is any unit vector obtained as above}\}$$

defines the *tangent cone* to S at X_0.

It follows as in §5.6, that $T \in \mathscr{K}_0 \Rightarrow G'(X_0)$, $T \leq \mathcal{O}$, where G' is the Jacobian matrix of G having as its *rows* the gradients $\nabla g_i(X_0) = A_i$, say, $i = 1, 2, \ldots, m$. Indeed, for $X_n \in S$; $X \neq X_0$:

$$0 \geq \frac{g_i(X_n) - g_i(X_0)}{|X_n - X_0|} = \nabla g_i(X_0) \cdot \frac{(X_n - X_0)}{|X_n - X_0|} + 3(X_n - X_0),$$

so that as $n \to \infty$: $\nabla g_i(X_0) \cdot T \leq 0$, $i = 1, 2, \ldots, m$. We can now obtain a converse assertion of the type alluded to in the footnote at the end of §5.7.

(A.27) **Proposition.** *If G is C^1 near X_0 and $G'(X_0)$ is of maximal row rank $m \leq d$, then $T \in \mathscr{K}_0$ iff $G'(X_0)T \leq \mathcal{O}$.*

PROOF. We know that if $T \in \mathscr{K}_0$, then $G'(X_0)T \leq \mathcal{O}$. Conversely, suppose that $T' \overset{\text{def}}{=} G'(X_0)T \leq \mathcal{O}$, for some T with $|T| = 1$, and let $l_i(X) = \nabla g_i(X_0) \cdot X$, $i = 1, 2, \ldots, m$. As X ranges over $\mathbb{R}^d$, $L(X) = (l_1(X), \ldots, l_m(X))$ ranges over $\mathbb{R}^m$, since the l_i are linearly independent by hypothesis. Thus there exist $X_j \in \mathbb{R}^d$, $j = 1, 2, \ldots, m$ for which the vectors $L(X_j)$ are linearly independent in $\mathbb{R}^m$, so that the $m \times m$ matrix $\mathbb{G}$ with elements $g_{ij} = \nabla g_i(X_0) \cdot X_j$, $i, j = 1, 2, \ldots, d$, is invertible.

Next, observe that $(0, \mathcal{O}) \in \mathbb{R}^{m+1}$ is in the $\mathcal{O}$-level set of

$$F(t, V) \overset{\text{def}}{=} G\left(X_0 + \sum_{j=1}^{m} v_j X_j + tT\right) - tT' \tag{39}$$

(since $G(X_0) = \mathcal{O}$), and the Jacobian matrix $F_V(0, \mathcal{O}) = \mathbb{G}$, (since $(\partial f_i / \partial v_j)(0, \mathcal{O}) = \nabla g_i(X_0) \cdot X_j$). Hence by implicit function theory [Ed], we can represent a neighborhood of this level set as the *graph* of a unique C^1 function $V(t)$ defined for $|t| < \varepsilon$, say; i.e., $F(t, V(t)) \equiv \mathcal{O}$ for $|t| < \varepsilon$, and $V(0) = \mathcal{O}$. Differentiating and evaluating at $t = 0$ we obtain

$$\mathcal{O} = G'(X_0)\left(\sum_{j=1}^{m} v_j'(0) X_j + T\right) - T' = \sum_{j=1}^{m} v_j'(0) G'(X_0) X_j = \overline{\mathbb{G}} V'(0)$$

(since $T' = G'(X_0)T$). But $\overline{\mathbb{G}}$, the transpose of $\mathbb{G}$, is nonsingular by construction so that $V'(0) = \mathcal{O}$.

Finally, for $0 \leq t < \varepsilon$:

$$X(t) \overset{\text{def}}{=} X_0 + \sum_{j=1}^{m} v_j(t) X_j + tT \in S,$$

(since $G(X(t)) \equiv tT' \leq \mathcal{O}$) and as $t \searrow 0$,

$$\frac{X(t) - X_0}{t} = T + \sum_{j=1}^{m} (v_j(t)/t) X_j \to T + \sum_{j=1}^{m} v_j'(0) X_j = T.$$

But this means that $(X(t) - X_0)/|X(t) - X_0| \to T/|T| = T \in \mathscr{K}_0$. $\qquad\square$

These results enable us to relax the hypotheses on the Lagrange multiplier rule in $\mathbb{R}^d$. (See Theorem 5.16.)

(A.28) **Theorem.** *Suppose G is C^1 and $G'(X_0)$ is of maximal row rank $m \leq d$, while f is real valued and differentiable at $X_0 \in \mathbb{R}^d$. If $f(X) \geq f(X_0)$, when $G(X) \leq G(X_0) \in \mathbb{R}^m$, then there exist unique Lagrangian multipliers, $M = (\mu_1, \ldots, \mu_m) \geq \mathcal{O}$, such that $\nabla \tilde{f}(X_0) = \mathcal{O}$, where $\tilde{f} = f + M \cdot G$, and $\mu_i g_i(X_0) = 0$, $\forall i = 1, 2, \ldots, m$.*

PROOF. Assume that $G(X_0) = \mathcal{O}$, let L and S be as in the proof of the preceding proposition, and set $l(X) = \nabla f(X_0) \cdot X$. Then from the proposition, $L(T) \leq \mathcal{O} \Rightarrow T \in \mathscr{K}_0$, and there is a sequence $X_n \in S$, $X_n \neq X_0$, for which $(X_n - X_0)/|X_n - X_0| \to T$ as $n \to \infty$, if $|T| = 1$. By hypothesis, $f(X_n) \geq f(X_0)$, so that as $n \to \infty$, $0 \leq (f(X_n) - f(X_0))/|X_n - X_0| \to \nabla f(X_0) \cdot T$; i.e., $L(T) \leq \mathcal{O} \Rightarrow l(T) \geq 0$, $\forall$ unit vectors T, and hence by homogeneity, for all $T \in \mathbb{R}^d$.

Thus by Proposition A.26(b) there exist unique $M \in \mathbb{R}^m$ for which $M \geq \mathbb{O}$ and

$$\nabla f(X_0) = -\sum_{i=1}^{m} \mu_i \nabla g_i(X_0),$$

or

$$\nabla \tilde{f}(X_0) = \mathbb{O}, \quad \text{if} \quad \tilde{f} = f + M \cdot G.$$

Now, were some $g_j(X_0) < 0$, then by continuity we should have $g_j < 0$ near X_0, and such constraints would be inactive in the minimization. It follows that we can restrict the g_i to those for which $g_i(X_0) = 0$, and apply that part already established to these g_i. But then, by uniqueness, we can simply set $\mu_j = 0$ when $g_j(X_0) < 0$, and the conclusion follows. $\square$

Bibliography

[AK]   N.I. Akhiezer
         The Calculus of Variations. Blaisdell Publishing Company: Boston, 1962.

[A–F]   M. Athans and P. Falb
         Optimal Control. McGraw-Hill Book Company: New York, 1966.

[A–S]   P. Aris and W.C. Strieder
         Variational Methods Applied to Problems of Diffusion and Reaction. Springer
         Tracts in Natural Philosophy, Vol. 24, Springer-Verlag: New York, 1973.

[Ar]   A.M. Arthurs
         Complementary Variational Principles, 2nd ed. Oxford University Press:
         Oxford, 1980.

[Bar]   S. Barnett
         Introduction to Mathematical Control Theory. Clarendon Press: Oxford,
         1975.

[B–P]   V. Barbu and T. Precupanu
         Convexity and Optimization in Banach Spaces. Sijthoff and Nordhoff:
         Bucuresti, 1978.

[Ba]   R. Pallu de la Barriere
         Optimal Control Theory. W.B. Saunders Company: Philadelphia, 1967.

[B–J]   D.J. Bell and D.H. Jacobson
         Singular Optimal Control Problems. Academic Press: New York, 1975.

[Be]   R.E. Bellman
         Introduction to the Mathematical Theory of Control Process. Academic
         Press: New York, 1967.

[Ber]   L.D. Berkovitz
         An Introduction to Optimal Control Theory, Springer-Verlag: New York,
         1974.

[Bi]   M.A. Biot
         Variational Principles in Heat Transfer. Clarendon Press: Oxford, 1972.

[Bl]   G.A. Bliss
         Calculus of Variations. Open Court Publishing Company: Chicago, 1925.
         Lectures on the Calculus of Variations. University of Chicago Press:
         Chicago, 1946.

[Bo]   O. Bolza
          Lectures on the Calculus of Variations, 3rd ed. Chelsea Publishing Company: New York, 1973.
          Vorlesungen uber Variationsrechnung. B.G. Teubner: Leipzig and Berlin, 1909.

[B–dP]   W. Boyce and R. DiPrima
          Elementary Differential Equations and Boundary Value Problems, 2nd ed. John Wiley and Sons: New York, 1969.

[CCV]   *Contributions to the Calculus of Variations*, 4 vols. University of Chicago Press: Chicago, 1930–1941.

[C]   C. Carathèodory
          Calculus of Variations and Partial Differential Equations of the First Order: Vols. I, II. Holden-Day: San Francisco, 1965, 1967.

[Ca]   L.B. Carll
          A Treatise on the Calculus of Variations. John Wiley and Sons: New York, 1890.

[Ce]   L. Cesari
          Optimization: Theory and Applications. Springer-Verlag: New York, 1983.

[Cla]   F. Clarke
          Optimization and Nonsmooth Analysis. John Wiley and Sons: New York, 1983.

[Cl]   J.C. Clegg
          Calculus of Variations. Oliver and Boyd: Edinburgh, 1968.

[C–L]   E. Coddington and N. Levinson
          Theory of Ordinary Differential Equations. McGraw-Hill Book Company: New York, 1955.

[Co]   R. Courant
          Calculus of Variations and Supplementary Notes and Exercises. New York University Lecture Notes, 1957.

[C–H]   R. Courant and D. Hilbert
          Methods of Mathematical Physics, Vol. I. Interscience Publishers: New York, 1953.

[Cr]   B.D. Craven
          Mathematical Programming and Control Theory. Chapman and Hall: London, 1978.

[Ed]   C.H. Edwards, Jr.
          Advanced Calculus of Several Variables. Academic Press: New York, 1973.

[El]   L.E. El'sgol'c
          Calculus of Variations. Addison-Wesley Publishing Company: Reading, MA, 1962.

[E–T]   I. Ekeland and R. Temam
          Convex Analysis and Variational Problems, North-Holland: Amsterdam, 1977.

[Ew]   G.M. Ewing
          Calculus of Variations with Applications. W.W. Norton: New York, 1969.

[Fl]   W.D. Fleming
          Functions of Several Variables, 2nd ed. Springer-Verlag: New York, 1977.

[F–R]   W.D. Fleming and R. Rishel
          Deterministic and Stochastic Optimal Control. Springer-Verlag: New York, 1975.

[F]   M.J. Forray
          Variational Calculus in Science and Engineering. McGraw-Hill Book Company: New York, 1968.

[Fo]   A.R. Forsyth
          Calculus of Variations. Dover Publications: New York, 1960.
[Fox]   C. Fox
          An Introduction to the Calculus of Variations. Oxford University Press:
          Fair Lawn, NJ, 1950.
[Fu]   P. Funk
          Variationsrechnung und ihre Anwendung in Physik und Technik. Springer-
          Verlag, OHG.: Berlin, 1962.
[G–F]   I.M. Gelfand and S.V. Fomin
          Calculus of Variations. Prentice-Hall: Englewood Cliffs, NJ, 1963.
[G–L]   J. Gregory and C. Lin
          *Constrained Optimization in the Calculus of Variations and Optimal Control
          Theory.* van Nostrand Reinhold: New York, 1992.
[G–T]   D. Gilbarg and N. Trudinger
          Elliptic Partial Differential Equations of Second Order. Springer-Verlag:
          New York, 1977.
[G]   S.H. Gould
          Variational Methods for Eigenvalue Problems, University of Toronto Press:
          Toronto, 1957.
[H]   J.S. Hadamard
          Leçons sur le Calcul de Variations, A. Herman et Fils: Paris, 1910.
[H–K]   G. Hadley and M. Kemp
          Variational Methods in Economics. North-Holland/Elsevier: Amsterdam,
          1971.
[H–L–P]   G.H. Hardy, J.E. Littlewood, and G. Polya
          Inequalities. Cambridge University Press: Cambridge, 1934.
[He]   M.R. Hestenes
          Calculus of Variations and Optimal Control Theory. John Wiley and Sons:
          New York, 1966.
[I–K]   E. Isaacson and H.B. Keller
          Analysis of Numerical Methods. John Wiley and Sons: New York, 1966.
[I–T]   A.D. Ioffe and V.M. Tihomirov
          Theory of Extremal Problems. North-Holland: Amsterdam, 1979.
[Ka–S]   M. Kamien and N. Schwartz
          Dynamic Optimization: The Calculus of Variations of Optimal Control in
          Economics and Management. North-Holland: New York, 1981.
[Ke]   H.B. Keller
          Numerical Solution of Two Point Boundary Value Problems. SIAM: Phila-
          delphia, PA, 1976.
[K–P]   P. Kosmol and M. Pavon
          Lagrange approach to the optimal control of diffusions. *Acta Appl. Math.*
          32 (1993), 101–122.
[K–S]   D. Kinderlehrer and G. Stampacchia
          An Introduction to Variational Inequalities and Their Application. Aca-
          demic Press: New York, 1980.
[K]   A. Kneser
          Lehrbuch der Variationsrechnung. Friedr. Vierweg & Sohn: Braunschweig,
          1900, 1925.
[L]   C. Lanczos
          The Variational Principles of Mechanics. University of Toronto Press:
          Toronto, 1949.
[L–M]   E.B. Lee and L. Markus
          Foundations of Optimal Control Theory. John Wiley and Sons: New York,
          1967.

[Le]   G. Leitmann
           The Calculus of Variations and Optimal Control. Plenum Press: New York,
           1981.
[Li]   L.A. Liusternik
           Shortest Paths: Variational Problems. Macmillan: London, 1964.
[Lo]   P. Loewen
           Optimal Control via Nonsmooth Analysis. American Mathematical Society,
           Providence, RI, 1991.
[M]   K. Maurin
           Calculus of Variations and Classical Field Theory, I. Lecture Notes Series,
           #34. Aaarhus: Danemark, 1972.
[M–S]   J. Macki and A. Strauss
           Introduction to Optimal Control Theory. Springer-Verlag: New York, 1982.
[Mi]   S.G. Mikhlin
           Variational Methods in Mathematical Physics. Macmillan: New York, 1964.
           The Problem of the Minimum of a Quadratic Functional. Holden-Day: San
           Francisco, 1965.
           The Numerical Performance of Variational Methods. Walters-Noordhoff:
           Groningen, 1971.
[Mo]   C.B. Morrey, Jr.
           *Multiple Integral Problems in the Calculus of Variations and Related
           Topics.* University of California Press: Berkeley, 1943.
[Mor]   M. Morse
           The Calculus of Variations in the Large. American Mathematical Society:
           Providence, RI, 1934.
[Mu]   F.D. Murnaghan
           The Calculus of Variations. Spartan Books: Washington, DC, 1962.
[N]   I. Nering
           Linear Algebra and Matrix Theory, 2nd ed., John Wiley and Sons: New
           York, 1970.
[Ne]   L. Neustadt
           Optimization: A Theory of Necessary Conditions. Princeton University
           Press: Princeton, NJ, 1976.
[No]   A.R.M. Noton
           Variational Methods in Control Engineering. Pergamon Press: New York,
           1965.
[O–R]   J.T. Oden and J.R. Reddy
           Variational Methods in Theoretical Mechanics. Springer-Verlag: Berlin,
           1976.
[Os]   R. Osserman
           A Survey of Minimal Surfaces. Van Nostrand Reinhold: New York, 1969.
[P]   L.A. Pars
           An Introduction to the Calculus of Variations. Heinemann: London, 1962.
[Pe]   I.P. Petrov
           Variational Methods in Optimal Control Theory. Academic Press: New
           York, 1968.
[Po]   L.S. Pontjragin with V.G. Boltjanskii, R.S. Gamkrelidze, and E.F.
           Mischenko
           The Mathematical Theory of Optimal Processes. Pergamon–Macmillan:
           New York, 1964.
[P–S]   G. Polya and M. Schiffer
           Convexity of functionals by transplantation. *J. d'Anal. Math.*, 1953/54,
           247–345.

[R]   K. Rektorys
        Variational Methods in Mathematics, Science and Engineering. Reidel
        Publishing Company: Boston, 1975.
[Ro]   R.T. Rockafellar
        Convex Analysis. Princeton University Press: Princeton, NJ, 1970.
[Rud]   W. Rudin
        Principles of Mathematical Analysis, 2nd ed., McGraw-Hill Book Com-
        pany: New York, 1964.
[Ru]   H. Rund
        The Hamilton–Jacobi Theory in the Calculus of Variations. D. Van
        Nostrand Company: Princeton, 1966.
[Rus]   J.S. Rustagi
        Variational Methods in Statistics. Academic Press: New York, 1979.
[S]   H. Sagan
        Introduction to the Calculus of Variations. McGraw-Hill Book Company:
        New York, 1969.
[S–S]   A. Seierstad and K. Sydsaeter
        Optimal Control with Economic Applications. North-Holland: New York,
        1987.
[Se]   L.A. Segel
        Mathematics Applied to Continuum Mechanics. Macmillan: New York,
        1977.
[Sew]   M.J. Sewell
        Maximum and Minimum Principles. Cambridge University Press, New
        York, 1987.
[Sm]   D.R. Smith
        Variational Methods in Optimization. Prentice-Hall: Englewood Cliffs, NJ,
        1974.
[Smi]   P. Smith
        Convexity Methods in Variational Calculus. Research Studies Press Ltd.:
        Letchworth, Hertfordshire, England, 1985.
[St]   W. Stacey
        Variational Methods in Nuclear Reactor Physics. Academic Press: New
        York, 1974.
[S–F]   G. Strang and G.J. Fix
        An Analysis of the Finite Element Method. Prentice-Hall: Englewood
        Cliffs, NJ, 1973.
[T]   L. Tonelli
        Fondamenti di Calcolo delle Variazione, I. II, N. Zanichelli: Bologna, 1921,
        1923.
[Ti]   V.M. Tikhomirov
        Fundamental Principles of the Theory of Extremal Problems, John Wiley &
        Sons: New York, 1986.
[TO]   G. Leitmann, Ed.
        Topics in Optimization. Academic Press: New York, 1967.
[Tr]   J.L. Troutman
        Boundary Value Problems of Applied Mathematics. PWS Publishing Com-
        pany, Boston, 1994.
[V]   M.M. Vainberg
        Variational Methods for the Study of Nonlinear Operations. Holden-Day:
        San Francisco, 1964.
        *Variational Method and Method of Monotone Operators in the Theory of
        Nonlinear Equations.* John Wiley and Sons: New York, 1973.

[W]   K. Washizu
           Variational Methods in Elasticity and Plasticity. Pergamon Press: New
           York, 1967.
[Wa]   J. Warga
           Optimal Control of Differential and Functional Equations. Academic Press:
           New York, 1972.
[W–S]   A. Weinstein and W. Stenger
           Methods of Intermediate Problems for Eigenvalues. Academic Press: New
           York, 1972.
[Wei]   H. Weinberger
           A First Course in Partial Differential Equations. Blaisdell Publishers: New
           York, 1965.
[We]   R. Weinstock
           Calculus of Variations with Applications to Physics and Engineering.
           McGraw-Hill Book Company: New York, 1952.
[Y]   L.C. Young
           Lectures on the Calculus of Variations and Optimal Control Theory. W.B.
           Saunders: Philadelphia, PA, 1969.
[Y–M]   W. Yourgrau and S. Mandelstam
           Variational Principles in Dynamics and Quantum Theory. Dover Publica-
           tions: New York, 1979.
[Z]   E. Zeidler
           Nonlinear Functional Analysis and its Applications, III: *Variational Methods
           and Optimization.* Springer-Verlag: New York, 1985.

Historical References

In addition to the appropriate comments and surveys to be found in the previous
works, the following references are of particular interest:

C. Carathèodory
The beginning of research in the calculus of variations, (and) Basel und der Beginn der
Variationsrechnung. *Ges. Math. Schriften*, II. C.H. Becksche Verlagsbuchhandlung:
München, 1955.
H. Goldstine
A History of the Calculus of Variations from the 17th through the 19th Centuries.
Springer-Verlag, New York, 1980.
M. Kline
Mathematical Thought from Ancient to Modern Times. Oxford University Press:
New York, 1972.
D.J. Struik
A Source Book in Mathematics 1200–1800. Harvard University Press: Cambridge,
MA: 1969.
G.F. Temple
100 Years of Mathematics. Springer-Verlag: New York, 1981.
I. Todhunter
A History of Progress of the Calculus of Variations in the Nineteenth Century.
Cambridge University Press: Cambridge, 1862.
R. Woodhouse
A Treatise of Isoperimetrical Problems and the Calculus of Variations. Cambridge
University Press: Cambridge, 1810.

Jean d'Alembert (1717–1783)

Daniel Bernoulli (1700–1782)
Jakob Bernoulli (1654–1705)
Johann Bernoulli (1667–1748)
Gilbert A. Bliss (1876–1951)
Oskar Bolza (1857–1942)

Augustin L. Cauchy (1789–1857)
Richard Courant (1888–1972)

P.G.L. Dirichlet (1805–1859)
Paul du Bois-Reymond (1831–1889)

Albert Einstein (1879–1955)
Leonhard Euler (1707–1783)

Pierre de Fermat (1601–1665)
Maurice Fréchet (1878–1973)

Galileo Galilei (1564–1642)
Carl F. Gauss (1777–1855)
George Green (1793–1841)

William R. Hamilton (1805–1865)
David Hilbert (1862–1943)

Carl Jacobi (1804–1851)

Adolf Kneser (1862–1930)

Joseph Louis Lagrange (1736–1813)
Pierre-Simon Laplace (1749–1827)
Adrien-M. Legendre (1752–1833)
Gottfried Wilhelm Leibniz
  (1646–1716)
Joseph Liouville (1809–1882)

Isaac Newton (1642–1727)
Emmy Noether (1882–1935)

Lord Rayleigh
  (John William Strutt)
  (1842–1919)
Bernhard Riemann (1826–1866)

Erwin Schrödinger (1887–1961)
Charles Sturm (1803–1855)

Karl Weierstrass (1815–1897)

Ernst Zermelo (1871–1953)

Answers to Selected Problems

Chapter 0

0.3. Maximum occurs at $(-2, 2)$; $f(-2, 2) = \frac{28}{3}$. Minimum occurs at $(-1, -2)$; $f(-1, -2) = -\frac{7}{3}$.

0.5. (a), (e) not convex.
(b) convex, but not strictly so.
(c), (d) strictly convex.

Chapter 2

2.1. Only (a), (c), (f), are *not* subspaces.

2.5. (a) $\delta J(y; v) = 3y(a)^2 v(a)$.
(c) $\delta J(y; v) = -\frac{1}{2}\int_a^b (2 + x^2 - \sin y'(x)) \cos y'(x)v'(x)\, dx$.
(e) $\delta J(y; v) = \int_a^b [2x^2 y(x)v(x) + e^{y'(x)}v'(x)]\, dx$.
(g) $\delta J(y; v) = 2(\int_a^b [2y'(x) + x^2 y(x)]\, dx)(\int_a^b [1 + y'(x)]v(x)\, dx) + (\int_a^b [2v'(x) + x^2 v(x)]\, dx)(\int_a^b [1 + y'(x)]^2\, dx)$.

Chapter 3

3.6. $y_0(x) = x^2 - 1$; $y_1(x) = 0$; $y_2(x) = $ const.

3.7. $y(x) = -1 + (2 - e)x + e^x$; $y_1(x) = e^x - ex - 1$;   incompatible.

3.9. $y(x) = \frac{1}{7}(19x - 12x^{-2})$; $y_1(x) = \frac{1}{5}(x + 4x^{-2})$; $y_2(x) = 0$.

3.11. $y(x) = \sqrt{9 - x^2}$.

3.13. $y(x) = -\sqrt{1 - x^2}$.

3.15. $y(x) = \frac{1}{4}(11 - 3x^{4/3})$.

3.19. $y(x) = x^3 - 3x$.

3.25.   (b) $\dfrac{wL}{2\mu}\left(\dfrac{x_1^4}{12} - \dfrac{x_1^3}{3} + \dfrac{x_1^2}{2}\right)$, where $x_1 = x/L$;

        (d) $\dfrac{M}{4\mu}x^2(1 - x_1)$ with max. at $x = 2L/3$.

3.27. $y(x) = \dfrac{\sqrt{5}}{2}(x^3 - x)$.

3.32.   (c) $A(\beta) = l^2(\beta - \sin \beta)/2\beta^2$.

3.38.   (c) $F(u_0) = \dfrac{2[h + g(T - 1 + e^{-T})]^2}{2T - 3 + 4e^{-T} - e^{-2T}}$.

3.39.   (c) They are the same.

3.41.   (d) $F_C = \frac{9}{20} > F_P = \frac{9}{8}\log(\frac{13}{9}) > F_T = \frac{5}{13}$.

Chapter 5

5.25.   (b) Each $y \in \mathscr{D}_1$.

5.27. $y(x) = 7c(x^{3/2} - 1) + 1$ with $c = (2^{3/2} - 1)^{-1}$.

5.29. $y(x) = 1 + \sin x$.

5.31. $y(x) = x(\ln x - 1)$.

5.37. $y(x) = (\frac{7}{2})^{1/5}\sqrt{x}$.

5.39. $y(x) = \frac{5}{4}(5x^4 - 1)$.

Chapter 6

6.1.   (a) $\cos y'(x) = c$.

        (c) $\dfrac{y'(x)}{x\sqrt{1 + y'(x)^2}} = c$.

        (e) $x - y(x) = 0$.

6.3. $y(x) = c_1 \sin x$.

6.5. $y(x) = 2x^2 - 1$.

6.7.   (a) $y(x) = x$.
        (b) None.

6.9. $y(x) = -\frac{1}{6}(x^3 + 5x)$.

6.11. $y(x) = 2 \cos\left(\dfrac{x}{2}\right)$.

6.17. (a) $y(x) = c_1 \sin x$.
(b) None.

6.19. $y(x) = n\pi x, n = 0, \pm 1, \pm 2, \ldots$.

6.21. (c) $y = \pm x/\sqrt{2}$.

6.23. (a) $y(x) = \pm \dfrac{\sqrt{2}}{n\pi} \sin(n\pi x), n = 1, 2, 3, \ldots$.

6.29. (d) $y(x) = \dfrac{l_1}{2}\left(\dfrac{x}{l}\right)^2 (3l - x)$.

6.31. $y(x) = \cos x$.

6.33. $y(x) = \frac{1}{2}x^2 + \ln x + \frac{1}{3}(4x^{-1} + 7x - 8)$.

6.34. (b) $\dfrac{\mu y''^2}{(1 + y'^2)^{5/2}} + P(\sqrt{1 + y'^2}) = cy' + c_0$ for constants c_0, c.

6.41. (a) $u_{xx} + u_{yy} + x^2 + y^2 = 0$.
(c) $3u_x u_{xx} - 6u_y^2 u_{yy} - u = 0$.

Chapter 7

7.5. (a) $c = 0$ with (i) or (ii).
(d) Any point c with (i); only when $\cos c = 0$, with (ii).

7.6. (a) $\hat{y}(x) = \pm \begin{cases} 1 - x, & 0 \le x \le 1, \\ x + 1, & -1 \le x < 0. \end{cases}$
(c) Yes, many.

7.7. (a) Only two possible solutions, each with $\hat{y}'(x) = \pm 2$.
(c) No.

(Note: For 7.6 and 7.7, express the Weierstrass–Erdmann conditions in terms of m and n, the *distinct* limiting values of $\hat{y}'$ at a potential corner point, and determine these values.)

7.9. (c) $\hat{y}(x) = \begin{cases} x, & x \le c, \\ e^x - 1, & x \ge c, \end{cases}$ where $c = e^{c-1}$.

7.23. (d) $\tau = T$ if $T^2 \le 2h$ and $\tau = T - \sqrt{T^2 - 2h}$ otherwise; (e) $T_0 = 3\sqrt{2h/5}$.

Chapter 8

8.3. $T = \frac{1}{2}ml^2\dot{\theta}^2 + \frac{1}{2}m_1 l^2(\dot{\theta}^2 + \dot{\theta}_1^2) + m_1 l^2\dot{\theta}\dot{\theta}_1 \cos(\theta - \theta_1)$,
$U = mgl(1 - \cos\theta) + m_1 gl(2 - \cos\theta - \cos\theta_1)$.
(c) Linearized equations are

$$\ddot{\theta} + \frac{m_1}{m}\ddot{\theta}_1 = -g\theta/l$$

and
$$\ddot{\theta} + \ddot{\theta}_1 = -g\theta_1/l.$$

8.5. $T = \frac{1}{2}ml^2(\dot{\varphi}^2 + \sin^2 \varphi\dot{\theta}^2),$
     $U = mgl(1 + \cos \varphi).$
     (f) $S_t + \dfrac{ml^{-2}}{2}[S_\varphi^2 + \sin^{-2} \varphi S_\theta^2] = -mg(1 + \cos \varphi).$

8.6. $S = c_1 t - (c_1 e^q)^{1/3} + c_0.$

Chapter 10

10.1. $y_0(t) = \dfrac{\sinh 2(1 - t)}{\sinh 2}$ gives unique minimum.

10.3. Equality holds iff $y_0(t) \equiv t$ or $y_0(t) \equiv -t.$

10.5. Take $u_0(t) = 1, t < 60;$ and $u_0(t) = 0, t > 60.$

10.12. (d) The equation $\ddot{p} - (1 + \sin t)^2 p = -2e^{-t}$ is not elementary.

10.14. $u_0(t) = 15(1 + 4e^{t-2})/(8e^{-4} - 11).$

10.16. $P(2) = 2(Y_0(2) - (5, 2)) = (-22/7, 16/21); u_0(t) = -11t/7 + 58/21.$

10.19. $U_0(t) = 0$ is the unique optimal control, but $Y_0(t) = 0$ might not be the only minimizing trajectory.

10.21. (a) $y_0(t) = 1 - t.$
     (b) If $T = 2,$ solution is not unique;
        if $T = \frac{1}{2},$ solution does not exist.

10.23. Minimum is $-\frac{2}{3}$ which occurs if $u_0(t) = 1 - t.$

10.26. (a) Minimum occurs where $y_1(2) = -y^2(2) = e^{-y_1^2(2)} = a,$ say. The minimum value is $a/\sqrt{1 + a^2},$ but neither optimal control nor optimal state is unique.

Chapter 11

11.1. (a) $y_0(t) = [\sinh 2(1 - t)]/\sinh 2.$
     (e) $\lambda_0 = 1, u_0(t) = \text{sgn}[1 - 3^{-1/2}(2 - t)], \dot{y}_0(2) = 2 - 2\sqrt{3}.$
     (f) $u_0(t) = \text{sgn}(5 - 2t),$ so
     $$y_0(t) = -5te^{2t}, \qquad t \le \tfrac{2}{5},$$
     $$= (5t - 4)e^{2t}, \quad t \ge \tfrac{2}{5}.$$
     (h) Optimal trajectories occur only along primary parabolas.

11.5. (b) $T_0 = \frac{1}{2} \log(1 + 2/\sin 1).$

11.11. $u_0(t) = +1$ or -1 with at most two switching points, will give the unique optimal control from any reachable $A.$

11.12. Optimal controls are unique and vertex-valued, but cannot bound number of switching points.

Index

Undergraduate Texts in Mathematics

(continued from page ii)

Lidl/Pilz: Applied Abstract Algebra.

Macki-Strauss: Introduction to Optimal Control Theory.

Malitz: Introduction to Mathematical Logic.

Marsden/Weinstein: Calculus I, II, III. Second edition.

Martin: The Foundations of Geometry and the Non-Euclidean Plane.

Martin: Transformation Geometry: An Introduction to Symmetry.

Millman/Parker: Geometry: A Metric Approach with Models. Second edition.

Moschovakis: Notes on Set Theory.

Owen: A First Course in the Mathematical Foundations of Thermodynamics.

Palka: An Introduction to Complex Function Theory.

Pedrick: A First Course in Analysis.

Peressini/Sullivan/Uhl: The Mathematics of Nonlinear Programming.

Prenowitz/Jantosciak: Join Geometries.

Priestley: Calculus: An Historical Approach.

Protter/Morrey: A First Course in Real Analysis. Second edition.

Protter/Morrey: Intermediate Calculus. Second edition.

Ross: Elementary Analysis: The Theory of Calculus.

Samuel: Projective Geometry. *Readings in Mathematics.*

Scharlau/Opolka: From Fermat to Minkowski.

Sigler: Algebra.

Silverman/Tate: Rational Points on Elliptic Curves.

Simmonds: A Brief on Tensor Analysis. Second edition.

Singer/Thorpe: Lecture Notes on Elementary Topology and Geometry.

Smith: Linear Algebra. Second edition.

Smith: Primer of Modern Analysis. Second edition.

Stanton/White: Constructive Combinatorics.

Stillwell: Elements of Algebra: Geometry, Numbers, Equations.

Stillwell: Mathematics and Its History.

Strayer: Linear Programming and Its Applications.

Thorpe: Elementary Topics in Differential Geometry.

Troutman: Variational Calculus and Optimal Control. Second edition.

Valenza: Linear Algebra: An Introduction to Abstract Mathematics.

Whyburn/Duda: Dynamic Topology.

Wilson: Much Ado About Calculus.